IET TELECOMMUNICATIONS SERIES 105

Non-Geostationary Satellite Communications Systems

Other volumes in this series:

Volume 9 **Phase noise in signal sources** W.P. Robins
Volume 12 **Spread spectrum in communications** R. Skaug and J.F. Hjelmstad
Volume 13 **Advanced signal processing** D.J. Creasey (Editor)
Volume 19 **Telecommunications traffic, tariffs and** costs R.E. Farr
Volume 20 **An introduction to satellite communications** D.I. Dalgleish
Volume 26 **Common-channel signalling** R.J. Manterfield
Volume 28 **Very small aperture terminals (VSATs)** J.L. Everett (Editor)
Volume 29 **ATM: the broadband telecommunications solution** L.G. Cuthbert and J.C. Sapanel
Volume 31 **Data communications and networks, 3rd edition** R.L. Brewster (Editor)
Volume 32 **Analogue optical fibre communications** B. Wilson, Z. Ghassemlooy and I.Z. Darwazeh (Editors)
Volume 33 **Modern personal radio systems** R.C.V. Macario (Editor)
Volume 34 **Digital broadcasting** P. Dambacher
Volume 35 **Principles of performance engineering for telecommunication and information systems** M. Ghanbari, C.J. Hughes, M.C. Sinclair and J.P. Eade
Volume 36 **Telecommunication networks, 2nd edition** J.E. Flood (Editor)
Volume 37 **Optical communication receiver design** S.B. Alexander
Volume 38 **Satellite communication systems, 3rd edition** B.G. Evans (Editor)
Volume 40 **Spread spectrum in mobile communication** O. Berg, T. Berg, J.F. Hjelmstad, S. Haavik and R. Skaug
Volume 41 **World telecommunications economics** J.J. Wheatley
Volume 43 **Telecommunications signalling** R.J. Manterfield
Volume 44 **Digital signal filtering, analysis and restoration** J. Jan
Volume 45 **Radio spectrum management, 2nd edition** D.J. Withers
Volume 46 **Intelligent networks: principles and applications** J.R. Anderson
Volume 47 **Local access network technologies** P. France
Volume 48 **Telecommunications quality of service management** A.P. Oodan (Editor)
Volume 49 **Standard codecs: image compression to advanced video coding** M. Ghanbari
Volume 50 **Telecommunications regulation** J. Buckley
Volume 51 **Security for mobility** C. Mitchell (Editor)
Volume 52 **Understanding telecommunications networks** A. Valdar
Volume 53 **Video compression systems: from first principles to concatenated codecs** A. Bock
Volume 54 **Standard Codecs: image compression to advanced video coding, 3rd edition** M. Ghanbari
Volume 59 **Dynamic Ad Hoc Networks** H. Rashvand and H. Chao (Editors)
Volume 60 **Understanding Telecommunications Business** A Valdar and I Morfett
Volume 65 **Advances in Body-Centric Wireless Communication: Applications and State-of-the-art** Q. H. Abbasi, M. U. Rehman, K. Qaraqe and A. Alomainy (Editors)
Volume 67 **Managing the Internet of Things: Architectures, Theories and Applications** J. Huang and K. Hua (Editors)
Volume 68 **Advanced Relay Technologies in Next Generation Wireless Communications** I. Krikidis and G. Zheng
Volume 69 **5G Wireless Technologies** A. Alexiou (Editor)
Volume 70 **Cloud and Fog Computing in 5G Mobile Networks** E. Markakis, G. Mastorakis, C. X. Mavromoustakis and E. Pallis (Editors)
Volume 71 **Understanding Telecommunications Networks, 2nd edition** A. Valdar
Volume 72 **Introduction to Digital Wireless Communications** Hong-Chuan Yang
Volume 73 **Network as a service for next generation internet** Q. Duan and S. Wang (Editors)
Volume 74 **Access, Fronthaul and Backhaul Networks for 5G & Beyond** M. A. Imran, S. A. R. Zaidi and M. Z. Shakir (Editors)
Volume 76 **Trusted Communications with Physical Layer Security for 5G and Beyond** T. Q. Duong, X. Zhou and H. V. Poor (Editors)
Volume 77 **Network Design, Modelling and Performance Evaluation** Q. Vien
Volume 78 **Principles and Applications of Free Space Optical Communications** A. K. Majumdar, Z. Ghassemlooy, A. A. B. Raj (Editors)
Volume 79 **Satellite Communications in the 5G Era** S. K. Sharma, S. Chatzinotas and D. Arapoglou
Volume 80 **Transceiver and System Design for Digital Communications 5th Edition** Scott R. Bullock
Volume 81 **Applications of Machine Learning in Wireless Communications** R. He and Z. Ding (Editors)
Volume 83 **Microstrip and Printed Antenna Design 3rd Edition** R. Bancroft
Volume 84 **Low Electromagnetic Emission Wireless Network Technologies: 5G and beyond** M.A. Imran, F. Héliot and Y.A. Sambo (Editors)
Volume 86 **Advances in Communications Satellite Systems Proceedings of The 36th International Communications Satellite Systems Conference (ICSSC-2018)** I. Otung, T. Butash and P. Garland (Editors)
Volume 87 **Real Time Convex Optimisation for 5G Networks and Beyond** T.Q. Duong, L. D. Nguyen and H. D. Tuan
Volume 89 **Information and Communication Technologies for Humanitarian Services** M. N. Islam (Editor)
Volume 90 **Communication Technologies for Networked Smart Cities** S.K. Sharma, N. Jayakody, S. Chatzinotas and A. Anpalagan (Editors)
Volume 91 **Green Communications for Energy-Efficient Wireless Systems and Networks** Himal Asanga Suraweera, Jing Yang, Alessio Zappone and John S. Thompson (Editors)

Volume 92 **Flexible and Cognitive Radio Access Technologies for 5G and Beyond** H. Arslan and E. Başar (Editors)
Volume 93 **Antennas and Propagation for 5G and Beyond** Q. Abbasi, S.F. Jilani, A. Alomainy and M. A. Imran (Editors)
Volume 94 **Intelligent Wireless Communications** G. Mastorakis, C. X. Mavromoustakis, J. M. Batalla and E. Pallis (Editors)
Volume 95 **ISDN applications in education and training** R. Mason and P.D. Bacsich
Volume 96 **Edge Caching for Mobile Networks** H. Vincent Poor and Wei Chen (Editors)
Volume 98 **Artificial Intelligence Applied to Satellite-based Remote Sensing Data for Earth Observation** M. P. Del Rosso, A. Sebastianelli and S.L. Ullo (Editors)
Volume 99 **Metrology for 5G and Emerging Wireless Technologies** T. H. Loh (Editor)
Volume 101 **Wireless Mesh Networks for IoT and Smart Cities: Technologies and applications** L. Davoli and Gi. Ferrari (Editors)
Volume 104 **Enabling Technologies for Social Distancing: Fundamentals, concepts and solutions** D. N. Nguyen, D.T. Hoang, T. X. Vu, E. Dutkiewicz, S. Chatzinotas and B. Ottersten

Non-Geostationary Satellite Communications Systems

Edited by
Eva Lagunas, Symeon Chatzinotas, Kang An and Bassel F. Beidas

The Institution of Engineering and Technology

Published by The Institution of Engineering and Technology, London, United Kingdom

The Institution of Engineering and Technology is registered as a Charity in England & Wales (no. 211014) and Scotland (no. SC038698).

First published 2022

The Institution of Engineering and Technology
Futures Place
Kings Way, Stevenage
Hertfordshire, SG1 2UA, United Kingdom

www.theiet.org

British Library Cataloguing in Publication Data
A catalogue record for this product is available from the British Library

ISBN 978-1-83953-566-6 (hardback)
ISBN 978-1-83953-567-3 (PDF)

Typeset in India by Exeter Premedia Services Private Limited

Cover Image: Andriy Onufriyenko / Moment via Getty Images

Contents

About the Editors

Eva Lagunas is a research scientist in the Interdisciplinary Centre for Security, Reliability and Trust (SnT) at the University of Luxembourg. Her research interests include radio resource management and general wireless networks optimization. She is a senior member of IEEE and a member of EURASIP, where she contributes to the Technical Area Committee (TAC) of Theoretical and Methodological Trends in Signal Processing. She received her PhD degree in Telecommunications Engineering from the Polytechnic University of Catalonia (UPC), Barcelona, Spain. She conducted research stays at the Department of Information Engineering, Pisa, Italy, and at the Center for Advanced Communications (CAC), Villanova University, PA, USA.

Symeon Chatzinotas is Full Professor and Head of the SIGCOM Research Group at SnT, University of Luxembourg. He is coordinating the research activities on communications and networking across a group of 80 researchers, acting as a PI for more than 40 projects and main representative for 3GPP, ETSI, DVB. He is currently serving in the editorial board of the IEEE Transactions on Communications, IEEE Open Journal of Vehicular Technology and the International Journal of Satellite Communications and Networking. In the past, he has been a Visiting Professor at the University of Parma, Italy and was involved in numerous R&D projects for NCSR Demokritos, CERTH Hellas and CCSR, University of Surrey. He is an IEEE Fellow and the co-recipient of the 2014 IEEE Distinguished Contributions to Satellite Communications Award and Best Paper Awards at WCNC, 5GWF, EURASIP JWCN, CROWNCOM, ICSSC. He has (co-)authored more than 700 technical papers in refereed international journals, conferences and scientific books.

Kang An is a senior engineer with the Sixty-third Research Institute, National University of Defense Technology, Nanjing, China. His research interests include cooperative and cognitive communications, radio resource management and optimization in wireless networks. He is a member of the IEEE. He received his PhD degree from the Army Engineering University, Nanjing, China.

Bassel F. Beidas is a scientist with the Advanced Development Group at Hughes, USA. He is responsible for the research and development in advanced transmission technologies, which have been successfully incorporated into premier product lines in cellular and satellite communications. His research interests include signal classification, interference cancellation, adaptive signal processing, MIMO, synchronization, and nonlinear systems. He holds over 35 US patents on digital communications techniques and has several patents pending. He was the winner of the prestigious 2021 Satellite Communications Technical Recognition Award, from the IEEE ComSoc Satellite and Space Communications Technical Committee. He received the M.S. degree (Hons.) in electrical engineering from the California Institute of Technology, USA, and the Ph.D. degree in electrical engineering from the University of Southern California, USA.

Preface

With the aim of boosting connectivity across the globe, lower orbit satellites are forecast to significantly increase the broadband coverage in isolated territories where terrestrial infrastructure is too expensive or unfeasible to deploy. Thanks to the recent advances in spacecraft manufacturing, multiple private sector satellite companies are showing interest in rapidly launching several conveyor-belt manufactured satellites, to create a dense "net" around Earth. Based on recent studies, 7,000 small satellites are likely to be launched between 2018 and 2027 for a variety of missions, where 82% are associated with the roll-out of constellations planned by private companies (e.g., SpaceX) [1].

While the launch of thousands of new satellites will definitely boost the space economy and help bridging the Digital Divide, Non-Geostationary Satellite Orbit (NGSO) constellation success will only be possible by addressing critical technical and regulatory challenges. Furthermore, to unleash full potential of NGSO constellations, stakeholders need to achieve a seamless integration with the existing ecosystem, including the cellular communications network and the existing space-based communications systems.

This book aims at reviewing these critical challenges and at shedding light into the potential technical solutions and guidelines for addressing these challenges. In particular, the book is divided into three main parts as listed below:

- Part I: NGSO basic concepts: This part provides a general overview of NGSO systems and their main challenges; an overview on the spectrum regulations and discussion about the NGSO role within the next generation of wireless cellular communications. This part is composed by Chapter 1 until Chapter 3.
- Part II: Technological Enablers: This part goes into the details of different technological enablers such as flat antenna arrays, payload design, radio-frequency impairments compensation, radio resource and interference management, multiple access schemes, constellation design, inter-satellite links and massive multiple-input multiple-output (MIMO) framework. This part is composed by Chapter 4 until Chapter 12.
- Part III: System Level Operations: This part discusses Software-Defined Networking (SDN) applied to NGSO, network security aspects, and the on-going 3GPP integration of Non-Terrestrial Networks (NTN) into the 5G standards. To conclude, an overview of NTN testbeds for 5G is presented. This part is composed by Chapter 13 until Chapter 16.

NGSO satellite communications have recently entered a period of renewed interest motivated by technological advances and nurtured through private investment and ventures. The writing of this book was mainly prompted by the fast developments in NGSO satellite communications in the past decade. The primary aim has been to include a maximum of useful information, with particular attention to the needs of researchers, scientists, or engineers who would like to delve deeper into the technical aspects of NGSO satellite communication systems design.

The Editors would like to acknowledge the contributions of the many reputed experts that have made this book possible. We are most grateful to them for their technical input and time devoted to this book.

Eva Lagunas, Symeon Chatzinotas
Kang An, Bassel F. Beidas

Reference

[1] Prospects for the Small Satellite Market (Euroconsult), online; 2018. Available from: https://digital-platform.euroconsult-ec.com/product/ prospects-for-the-small-satellite-market/.

Chapter 1

Non-geostationary orbit systems introduction and challenge identification

Hayder Al-Hraishawi[1], Houcine Chougrani[1], Steven Kisseleff[1], Eva Lagunas[1], and Symeon Chatzinotas[1]

Non-geostationary orbit (NGSO) satellites are anticipated to support various new communication applications from different industries. NGSO communication systems are known for a number of key features such as lower propagation delay, smaller size, and lower signal losses in comparison to the conventional geostationary orbit (GSO) satellites, which can potentially enable latency-critical applications to be provided through satellites. NGSO promises a significant boost in communication speed, and thus, tackling the main inhibiting factors of commercializing GSO satellites for broader utilisation.

However, there are still many NGSO deployment challenges that need to be adequately addressed in order to ensure seamless integration not only with GSO systems but also with terrestrial networks. These unprecedented challenges are identified in this chapter, including coexistence with GSO systems in terms of spectrum access and regulatory issues, satellite constellation and architecture designs, resource management problems, and user equipment requirements. Furthermore, future research challenges inspired by utilising NGSO systems to advance satellite communications within versatile applications are also provided.

1.1 Introduction

Satellites have a distinctive ability to cover wide geographical areas through a minimum amount of infrastructure on the ground, which qualifies them as an appealing solution to fulfil the growing number of diverse applications and services either as a stand-alone system, or as an integrated satellite-terrestrial network [1]. Currently, the field of satellite communications is drawing increased attention in the global telecommunications market as several network operators start using satellites in

[1]Interdisciplinary Centre for Security, Reliability and Trust (SnT), University of Luxembourg, Luxembourg

backhauling infrastructures for connectivity and for 5G system integration [2]. Recently, due to the swift rise of "NewSpace" industries that are developing small satellites with new low-cost launchers, a large number of satellite operators have started planning to launch thousands of NGSO satellites to satisfy the burgeoning demand for global broadband, high-speed, heterogeneous, ultra-reliable and low latency communications. For instance, the emerging NGSO satellites and mega constellations such as SES O3b, OneWeb, Telesat, and Starlink have a system capacity reaching the terabits-per-second level [3].

In the last few years, the notion of utilising large fleets of NGSO satellites, especially in the low Earth orbit (LEO), to provide reliable, low-latency, and high-speed Internet from space has re-gained popularity and experienced a tremendous growth. This trend is rather surprising given the unfortunate faring of past NGSO constellations, but it appears that both technological and business momenta are favourable with the achievements of SpaceX, SES O3B, and OneWeb. In fact, between 2014 and 2016, a new wave of proposals for large LEO constellations emerged with the target of providing global broadband services [4]. Specifically, the number of satellites launched into space has substantially increased according to the recent satellite database released by the Union of Concerned Scientists [5]. This database has listed more than 4,000 operational satellites currently in orbit around Earth with a big difference between the number of GSO and NGSO satellites in favour of the latter as depicted in Figure 1.1. In detail, approximately 90% of the total number of operational satellites are within the NGSO constellations.

NGSO satellites on a geocentric orbit include LEO, medium Earth orbit (MEO) and highly elliptical orbit satellites, which are orbiting constantly at a lower altitude than that of GSO satellites, and thus, their link losses and latency due to signal propagation are lower [6]. These intrinsic features of NGSO systems besides their fairly

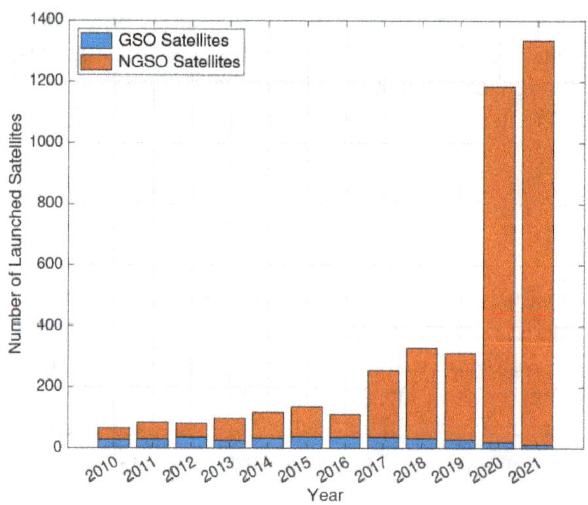

Figure 1.1 A comparison between the number of launched GSO and NGSO satellites per year

large footprints and fast deployment offer an interesting set of advantages for the high-speed interactive broadband services [7]. Furthermore, the most recent developments in NGSO systems empower satellites to manage narrow steerable beams covering a relatively broad area, which facilities the use of smaller and lower-cost equipment at the user terminals [8]. Specifically, the offered capacities by NGSO satellites can be further increased by utilizing high-frequency bands along with throughput enhancement techniques such as spectrum sharing, cooperative gateway diversity, interference mitigation, large antenna array for distributed beamforming, and spatial multiplexing [9, 10].

Furthermore, satellite systems have been contributing to telecommunication services in a wide range of sectors such as aeronautical, maritime, military, rescue and disaster relief [11]. Beyond this, NGSO systems are envisaged to be an appealing solution for future non-terrestrial networks (NTN) to meet the demanding 6G system requirements in terms of both large throughput and global connectivity [12]. In this direction, the third generation partnership project (3GPP) standards group has been defining the use of satellite communication networks for its integration with terrestrial communication networks in order to support future wireless ecosystems [13]. Moreover, by harnessing the satellite's geographical independence, wireless connectivity can be extended to the underserved and unserved areas, where NGSO systems can be an efficient solution for viable deployments of 5G and beyond networks. NGSO satellite capabilities of ubiquitous coverage and connectivity can also be leveraged for provisioning resiliency and continuity to mobile platforms such as on-board aircraft, high-speed trains, sea-going vessels, and land-based vehicles that are beyond the reach of a terrestrial cell site [14].

In addition to the NGSO's unique capabilities in providing global coverage, low-latency communication, and high-speed Internet access point, these systems can essentially change the way satellite missions are designed and operated in the near future [15]. In particular, the recent technological progress has evolved the possibility of constructing a chain production of cheaper NGSO satellites with very short lifespans. Accordingly, the satellite infrastructure will be more regularly upgraded, and thus, the payload design can be more innovative in terms of on-board technologies. Evidently, the NGSO system can help bridge the digital divide across the globe and can create new capabilities and services for different enterprise verticals [16]; however, that comes with some important questions about their operations and the required developments. The next sections explore the NGSO system characteristics and classification, the key challenges faced in this rising field, and the promising future research directions for NGSO systems.

1.2 NGSO system characteristics and classification

NGSO satellites have been already used in numerous applications, such as telecommunications, Earth and space observation, asset tracking, meteorology, and scientific projects. Depending on the provided services, NGSO systems can be classified into two categories: space-based Internet providers and small satellite missions.

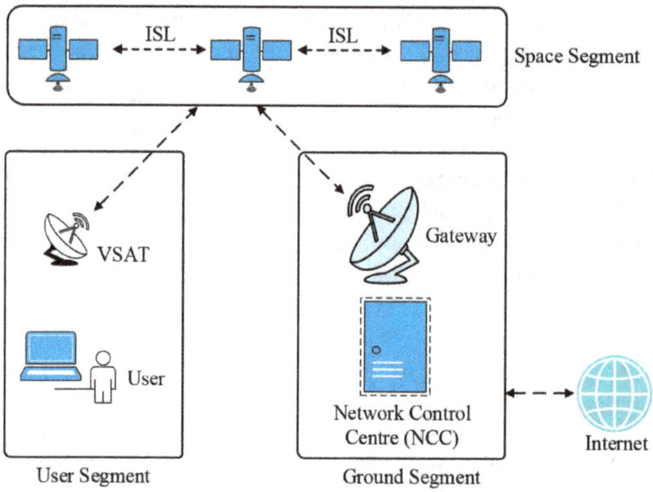

Figure 1.2 Schematic diagram for a space-based Internet system

1.2.1 Space-based Internet providers

The space-based Internet services have been provided by multiple companies such as Hughes, Eutelsat, Viasat, and Gilat since the 1970s to regions with underdeveloped infrastructure. However, most of the existing systems utilize GSO satellites that are 35,786 km above Earth, resulting in slow and expensive Internet connections. Consequently, the use of GSO-based Internet systems has been limited to latency-tolerant applications. In contrast, the emerging NGSO mega-constellations will operate from lower altitudes, between 160 km and 2000 km above Earth, which lowers signal propagation loss and latency, and reduces the hardware complexity of user equipment. Several private sector companies are on their way to provide space-based Internet services in the upcoming few years, such as SpaceX, OneWeb, and SES. They have obtained licenses, launched many satellites and successfully performed initial tests. Internet giants are also foreseeing market opportunities to extend their services via NGSO systems. For example, Amazon introduced the Kuiper project to offer high-speed broadband connectivity to people globally. Likewise, Google has invested in Starlink and supported the Loon project.

Generally, a space-based Internet system consists of three main components: the space segment, ground segment, and user segment (see Figure 1.2). The space segment can be a satellite or a constellation of satellites, while the ground segment involves a number of ground stations/gateways that relay Internet data to and from the space segment, and the user segment includes a small antenna at the user location, often a very small aperture terminal (VSAT) antenna with a transceiver. Additional critical entities within this structure are (i) network management centre (NMC) and (ii) network control centre (NCC). The centralized NMC is the functional entity in charge of the management of all the system elements such as fault, configuration, performance, and security management. The NCC is the functional entity that provides real-time control signalling such as session/connection control, routing, access control to satellite resources, etc. [17].

1.2.2 Small satellite missions

The space industry is experiencing a profound change due to the miniaturization of electronic equipment to manufacture satellites leading to the emergence of new low-cost small satellites. The miniaturization of satellites is making space more affordable and accessible than ever, which will enable any country, university, startup or even school to reach space in an affordable way within a reasonable time period. Thus, these developments have unlocked the missions that satellites can carry and execute for different applications. In particular, the most relevant small satellite missions in this context include but are not limited to:

- Earth and space observation: this is one of the widespread uses of satellite constellations in different orbits including capturing high-resolution images of Earth and outer space, remote sensing in various frequencies, RF monitoring, global navigation satellite system reflectometry, etc.
- Asset tracking: satellite payload in asset tracking projects consists of a device equipped with communication components to collect information sent from objects on the ground and transmit it back to ground stations.
- Meteorology: small satellites are able to play an important role in storm detection and in the development of climate and weather models that enhance weather forecasts. For instance, NASA RainCube project has started the testing phase for the location, tracking and analysis of rain and snowstorms over the entire Earth.
- Agriculture: crop monitoring is another potential use of nanosats, where a better control of harvests, the improvement of the quality of agricultural products, the finding of diseases in crops and analysis of the ramifications derived from the periods of drought can be facilitated by using nanosats.
- Educational activities: the development of scientific experiments outside the Earth has become another common application of small satellites, which are unprecedented opportunities brought up by nanosats with their myriad possibilities.
- Government space programs: the goals of these government programs vary from national security to emergency response. Some other useful applications can be for protecting the environment through the detection of forest fires, studying the progress of melting ice, fighting against ocean pollution, detection of oil spills, monitoring marine life, controlling desertification, etc.

With these diverse applications and the rapid developments in mind, there is definitely an exciting future ahead for small satellite missions in many fields but the advancement of future cooperative distributed space systems will probably require a high degree of operational autonomy.

1.3 NGSO deployment challenges

Notwithstanding the growing interest in NGSO satellites due to their essential feature of providing high-speed pervasive connectivity for a wide variety of use cases and applications, there are still many daunting challenges in the NGSO satellite evolution to be addressed in order to achieve high quality communications [18]. This section presents several key challenges (see Figure 1.3) including satellite constellation and architecture designs, coexistence with GSO and other NGSO systems in terms of spectrum access and regulatory issues, system control and operation, as well as user equipment requirements. In the following, the related critical challenges of NGSO systems development and integration are discussed with highlighting the most relevant solutions.

1.3.1 Regulatory and coexistence issues

According to the international telecommunication union (ITU) regulations, the interference inflicted on GSO satellites from NGSO satellite systems shall not degrade GSO satellite's performance and shall not claim protection from GSO systems in the fixed-satellite and broadcasting-satellite services [6]. Specifically, the effective power flux density (EPFD) within the frequency bands that are allocated to GSO systems and at any point on the Earth's surface visible from the GSO satellite orbit shall not exceed the given predefined limits in the ITU regulations. Although NGSO systems have the potential for global coverage and high performance, many

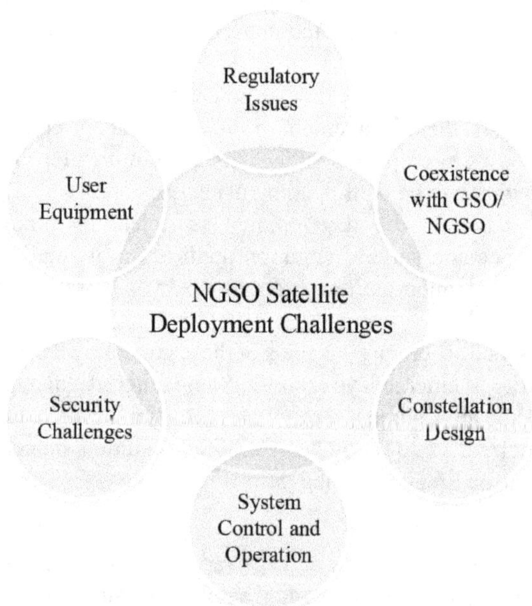

Figure 1.3 NGSO satellites deployment challenges

of their regulatory rules were coined nearly two decades ago based on the proposed technical characteristics of NGSO satellites at the time. This is very challenging from a spectral coexistence viewpoint, and it will require much more agile systems. Moreover, the deployment of NGSO satellites is undergoing significant densification compared to existing GSO systems, which is leading to unprecedented inter-satellite coexistence challenges. The high interference levels will not only result from a large number of operating satellites but also from the expected high heterogeneity of the NGSO systems. Therefore, it is imperative to scrutinize the interference interactions between different GSO and NGSO systems to ensure a consistent hybrid deployment landscape.

The recent surging activities concerning the use of NGSO satellite constellations have propelled the regulatory environment towards adapting and extending their rules to ensure the safe and efficient deployment of NGSO operations. International regulators have the challenging task to establish a fair and transparent competitive framework for all satellite broadband players while prioritising socio-economic growth. Specifically, during the world radio communications conference in 2015 (WRC-15), different national delegates have expressed their concerns about the increasing number of requests submitted for NGSO satellite systems operating in the Fixed-Satellite Service (FSS) subject to the EPFD limits in Article 22 and to coordination under no. 9.7B of the Radio Regulations (RR). Furthermore, the global satellite coalition (GSC) during WRC-19 has agreed on defining a regulatory framework for NGSO satellites to operate in the Q/V bands. They also have planned a new agenda item for WRC-23 to further study a number of issues including technical considerations related to space-to-space links, which will be important for global NGSO and hybrid NGSO-GSO networks. Moreover, the ITU vision for the next WRC-23 aims to bring the satellite industry forward to work together with governments to shape a global perspective on connectivity that also addresses national and regional requirements.

At this point, some aspects and scenarios need further investigations in this direction, which are enumerated and briefly described in the following.

- NGSO and GSO coexistence: NGSO single-entry power flux density (PFD) limits in certain parts of the frequency range 10.7–30 GHz are included in Article 22 of the RR since 2000, with the main goal to protect GSO systems operating in the same frequency bands. Later, the single-entry PFD limit was found to be not enough as the number of NGSO satellites was growing at a rapid pace. This led to the definition of EPFD which takes into account the aggregate of the emissions from all NGSO satellites. An example of multiple NGSO systems causing interference to a GSO receiver is shown in Figure 1.4. In this direction, a specific software tool has been made available for operators and regulators to check these limits for specific NGSO satellites [19]. The European Space Agency has also launched a separate activity to build its own simulator [20]. Moreover, a feasible solution can be proposed by constructing a large discrimination angle and exclusion zones are typically considered to limit interference with GSO communications systems [21].

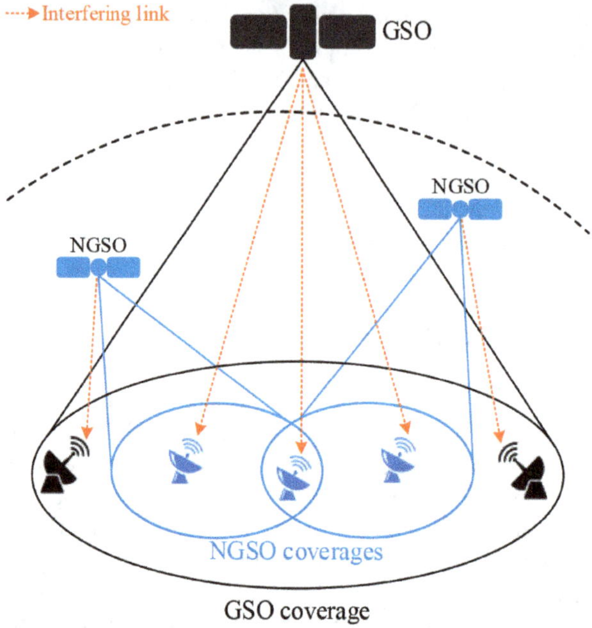

Figure 1.4 *Aggregated inter-satellite interference between GSO and NGSO systems*

- NGSO Earth station operations: the ground infrastructure required to operate an NGSO constellation is significantly more complex than that of a single GSO satellite. Therefore, the impact of deploying multiple NGSO Earth stations distributed over the coverage area has to be carefully designed to ensure minimal impact on other users within the shared spectrum. However, from the regulators' perspective, there is no individual licensing of Earth stations because the general impression is that mitigation techniques can be employed by the operators to avoid detrimental interference, e.g., switching to alternative frequencies, as elaborated in Federal Communications Commission (FCC) documentations [22].
- NGSO FSS user terminals: in general, and excluding large latitudes, GSO FSS user terminals have a significant gain in high elevation directions with limited gain towards the horizon, as the satellite is usually placed above the region of interest. Recently, advocates of a new generation of NGSO FSS systems have sought after the FCC authority to modernize the relevant regulations, and consequently, the FCC has proposed to update certain frequency allocations in the Ka-band, power limits, and service rules to facilitate these emerging systems [22].
- Coordination with other NGSO networks: in view of the constellation and orbital overcrowding, it is very likely that large NGSO constellations will cause interference to other NGSO systems. However, the preliminary interference

risk analysis carried out in Reference 23 considering both Ka-band and V-band suggests that the risk is relatively low, concluding that the need for interference mitigation might be limited. In case of unacceptable interference situations, the mitigation techniques described in Annex 1 of Reference 24 should be considered in order to achieve satisfactory sharing between different NGSO systems, although other techniques are not excluded.

1.3.2 Satellite constellation design

Generally, satellite orbit constellation design is a key factor that directly affects the overall system performance. The key constellation parameters include the type of orbit, altitude of the orbit, number of orbits, number of satellites in each orbit, and satellite phase factor between different orbit planes [25]. Several earlier studies have considered systematic constellation patterns of satellites such as polar constellations and Walker-Delta patterns [26], which are formulated based on the relative positions of the satellites in the Earth-centred inertial frame. Additionally, in Reference 27, the concept of flower constellations has been proposed to put all satellites in the same 3D trajectory in the Earth-centred Earth-fixed frame. However, these design approaches do not take into consideration the demand characteristics on Earth, which makes them inefficient strategies when bearing in mind the non-uniform and uncertain demand over the globe. Accordingly, a more competent strategy would be a staged flexible deployment that adapts the system to the demand evolution and begins covering the regions that have high-anticipated demands.

In Reference 28, another constellation design concept is proposed that can be applied to NGSO systems in order to constitute reconfigurable satellite constellations where satellites can change their orbital characteristics to adjust global and regional observation performance. This concept allows for establishing a flexible constellation for different areas of interest. However, introducing reconfigurability feature to the constellation requires a higher maneuvering capability of the satellites and more energy consumption and that can be a deterrent factor when multiple successive reconfigurations are needed over the life cycle. On the other hand, a hybrid constellation design is proposed in Reference 29 to utilize multiple layers and mixed circular-elliptical orbits, thus, accommodating the asymmetry and heterogeneity of the traffic demand. Nonetheless, the optimization of adapting the constellation to growing demand areas is a challenging issue to be addressed in the context of integrating an entire hybrid model. Moreover, an integrated framework that accounts for the spatial-temporal traffic distributions and optimizes the expected life cycle cost over multiple potential scenarios can be an initial plan to surmount the constellation design hurdles.

Traditional global constellation systems are no longer valid solutions for NGSO systems due to high cost and inflexibility to react to uncertainties resulting from market demands and administrative issues. Therefore, regional coverage constellations are promising solutions for satellite operators as they will be able to tackle the economic and technical issues in a flexible manner [30]. Regional constellations focus on the coverage over a certain geographical region by using a small number of satellites in the system and they can achieve the same or better performance compared to global-coverage constellations. Regional coverage constellations can also provide sufficient redundancy by deploying multiple NGSO satellites in lieu of a single GSO satellite, and thus, operators can hand off traffic to satellites that avoid beam overlapping, and therefore interference [31]. However, designing an optimal regional constellation is a complicated process, which requires optimizing the orbital characteristics (e.g., altitude and inclination) while considering asymmetric constellation patterns, especially for complex time-varying and spatially-varying coverage requirements. This research area has not been well investigated in the open literature, and thus, new sophisticated approaches to design optimal constellation patterns are needed to be tailored to different orbital characteristics and the NGSO environments.

1.3.3 System control and operation

Satellite systems are complex cyber-physical systems that are notoriously difficult to operate owing to the extensive physical distance with the asset. Basically, GSO satellites can be operated individually, since each asset occupies a specific orbital slot and provides service over a specific coverage area. The operation is usually split between two main functions NMC and NCC [17], as presented in Figure 1.5. The two types of operations are tightly linked and there are strict coordination procedures between them, especially when the communication payload has to be reconfigured (e.g., carrier switching, power control). Furthermore, the relevant hardware

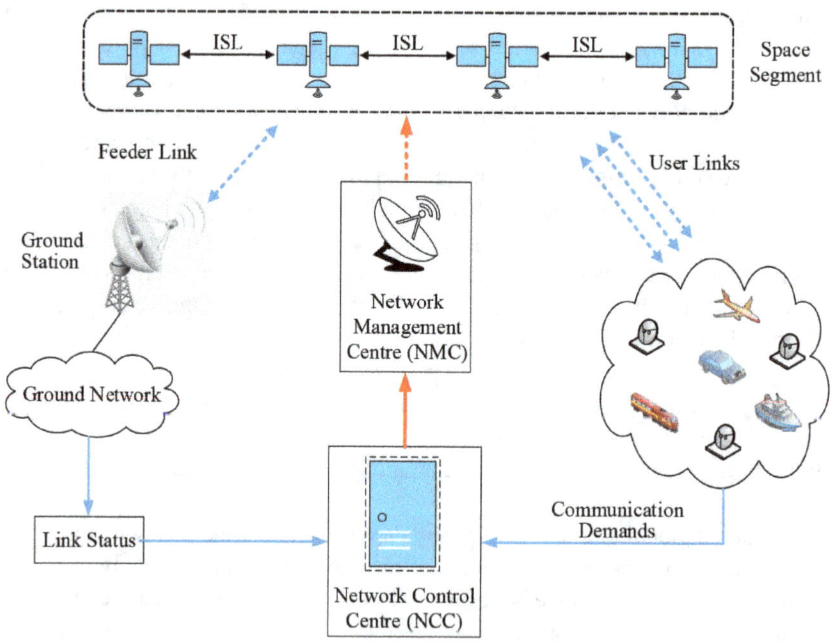

Figure 1.5 Diagram of a satellite communication system architecture

and software for NMC and NCC are usually replicated over multiple geographically distanced sites on the globe to avoid single points of failure on the ground.

For NGSO systems, it is apparent that these operations become even more involved for two main reasons: (i) a large number of gateways is required and (ii) there are multiple satellites that have to be jointly operated/configured so that they optimize the performance of the communication service as the constellation rotates. The former reason is currently a large capital expenditures (CAPEX) for the deployment of mega-constellations, which can be partially mitigated by deploying inter-satellite links for routing communication data in space. The latter reason is mainly driven by the relative motion between the constellation and user terminals, and unbalance of data traffic/demand depending on the geographical location of the users, which requires the constant reconfiguration of satellites in terms of resource allocation.

The control and operation mechanisms are fundamental issues for the NGSO satellites. These issues can be settled by operating NGSO system in either centralized or decentralized manners. In centralized architectures, highly efficient network management can be achieved but that comes at the expense of incurring a non-negligible complexity and an increased operating expenditures (OPEX). Specifically, network controllers in the centralized architectures typically execute in servers located at a terrestrial network. The control channels between a controller and each node (satellite or ground station) will require additional bandwidth resources in addition to the resource allocation burden. On the other hand, in decentralized architectures, each NGSO satellite independently regulates its operating parameters such as power allocation and topology management. It is critical for this architecture to develop energy-efficient and delay-sensitive distributed algorithms that are able to run in the on-board units of satellites such that the number of messages that need to be exchanged among satellites and their neighbours is minimized. However, global optimal control and operation policies are difficult to achieve in this decentralized setup.

Far from the technical aspects, other NGSO operational challenges/concerns are raised by the astronomy community as some rough estimates suggest there could be more than 50 000 satellites in total added to Earth orbits in the near future, which will make our planet blanketed with satellites. Therefore, some experts are alarmed by the plans of mega-constellation companies and raised many concerns specifically about the defunct satellites and smaller pieces of space debris. Additionally, astronomers have already expressed their disquiet about the resulting light pollution from the massive number of visible satellites, which will probably affect their scientific observations of the Universe. Thus, these concerns are briefly discussed next.

- Light pollution: the proliferation of LEO satellites at altitudes less than 2,000 km will jeopardize the ability to observe, discover and analyse the cosmos from the Earth's surface. The astronomy community claims that the number of visible satellites will outnumber the visible stars and that their brightness in both optical and radio wavelengths will significantly influence their scientific research [32]. A major issue with commercial satellite constellations is their visibility from the ground, where the prime contributing factor to light pollution from satellite constellations is the satellites' size. However, currently there are a few mitigating options that can be considered to alleviate these concerns, which are presented in Reference [33]. For instance, making satellites as small as possible, minimizing the reflectivity of satellites, and providing the most accurate satellite orbits to understand observational "avoidance zones" by time or location for astronomy. Venkatesan *et al.* [34] have called this issue an "unfortunate irony" because the technology indebted to centuries of study of orbits and electromagnetic radiation from space now holds the power to prevent the astronomical community from further exploration of the Universe. To this direction, the international astronomical research community has been active seeking a seat at decision-making tables to mitigate the impact of satellite mega-constellation on astronomical research.

- Space debris: since the commercialization of NGSO satellites enters the realm of technical feasibility, many orbital debris concerns have been raised due to the long-term impact that results from placing thousands of satellites in orbits and the risk of causing satellite collisions. Moreover, the advent large constellations of NGSO satellites have been added to the existing debate about the long-term impact of distributed spacecraft missions on orbital debris propagation. Thus, the field of studying the orbital debris is evolving in order to examine the potential debris mitigation strategies. For example, the work in Reference 35 investigates the impact of large satellite constellations on the orbital debris environment and uses OneWeb, SpaceX, and Boeing proposals as case studies. Kelly and Bevilacqua [36] study retrieving and relocating large debris for placement into the "graveyard" orbit above the geostationary regime as a way to mitigate orbital debris congestion. This work derives an analytical deorbit solution based on Lyapunov control theory combined with the calculus of variations. Another cost-effective way to diminish satellite debris has proposed to use a high power pulsed laser system on Earth to make plasma jets on the objects, slowing them slightly, and causing them to re-enter and burn up in the atmosphere [37].

1.3.4 User equipment

Reducing communication latency in satellite systems can only be achieved by moving satellites closer to Earth, namely, the low altitude NGSO satellites offer much lower latency compared with GSO. The closer a satellite is placed, the faster its movement is perceived from the user terminals on Earth, which imposes additional challenges to the user terminal equipment because it has to be able to track the satellite movement and perform handover from one satellite to another. The complexity of user equipment has an impact on its cost, which has been identified as a potential barrier for the commercial success of NGSO satellite communication systems. Previously, broadband LEO networks required expensive user equipment composed of mechanical gimbaled antennas, which has narrowed their roll out to only the customers with high purchasing power mainly within the enterprise market [4]. Thus, a new generation of an antenna and terminal technology was needed that should be affordable, easy to use, and adaptive to the increasingly complex space ecosystem. In other words, inexpensive user equipment capable of tracking LEO satellites is a significant component for widespread adoption and crucial to the business success of NGSO systems. In this context, AST & Science initiative envisions building a space-based cellular broadband network to be accessible by standard smartphones where users will be able to automatically roam from land networks to a space network [38].

Conventional parabolic antennas provide good directivity at the expense of costly mechanical steering [39]. The continuous narrow beam pointing is a challenging task, which has led the ground equipment developers to fight in the battle of technical innovations. Electronic beam steering via antenna arrays, which have thus far been mainly used for military applications, are gaining momentum not only for NGSO satellites but also for moving platforms [40]. Low-cost and high-performance beam-tracking antennas are considered as a game-changer for the satellite community, and several companies are in the final stages of sending their products to the market, e.g., C-ComSat Inc, Kymeta, and ViaSat. Other antenna manufacturers are developing advanced silicon chips that can be used as building blocks of smart digital antennas to create electronically steered multi-beam array antennas [41]. For instance, the startup Isotropic Systems has been working on developing modular antenna systems that are able to track more than one satellite at a time with a single antenna, which will enable multi-orbit operations and reduce the cost by combining their assets into a single integrated terminal without needing to duplicate circuity [42].

Parabolic antennas are difficult to install, configure and operate, but they will still be dominant in governmental institutions and big moving platforms like cruise ships. Nevertheless, electronically steerable flat panel antennas are an imperative ground segment innovation offering a more agile, affordable and scalable antenna product capable of performing the same function as parabolic antennas, opening the door to the NGSO services to also small user terminals. User mobility is another challenge to be addressed using inexpensive antennas. Interestingly, manufacturing a small, low-cost, flat-panel antenna that can be installed on various mobile assets seems feasible with employing the electrically steerable flat panel antennas. Moreover, ground equipment can benefit from satellites that have more flexibility and on-board processing capabilities that allow creating small and high power beams over certain regions or assets, and that will change how the landscape leverages the assets in the sky to facilitate user connectivity on the ground.

Furthermore, the engagement of the satellite industry with the 3GPP to integrate satellite networks into the 5G ecosystem yields an outcome that handheld users can be served by LEO and GSO in S-band with appropriate satellite beam layouts. Besides, other users with high transmit and receive antenna gains (e.g., VSAT and proper phased array antenna) can be served by LEO and GSO in both S-band and Ka-band [13]. This also requires 5G functionalities to take into account the issues of long propagation delays, large Doppler shifts, and moving cells in NTN, and to improve timing and frequency synchronization. The characteristics of this user equipment are specified in Reference 13. In particular, the VSAT user equipment consists of a directional antenna (i.e., phased array antenna) with circular polarization and 60 cm equivalent aperture diameter, whereas the handheld user has an omnidirectional antenna element (e.g., a dipole antenna) with linear polarization.

1.3.5 Security challenges

Proper security mechanisms are essential for NGSO communication systems because they are susceptible to security threats such as eavesdropping, jamming, and spoofing. For instance, any sufficiently well-equipped adversary can send spurious commands to the satellite and gain full access to satellites as well as data, enabling them to cause serious damage. In addition to the blind jamming [43], intelligent jamming exploiting the communication protocols can be used [44]. In this framework, applications of satellite-aided massive uncoordinated access are very vulnerable to such intelligent jamming due to the reduced coordination, i.e., increased uncertainty related to the structure of the received signal. Another example of potentially malicious activity that requires additional security measures is related to denial-of-service attacks, which can be conducted by adversaries via sending a large number of spurious messages to the satellite [45]. Thus, satellites under this attack will spend significant computational processing power and time on spurious messages, which degrades the quality of service for the legitimate users. NGSO satellites can be particularly susceptible to this kind of attacks due to rather limited computational power, such that the satellite can be easily overloaded with processing tasks and may not be able to provide the requested service within the short visibility window.

Security of satellite communication is traditionally provisioned via cryptography-based techniques on the upper layers. The drawback of these techniques is their high computational complexity. Thus, more efficient and sophisticated methods from the areas of quantum key distribution (QKD), blockchain technology (BCT), and physical layer security have been proposed in References 46–48. QKD provides means to detect if the transmission has been eavesdropped or modified. For this, quantum coherence or entanglement is employed, which is based on a unique connection between the transmitter and the receiver. The disadvantage of this scheme is the need to exchange the keys, which may need time since entangled particles need to be produced and sent. Hence, this approach may not always be suitable for NGSO and especially LEO satellites due to the fast passage of the satellite. However, free space optical (FSO) communication technology is an interesting alternative to RF inter-satellite-links owing to the wide bandwidth and high data rate that an FSO system can offer, where the optical technologies are foreseen as a key enabler for ultra-secure communications with the use of QKD.

The communications between ground stations and NGSO satellite constellations require decentralized tracking and monitoring of active and inactive space assets. In addition, it requires assessing the space environment through a network of multi and heterogeneous satellite nodes in different orbits. In this respect, BCT can be utilized for securing satellite communications and authenticating space transactions between the NGSO constellations and ground stations [47]. The key feature of BCT is to authenticate the satellite's identity, ground station's identity, or communication pattern validity through a history record of changes such as the configuration and re-configuration history of the satellite and space information network. Therefore, BCT can be beneficial to protect satellite communication against denial-of-service, distributed denial-of-service, and insider attacks. However, BCT challenges should be scrutinized as well, such as the BCT database storage and distribution for all satellite nodes in a network.

Physical layer security is known to be an effective approach to achieve reasonable levels of security without imposing additional computational complexity for data encryption/decryption [49]. This approach is very popular in terrestrial networks, where the spatial filters are designed with respect not only to the user demands but also to the secrecy against an eavesdropper with a partially known or unknown location. Nevertheless, the satellite-terrestrial communication link usually does not have enough spatial diversity to distinguish between the intended users and eavesdroppers. Interestingly, the joint precoding over multiple NGSO satellites with overlapping coverage areas may solve this issue under some conditions, since the spatial diversity associated with the antennas of the adjacent satellites can be exploited to enhance the secrecy performance. A physical layer security technique can be introduced as an added layer of defence into NGSO satellites but more research efforts are required in this area for further development.

1.4 Future research challenges

Apparently, NGSO satellites are going to be an important part of future wireless communication networks, where they will converge with other wireless systems to achieve ubiquitous coverage, hybrid connectivity, and high capacity. Satellite technologies are under constant development to respond to the fast-changing demands of contemporary commercial and governmental systems through significantly higher capabilities and in a cost-effective manner. The disruptive potential of NGSO satellites does not lay only in serving the poorly connected areas but it also promises to open new frontiers for digital innovation. In this section, we will present some future research directions inspired by utilizing NGSO systems in various applications.

1.4.1 Open RAN

Open Radio Access Network (ORAN) initiatives were developed to split the RAN into multiple functional parts thereby enabling the interoperability of the vendor-independent off-the-shelf hardware, openness of the software and the interfaces. Furthermore, the movement of ORAN actively promotes disaggregated RAN architectures enabled by standardized communication and control interfaces among the constituent components. The goal is to empower innovation, enhance security and increase sustainability. The ORAN Alliance [50] actively promotes these initiatives.

All these aspects are very beneficial for satellite communication systems. For comparison, current satellite networks mainly rely on the implementation by a single manufacturer. This sole manufacturer usually provides all necessary network components, which are 'hard-wired' within the system without any possibility to reconfigure. Hence, such vendor-dependent satellite networks lack flexibility and adaptability, especially for longer missions of more than 10 years, since the satellite hardware components can hardly be replaced while the software can hardly be updated. On the other hand, the persistent growth of the traffic demand and the number of services with varying requirements, demand timely updates of the network configuration. In this context, ORAN offers the possibility to easily exchange the components with more advanced ones or extend the network by incorporating additional infrastructure. In addition, a novel strategy for network management has been proposed for the ORAN architecture, which is based on artificial intelligence (AI) and machine learning (ML)-driven policy definitions and resource management [51, 52]. This strategy enables the AI/ML-based solutions to perform computationally intense tasks and the decision-making triggered by the network itself.

For NGSO satellite networks, the reconfiguration capability and vendor independence of ORAN are of special interest, since they allow a flexible extension of the constellation by adding more satellites or replacing their hardware and software with non-proprietary updates, which may work more efficiently in future. In this context, there are various challenges, since the compatibility of such diverse hardware may require a careful system design. In particular, the availability of data and the way how it is processed in different satellites needs to be taken into account. The most affected use cases for the application of ORAN seem to be resource management, carrier planning, and network adaptation. In addition, multi-layer mega-constellations seem to be the most demanding scenario for such an architecture. These use cases need to be analysed in order to determine the price that needs to be paid for the enhanced flexibility of ORAN.

1.4.2 Broadband connectivity for space missions

Space-based Internet systems emerge as solutions to provide Internet access through a large number of LEO or MEO satellites. In addition to their unique capabilities in providing global coverage, low-latency communication, and high-speed Internet access point, they can notably change the way satellite missions are designed and operated in the near future. More specifically, the number of small satellite constellations in lower orbits for space downstream applications, such as Earth observation, remote sensing, and Internet of Things (IoT) collection, is constantly increasing. Currently, downstream mission operators heavily depend on a network of ground stations distributed across the globe for the purpose of downlinking data and controlling small satellites through telemetry and telecommand. Therefore, one of the key challenges for future space missions is providing real-time uninterrupted connectivity, which is fairly infeasible in the current satellite system infrastructure due to the magnitude and cost of the needed gateway network on the ground. Even though some innovative concepts towards ground network sharing have recently appeared, such as Amazon AWS ground station [53] and Microsoft Azure Orbital [54], the number and duration of ground access sessions are most of the times limited, preventing real-time mission operation and continuous high-throughput downstreaming data.

Assuming a scenario where small satellites for downstream applications can directly access the Internet via a space-based Internet provider in a higher orbit, the small satellites can be constantly connected to the network without depending on a private or shared distributed network of ground stations [55]. This is certainly a game changer for the design and operation of future downstream satellite missions since the communication link has to be pointing towards the sky instead of the Earth. This approach can be also replicated for the space-based Internet providers to enable a larger degree of connectivity in space network topologies. Further, this structure can lead to more inexpensive and sustainable space systems by reducing the number of required ground stations, while achieving real-time and reliable space communications. Employing the space-based Internet systems in this context can provide coordination of multiple constellations and awareness of the operational characteristics of each counterpart system. Additionally, space-based Internet systems will allow a satellite system to function strategically by transmitting telemetry, tracking, and command data between small satellite terminals and satellite control centres on the ground. However, the expected connectivity improvement will be achieved at the cost of higher complexity which is essential for load balancing between satellite links and for finding paths with the shortest end-to-end propagation delay, as well as tackling the dynamicity of the nodes (e.g., high relative speeds and frequent handovers), which are yet unexplored areas in the literature.

1.4.3 Edge computing

One of the main challenges for the operation of satellites in general and especially NGSO satellites is the rather low information processing capabilities of the on-board processors [56]. Consequently, complex processing tasks, such as online optimization of the resource allocation strategy, data processing for Earth observation applications, data aggregation for IoT, etc., can hardly be executed using a single satellite processor. Instead, the processing can be done in a distributed manner by pushing it from the central unity, e.g., GEO satellite, to the edge, e.g., NGSO satellites [57, 58]. Besides that, computation offloading via NGSO satellites has been proposed in various works, e.g., [59]. Moreover, edge computing has emerged as a promising solution to alleviate the high latency issue by deploying processing and storage resources closer to users, especially for resource-hungry and delay-sensitive applications. Thus, integrating edge computing into NGSO networks can improve the performance of satellite networks by providing near-device processing capability. In this system, large amount of data generated by users can be processed through NGSO satellites instead of redirecting it to other servers, which will reduce network traffic load and the processing delay. While this application seems very promising, its practical limitations and requirements are not yet fully understood as it has started to attract the attention of researchers only in the last few years.

1.4.4 Space-based cloud

Far from the common use of satellites as relay devices, the space-based cloud concept has emerged as a promising and secured paradigm for data storage over NGSO satellites, particularly in the context of big data technologies and applications [60]. The key advantage of space-based data storage is providing complete immunity from natural disasters occurring on Earth. Furthermore, utilizing NGSO satellites for data storage can offer more flexibility to some cloud networks that are designed to transfer data globally regardless of the geographical boundaries and terrestrial obstacles [61]. For instance, mega-corporations and large organizations that are located at different global sites can share big data through a space-based cloud and benefit from the faster transfer rate compared to the traditional terrestrial cloud networks, especially for delay-sensitive services.

In this perspective, a startup company named Cloud Constellation is planning to establish a space-based datacentre platform, named SpaceBelt that is offering secure data storage through LEO satellites and well-connected secure ground networks. In this infrastructure, the data-storage system is built upon multiple distributed satellites equipped with data-storage servers. However, the communication window between a ground station and an NGSO satellite is sporadic and the power budget in satellites is limited. Hence, this infrastructure imposes a significant challenge on developing scheduling algorithms for energy-efficient downloading files from the space-based datacentres to meet the dynamic demands of users under time-varying channel conditions. Besides, the existing operational algorithms for task scheduling in terrestrial cloud datacentres are not applicable to the space-based cloud infrastructures [62].

1.4.5 IoT via NGSO satellites

The flexibility and scalability properties of NGSO satellites make their employment within the IoT ecosystem more appealing to shape novel architectures that uplift the interoperability among a plethora of applications and services [25]. Thus, by exploiting the relatively short propagation distances of NGSO satellite constellations, IoT terminals can be designed to be small-sized, long-life, and low-power, which is ideal for IoT operation. Moreover, the reduced OPEX and CAPEX of NGSO satellites compared to GSO ones render them into efficient facilitators for the deployment of efficient IoT services over wide geographical areas [63]. Hence, these exceptional features of NGSO satellites can unleash the full potential of IoT, which will establish a universal network with billions of worldwide interconnected devices.

In this direction, the 3GPP organization in its release 17 [64] has studied the necessary changes to support Narrow-Band IoT (NB-IoT) over satellites, including both GSO and NGSO systems. The objective here is to identify a set of features and adaptations enabling the operation of NB-IoT within the NTN structure with a priority on satellite access. In this context, some works have already started to adapt and evaluate these protocols under the NGSO system constraints specifically the relative satellite motion [65]. Nevertheless, the progress is still in its infancy and more research efforts are required for a seamless integration, particularly in connecting NGSO satellites to mobile or stationary IoT devices and supporting ultra reliable low latency communications.

1.4.6 Caching over NGSO satellites

Benefiting from the high-capacity backhaul links and ubiquitous coverage, NGSO satellites can help bring content closer to the end users, and thus, these satellites can be considered as an option for data caching. NGSO satellites also have the ability to multi-cast data and quickly update the cached content over different locations. Additionally, the symbiotic relationship between satellite and terrestrial telecommunication systems can be exploited to create a hybrid federated content delivery network, which will substantially ameliorate user experience. Therefore, integration of NGSO satellites into future Internet with enabling in-network caching makes traffic demands from users for the same content to be easily accommodated without multiple transmissions, and thereby, more spectral resources can be saved along with reducing transmission delay. Further, a promising strategy in this context is the combination of caching with edge computing over NGSO satellites, such that data processing, content analysis and caching are seamlessly integrated and harmonized. However, the time-varying network topology and limited on-board resources in NGSO satellites have to be taken into account when designing caching placement algorithms alongside their fast convergence and low complexity.

1.4.7 UAV/HAP and NGSO coordination

The use cases of low-cost unmanned aerial vehicles (UAVs) as flying mobile base-station are rapidly growing to expand wide-scale coverage range and improve wireless network capacity. Integrating terrestrial, airborne, and satellite networks into a single wireless system could provide comprehensive and efficient services. Moreover, UAVs and high-altitude platforms (HAPs) offer a high degree of mobility and a high chance for LoS connectivity, which makes them perfect mobile relays for satellite-terrestrial links. The use of NGSO and especially LEO satellites seems very promising due to a much smaller latency compared to GSO satellites, which is a necessary condition for the proper functioning and autonomous operation of the UAVs.

By introducing UAVs as part of the integrated space-air-ground system novel types of networks have been envisioned [66], such as UAV-aided cognitive satellite-terrestrial networks [67], cell-free satellite-UAV networks as part of future 6G systems [68], etc. Specifically, massive integrated networks are envisioned with multiple satellite orbits as part of NGSO mega-constellations, multiple UAVs and HAPs. Such networks pose many challenges for the coordination, navigation and synchronization. Some of the typical impairments to be considered in this context are high Doppler shift, pointing errors and outdated channel state information. Another challenge is the topology control and multi-hop signal routing for such dynamic networks.

1.5 Conclusions

The deployment of NGSO satellites has been trending over recent years owing to their low free-space attenuation, low-profile antenna, small propagation delay, and reduced orbital injection cost per satellite. The successful realization of NGSO communication systems is being achieved by the ongoing development of new technologies and the growing interest and investments, which have indeed pushed the satellite communication potentials towards higher bounds that need to be explored to support the rapid proliferation of various space-based applications and services. In addition, NGSO systems can be employed to support the terrestrial networks and facilitate matching the rapid 5G ecosystem evolution by increasing the offered coverage and network capacity.

This chapter has presented the uprising technologies and research outlook in the realm of NGSO satellite systems along with the key technical challenges to integrate NGSO satellites into the global wireless communication platforms. In particular, the state-of-the-art in NGSO systems has been discussed first via exploring the attributes of both NGSO space-based Internet providers and small satellite missions. Next, in addition to studying the restrictions due to the coexistence with GSO systems, constellation design and resource management challenges, and user equipment requirements have been explored as well. Finally, several future research directions to deliver highly reliable and efficient global satellite communications for various applications were discussed.

References

[1] Perez-Neira A.I., Vazquez M.A., Shankar M.R.B., Maleki S., Chatzinotas S. 'Signal processing for high-throughput satellites: challenges in new interference-limited scenarios'. *IEEE Signal Processing Magazine*. 2019, vol. 36(4), pp. 112–31.

[2] Giambene G., Kota S., Pillai P. 'Satellite-5G integration: a network perspective'. *IEEE Network*. 2019, vol. 32(5), pp. 25–31.

[3] Su Y., Liu Y., Zhou Y., Yuan J., Cao H., Shi J. 'Broadband LEO satellite communications: architectures and key technologies'. *IEEE Wireless Communications*. 2019, vol. 26(2), pp. 55–61.

[4] del Portillo I., Cameron B.G., Crawley E.F. 'A technical comparison of three low earth orbit satellite constellation systems to provide global broadband'. *Acta Astronautica*. 2019, vol. 159, pp. 123–35.

[5] *UCS satellite database 2021* [online]. Bethesda (MD): Union of Concerned Scientists. 2021. Available from https://www.ucsusa.org/resources/satellite-database

[6] 'Simulation methodologies for determining statistics of short-term interference between co-frequency, codirectional non-geostationary-satellite orbit fixed-satellite service systems in circular orbits and other non-geostationary fixed-satellite service systems in circular orbits or geostationary-satellite orbit fixed-satellite service networks'. *ITU-R S.1325-3*. 2003.

[7] Guidotti A., Vanelli-Coralli A., Conti M, *et al.* 'Architectures and key technical challenges for 5G systems incorporating satellites'. *IEEE Transactions on Vehicular Technology*. 2019, vol. 68(3), pp. 2624–39.

[8] Guan Y., Geng F., Saleh J.H. 'Review of high throughput satellites: market disruptions, affordability-throughput map, and the cost per bit/second decision tree'. *IEEE Aerospace and Electronic Systems Magazine*. 2019, vol. 34(5), pp. 64–80.

[9] Al-Hraishawi H., Maturo N., Lagunas E., Chatzinotas S. 'Scheduling design and performance analysis of carrier aggregation in satellite communication systems'. *IEEE Transactions on Vehicular Technology*. 2021, vol. 70(8), pp. 7845–57.

[10] Kibria M.G., Lagunas E., Maturo N., Al-Hraishawi H., Chatzinotas S. 'Carrier aggregation in satellite communications: impact and performance study'. *IEEE Open Journal of the Communications Society*. 2020, vol. 1, pp. 1390–402.

[11] Al-Hraishawi H., Lagunas E., Chatzinotas S. 'Traffic simulator for multibeam satellite communication systems'. *2020 10th Advanced Satellite Multimedia Systems Conference and the 16th Signal Processing for Space Communications Workshop (ASMS/SPSC)*; Graz, Austria, 2020. pp. 1–8.

[12] Giordani M., Zorzi M. 'Non-terrestrial networks in the 6G era: challenges and opportunities'. *IEEE Network*. 2021, vol. 35(2), pp. 244–51.

[13] '3GPP TR 38 821 V16 0 0. 3rd generation partnership project; technical specification group radio access network; solutions for NR to support non-terrestrial networks (NTN)'. *3rd Generation Partnership Project*. 2019, vol. 16.

[14] Di B., Song L., Li Y., Poor H.V. 'Ultra-dense LEO: integration of satellite access networks into 5G and beyond'. *IEEE Wireless Communications*. 2019, vol. 26(2), pp. 62–69.

[15] Babich F., Comisso M., Cuttin A., Marchese M., Patrone F. 'Nanosatellite-5G integration in the millimeter wave domain: A full top-down approach'. *IEEE Transactions on Mobile Computing*. 2020, vol. 19(2), pp. 390–404.

[16] Latio R.G.W. 'Social and cultural issues: the impact of digital divide on development and how satellite addresses this problem'. *Online Journal of Space Communication*. 2021, vol. 2(5), pp. 1–17.

[17] *ETSI TS 101 545-3. DVB-RCS2 higher layer satellite specification* [online]. European Telecommunications Standards Institute. 2020. Available from http://www.etsi.org

[18] Al-Hraishawi H., Chatzinotas S., Ottersten B. 'Broadband non-geostationary satellite communication systems: research challenges and key opportunities'. *IEEE International Conference on Communications Workshops (ICC Workshops)*; Montreal, QC, 2021. pp. 1–6.

[19] *Functional description to be used in developing software tools for determining conformity of non-geostationary-satellite orbit fixed-satellite service systems or networks with limits contained in article 22 of the radio regulations*. International Telecommunication Union-Recommendation (ITU-R) S.1503-3; 2018.

[20] European Space Agency (ESA). *CONSTELLATION - Simulator for managing interference between NGSO constellations and GSO*. 2019. Available from https://artes.esa.int/projects/constellation

[21] Higgins R.P., Boeing C. *Method for limiting interference between satellite communications systems*. US Patent No. 6,866,231. 2005.

[22] Federal Communications Commission. *Updated rules to facilitate non-geostationary satellite systems* [online]. 2017. Available from Https://www.fcc.gov/document/updated-rules-facilitate-non-geostationary-satellite-systems

[23] Tonkin S., De Vries J.P. 'NewSpace spectrum sharing: assessing interference risk and mitigations for new satellite constellations'. *SSRN Electronic Journal*. 2018, pp. 1–108.

[24] *Methods to enhance sharing between NGSO FSS systems (except MSS feeder links) in the frequency bands between 10-30 GHz*. International Telecommunication Union-Radiocommunication (ITU-R) S.1431; 2020.

[25] Qu Z., Zhang G., Cao H., Xie J. 'LEO satellite constellation for internet of things'. *IEEE Access*. 2017, vol. 5, pp. 18391–401.

[26] Walker J.G. 'Satellite constellations'. *Journal of the British Interplanetary Society*. 1984, vol. 37, pp. 559–72.

[27] Mortari D., Wilkins M.P., Bruccoleri C. 'The flower constellations'. *The Journal of the Astronautical Sciences*. 2004, vol. 52(1–2), pp. 107–27.

[28] Paek S.W. 2012. *Reconfigurable Satellite Constellations for GEO-spatially Adaptive Earth Observation Missions*. [dissertation]. Massachusetts Institute of Technology.

[29] Chan S., Samuels A., Shah N., Underwood J., de Weck O. 'Optimization of hybrid satellite constellations using multiple layers and mixed circular-elliptical orbits [online]'. *22nd AIAA International Communications Satellite Systems Conference & Exhibit (ICSSC)*; Monterey, CA, Reston, Virigina, 2004. pp. 1–15. Available from http://arc.aiaa.org/doi/book/10.2514/MICSSC04

[30] Lee H.W., Shimizu S., Yoshikawa S., Ho K. 'Satellite constellation pattern optimization for complex regional coverage'. *Journal of Spacecraft and Rockets*. 2004, vol. 57(6), pp. 1309–27.

[31] Lee H.W., Jakob P.C., Ho K., Shimizu S., Yoshikawa S. 'Optimization of satellite constellation deployment strategy considering uncertain areas of interest'. *Acta Astronautica*. 2018, vol. 153, pp. 213–28.

[32] McDowell J.C. 'The low earth orbit satellite population and impacts of the spacex starlink constellation'. *The Astrophysical Journal*. 2018, vol. 892(2), L36.

[33] Constance W., Jeffrey H., Lori A, *et al*. 'Impact of satellite constellations on optical astronomy and recommendations toward mitigations'. *Bulletin of the AAS*. vol. 52(2). n.d.

[34] Venkatesan A., Lowenthal J., Prem P., Vidaurri M. 'The impact of satellite constellations on space as an ancestral global commons'. *Nature Astronomy*. 2018, vol. 4(11), pp. 1043–48.

[35] Foreman V.L., Siddiqi A., De Weck O. 'Large satellite constellation orbital debris impacts: case studies of oneweb and spacex proposals'. *AIAA SPACE*

and Astronautics Forum and Exposition; Orlando, FL, Reston, Virginia, 2018. Available from https://arc.aiaa.org/doi/book/10.2514/MSPACE17

[36] Kelly P., Bevilacqua R. 'An optimized analytical solution for geostationary debris removal using solar sails'. *Acta Astronautica*. 2019, vol. 162, pp. 72–86.

[37] Phipps C.R., Baker K.L., Libby S.B, *et al.* 'Removing orbital debris with lasers'. *Advances in Space Research*. 2019, vol. 49(9), pp. 1283–300.

[38] *AST mobile space: transforming connectivity eliminating coverage gaps for billions* [online]. AST and Science. 2021. Available from hhttps://ast-science.com/

[39] Cheng Y., Song N., Roemer F, *et al. 2012 IEEE first AESS European conference on satellite telecommunications (ESTEL)*; Rome, Italy, 2019. pp. 1–7.

[40] Guidotti A., Vanelli-Coralli A., Foggi T, *et al.* 'LTE-based satellite communications in LEO mega-constellations'. *International Journal of Satellite Communications and Networking*. 2019, vol. 37(4), pp. 316–30. Available from https://onlinelibrary.wiley.com/toc/15420981/37/4

[41] Sadhu B., Gu X., Valdes-Garcia A. 'The more (antennas), the merrier: A survey of silicon-based mm-wave phased arrays using multi-IC scaling'. *IEEE Microwave Magazine*. 2016, vol. 20(12), pp. 32–50.

[42] Isotropic Systems Inc. *The world's first multi-service high throughput terminals* [online]. Available from https://www.isotropicsystems.com

[43] Lichtman M., Poston J.D., Amuru S, *et al.* 'A communications jamming taxonomy'. *IEEE Security & Privacy*. 2016, vol. 14(1), pp. 47–54.

[44] Lichtman M., Reed J.H. 'Analysis of reactive jamming against satellite communications'. *International Journal of Satellite Communications and Networking*. 2016, vol. 34(2), pp. 195–210. Available from http://doi.wiley.com/10.1002/sat.v34.2

[45] Roy-Chowdhury A., Baras J.S., Hadjitheodosiou M., Papademetriou S. 'Security issues in hybrid networks with a satellite component'. *IEEE Wireless Communications*. 2016, vol. 12(6), pp. 50–61.

[46] Bonato C., Tomaello A., Da Deppo V., Naletto G., Villoresi P. 'Feasibility of satellite quantum key distribution'. *New Journal of Physics*. 2009, vol. 11(4), 045017.

[47] Xu R., Chen Y., Blasch E., Chen G. 'Exploration of blockchain-enabled decentralized capability-based access control strategy for space situation awareness'. *Optical Engineering*. 2019, vol. 58(4), p. 1.

[48] Zheng G., Arapoglou P.D., Ottersten B. 'Physical layer security in multibeam satellite systems'. *IEEE Transactions on Wireless Communications*. 2011, vol. 11(2), pp. 852–63.

[49] Al-Hraishawi H., Baduge G.A.A., Schaefer R.F. 'Artificial noise-aided physical layer security in underlay cognitive massive MIMO systems with pilot contamination'. *Entropy*. 2017, vol. 19(7), p. 349.

[50] Alliance O. 'ORAN: towards an open and smart RAN'. *White Paper*. 2018.

[51] Gavrilovska L., Rakovic V., Denkovski D. 'From cloud RAN to open RAN'. *Wireless Personal Communications*. 2020, vol. 113(3), pp. 1523–39.

[52] Alliance O. 'ORAN working group 2: AI/ML workflow description and requirements'. *Technical Report*. 2019.

[53] *AWS ground station* [online]. Amazon.Com. 2021. Available from https://aws.amazon.com/ground-station

[54] *Azure orbital* [online]. Microsoft.Com. 2021. Available from https://azure.microsoft.com/en-us/services/orbital

[55] Al-Hraishawi H., Minardi M., Chougrani H., Kodheli O., Montoya J.F.M., Chatzinotas S. 'Multi-layer space information networks: access design and softwarization'. *IEEE Access*. 2020, vol. 9, pp. 158587–98.

[56] Lovelly T.M., Bryan D., Cheng K, *et al.* 'A framework to analyze processor architectures for next-generation on-board space computing'. IEEE Aerospace Conference, 2014. pp. 1–10.

[57] Zhang Z., Zhang W., Tseng F.H. 'Satellite mobile edge computing: improving qos of high-speed satellite-terrestrial networks using edge computing techniques'. *IEEE Network*. 2019, vol. 33(1), pp. 70–76.

[58] Li C., Zhang Y., Xie R., Hao X., Huang T. 'Integrating edge computing into low earth orbit satellite networks: architecture and prototype'. *IEEE Access*. 2021, vol. 9, pp. 39126–37.

[59] Tang Q., Fei Z., Li B., Han Z. 'Computation offloading in LEO satellite networks with hybrid cloud and edge computing'. *IEEE Internet of Things Journal*. 2021, vol. 8(11), pp. 9164–76.

[60] Jia X., Lv T., He F., Huang H. 'Collaborative data downloading by using inter-satellite links in LEO satellite networks'. *IEEE Transactions on Wireless Communications*. 2017, vol. 16(3), pp. 1523–32.

[61] Huang H., Guo S., Wang K. 'Envisioned wireless big data storage for low-earth-orbit satellite-based cloud'. *IEEE Wireless Communications*. 2018, vol. 25(1), pp. 26–31.

[62] Huang H., Guo S., Liang W., Wang K., Okabe Y. 'CoFlow-like online data acquisition from low-earth-orbit datacenters'. *IEEE Transactions on Mobile Computing*. 2020, vol. 19(12), pp. 2743–60.

[63] Bacco M., Cassara P., Colucci M., Gotta A. 'Modeling reliable M2M/IoT traffic over random access satellite links in non-saturated conditions'. *IEEE Journal on Selected Areas in Communications*. 2018, vol. 36(5), pp. 1042–51.

[64] Liberg O., Lowenmark S.E., Euler S, *et al.* 'Narrowband internet of things for non-terrestrial networks'. *IEEE Communications Standards Magazine*. 2020, vol. 4(4), pp. 49–55.

[65] Chougrani H., Kisseleff S., Martins W.A., Chatzinotas S. 'NB-IoT random access for non-terrestrial networks: preamble detection and uplink synchronization'. *IEEE Internet of Things Journal*. 2021, pp. 1–1.

[66] Hosseini N., Jamal H., Haque J., Magesacher T., Matolak D.W. 'UAV command and control, navigation and surveillance: A review of potential 5G and satellite systems'. *IEEE Aerospace Conference*; Big Sky, MT, 2019. pp. 1–10.

[67] Hua M., Wang Y., Lin M., Li C., Huang Y., Yang L. 'Joint comp transmission for UAV-aided cognitive satellite terrestrial networks'. *IEEE Access*. 2019, vol. 7, pp. 14959–68.
[68] Liu C., Feng W., Chen Y., Wang C.-X., Ge N. 'Cell-free satellite-UAV networks for 6G wide-area internet of things'. *IEEE Journal on Selected Areas in Communications*. 2021, vol. 39(4), pp. 1116–31.

Chapter 2

Spectrum regulation for non-geostationary orbit satellite systems

Jesús Arnau[1]

2.1 Introduction

The last few years have witnessed the development of several non-geostationary orbit (NGSO) satellite constellations that aim to provide global broadband coverage, be it for professional communications or for direct-to-home Internet connectivity [1–3]. In total, they consist of several thousand of satellites, all of them operating in bands already in use by geostationary (GSO) satellite networks. The goal of this chapter is to review, at a tutorial level, the existing international regulations surrounding the use of spectrum by these NGSO constellations, with a particular focus on how they share spectrum with GSO networks.

Satellite communications are international in nature and therefore need some form of global regulation regarding their access to the radio spectrum. The global treaty fulfilling this function is the International Telecommunications Union (ITU) Radio Regulations (RR) [4]. The principle underlying this treaty is that Member States that sign it must operate their radio stations—whatever their purpose—so that they do not cause harmful interference to the radio services of other Member States that operate in accordance with the provisions in the treaty.

The RR establishes the procedure that satellite services must follow to obtain international recognition for the frequency and orbital resources they plan to use. Such recognition is obtained on a first-come, first-served basis by submitting a request to the ITU. For several types of satellite services, including NGSO satellite systems, the underlying principle is that international recognition is acquired through successful negotiations with affected administrations (i.e., through coordination agreements).

Furthermore, according to the RR, and unless otherwise stated, NGSO systems shall not cause unacceptable interference to, and shall not claim protection from, GSO networks in the fixed-satellite service (FSS) and the broadcasting-satellite service (BSS) (see Article 22.2 of the RR). The RR also defines ways to

[1]Spectrum Group, Ofcom, London, United Kingdom The views expressed herein are those of the author and do not necessarily reflect those of his employer

quantify what unacceptable interference is in this case, although two countries can always reach a bilateral agreement allowing them to cause more interference to the services of the other; in such a case, both countries still need to protect services of third countries according to the RR.

In this chapter, we provide a tutorial introduction to spectrum regulation for NGSO satellite systems. While there are NGSO systems operating in the mobile-satellite service (MSS), the scope of this chapter is limited to the regulations surrounding NGSO systems operating in the FSS, which mostly use K_a and K_u band for their links. We will briefly touch upon their planned use of higher bands, too.

The remainder of this chapter is structured as follows. Section 2.2 provides an introduction to the basic aspects of national and international regulation of satellite services, including an introduction to the RR. Section 2.3 summarises the spectrum usage and characteristics of some of the main NGSO FSS satellite systems. Section 2.4 explains in more detail the process by which NGSO systems obtain international recognition for their frequency assignments. Section 2.5 provides further detail into the way NGSO systems and GSO networks share spectrum. The chapter ends by listing some open challenges in NGSO spectrum management in section 2.6.

2.2 The basics of spectrum regulation for satellite services

Satellite services are amongst the most international radio communications services by their very nature; transmissions from a single GSO satellite can reach almost one third of the globe if using a global antenna beam, and constellations of NGSO satellites can cover the whole Earth. For this reason, international coordination of satellite systems is crucial. Moreover, operators will want to place earth station in several countries and have the right to transmit and, if needed, to receive signals in specific frequency bands. In other words, satellite operators need to obtain recognition for the spectrum used by both space stations (satellites) and earth stations (user terminals and gateways), as follows:

1. For the space segment, operators need to obtain international recognition for the frequency and orbital resources they plan to use. These are obtained on a first-come, first-served basis by submitting a request to the ITU in what is called a satellite filing. Such recognition is conditional on certain requirements being met as specified by the RR, the global treaty that governs the use of radiocommunication spectrum.
2. For the earth segment, operators need to obtain licences from individual countries, unless there is an explicit exemption in place. Licences provide national recognition for the use of certain frequencies within a country. Licences are awarded to stations only if they meet certain requirements that facilitate their coexistence with other stations—hence their importance from the point of view of national spectrum management.

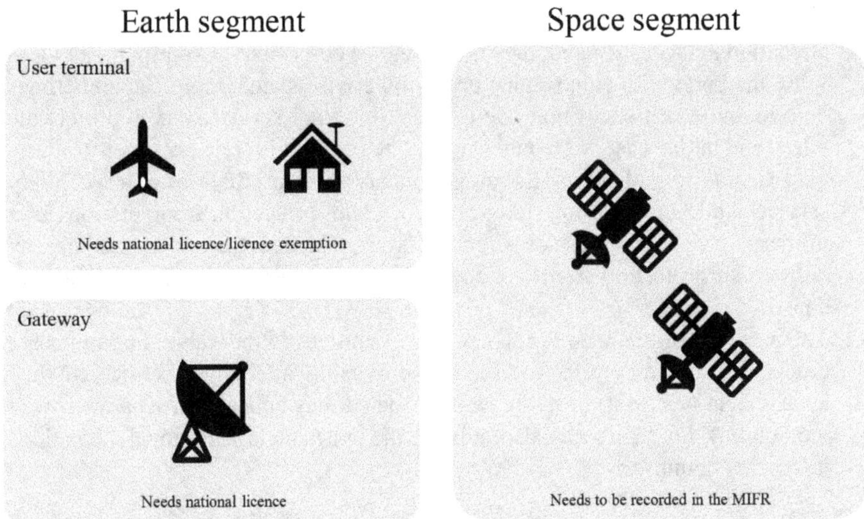

Figure 2.1 *Regulatory framework for the different parts of a NGSO satellite system*

2.2.1 Introduction to the radio regulations

A commonplace statement in radio communications is to say that radio waves do not stop at country borders. It is very true, and because of this, a global agreement is needed to regulate the use of the radio spectrum. As already mentioned, the RR [4] is the global treaty fulfilling this function.

The RR is an international treaty that is binding to ITU Member States. It is revised by administrations and members of the ITU during World Radio Conferences (WRC), which take place roughly every 4 years. It consists of four volumes containing articles, appendices, Resolutions and recommendations, respectively [5].

The founding principles of these regulations are outlined in their preamble. Notably, Members States shall bear in mind that radio frequencies and orbits are limited natural resources and that they must be used rationally, efficiently and economically. Member States shall operate their radio stations—whatever their purpose—so that they do not cause harmful interference to the radio services of other Member States operating in accordance to the provisions of this treaty.

2.2.1.1 Article 5

One of the key pieces of the RR is the Table of Frequency Allocations contained in Article 5, which currently covers the radiocommunications spectrum from 8.3 kHz to 275 GHz. It details which services can be operated in each band and each ITU Region (see RR No. 5.2), and under which conditions. A radiocommunications service involves the transmission, emission and/or reception of radio waves for specific telecommunication purposes (c.f. RR No. 1.19).

1. Services in CAPITALS* are primary services. If several services have a primary allocation in the same band, they are said to be co-primary. Co-primary services share the band with equal rights unless otherwise stated: there can sometimes be a footnote that states that one of the co-primary services has to protect the other, and in that case we would say that the protected service is super-primary.
2. Services in 'normal characters' are secondary services. Stations of a secondary service shall not cause interference to, or claim protection from, stations of a primary service regardless of when they were assigned. Secondary stations can only claim protection from stations of secondary services assigned at a later date.
3. The numbers of the form 5.XXX refer to footnotes to the Table. Footnotes are crucial because they refer to essential provisions affecting a service (if they appear right next to it) or to the whole band (if they appear in a separate row at the bottom). Footnotes can also indicate the existence of additional allocations in specific countries.

It shall be remarked that allocations are to radiocommunication services, and not to applications or technologies: the RR are technically neutral. However, it can sometimes be seen that a band has been identified or designated for a specific use. These concepts are not explicitly defined and have no regulatory implication, but express the interest or intention of some administrations on a future use of that band for a specific technology or application. A common example is No. 5.516B, which identifies several bands for high-density applications in the fixed-satellite service†; another example is the identification of the bands 24.25–27.5 GHz, 42.5–43.5 GHz, 47.2–48.2 GHz, 45.5–47 GHz and 66–71 GHz for use by administrations wishing to deploy the terrestrial component of International Mobile Telecommunications (IMT) achieved during WRC-19.

It is also important to remark that if administrations assign frequencies to a station in violation of the Table of Frequency Allocations (or of other provisions of the RR), they must do so under the express condition that such station will be operated on a non-interference, non-protection basis from stations operating in accordance with the regulations (RR No. 4.4). Each Member State is thus sovereign over the radio spectrum in its own territory, subject to the respect of the RR. In use of their sovereignty, administrations create their own national frequency allocation tables in reflection of the RR's. Examples are the USA table of frequency allocations [6], the UK FAT [7] and the Tableau national de répartition des bandes de fréquences of France [8]; the European Communications Office keeps links to the national table of allocations of 48 countries on the European continent [9].

*Of the six languages of the ITU, this applies to the English, French, Russian and Spanish versions. Bold characters are used instead in the Arabic and Chinese versions.

†According to Resolution 143 (rev. WRC-19), HDFSS systems are characterised by the flexible, rapid and ubiquitous deployment of large numbers of cost-optimised earth stations employing small antennas and having common technical characteristics.

2.2.1.2 Article 9

This article details the procedure for effecting coordination with or obtaining agreement of other administrations. We will discuss coordination in more detail later on.

2.2.1.3 Article 11

Another key piece of the RR is Article 11, which deals with the notification and recording of frequency assignments. The notification process is crucial because administrations shall notify to the Bureau all frequency assignments capable of causing harmful interference to any service of another administration (No. 11.3). Moreover, administrations can also choose to notify any other frequency assignments for which they wish to obtain international recognition (No. 11.7). Upon reception of a notification, the Bureau will examine it with respect to its conformity with the relevant provisions of the RR. Frequency assignments that receive a favourable finding are recorded in the master register. According to No. 8.3, any frequency assignment recorded in the master register with a favourable finding shall have the right to international recognition, i.e. other administrations shall take it into account when making their own assignments, in order to avoid harmful interference.

2.2.1.4 Article 21

This article sets power limits that terrestrial and space services shall respect in some of the bands they share on a co-primary basis in order to facilitate coexistence. Such limits are examined under the relevant provisions of Article 11 (see previous item) before granting international recognition to frequency assignments.

2.2.1.5 Article 22

This article is of special interest for the purposes of this chapter, as it is devoted to space services. In particular, Section II of Article 22 details how space services shall control interference to GSO satellite systems and includes the notion that, unless otherwise stated, NGSO systems shall not cause unacceptable interference to, and shall not claim protection from, GSO networks in the fixed-satellite service and the broadcasting-satellite service (RR No. 22.2). We shall explore the implications of this article much more in detail in sections 2.4.2 and 2.5.

2.2.1.6 Appendices

The appendices to the RR contain additional provisions and information needed to apply the RR. Examples of appendices of interest in satellite communications are:

- Appendix 4, defining the data that must be provided to the ITU, when filing for a frequency assignment.
- Appendix 5, defining when coordination is required under the provisions of Article 9.
- Appendix 7, containing methods to determine the coordination area around earth stations.

- Appendix 8, containing a method to determine if coordination is needed between two GSO satellites.
- Appendices 30, 30A and 30B. These appendices describe the rules around the broadcasting-satellite and fixed-satellite service plans. These are quite different rules that apply to some specific frequency bands. In planned bands, orbital slots and frequency resources are split into allotments and distributed equitably among Member States. Operations in planned bands are protected from harmful interference from other networks even if they are not currently in operation, thus ensuring capacity for future use by Member States. Further information can be found in Reference 10.

2.2.1.7 Resolutions

Resolutions can be seen as a type of decisions adopted by WRC. They can be used to invite the ITU-R or the Radiocommunications Bureau to carry out certain actions, like when setting the agenda of the next WRC. An example of this is Resolution 811 (WRC-19), which sets the agenda for WRC-23.

More commonly, Resolutions are used to state rules that administrations shall comply with when using certain bands or services. For example, Resolution 750 (rev. WRC-19) deals with compatibility between the Earth exploration-satellite service (passive) and relevant active services, and identifies maximum recommended unwanted emission powers that shall not be exceeded in order to ensure such compatibility. Other examples of particular relevance for NGSO systems are the following:

1. Resolution 32 (WRC-19), on regulatory procedures for short-duration NGSO missions.
2. Resolution 35 (WRC-19), on milestones for NGSO constellations deployment, see section 2.4.5.
3. Resolution 76 (rev. WRC-15), on the protection of GSO satellite networks from the aggregate interference from several NGSO systems in K_u and K_a band. We will discuss it in more detail in section 2.5.2.
4. Resolution 769 (WRC-19), on the protection of GSO satellite networks from the aggregate interference from several NGSO systems in Q/V band. We will discuss it in more detail in section 2.5.3.1.
5. Resolution 770 (WRC-19), on the protection of GSO networks from single-entry interference from NGSO systems in Q/V band, as explained further in section 2.5.3.

2.2.1.8 Rules of procedure

The Rules of Procedure [11] are a separate document approved by a separate entity, the Radio Regulations Board. They provide clarifications to the application of particular regulations, or establish practical procedures that are sometimes needed to properly apply the RR. The rules of procedure shall be used by administrations and the Radiocommunication Bureau in the application of the RR.

2.2.1.9 Recommendations

Another relevant set of documents that are not part of the RR (although they can sometimes be incorporated by referencing them) are the ITU-R recommendations.

Recommendations provide advice on a range of issues, including how to model radio propagation and antenna radiation patterns, the best methodologies to carry out certain calculations, or the amount of interference or degradation that certain links should be designed to tolerate.

Some recommendations of interest for GSO and NGSO systems are listed below.

- ITU-R P.618, which contains propagation data and prediction methods required for the design of Earth-space telecommunication systems.
- ITU-R S.465, which contains a reference radiation pattern of Fixed Satellite Service (FSS) earth station antennas for use in coordination and interference assessment in the frequency range 2–31 GHz.
- ITU-R S.1323, which recommends maximum permissible levels of interference in an FSS satellite network caused by other codirectional FSS networks below 30 GHz.
- ITU-R S.1325, describing simulation methodologies for determining statistics of short-term interference between certain NGSO systems and GSO networks.
- ITU-R S.1432, on the allowable error performance degradations to FSS hypothetical reference digital paths arising from time invariant interference below 30 GHz.
- ITU-R S.1503, containing a functional description to be used in developing software tools for determining conformity of NGSO FSS systems with certain limits in Article 22 of the RR. We describe this recommendation in greater detail in sections 2.5.2.2 and 2.6.1.

2.2.2 National licensing

A radio licence gives national recognition for the use of certain frequencies within a country. It limits the technical and operational characteristics of the concerned radio station (or stations, as we will see) so that it can coexist with other spectrum users.

According to RR No. 18.1, all transmitting stations established or operated by private persons or enterprises must be licensed by or on behalf of the government of the country to which it is subject. Other provisions of RR Article 18 deal with the essential responsibilities of the licensee and provide guidance on licensing mobile stations that may move across administrations. When translated into national terms, No. 18.1 implies that the operation of a transmitting station on Earth requires either:

1. **An individual licence.** For example, if an administration has deployments of two co-primary services in a band, it will often need to resort to individual licences to ensure interference is managed.
2. **A network licence.** These can apply to certain geographic areas or to whole countries. For example, mobile stations (terrestrial or satellite) are often licensed in this way.
3. **A licence exemption.** That is, a national decision that allows operation without any form of licence on a non-interference basis, provided that the equipment meets certain technical parameters that limit the interference it can cause. Wi-Fi and Bluetooth devices are examples of this.

Licences often come with a fee, and may take some time for the administration to process depending on their complexity and on whether the process can be automated or not [12].

It is worth remarking again that nations have sovereignty over the radio spectrum in their territory, subject to respecting the RR treaty that binds them.

2.3 Summary of bands used by NGSO systems in the fixed-satellite service

As explained in section 2.1, in this chapter we focus on NGSO systems operating in the FSS in K_u and K_a band. Before explaining the international regulatory framework in more detail, we provide a summary of the characteristic and spectrum usage of some of those systems in Table 2.1.

It is worth mentioning that several of these systems also plan to make use of higher bands (around 50 GHz and 40 GHz) in future evolutions; while this is out of the scope of this chapter, we will introduce some of the NGSO regulatory framework around these bands later in section 2.5.3. Finally, it is also worth remarking that NGSO constellations operating in other radiocommunication services (like the MSS) operate in other bands.

Table 2.1 Summary of constellations and their band use

	Frequency bands (GHz)			
	Feeder link	**User link**	**#sat.**	**Altitude (km)**
Kuiper	↓ 17.7 -18.6, 18.8 – 20.2 ↑27.5 – 29.1, 29.5 – 30.0		3236	590, 610, 630
OneWeb	↓ 17.8 – 18.6, 18.8 – 19.3	10.7 – 12.7	650	1100, 1200
SpaceX	↑27.5 – 29.1, 29.5 – 30.0	14.0 – 14.5	4408	590, 540, 570
Telesat	↓ 17.8 – 18.6, 18.8 – 19.3, 19.7 – 20.2 ↑27.5 – 29.1, 29.5 – 30.0		298	1015, 1325

Source: [13, 14] and references therein.

Table 2.1 shows how existing and planned constellations target most of the FSS allocations in K_a band, and sometimes also in K_u band. From a regulatory perspective these bands have different characteristics, as we explain below.

2.3.1 K_u band downlink (10.7–12.7 GHz)

In this band, the FSS is co-primary with the terrestrial fixed and mobile services. The part 10.7–11.7 GHz has an FSS allocation across all regions, but this is not the case above 11.7 GHz (e.g. there is no FSS allocation in region 1 in the part 11.7–12.5 GHz, see Chapter II of Article 5 of the RR).

2.3.2 K_u band uplink (14.0–14.5 GHz)

Within this band, in the part 14.0–14.25 GHz the FSS shows as co-primary with the radionavigation service only. However, the radionavigation service shall protect the FSS in that band as per No. 5.504 of the RR.

2.3.3 K_a band downlink

- **17.7–18.6 GHz**. In this band, the FSS is co-primary with the terrestrial fixed and mobile services. This means that, depending on the use in each country, operators might need to coordinate their earth stations with e.g. fixed links, or to accept interference from them.
- **18.6–18.8 GHz**. This band does not show up in Table 2.1 because it is limited to systems with an orbit of apogee greater than 20 000 km.
- **18.8–19.3 GHz**. In this band, the FSS is also co-primary with the terrestrial fixed and mobile services. Within this band, NGSO systems are subject to coordination with respect to GSO networks, i.e. Article 22.2 of the RR does not apply.
- **19.3–19.7 GHz**. In this band, the FSS is also co-primary with the terrestrial fixed and mobile services.
- **19.7–20.2 GHz**. In this band, only the FSS and MSS have a primary allocation, which makes it generally easier for operators to place earth stations. The whole band is identified for high-density applications across all regions (see RR 5.516B).

2.3.4 K_a band uplink

- **27.5–29.1 GHz**. In this band, the FSS is co-primary with the terrestrial fixed and mobile services. This means that, depending on the use in each country, operators may need to coordinate their earth stations with e.g. fixed links. Within this band, NGSO systems in the segment 28.6–29.1 GHz are subject to coordination with respect to GSO networks and Article 22.2 does not apply.
- **29.1–29.5 GHz**. This band does not show up in Table 2.1 because its use by NGSO systems is limited to feeder links for those operating in the MSS.

- **29.5–30 GHz**. Like in the downlink segment 19.7–20.2 GHz mentioned above, in this band, only the FSS and MSS have a primary allocation, which makes it generally easier for operators to place earth stations. The whole band is identified for high-density applications across all regions.

2.4 Efficient use of spectrum and orbital resources: the filing process

Radiocommunication spectrum is considered to be a scarce resource, and so are orbital resources—and very especially orbital slots in the GSO arc. This is why, according to the RR, Member States must bear in mind that radio frequencies and satellite orbital locations are limited natural resources and strive to use them rationally, efficiently and economically, in such a way that countries may have equitable access to both (see Article 0.3 of the RR).

As hinted in section 2.2, international recognition of a satellite frequency assignment is obtained through submitting a satellite filing to the ITU. Simplifying a bit, the process goes as follows:

1. Operators submit a filing through an administration
2. The Radiocommunications Bureau checks that the filing is compliant with existing regulation. If so:
3. Operators need to coordinate with previously filed systems, as appropriate.
4. The filing is recorded in the Master International Frequency Register.
5. The filing must be brought into use before a certain deadline to retain its recognition.

Below we describe each of these steps more exhaustively. Do note that we must omit some details for ease of exposition; see e.g. [15] and [16] for a more comprehensive account.

2.4.1 Initial submission of a satellite filing

A satellite filing is a description of the frequency assignments of a satellite network in terms of its characteristics, including orbital parameters, space station transmission and reception parameters (including frequency bands, emission bandwidth, power, antenna gain and receiver noise temperature), earth station parameters, type of service and service area [15]. A filing can only be submitted to the ITU by an administration of a Member State, and not directly by a private company.

2.4.2 First examination of a filing by the Radiocommunication Bureau

When the Radiocommunication Bureau receives a filing, it then examines it in accordance with the RR and the rules of procedure. An NGSO filing must be compliant with the RR, and among other things this means it must

- Be consistent with the international table of frequency allocations (see Article 5 of the RR).
- Respect the satellite emission limits that ensure protection of co-frequency terrestrial services, if applicable (see Article 21 of the RR).
- In K_u and K_a, respect the equivalent power-flux density (EPFD) limits that ensure protection of GSO networks both from earth stations and space stations, except in the bands 18.8–19.3 GHz and 28.6–29.1 GHz where coordination applies instead (see Article 22 of the RR).
- In Q/V bands, respect maximum degradation limits defined over a set of hypothetical GSO reference links (see Article 22 of the RR and Resolution 770).

The last two bullet points are quite important in our exposition. As explained, No. 22.2 of the RR says that, unless otherwise stated, NGSO systems must not cause interference to, or claim protection from, GSO networks in the FSS and BSS that operate in accordance with the RR. This principle has resulted in slightly different conditions in the K_u and K_a bands as compared to the Q/V bands. In both cases, however, operators will have to put technical mitigations in place in order to ensure protection of GSO networks.

Coexistence between NGSO systems and GSO networks is a crucial topic in spectrum management nowadays, and we will revisit it in a dedicated section (section 2.5). Before moving on, we shall reiterate that No. 22.2 does not apply in two parts of K_a band, one in the uplink and one in the downlink, where NGSO systems and GSO networks must complete coordination on a first-come, first-serve basis instead. We explain the principles of coordination in more detail in the paragraphs below.

2.4.3 Seeking agreement from other administrations

Filings are processed differently depending on whether the requested bands are planned or unplanned.

In bands subject to a plan, equitable access to spectrum is guaranteed by *a priori* planning, e.g., by creating allotments with certain orbital locations and technical parameters, and ensuring their protection regardless of whether the allotment is in use or not.

In bands not subject to a plan, recognition is obtained on a first-come, first-served basis. More specifically, international recognition is acquired through successful negotiations with affected administrations; in other words, by reaching a coordination agreement. This is the case for the NGSO systems under discussion in this chapter: NGSO systems operating in the FSS need to seek coordination agreements with previously filed NGSO systems operating or planning to operate in the same bands. Also, and as already explained, NGSO systems and GSO networks in the FSS must reach coordination agreements in the bands 18.8–19.3 GHz and 28.6–29.1 GHz.

Coordination enables parties to discuss and find solutions that will allow their satellite systems to coexist. This often requires the use of technical mitigation methods, but agreements beyond technical parameters are also possible [17].

Coordination is meant to allow equitable access to bands and orbital resources while protecting existing systems. However, it requires time and effort to complete, especially in congested bands where the list of existing networks may be very long. It can also introduce uncertainty in the business plan because it introduces operational constraints that cannot be accurately predicted at the start of the filing process.

2.4.4 Recording in the master international frequency register

An NGSO system which has completed coordination can then be notified, meaning that it is recorded in the Master International Frequency Register. Recording in this register gives international recognition to the system, which can then have regulatory confidence over the parameters it has filed for and coordinated. In regulatory terms, this international recognition means other administrations shall take these assignments it into account when making their own, in order to avoid harmful interference–see No. 8.3 of the RR.

Even if coordination is not completed with some previous systems, frequency assignments could still be recorded in the master register with the important caveat that they shall not cause interference to, or claim protection from, those previous systems with which coordination was not achieved.

It shall be remarked that the notification process is not exclusive of frequency assignments of space services. According to the RR, administrations shall notify to the Bureau all frequency assignments capable of causing harmful interference to any service of another administration (No. 11.3). Administrations can also choose to notify any other frequency assignments for which they wish to obtain international recognition (No. 11.7).

2.4.5 Bringing into use

The international recognition given by the recording in the master register is conditional on several things. To start with, the frequency assignment must be brought into use within a defined regulatory time limit, i.e., a station that is capable of using such frequency assignment must be put in place before a certain deadline. A bringing into use deadline helps prevent spectrum warehousing and improves the overall efficiency of the process.

For a GSO satellite, e.g., the deadline is 7 years from the date of receipt of the filing. This concept has been recently adapted to NGSO systems, too: while in the past a full constellation of hundreds of satellites could be brought into use by putting into orbit a single satellite, a series of completion milestones must be fulfilled now—see Resolution 35 (WRC-19).

In particular, operators must deploy:

- 10% or more of the total number of satellites within 2 years after the end of the 7-year period.
- 50% or more of the total number of satellites within 5 years after the end of the 7-year period.
- 100% within 7 years after the end of the 7-year period.

Finally, it is worth remarking there are other changes that could affect an NGSO system once it has been correctly recorded in the master register. For example, there are limits to how much degradation all NGSO systems can cause to GSO networks. If such limits are ever exceeded, administrations shall remediate the situation by appropriate modification of their systems; see Resolution 76 (Rev. WRC-15) and Resolution 769 (WRC-19) [4].

2.5 Sharing between NGSO systems and GSO networks

As already mentioned, coexistence between NGSO systems and GSO networks is a crucial topic in spectrum management. The underlying principle is that, unless otherwise stated, NGSO systems must not cause interference to, or claim protection from, GSO networks in the FSS and BSS that operate in accordance with the RR. To achieve this, NGSO systems need to resort to several types of mitigation techniques that make this coexistence possible. Additionally, this principle has resulted in different implementations across different spectrum bands, as we explain in the following paragraphs.

2.5.1 Technical mitigation techniques to facilitate coexistence

Non-GSO systems need to resort to different mitigation techniques in order to be able to share the spectrum, both with other NGSO systems and with GSO networks [18, 19]. The most basic techniques are frequency planning and coverage design that are aware of other spectrum users, but other more sophisticated options are also common. For example, earth stations may leverage satellite diversity and communicate only with satellites that require pointing far from the GSO arc. Satellites may also change their parameters, e.g., by ceasing transmission when passing through the main beam of a GSO satellite link, or by tilting the satellite in order to increase discrimination.

2.5.2 Regulatory framework in K_u and K_a band

2.5.2.1 Description

As we explained in section 2.4.3, in K_u and K_a bands NGSO systems shall respect EPFD limits contained in Article 22 of the RR, except in the bands 18.8–19.3 GHz and 28.6–29.1 GHz, where coordination applies instead. An NGSO system that complies with these limits is deemed to be in compliance with No. 22.2 of the RR, i.e., it is deemed not to cause harmful interference to GSO networks—including future networks. As mentioned, these EPFD limits may be exceeded on the territory of countries that have so agreed. These limits apply to each NGSO system, but aggregate limits for all co-frequency NGSO systems have also been specified.

The EPFD is defined as the sum of power-flux densities produced, by all stations of an NGSO system, at a point on the Earth's surface or in the geostationary orbit, as appropriate:

$$\text{epfd} = 10\log_{10}\left(\sum_{l=1}^{N} 10^{\frac{P_l}{10}} \cdot \frac{G_t(\theta_l)}{4\pi d_l^2} \cdot \frac{G_r(\phi_l)}{G_{r,max}}\right)$$

where P_l is the transmit power of the l-th station, d_l is the distance to the point under consideration, $G_t(\theta_l)$ is the antenna gain of the transmitting station in the direction of the receive station, $G_r(\phi_l)$ is the receive antenna gain in the direction of the transmit station, and $G_{r,max}$ is the maximum gain of that same receive station.

2.5.2.2 Implementation of the framework

To determine whether an NGSO system complies with the single entry EPFD limits in Article 22, the ITU performs a check at the filing stage using the methodology described in ITU-R S.1503 [20]. This methodology plays a crucial role in assessing whether NGSO FSS systems meet the EPFD limits not only in Ku and Ka band but also in parts of C band; it is also a component of the methodology to assess whether NGSO FSS systems meet limits on unavailability and throughput that exist for higher bands, as we will explain in Section 2.5.3.

The methodology in ITU-R S.1503 takes as input the NGSO system characteristics as specified in the ITU filing. Each of these parameters should be measurable, thus enabling administrations to verify they are being complied with. They should also be understood as limits, i.e., an NGSO system shall operate under the envelope they describe and shall not exceed that envelope during its operation, but can otherwise modify its operating characteristics provided that the above-mentioned limits are not exceeded.

ITU-R S.1503 determines the geometry that would result in the highest short-term EPFD and uses it in all calculations. This is known as the worst-case geometry. If several geometries produce the same short-term EPFD, then the worst-case geometry is the one with the highest likelihood, typically the one for which the lowest elevation angle of the GSO earth station occurs. Using that worst-case geometry, the algorithm then computes the statistical distribution of the EPFD, and compares it with the limits in Article 22 of the RR.

This procedure deals with the emissions of one single NGSO constellation, but how to ensure and verify that the aggregate emissions from all co-frequency NGSO constellations do not exceed the aggregate limits in Article 22? Note that aggregate emissions are very important from the perspective of GSO links.

The rules to deal with aggregate emissions are set in Resolution 76 in the RR, which states that administrations operating or planning to operate NGSO FSS systems in Ku and Ka band shall take all possible steps, including, if necessary, by modifying their systems, to ensure that the aggregate interference into GSO FSS and GSO BSS networks caused by such systems operating co-frequency in these frequency bands does not exceed the levels indicated in that recommendation.

How to implement this in practice was largely left open and is currently a matter of discussion in the ITU. But finding a solution might be challenging: if many co-frequency NGSO systems were put in orbit, the first to arrive could subsequently need to constrain its operation as new systems come into fruition, in order to ensure compliance with the aggregate limits.

However, it is worth remarking that, according to technical studies, at least three systems would be needed to create a small risk of aggregate interference (see *considering d)* in Resolution 76, which states that the single-entry limits in the RR were derived assuming a maximum effective number of 3.5 co-frequency NGSO systems). This is partly because NGSO constellations already must avoid interfering one another in order to operate, and therefore have some inherent flexibility to adapt and reduce the chances of aggregate interference if needed.

2.5.3 Regulatory framework in higher bands

In section 2.3, we mentioned that NGSO constellations plan to start making use of higher frequency bands in the near future. For this reason, WRC-19 adopted regulatory provisions to facilitate coexistence between NGSO systems and GSO networks in the bands 37.5–39.5 GHz and 39.5–42.5 GHz for the space-to-Earth direction, and 47.2–50.2 GHz and 50.4–51.4 GHz in the Earth-to-space direction. These are colloquially said to be part of the Q/V bands.

The thinking underlying the framework in these bands is slightly different to the one in K_u and K_a bands. In this case, NGSO systems must comply with limits in terms of increase in unavailability and decrease in average throughput that they cause to a set of hypothetical reference GSO links. Simplifying, these are (see Article 22.5 M):

- A single-entry increase of 3% in the unavailability of the GSO link, and
- A single-entry reduction of at most 3% in terms of spectral efficiency averaged over a year.

For each frequency, only reference GSO links whose original availabilities are in a certain range, i.e. which enjoy a certain range of link margin, are considered for protection and used in the calculations. Naturally, these calculations require accounting for the impact of rain fading, which is of paramount importance in Q/V bands.

In addition to the limits above, the aggregate emissions from all NGSO systems in the band shall not create more than a 10% increase in unavailability nor more than 8% decrease in average spectral efficiency to any reference link (see Article 22.5L). Aggregate limits are also specified for a set of so-called supplementary links, though these have not been fully defined yet—we will come back to this in later sections.

2.5.3.1 Implementation of the framework

Like in K_u and K_a bands, compliance with single-entry limits is checked through computer calculations by the ITU when the filing is received‡.

Also like in K_u and K_a bands, the methodology in ITU-R S.1503 is used to obtain the statistics of the EPFD generated by the NGSO satellites. The key difference though is that, while in K_u/K_a band the EPFD statistics are directly compared with the limits in Article 22, in Q/V band they are used to calculate two other metrics (increase in unavailability and decrease in spectral efficiency) in a set of hypothetical reference links.

The compliance with aggregate limits is also different to K_u/K_a band, and in this case consultation meetings are mandated. The current regulatory text is also more prescriptive on the steps to be taken should aggregate emissions exceed the limits. Also, and as explained, compliance with aggregate limits will have to be checked also on a second set of links, called supplementary links, whose exact form is yet to be defined; we touch upon this again in section 2.6.1.

2.5.4 Comparison

The sharing frameworks in K_u/K_a and Q/V bands differ in both the underlying metrics and environment they consider. In K_u and K_a bands, NGSO systems must not exceed the EPFD limits on any location visible from the GSO arc, and the methodology in ITU-R S.1503 is used to ascertain this. On the other hand, in Q/V bands NGSO systems are evaluated by checking the degradation they would cause to a set of hypothetical reference GSO links; such degradation is measured in terms of reduction in spectral efficiency and increase in unavailability. It is also worth remarking that in Q/V band, rain attenuation is accounted for directly in the calculations.

2.6 Open challenges

We finish this chapter by summarising some open challenges in spectrum management of NGSO satellite systems.

2.6.1 Dealing with aggregate interference

We have already described how taking into account the limits to aggregate interference could be difficult in practice.

In K_u and K_a bands there are no clearly defined procedures on how to ensure that the aggregate limits are not exceeded. On the other hand, for Q/V bands the RR establish the creation of periodic consultation meetings among affected administrations in order to ensure that the aggregate emissions from all co-frequency NGSO systems do not cause unacceptable interference to GSO networks.

‡The ITU does not yet have a software to do this for Q/V bands, so a provisional favourable finding (called a qualified favourable finding) is given to all requests that are backed by the filing administration.

As we mentioned, for Q/V bands the concept of supplementary links was introduced to verify the aggregate limits. Supplementary links are to be provided directly by administrations and should be representative of real operational GSO links. Work is still underway to define which conditions should a link fulfil in order to be considered representative and worth protecting, and also whether there should be different checks to those applied to the hypothetical reference links. The outcome of this work will have a strong impact on balance between NGSO flexibility of operation and GSO protection in those bands.

2.6.2 Accurately modelling NGSO systems in operation

Modern NGSO satellite systems tend to have flexible payloads, which, among other things, allows them to adapt their emission characteristics over time and space. Thanks to this, satellite operators not only adapt to varying service demands but also to different interference environments by implementing interference mitigation techniques.

It is important that the regulatory framework be capable of reflecting such interference mitigation techniques while still providing effective protection to other users of the spectrum.

One example of this is the software described in ITU-R S.1503. Such software should not underestimate the amount of interference created by an NGSO system, but should also allow modelling their real operation and interference mitigation techniques. Because of this, ITU-R S.1503 is frequently under review to better model NGSO systems.

2.6.3 Spectrum monitoring for NGSO systems

Because of the use of small spot beams and the above-mentioned payload flexibility, the emission characteristics of NGSO systems are not geographically homogeneous. As a consequence, if interference occurs, it may occur in some areas only, and it may also fluctuate quickly over time.

This has implications on spectrum monitoring and if a case of interference is reported, because it means that measurements taken far away from the victim site, (e.g. at an international spectrum monitoring station) will not necessarily represent the same interference conditions. For this reason, in the future there may be an increased need for administrations and interested parties in general to have monitoring capabilities that allow to take measurements as close as possible to the victim stations.

References

[1] Harris M. 'Tech giants race to build orbital internet [news]'. *IEEE Spectrum*. 2018, vol. 55(6), pp. 10–11.

[2] del Portillo I., Cameron B.G., Crawley E.F. "A technical comparison of three low-Earth orbit satellite constellation systems to provide global broadband".

Proceedings of 69th International Astronautial Congress (IAC); Bremen, Germany, Oct 2018. n.d.

[3] Kodheli O., Guidotti A., Vanelli-Coralli A. 'Integration of satellites in 5G through Leo constellations'. IEEE Global Communications Conference (GLOBECOM 2017); Singapore, 2017.

[4] ITU-R *Radio regulations*. Geneva: International Telecommunications Union; 2020.

[5] Restrepo J. *"ITU radio regulations", presentation during the ITU world radiocommunication seminar* [online]. 2020. Available from https://www.itu.int/en/ITU-R/seminars/wrs/2020/Pages/presentations.aspx

[6] Federal Communications Commission. *FCC online table of frequency allocations* [online]. Available from https://transition.fcc.gov/oet/spectrum/table/fcctable.pdf [Accessed Feb 2021].

[7] Ofcom. *United Kingdom frequency allocation table* [online]. 2017 Jan. Available from https://www.ofcom.org.uk/__data/assets/pdf_file/0016/103309/uk-fat-2017.pdf

[8] *Agence nationale des fréquences, "Tableau national de répartition des bandes de fréquences"* [online]. 2021 May. Available from https://www.anfr.fr/fileadmin/mediatheque/documents/tnrbf/TNRBF_2021-05-04.pdf

[9] *European communications office, list of national frequency tables* [online]. Available from https://efis.cept.org/views2/national_frequency_table.jsp

[10] Wang J. *Introduction to Bss & FSS plans [presentation during the ITU World Radiocommunication Seminar* [online]. 2020. Available from https://www.itu.int/en/ITU-R/seminars/wrs/2020/Pages/presentations.aspx

[11] ITU-R *Rules of procedure*. Geneva: International Telecommunications Union; 2021.

[12] ITU-R *Handbook on national spectrum management*. Geneva: International Telecommunications Union; 2015.

[13] ITU Satellite Webinars. *Episode 2: non-geostationary satellite systems: entering into the era of broadband service delivery* [online]. 2020 Oct. Available from https://www.itu.int/en/ITU-R/space/workshops/sat-webinars/Pages/default.aspx

[14] Ofcom. *Non-geostationary satellite systems: licensing updates* [online] [public consultation document]. 2021 Jul. Available from https://www.ofcom.org.uk/consultations-and-statements/category-2/non-geostationary-satellite-systems

[15] Ofcom. *Procedures for the management of satellite filings* [online]. 2016 Mar. Available from https://www.ofcom.org.uk/__data/assets/pdf_file/0027/63495/new_procedures.pdf

[16] Sakamoto M. *Overview of coordination and notification procedures (non-planned space services* [online] [presentation during the 29th World Radiocommunication Seminar]. 2020 Dec. Available from https://www.itu.int/en/ITU-R/seminars/wrs/2020/Pages/presentations.aspx

[17] Tham D. *Satellite coordination meetings, presentation during the ITU regional radiocommunication seminar for Asia* [online]. 2014. Available from

https://www.itu.int/en/ITU-R/space/workshops/RRS-14-Asia/Documents/CoordinationMeetings.pdf

[18] Sharma S.K., Chatzinotas S., Ottersten B. 'In-line interference mitigation techniques for spectral coexistence of Geo and NGEO satellites'. *International Journal of Satellite Communications and Networking*. 2016, vol. 34(1), pp. 11–39.

[19] Tonkin S., De Vries J.P. 'New space spectrum sharing: assessing interference risk and mitigations for new satellite constellations'. *SSRN Electronic Journal Presented at TPRC46: Research Conference on Communications, Information and Internet Policy, Washington DC, USA, October 2018*. n.d.

[20] ITU-R S.1503 *Functional description to be used in developing software tools for determining conformity of non-geostationary-satellite orbit fixed-satellite service systems or networks with limits contained in article 22 of the radio regulations*. Geneva: International Telecommunications Union; 2018 Jan.

Chapter 3

The role of non-geostationary orbit satellite systems in 5G integration

Alessandro Guidotti[1] and Alessandro Vanelli-Coralli[2]

The last years have seen an unprecedented demand for improved broadband connectivity, near-zero latency services, and ultra-reliable and heterogeneous communications. Such a trend is expected to further increase in the near future, with forecasts of 5.3 billion Internet users and 14.7 billion machine-to-machine (M2M) connections by 2023 [1]. The evolution of 5G into beyond 5G (B5G) and 6G networks aims at responding to this increasing need for ubiquitous and continuous connectivity services in all areas of our life: from education to finance, from politics to health, from entertainment to environment protection.

In this context, the multifaceted applications and services pose a vast variety of requirements calling for a flexible, adaptable, resilient and cost-efficient network able to serve heterogeneous devices with different capabilities and constraints. Today, it is well understood that only a *network of networks*, integrating multiple access methods, i.e., terrestrial fixed and wireless, and non-terrestrial (NT), will be able to provide the required capabilities [2–24]. The importance of airborne and spaceborne nodes, as foreseen for the 5G network infrastructure, is expected to become even more critical in the future, as preliminary envisioned by International Telecommunications Union (ITU) and 5G Infrastructure Public Private Partnership (5G-PPP), which highlight that low earth orbit (LEO) constellations will play a pivotal role for the 6G ecosystem [25, 26]. An NT component is able to extend and complement the terrestrial network both in rural/remote areas, in which a terrestrial infrastructure might not be available, and in densely populated areas, in which the actual capacity availability might be limited due to the users' density. This is substantiated by the attention that NT systems have been receiving within 3GPP; in fact, the definition of a global 5G-based standard including non-terrestrial networks (NTN) for all platforms, orbits, bands and devices can be a key enabler for a smooth integration of NT components into the 5G ecosystem. The integration of satellite and aerial access networks in 5G is foreseen to bring manifold benefits

[1]Consorzio Nazionale Interuniversitario per le Telecomunicazioni (CNIT), Research Unit at the University of Bologna, Bologna, Italy

[2]Department of Electrical, Electronic, and Information Engineering, University of Bologna, Bologna, Italy

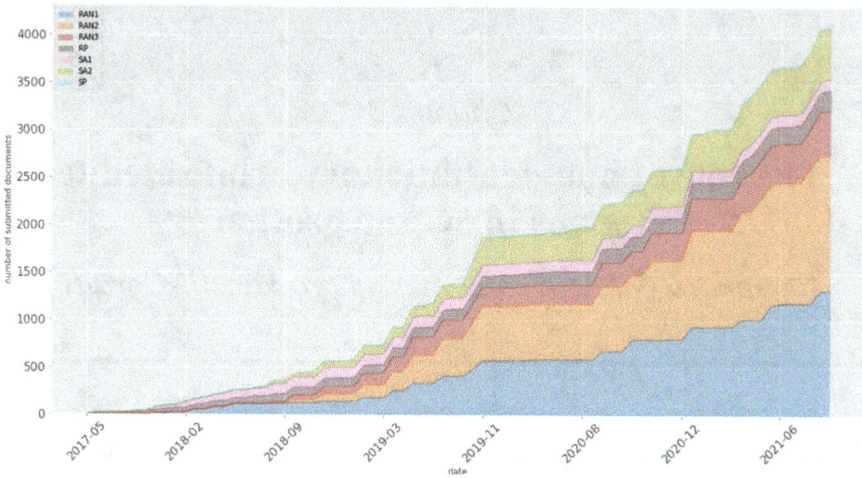

Figure 3.1 Number of documents submitted to 3GPP meetings related to NTN[*]

[3–9]: (1) complement 5G services in under- or un-served areas; (2) improve the 5G service reliability and continuity for M2M or Internet of Things (IoT) devices, or for mission critical services; and (3) enable the 5G network scalability by means of efficient multicast/broadcast resources for data delivery. Since Rel. 15, with the initial study items (SI) for 5G NTN on the definition of deployment scenarios and technical challenges, two work items (WI) were approved for Rel. 17 for the actual integration of the NT component in the 5G architecture aiming at [10]: (1) consolidating the preliminary performance assessment and potential impacts on the physical (PHY) and medium access control (MAC) layers; (2) analysing aspects related to Layers 2 and 3, including handover and dual connectivity (DC); and (3) identifying potential requirements for the upper layers. Moreover, from Rel. 17, narrow band-IoT (NB-IoT) has entered the normative phase as well. Between 2017 (when the first study proposals started) and 2021 (at the time of writing this chapter), 17 WIs and SIs have been defined exclusively for NTN, leading to a considerable amount of documents submitted to radio access network (RAN) and Service and system Aspects (SA) meetings, shown in Figure 3.1.

In the above landscape, an increasing interest is being received by non-geostationary orbit (NGSO) systems, mainly justified by the need to cope with the stringent low latency requirements for many of the services to be provided, e.g., mission critical, automotive, e-health, and by the technology innovations in the satellite payload miniaturisation, paving the way for nano-/pico-satellites and CubeSats. With respect to latency, it is worthwhile highlighting that LEO and very low earth

[*]The values reported in the figure have been obtained by web scraping of public information available on the 3GPP website on 28 October 2021.

orbit (vLEO) systems might provide a more favourable context even compared to terrestrial networks, over long distances. In fact, electromagnetic waves propagate at the speed of light in space, while in optic fibres they can travel at approximately 65–70% of that speed [11, 12]. In addition to that, NGSO nodes can be easily connected to any type of terminal on ground, from fixed dedicated ground stations to new radio (NR) gNode-B (gNBs), from moving platforms on ships or aircrafts to IoT devices. This is possible thanks to the manifold advantages that they have compared to Geosynchronous Orbit (GSO) systems: (1) reduced path loss, in particular for LEO or vLEO; (2) for a fixed antenna size and configuration, NGSO nodes produce smaller footprints on ground, allowing to increase the frequency reuse and ultimately the system throughput; (3) easier access to spectrum, since Ku-/Ka-band spectrum can be accessed without requests on a secondary basis, i.e., NGSO operators cannot claim protection from GSO systems and they need to guarantee that their operations will not affect GSOs; (4) the geometry between the nodes and the user terminals is varying over time, i.e., NGSO systems provide path diversity, which is particularly helpful in urban scenarios; and (5) the elevation angles at which NGSO nodes are seen by the user terminals are typically larger compared to legacy geostationary earth orbit (GEO) systems. It shall be noticed that we refer to the flying elements in the NGSO space segment as *NGSO nodes*; in fact, these systems consist in communication nodes that can be space- or air-borne, flying at different altitudes, and communicating among them, and with the terrestrial network if needed, through inter-node links.

To provide a truly global coverage, with at least one node covering any given terminal at any time, anywhere, a massive number of NGSO nodes is required; currently, there are many industrial and commercial endeavours in the new space era aiming at the realisation of NGSO mega-constellations, such as SpaceX Starlink, Amazon Kuiper, Oneweb and Telesat, to name a few. In this framework, many diverse technical challenges arise, including, but not limited to, constellation management, advanced resource management techniques and network orchestration. In this chapter, we review the architectures and related services for the integration of NGSO systems in the 5G ecosystem, also highlighting the main research and development challenges.

3.1 GPP standardisation

Currently, the activities within 3GPP are mainly aimed at completing the specifications for Rel. 17 and at identifying the technologies and network enhancements, if any, that shall be included in Rel. 18. In Guidotti *et al.* [10], a detailed overview of the status of 3GPP activities until Rel. 17 is provided and summarised below:

- **Rel. 15**. Two SIs started in 2017 on: (1) 'Study on NR to support Non Terrestrial Networks', with the RAN plenary as responsible group with the support of RAN1 (PHY layer); and (2) 'Study on using Satellite Access in 5G', under the supervision of SA1 (services). As for the former, the objective was to [4]: define the deployment scenarios and the related system parameters, identify and assess the potential key impact areas on the NR and identify the required adaptations of the 3GPP channel models for NTN; based on the outcome of these analyses, potential solutions for the identified key impacts on the RAN protocols and architectures were proposed. As for the latter SI, it was indeed initiated in 2017, but then moved to Rel. 16 and associated to the WI on 'Integration of Satellite Access in 5G', discussed below. It focused on the definition of a set of use cases for the integration of a satellite component in NR, together with the identification of the potential services, and a categorisation of use cases based on service continuity, ubiquity and scalability [5].

- **Rel. 16**. In this framework, the following activities were: (1) an SI on 'Study on architecture aspects for using satellite access in 5G', under SA2 (architecture) [6], (2) a WI on 'Integration of Satellite Access in 5G', under the supervision of SA1 [27]; and (3) an SI on 'Study on management and orchestration aspects with integrated satellite components in a 5G network', under SA5 (management) [7]. In these studies, the impacted areas related to the integration of an NT component into NR were identified, together with a set of potential solutions for two specific use cases: roaming between terrestrial and satellite networks, and 5G fixed backhaul between satellite enabled NR-RAN and the 5G Core; moreover, issues related to the interaction between the RAN and the core network were also discussed. The requirements of NR systems provided in TS 22.261 were extended with a satellite component in terms of multiple access technologies and connectivity models, with the addition of specific KPIs for NTN. Finally, aspects related to management and orchestration for NTN were also addressed, in particular identifying the most critical issues (and possible solutions) when including a satellite component in the NR network. With respect to the access technologies, RAN3 (interfaces) led the activities on the SI 'Study on solutions for NR to support NTN', which was completed at the end of 2019. Within this SI, building on the results obtained within Rel. 15 in TR 38.811, a set of required adaptations enabling NR technologies and operations in the NTN context was addressed, covering several issues in RAN1, RAN2 (Layers 2 and 3) and RAN3 (interfaces). In particular, the performance assessment of NR in scenarios including GEO and LEO satellites was provided at both system and link level, together with a preliminary set of potential solutions for NR adaptations at Layers 2 and 3. It shall be noticed that some architecture aspects were modified with respect to TR 38.811, which is superseded from this point of view by TR 38.821 [8].
- **Rel. 17**. At the end of 2019, two WIs were officially started for NTN: (1) 'Solutions for NR to support NTN', under RAN2 (Layers 2 and 3) activities, but covering RAN1, RAN2 and RAN3 technologies and techniques; and (2) 'Integration of satellite components in the 5G architecture', under SA2. The activities for the former focused on the: (1) consolidation of the performance assessment provided in TR 38.821 and of the potential impacts at PHY and MAC level; (2) analysis of aspects related to Layers 2 and 3, e.g., handover and DC; and (3) identification of the potential requirements for the upper layers

based on the considered architectures. The latter WI aims at extending the analysis provided in TR 23.737 related to: (1) the identification of impact areas of the satellite communications (SatCom) integration in NR systems, in particular aiming at minimising it; (2) the analysis of the issues related to the interaction between the RAN and the core network; and (3) the identification of solutions for the two use cases highlighted above (terrestrial/satellite networks roaming and 5G fixed backhaul).

In June 2021, a workshop was organised within RAN related to Rel. 18 and a few documents were submitted to the meeting related to NTN. Below, we report the potential techniques and technologies that have been preliminary identified as of interest for Rel. 18 NTN studies:

- **Coverage enhancements**. This is considered as a critical item for low data rate services, as messaging and voice for commercial smartphone services. The aspects on which the activities will focus are related to: (1) repetitions and diversity techniques for the relevant channels (including Physical Random Access Channel (PRACH) and techniques to enable full-power uplink transmission and reduced polarisation loss); (2) channel state information (CSI) aging mitigation; (3) the configuration of the demodulation reference signal PHY signals; (4) improvement of the performance of low-rate codecs in low link budget scenarios, including the reduction of the RAN protocol overhead (the initial work shall be performed in RAN1 and RAN2, with the possibility to then liaise with SA2 and SA4 if necessary); and (5) investigation of approaches to mitigate the packet interruption due to low uplink/downlink signal-to-noise ratio and beam/cell switching for NTN.
- **NR-NTN deployment above 10 GHz bands** and support for very-small-aperture terminals (VSAT) and earth station in motion (ESIM) NTN terminals. The overall recommendation is that the work shall focus on the general challenges related to NTN operations in the time division duplexing and frequency division duplexing (FDD) bands in FR2 (24.25–52.6 GHz), as well as the handling of the 7–24 GHz band. Moreover, the following list of specific technical items is currently considered as relevant: study and identify the NTN bands: (1) analysis of regulations, adjacent channel coexistence and future-proof protection of the terrestrial network; (2) specify the reception/transmission requirements for different VSAT/ESIM user equipment (UE) classes (not only 60 cm aperture antennas as in TR 38.821); (3) investigate and specify the UE timing and frequency pre-compensation accuracy requirements as needed; (4) specify the conformance testing; (5) specify the radio resource management requirements; (6) definition of the PHY layer parameters, such as the subcarrier spacing for the synchronisation signal block and data channels; and (6) beam management and bandwidth part operation/switching in NTN, considering the characteristics of the satellite beams (e.g., large beam foot print, multiple beams per satellite and FDD for FR2).

- **Non-Terrestrial Network-Terrestrial Network (NTN-TN) and NTN-NTN mobility and service continuity enhancements**. Further mobility enhancements are needed with respect to previous releases, in particular when referring to: (1) addressing handover interruption, handover signalling overhead and Random Access Channel (RACH) congestion; (2) addressing the radio link failure reduction issue for different delays and/or network topologies between the different access types/points/nodes; (3) NTN-TN and NTN-NTN measurement/mobility and service continuity enhancements; and (4) multi-connectivity (MC) for NTN.
- **Network-based UE location**. RAN1, with support from RAN2, RAN3 and RAN4, is focusing on the need to fulfil regulatory requirements for regulated services (e.g., lawful intercept, emergency communications and public warning service), as well as the handling of requirements where law enforcement applies that the network shall be able to provide a 'reliable' UE location (either network verified or network provided). Some other companies would like to treat this topic as second priority. The general recommendation is to start the Rel. 18 work with a SI to determine how the network can determine the UE location without relying on UE global navigation satellite system measurements or support.

To conclude the overview on Rel. 18, also aspects related to regenerative payloads, the reduction of the peak-to-average power ratio, and asynchronous MC and DC solutions are considered as of interest and will be addressed in the next meetings.

3.2 Enabled services

As discussed above, the role of satellites, and NGSO constellations in particular, is essential for the provisioning of 5G connectivity to fixed and mobile users anywhere on the globe, at any time. The definition of the NTN global standard based on the 5G technology can foster new service capabilities aimed at consistent service continuity, reliability and availability. Moreover, it can be expected that the overall cost of both the network infrastructure and devices can decrease, thanks to the economies of scale of the 5G ecosystem. Below, we provide an overview of the most relevant categories of services envisaged for NGSO-based 5G and B5G systems: broadband, backhauling, M2M/IoT and critical communications.

3.2.1 Broadband

Although GEO systems are typically the primary option, NGSO constellations have gained an ever increasing attention when aiming at providing broadband access at global scale, thanks to the specific advantages that lower altitudes provide, as previously discussed; this is demonstrated by the industrial endeavours from SpaceX Starlink, OneWeb, Amazon Kuiper, Telesat and LeoSat, to mention a few. Moreover, the availability of optical inter-satellite links (ISLs) will further boost the achievable capacity in the near future. In this context, the support of intelligent transport networks and connected vehicles (automotive, maritime and aircrafts) will have NGSO nodes as one of the key actors, complementing terrestrial networks where necessary or even being the only possible candidate where the terrestrial infrastructure is not available, as in maritime and aeronautical communications, or economically viable, as in rural areas. Focusing on the latter aspect, a mega-constellation of NGSO nodes can help to finally provide broadband connectivity to users in digital divide areas; in fact, in Europe, 20% of the European citizens never used the Internet, while 72% of them use Internet at least once a week [2]. In these scenarios, two types of broadband access can be foreseen: (2) fixed broadband, for households and premises; and (2) mobile, for outdoor users, pedestrian or on moving platforms.

3.2.2 Backhauling

Thanks to the reduced latency and global coverage, NGSO constellations can provide an efficient solution for the backhauling of both high-speed services or the aggregate traffic from a plethora of IoT devices. In this context, the NGSO nodes can be used as a single centralised backhaul for traffic off-loading, edge processing and resource sharing, in particular in those areas in which a terrestrial infrastructure is not feasible or available. In this framework, a networked constellation of LEO satellites interconnected by means of ISLs can lead to hybrid terrestrial-satellite routing algorithms, or satellite-only routing in case the terrestrial infrastructure is congested or temporarily/locally unavailable [25].

3.2.3 M2M/IoT

As introduced above, the market related to machine-type communications is expected to significantly grow in the next years. Applications can be foreseen in various vertical markets, including manufacturing, military, maritime, aviation, etc. It is expected that most of these services will be provided in S and L bands, with Ka and Ku bands exploited for the backhauling of the aggregate traffic generated by IoT devices.

Asset tracking is an example of service in this category, which can be applied to almost any vertical market. A moving platform carrying specific assets shall be continuously monitored, but it is likely to happen that, during its trip towards the destination, it will be moving through areas in which a terrestrial communication infrastructure is not present (either due to a low revenue for the operator or because PHY unfeasible, as in the sea or in the air). For instance: (1) a cargo flight will be able to connect to a terrestrial network while on ground, but not during the flight time; (2) a train or vehicle will move across rural areas with no or limited population, where terrestrial mobile network operators (MNOs) might not be encouraged to deploy an infrastructure; or (3) a ship can be connected to a terrestrial network when it is anchored or close to a harbour, but not off-shore. In this framework, the possibility to rely on satellite access through one or more satellite network operators (SNOs) managing NGSO constellations would actually guarantee service continuity. In this context, the 5G system should provide connectivity by means of both terrestrial and satellite radio access technologies (RAT). When both options are available, advanced management techniques can be implemented so as to either route the traffic towards the best performing network or to implement DC, as discussed below. In the latter case, it is worth mentioning that, when more than one satellite from the same network operator is visible, DC can be implemented as well. However, DC is a technique aimed at enhancing the capacity provided to the UEs; since in this scenario we are considering smart good tracking, which typically requires a limited amount of traffic per UE, it might be not necessary to implement advanced techniques. Even assuming that a single network element onboard the vehicle gathers the entire traffic before sending it through an MNO or SNO, the expected capacity requirements do not justify the implementation of advanced, and more complex, techniques.

3.2.4 Critical communications

When a natural or man-made (terrorist attack or mistake) disaster occur, it might happen that a part of the RAN or the core network become unavailable. Consequently, all of the services provided by one or more MNOs operating through the disrupted terrestrial infrastructure cannot be guaranteed anymore. Notably, the restoration of the communication infrastructure on the area is fundamental for both the population and, in particular, the first responders that need a telecommunication infrastructure to coordinate their efforts and to report to the command centre the evolution of the rescue operations. Similar necessities arise in case of extremely populated events (such as the Olympic Games or a concert), where many people are concentrated in a limited area, whereas densification alone is not sufficient to guarantee high-data rate connectivity to all users. On the contrary, users' demands will have to be met by complementing the terrestrial infrastructure with high altitude and satellite platforms. Their availability is not just aimed at introducing more capacity into the overall network, but also to provide ground for efficient off-loading techniques preventing congestion events. A constellation of NGSO nodes is also an excellent solution for the surveillance of critical infrastructure, in particular when the terrestrial network is not available (maritime) or too congested.

3.3 Radio access network

Based on the analyses performed within Rel. 15 and Rel. 16, a set of architecture options has been defined by 3GPP for NTN that applies to both GSO and NGSO systems [8]. The impact of the NTN environment on the technologies developed for terrestrial NR systems is strictly linked to the scenarios and architectures under consideration: the type of node and its capabilities, the constellation and the use cases. Moreover, this impact does not only depend on the specific implementation of the user, space and ground segments, but also on how these are interconnected and mapped to the 5G network elements. In terms of the space segment, two macro-categories have been identified: the space-borne, i.e., satellite-based communication platforms, and the air-borne, i.e., high-altitude platform systems (HAPS), devices.

There are several architectural options defined for NTN, which can be broadly classified depending on: (1) the type of payload, either transparent or regenerative, in which the node can contain a full gNB or part of it in case functional split solutions are implemented; and (2) the type of user access link, either direct or relay based, in which the UE is connected to an integrated access and backhaul (IAB) node and not directly to the satellite. Finally, with respect to the UE type, both handheld and dedicated satellite equipment, as VSATs, can be considered in the architectures described below.

3.3.1 Direct user access link

In this case, the RAN architecture can be based on both a transparent or a regenerative payload. Moreover, functional split solutions can also be implemented when regenerative payloads are considered, as detailed below.

Figure 3.2 shows the architecture with a transparent node, i.e., the payload only implements frequency conversion, filtering and amplification. Since the payload has no advanced processing capabilities, being basically equivalent to a Radio-Frequency (RF) repeater, the gNB is conceptually located at the system Gateway (GW). As a consequence, to be fully compatible with the 5G standard, both the feeder link and the user (service) link shall be implemented by means of the New Radio-Uu (NR-Uu) air interface; this is motivated by observing that the satellite, carrying a transparent payload, cannot terminate the NR-Uu procedures nor manage the quality of service (QoS) flows. Thus, as shown in Figure 3.2, the RAN consists of the gNB and the remote radio unit, which is given by the system GW and the transparent payload. It shall be noticed that, since the NR-Uu air interface is specifically designed for terrestrial systems, it is of paramount importance to properly assess the impact of the satellite channel impairments on the PHY and MAC procedures, as thoroughly discussed since Rel. 15 and Rel. 16, in particular in TR 38.811 [3] and TR 38.821 [8]. As for the connection towards the data network(s), everything is equivalent to a fully terrestrial system: the Next Generation (NG) air interface connects the gNBs and the next generation core network (NGC), which

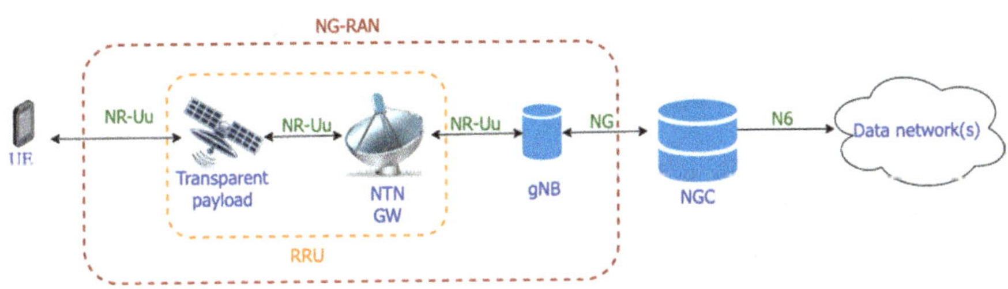

Figure 3.2 NTN-based RAN architecture with transparent payloads

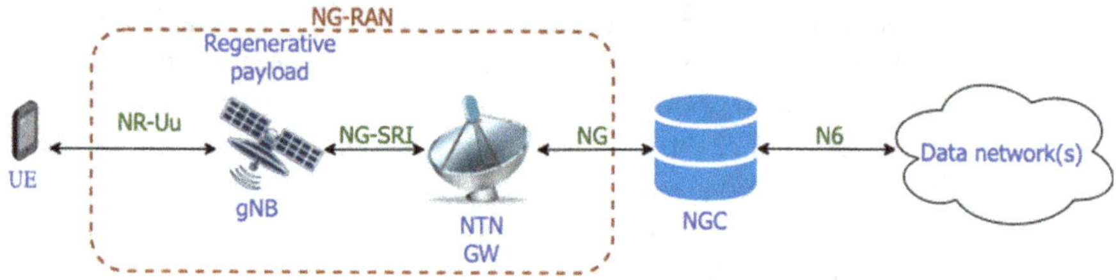

Figure 3.3 NTN-based RAN architecture with regenerative payloads

is then connected to the data network(s) by means of the N6 air interface†, as defined in TR 38.801 [14].

In Figure 3.3, the RAN solution with regenerative payloads is depicted. In this case, the gNB is fully implemented on-board and, thus, the NR-Uu protocols are terminated in the payload and the GW acts as a transport network layer node, terminating and supporting all transport protocols. In terms of the air interfaces to be used, this implies that the user (service) link is still implemented through the NR-Uu interface, while the feeder link can be operated through a NG interface. In this context, it shall be noticed that NG is a logical air interface, i.e., it can be implemented with any existing satellite radio interface (SRI), as long as specific signalling operations are guaranteed [13]. This means that the feeder link can be implemented with a modified version of the terrestrial air interface, or even by means of state-of-the-art solutions for SatCom, as the DVB-S2 [15], DVB-S2X [16] or DVB-RCS [17], air interfaces, subject to the satisfaction of the requirements posed by the NG interface. Notably, this solution allows to significantly reduce the latency for NR PHY and MAC procedures, thus easing the adaptations that might be needed for NTN. However, it is also more complex and the cost of the payload is increased.

In Figure 3.4, we report the option with regenerative payloads in which ISLs can be implemented between two or more nodes. The ISL is a transport link between the flying nodes, which can be implemented by means of the Xn air interface, another logical interface in 3GPP NR standardisation that allows to interconnect different gNBs. Thus, it can be implemented by means of any 3GPP or non-3GPP solution as the NG interface. It shall be mentioned that, while in the figure we are showing two on-board gNBs connected to two separate on-ground GWs and two separate NGCs, the latter can be the same. As long as ISLs are considered, it is worth to be mentioned that the latency with which the NR protocols have to cope is not only related to the user access and feeder links, but also to at least one ISL. Thus, depending on the number of hops between the flying nodes, the protocols timers might need to be adjusted.

Figure 3.5 shows an architecture option with direct access and regenerative payload, in which functional split concepts are applied. The functional split allows scalable solutions, a significant adaptability to the different use cases and vertical

†As per 3GGP specifications, N6 is the interface between the user plane function and any other external or internal data network(s) or service platform(s).

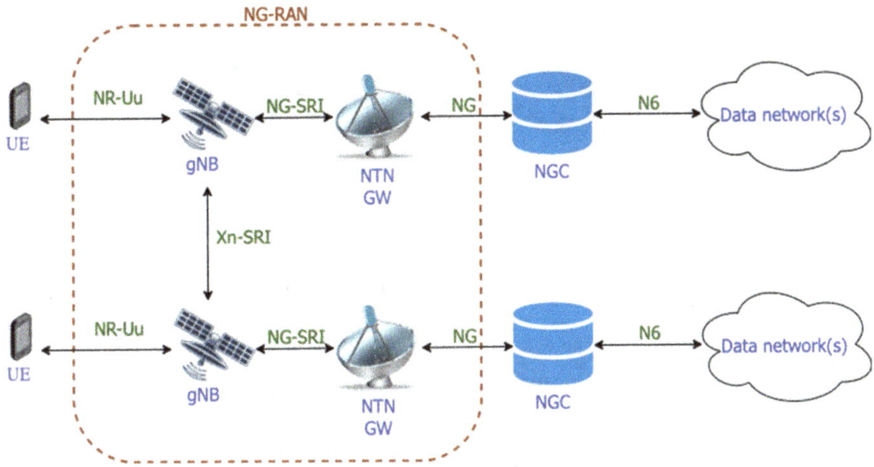

Figure 3.4 NTN-based RAN architecture with regenerative payloads and ISLs

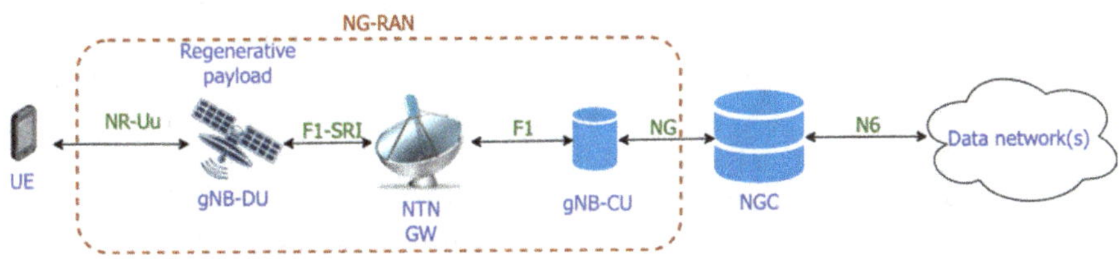

Figure 3.5 NTN-based RAN architecture with regenerative payloads and functional split

services, and enhanced performance in terms of load and network management; moreover, it is at the basis of network function virtualisation (NFV) and software-defined networks (SDN). Clearly, this solution also increases the overall system cost. As provided in TS 38.401 [18]: (1) a gNB can be split into a centralised unit (gNB-CU) and one or more distributed units (gNB-DU); (2) a gNB- DU can be connected to only one gNB-CU; (3) the air interface to be used between the gNB-CU and its gNB-DUs is the F1 air interface; (4) the F1 air interface is logical as the NG, i.e., as long as specific signalling operations are ensured, it can be implemented by means of any existing standard [19]. The split between CU and DUs can be implemented at different layers, and even within a given layer, as detailed in Reference 14; however, the most considered option in NR, which is also that considered for NTN for the moment being, is as follows: (1) PHY, MAC and radio link control (RLC) are implemented in the DU; and (2) the packet data convergence protocol (PDCP) and service data application layer (SDAP) for the user plane (UP) or the radio resource control (RRC) for the control plane are implemented in the CU. Finally, it is worth highlighting that intermediate solutions can also be envisaged: apart from a CU and the controlled DU, there can be an intermediate unit that further splits the gNB in

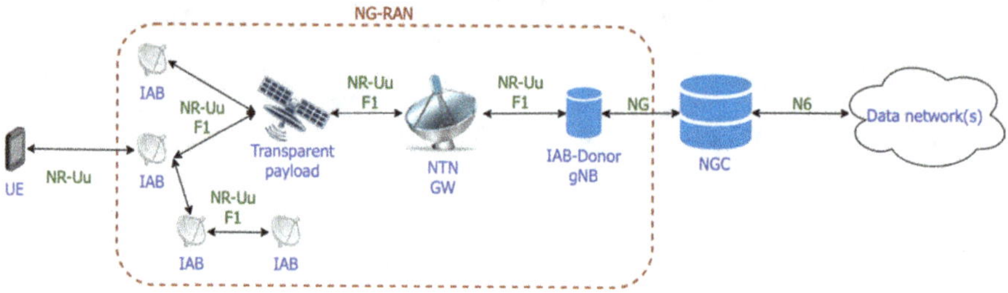

Figure 3.6 NTN-based RAN architecture with IAB-based access and transparent payloads

three entities and controls several DUs. This option is not yet considered in the current status of NTN systems.

3.3.2 IAB-based user access

The possibility to implement an indirect access, in which the platform is connected to an on-ground IAB then providing the user access link, is currently considered for further study within 3GPP. However, for the sake of completeness, it is worth to be introduced because it can provide a viable solution for backhauling and transport networks based on NTN. In this architecture, the UE is connected to an IAB; thus, from the users' perspective, the system is equivalent to a completely terrestrial 5G network. IABs are a new type of wireless backhaul elements introduced with 5G to address dense deployment scenarios, inspired by their LTE counterparts known as relay nodes, detailed in TR 38.809 [20] and TR 38.874 [21]. In the simplest implementation, an IAB-Donor gNB is connected, via the NG interface, to the NGC. The IAB-Donor gNB acts as a single logical entity that comprises a set of functions such as the distributed and centralised gNB units, gNB-DU and gNB-CU, the split of the latter in control and UP, gNB-CU-CP and gNB-CU-UP, and potentially other functions.

Based on these options, some of the potential architectures are shown in Figure 3.6 and for transparent and regenerative payloads, respectively. In both solutions, we can notice that the connection between the IAB-Donor gNB and the NGC is implemented by means of a NG air interface, while the user access and IAB-to-IAB links need a clarification. As per TR 38.874, each IAB element is composed by a DU that provides connectivity to the users and a mobile termination, which is the function terminating the radio interface protocols of the backhaul Uu interface towards the parent IAB. In fact, we can also notice the hierarchical structure that IABs allow to deploy, with a given IAB entity acting as controller of child IABs, although with limited functions compared to the donor IAB. Referring to the transparent payload architecture in Figure 3.6, this allows to deploy any number of on-ground IABs controlled by one or more nodes and also to have a hierarchical architecture on ground. In terms of the challenges, the same considerations on the NR-Uu and F1 logical interfaces on the user and feeder links discussed in the previous section hold.

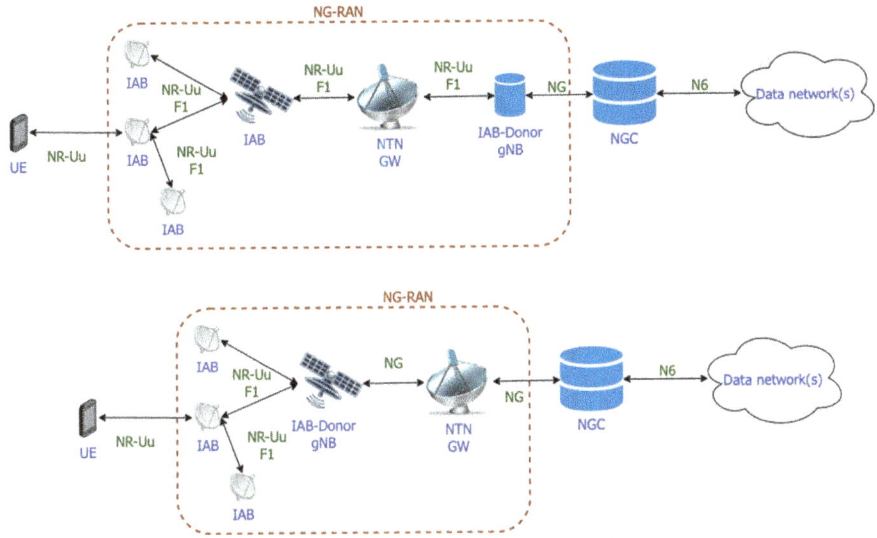

Figure 3.7 *NTN-based RAN architecture with IAB-based access and regenerative payloads*

When a regenerative payload is considered, we can implement on-board both an IAB or the IAB-Donor gNB elements, as shown in Figure 3.7. Notably, in the latter case, there is the need to implement the NG air interface on the feeder link, while the former option is very similar to the transparent case from the interface perspective. Clearly, regenerative platforms allow to reduce the latency and, in general, to better tailor the on-board functions to the specific service to be provided thanks to NFV/SDN.

3.3.3 Multi-connectivity

In MC, multiple transmitters can simultaneously configure the radio resources to be provided to a given terminal, introducing link diversity and enhancing the achievable capacity and reliability in many scenarios, e.g., residential homes, vehicles, high-speed trains and aircrafts, to name a few. Currently, the focus within NTN is directed towards DC solutions, in which two radio accesses can be used, both with transparent and regenerative payloads. In general, a single user terminal can be connected to: (1) an NTN-based RAN and a terrestrial RAN; (2) two NTN-based RANs. Moreover, in case of a regenerative payload, the gNB can be split into an on-board DU and an on-ground CU, as already discussed above. This leads to multiple architectural solutions, as shown in Figures 3.8 and 3.9 for Terrestrial-Non Terrestrial (T-NT) and NT-only DC RANs, respectively. The observations on the different air interfaces to be used provided above still apply. Clearly, in addition to them, the system complexity is increased due to the management and synchronisation of multiple transmission. In particular, in addition to supporting radio access with extended latencies as in the above scenarios, other adaptations might be needed

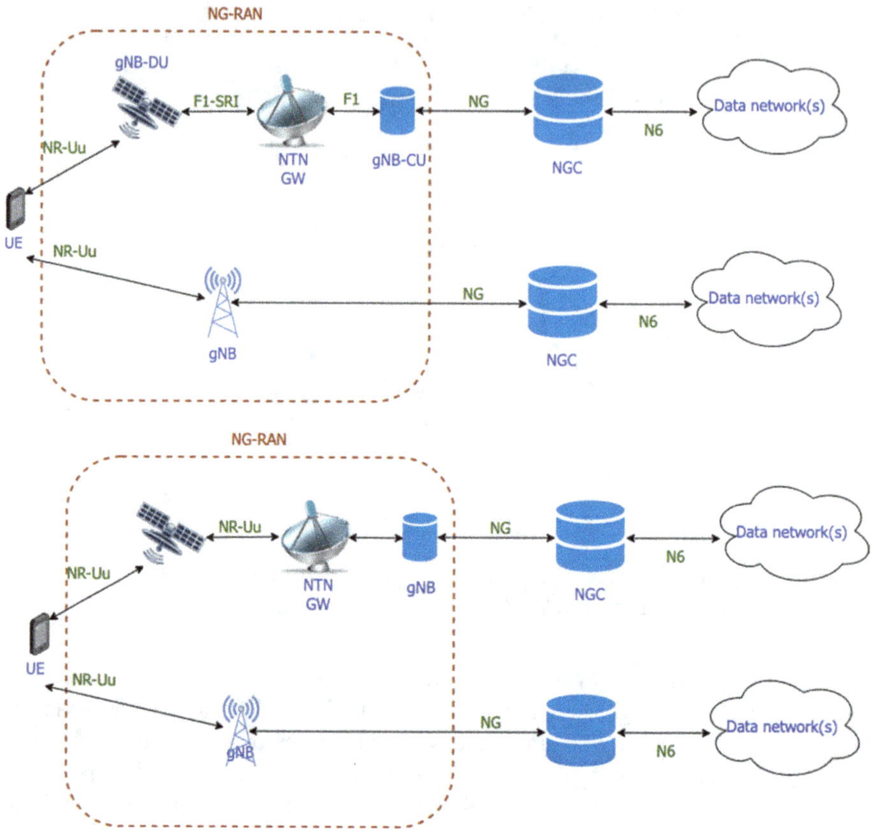

Figure 3.8 RAN architecture with DC provided through terrestrial and NT access with regenerative (top) and transparent (bottom) payloads

compared to terrestrial 5G to support: (1) an RAT potentially suffering from variable latency within the backhaul network, as in the case of a Xn-SRI interface crossing multiple nodes on different orbital planes; and (2) the potentially significant difference in delay when terrestrial and non-terrestrial access are merged into DC. Finally, it shall be mentioned that the RAN might flexibly select either the NT or the T gNB as master node.

3.3.4 NR adaptation challenges

In the framework of adapting the NR standard to NTN systems, several challenges arise related to the protocols and the PHY/MAC layers.

While one of the main 5G features is that of reducing the latency down to 1 ms, the propagation delay over NTN might be larger, depending on the orbit. In fact, while vLEO nodes can provide a better scenario ad outlined above, LEO or medium earth orbit (MEO) can pose several challenges. The extended latency has an impact on RRC, SDAP, PDCP, RLC and MAC layers, as well as PHY

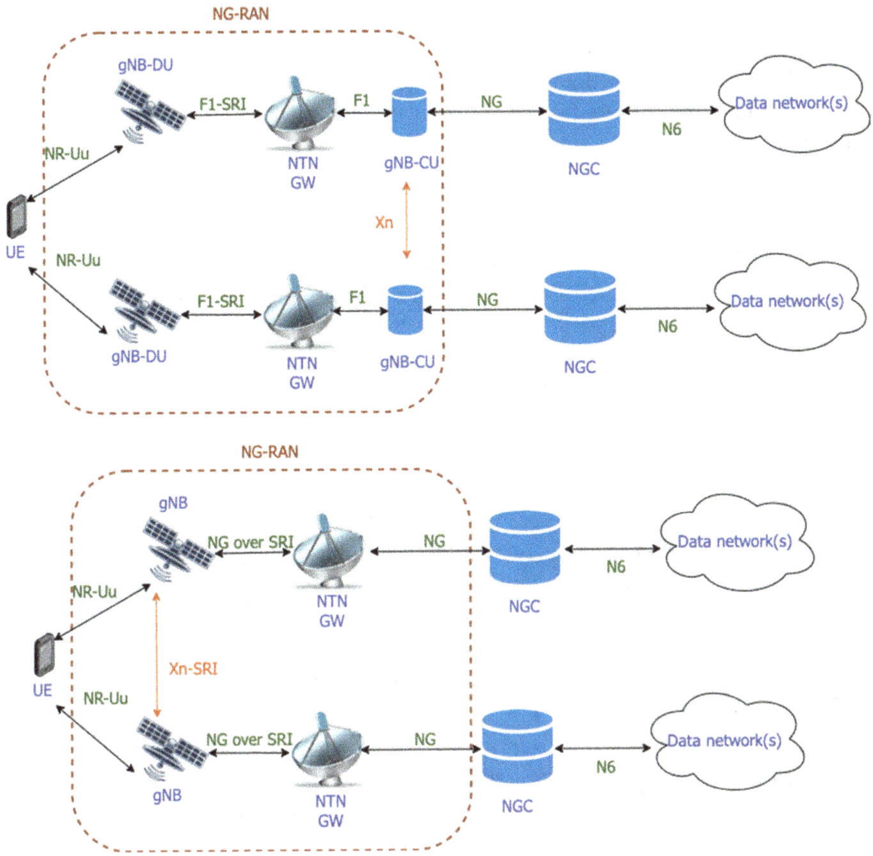

Figure 3.9 *RAN architecture with DC provided through NT access only with regenerative (top) and transparent (bottom) payloads*

procedures, including: (1) scheduling, in which the MAC scheduler shall allocate resources based on larger delays or even significantly different latency values in case DC is implemented with a terrestrial and an NT components, and link adaptation, in which latency might lead to suboptimal solutions due to outdated channel estimates; (2) random access, hybrid automatic repeat request and timing advance procedures; (3) tracking area management and handover procedures, due to the mobility of the NGSO nodes, in particular when moving beams platforms, i.e., antenna subsystems not allowing to mechanically steer the beams, are considered; and (4) timing and frequency acquisition and tracking, related to the potentially large speed of the NGSO nodes leading to variable differential delays and large Doppler shifts on a beam basis. A thoroughly detailed description of the above challenges and the related potential solutions can be found in both the literature [10, 22–26, 28], and in the 3GPP 38.811 and 38.821 TRs [4, 8]. These aspects are in line with the items considered to be relevant for Rel. 18, discussed above.

3.4 System architectures

The RAN options described above for the integration of NTN in the 5G ecosystem can be declined into different NGSO system architectures depending on the type of payload and design of the constellation [2, 25, 26]:

- **HTS broadband MEO**. MEO constellations are currently receiving an increased attention compared to the last few years. This is demonstrated, for instance, by O3b, which is aiming at providing connectivity to emerging and not sufficiently connected markets in Latin America, Africa, Middle East, Asia and Pacific with an MEO constellation. One of the main advantages is the possibility to provide data access at a reasonable latency, compared to their GEO counterparts. However, it is worth mentioning that a more advanced design of antennas is necessary to boost the performance.
- **LEO constellations**. As outlined in the introduction, LEO deployments require a large number of satellites to provide a complete global coverage, with LEO nodes ranging from some hundreds satellites up to thousands of satellites. The research and development in LEO constellations design is mainly directed towards mega-constellations in which the LEO nodes are equipped with optical ISLs, as demonstrated by the industrial endeavours from SpaceX Starlink, OneWeb, Amazon Kuiper, Telesat and LeoSat. In particular, Starlink has been allowed to deploy up to 12 000 satellites, currently focusing at an altitude of 550 km, but a request has been filed to bring the mega-constellation to 42 000 nodes. The satellite mass is relatively small (less than 260 kg) and the use of ISLs is foreseen; while preliminary operations in Ku-band are already possible, the full deployment is expected between 2027 and 2030. OneWeb is focusing on a 600 satellites (648 with spares) mega-constellation at 1200 km operating in Ku-band (feeder link) and Ka-band (user access link). Also in this case, the satellite mass is limited (150 kg at most) and the solar powered user terminals, both fixed and mobile, will provide 2G, 3G, LTE and Wi-Fi connectivity. Telesat is a mega-constellation of 298 larger satellites (700–750 kg) on hybrid orbits, in which both polar and inclined planes are included aiming at global coverage; the satellites operate in Ka-band and optical ISLs are foreseen (up to four links per satellite), with regenerative payloads and the possibility to implement on-board IP routing algorithms. LeoSat envisages even larger payloads (670 kg) for 108 satellites at 1 400 km on polar orbits; also in this case, Ka-band is foreseen for the user access link, together with up to four high-capacity ISLs creating a fully meshed inter-satellite network.

  LEO constellations mainly address Ka-/Ku-band Internet connection as a service, but also secure point-to-point communications. Notably, the large number of satellites will permit to achieve high-granularity coverage, thus providing higher capacity to the on-ground users; in fact, the LEO constellations basically work as a huge distributed network switch operating in the sky. In addition, the availability of optical ISLs can drastically reduce the latency of rerouting operations in space, taking also into account that speed of light in free space is higher than in terrestrial optic fibres (65–70% the speed of light, as mentioned above). As such, the transmission of large amounts of data between adjacent satellites can be performed in an almost negligible time, hence letting users experience a new Internet experience from the sky owing to the limited delay, coming from the low altitude of LEO satellites. In this framework, advanced and fast on-board processing of the signals is essential for efficiently routing the signals to the right destinations. Another aspect worth to be highlighted is that frequency coordination is of vital importance as they need to operate next to the already existing satellites in different orbits and in the same orbits.

- **vLEO constellations**. vLEO constellations at an altitude lower that 300 km are expected to play a pivotal role in future networks thanks to several technical benefits, e.g., de-orbiting time and lower radiation effects, to the increasing availability of low-cost launchers, and the reduced satellite manufacturing cost. They will be of particular interest in support of IoT services by means of improved on-board computation capabilities. Their function to extend the terrestrial network capacity will be carried out by means of nano-/pico-satellites, which, together with CubeSats, will make use of high data rate links in Ka-band or optical frequencies. In this context, it is worth to be mentioned that 3GPP included a scenario for 5G-based IoT via NTN with vLEO satellites; moreover, recently Sateliot launched the first nanosatellite in orbit aiming at providing IoT services.
- **Hierarchical aerial networks**. The integration of HAPS and unmanned aerial vehicles (UAVs) can be another key development for many different applications that provide connectivity for otherwise not connected areas, critical communications, environmental monitoring, massive machine-type communications, IoT and interplanetary communications. Clearly, exploiting multiple UAVs and HAPS requires a further level of cooperation and data exchange among the nodes, leading to the development of advanced multi-layered architectures as discussed below.

In addition to the above, it is worth to mention that fractioned cooperative constellations can be deployed as well. In this case, incomplete constellations, that would lead to intermittent connectivity if operating standalone, can cooperate and provide an equivalent global coverage. Such an approach requires advanced on-board processing capabilities and optical ISLs and leads to complex inter-constellation coordination and management from the ground segment.

Within the integration of NTN systems, in particular NGSO, into the 5G and beyond ecosystem, it is widely understood that a further level of integration is needed to reach a truly integrated architecture such as the multidimensional multilayer (MD-ML) system shown in Figure 3.10. In the envisaged network, the 2D horizontal terrestrial (T) deployment is complemented and enhanced by means of a third, vertical dimension, represented by an ML NT component consisting in communication nodes, space-/air-borne, flying at different altitudes and communicating among them and with the terrestrial element through inter-node links. We refer here to nodes, instead of satellites, because in the proposed architecture the flying nodes are either space- or air-borne. The access network is formed by the T and NT access networks, with the latter structured in the NTN ground segment and the NT-segment.

3.5 Research and development challenges

The complex, yet flexible and adaptable, architecture proposed in Figure 3.10, poses numerous challenges that need to be addressed to develop the required technical enablers, as outlined below.

3.5.1 Architecture design

The full integration of the NT component into the B5G infrastructure calls for a design of the overall architecture where there is no distinction between terrestrial and non-terrestrial network elements that can be therefore orchestrated in order to provide the optimal cost-efficient network configuration able to satisfy the traffic requirements. In this framework, three main enablers shall be developed: NT-access network softwarisation, virtualisation and disaggregation. This is expected to increase the flexibility and adaptability of the network, as well as the sharing between the terrestrial and NT components of architectural elements and of network

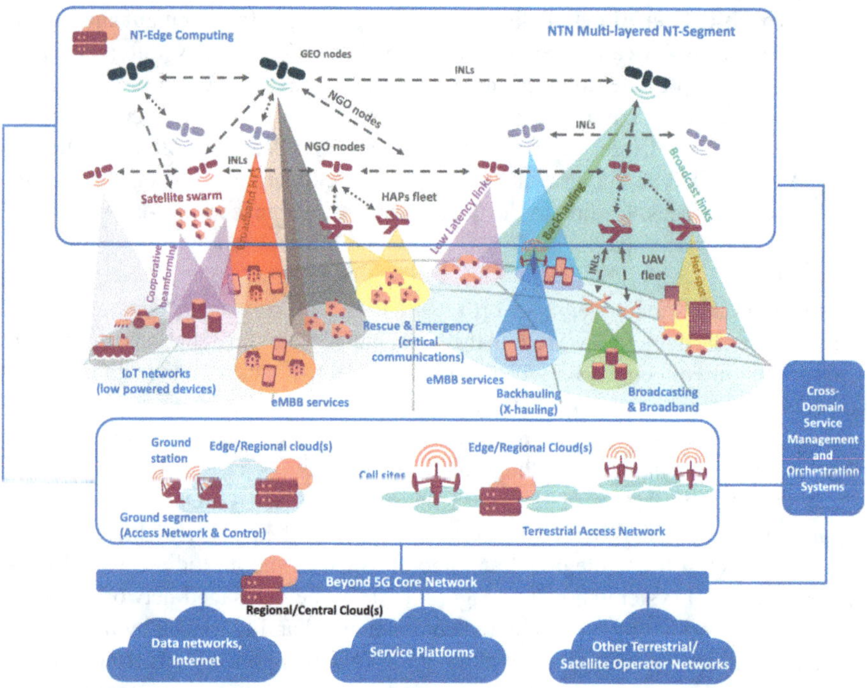

Figure 3.10 B5G multidimensional multilayer-integrated architecture [22]

functions, thus reducing the overall cost and opening new markets opportunities. Notably, the NT access network also entails the ML NT segment, i.e., what is called space segment in traditional SatCom architectures, thus bringing softwarisation, virtualisation and disaggregation in the sky, i.e., in the flying nodes. In this context, it shall be mentioned that there is a commonly shared vision for 6G architectures that shall be based on: (1) native programmability and soft re-architecting, so as to customise the behaviour of network resources through standardised programming interfaces for network control, management and servicing functionality; and (2) the introduction of native slices, to enable an easy and efficient execution of multiple and various types of services at a given time on the same infrastructure. These two directions are envisaged for the global 6G infrastructure and, clearly, they shall be applied to SatCom as well in order to foster a true and seamless integration.

3.5.2 Constellation: hierarchical design

The challenge is to go beyond the current single orbit (layer) constellation design towards ML and hierarchical constellations consisting of nodes flying at different altitudes and communicating among them through horizontal inter-node links, i.e., among nodes at the same altitude, and vertical inter-node links, i.e., among nodes flying at different altitudes, or even terrestrial nodes. Hierarchical constellations will be instrumental in ensuring flexibility and adaptability, and the possibility to provide cost-efficient progressive service coverage, e.g., NB-IoT services can be provided through incomplete vLEO constellations of low-cost platforms relying on GEO/GSO large platforms to ensure continuous core network connection to the on-ground devices. In this context, as previously outlined, there are already industrial endeavours considering hybrid constellations.

3.5.3 Resource optimisation: infrastructure as a resource

Beyond the bandwidth, time, power and space domains, resource optimisation shall address the infrastructure as a resource to be configured according to the service requirements. The flexibility of the MD-ML architecture allows the definition of a network that evolves into new forms that are better able to meet the traffic requirements. Autonomous and intelligent predictive optimisation of the infrastructure shall be developed to enable a timely and dynamic reconfiguration, thus calling for the application of artificial intelligence (AI) concepts. For instance, neural networks were exploited in de Cola and Bisio [29] to achieve the optimal resource allocation for eMBB services, taking into account the requested QoS and integrating a satellite and a terrestrial 5G link. Such possibilities were from a networking perspective, smart NGSO nodes can be designed as edge nodes, requiring the autonomous management of diverse requirements originated in different networks [30]. In terms of network management, the definition of the type of functional split and where the different elements in the RAN shall be implemented, thoroughly discussed above, is fundamental. In NFV/SDN, the research and innovation challenges are related to the design and implementation of a number of functions also realising a service chaining by constructing a forwarding graph interconnecting network functions in the

NFV [2]. As mentioned discussing the RAN challenges above, the need to cope with mobility implies to also develop efficient and seamless handover at beam, gateway and satellite level, considering also the advent of mega-constellations. As a consequence, suitable orchestration schemes, fully centralised or distributed to better cope with the increase of processing delay, shall be developed. The heterogeneity of the involved technologies leads to the need for a unified network management model to get a flexible and consistent network management plane, with consequent implications also at system orchestration and security level.

Dynamic network management algorithms shall take into account the multi-provider MD-ML architecture in identifying the optimal routing algorithm. In this framework, delay tolerant network (DTN) technologies, currently applied to deep space environments, are becoming of interest also for inter-node communications, thanks to their good handling of network partitions and the underlying store-and-forwarding principle. Nodes persistently store the received data and then send information without a full knowledge of the topology, acting also in case of intermittent connectivity, which can be opportunistic or scheduled, thus making DTN relevant in the context of future smart NGSO constellation communication. In the framework of space networking, ITU-T FG-NET-2030 highlighted the importance of fostering a vision consisting of LEO mega-constellations for 2030 networks, with satellites that can be interconnected to form a network infrastructure in space [25]; this infrastructure will then be further integrated with the network infrastructures on the ground. Notably, the key challenge here is the frequent handover between the satellite and terrestrial networks, caused by the large velocity the LEO satellites on their orbit. In the foreseen integrated infrastructure, two new key components of the space networking architecture are proposed: (1) an SDN-based controller, which shall be deployed on-board an MEO or GEO satellite so as to guarantee the management of a large number of LEO nodes and in charge of forwarding the requested data; and (2) a mobile edge computing server, located on-board the LEO satellites and embedding local computing and storage capabilities, so as to provide lower latency services to the locally attached users. Aspects related to routing and caching in space networks are also discussed in Di *et al.* [31] related to dense LEO constellations.

3.5.4 Dynamic spectrum management: coexistence and sharing

Spectrum usage maximisation shall be sought at all levels of the system, from spectrum efficiency, i.e., bit/s/Hz, through system throughput, to the full utilisation of the allocated bandwidth. To this aim, dynamic spectrum coexistence and sharing between the terrestrial and NT segments, as well as among the different layers of the NGSO architecture shall be developed. In this context, sharing can happen at various levels: (1) intra-constellation, i.e., among nodes of the same constellation; (2) inter-constellation, i.e., among nodes of different constellations; and (3) between terrestrial and NT networks. AI might provide a valuable resource in predictively define the spectrum sharing environment taking into account the flying patterns of the NGSO nodes.

Alongside spectrum sharing, in order to support the ever-increasing capacity demand, the use of new spectrum, up to optical communications, shall be investigated, in particular for the feeder and the inter-node links. In this case, channel characterisation and measurements from different orbits and altitudes are required to provide accurate models to be used for the system design.

3.5.5 RAT: flexibility and adaptability

It has been shown that, with limited and acceptable modifications, the NR air interface can be adapted to the NTN case, thus introducing a completely new dimension in the overall 5G ecosystem. However, to enable the full exploitation of the NT component, the 3GPP waveform design shall address the NT channel characteristics and in particular, the Doppler effects introduced by the NGSO nodes, the delay and latency aspects characteristics of the higher altitude nodes, and the need for efficient support of channel estimation procedures since CSI is fundamental for the beamless communication approach. At the same time, new numerologies shall be investigated to enable the PHY layer flexibility and adaptability in support of the service heterogeneity.

3.5.6 Antennas and user-centric coverage

The extreme variety of traffic request and of user density through the services typologies and covered areas requires the coverage design to go beyond the current geographical approach to provide a user-centric communication in which links are dynamically created to follow the served user. To this aim, new antenna designs, providing flexible beamforming for beamless systems as well as support for new spectrum frequencies, shall be addressed. Going beyond the large scale antenna array concept, distributed antenna systems in the sky, supported by node communication and cooperation to implement multiple input multiple output solutions, shall be researched and developed. The technology innovations are expected to enable the generation of narrower beams for a more efficient use of power and spectral resources. Moreover, the possibility to steer the beams will also play a key role to cope with the dynamicity of the traffic demands, avoid interference between different constellations/systems and tackle the challenges related to mobility management in 5G systems. Advances in antenna design will thus be of paramount importance to make satellite systems ready for very high throughput requirements and fulfil the challenges of multi-orbit and mobility scenarios. In this framework, active-phased antenna arrays and meta-surface antennas, characterised by low power consumption and cost, are foreseen to be critical to provide the required flexibility. Active antenna technology will enable the NGSO nodes to create ad hoc beam responses characterised by null-steering in any direction at any time, leading to the envisaged user-centric coverage discussed above. It shall be mentioned that, for an optimal compromise between the flexibility and the power and mass requirements of the payload, hybrid analogue/digital beamforming architectures will play here a central role.

Finally, also in the ground segment, it is worth to mention that low-profile electronically steerable antennas provide benefits for mobile platforms, as cars, planes, trains and ships. This will allow to connect the devices on moving platforms to different NGSO constellations.

3.5.7 AI: exploitation of NT dynamics

To address the complexity of the MD-ML architecture autonomous and intelligent network management system shall be appropriately developed, in particular taking into account the predictable dynamic of the NT segment, where significant periodicity of the infrastructure characteristics can be exploited, e.g., the periodic repetition of channel conditions due to the node position in the sky. AI solutions can provide the means for fast decision-making in order to meet the full potential of integrated systems consisting of space-borne, air-borne and terrestrial components. This could include predictive spectrum allocations, predictive routing and even autonomous replanning at the satellite without the delays introduced by the decision-making loops on ground.

For instance, deep learning algorithms have been implemented for detection problems in spectrum monitoring scenarios [32], allowing the satellites to quickly react to the interfering environment rearranging the frequency allocation, transmit power and beamforming configurations of the payload. The reconfiguration of the payload has also been addressed based on genetic algorithms in References [33, 34]. Other application of AI is related to the PHY layer, aiming at enhancing the signal processing and channel estimation performance [35].

3.5.8 Security

The provision of an end-to-end security at application layer, e.g., encrypted video streams, is not sufficient anymore. Within 3GPP, novel security schemes are incorporated and they will need to be adjusted so as to operate in the context of mixed delivery systems. Thus, there is need to address security problems for the future MD-ML networks and the new services applications. Moreover, more emphasis is being placed on interplanetary systems as we focus on lunar and then mars missions where again security in these environments needs to be visited. In this framework, the following topics are of interest for near-term research and development endeavours: (1) security integration between satellite and terrestrial systems; (2) blockchain technology for SatCom; (3) quantum key distribution and key management over satellite; (4) secure multicasting over satellite; (5) secure multiple access; and (6) RF-level security.

References

[1] Cisco White Paper. *Cisco Annual Internet Report (2018-2023)* [online]. 2020 Mar. Available from https://www.cisco.com/c/en/us/ index.html

[2] Networld 2020. *Strategic research and innovation agenda 2021-27* [online]. 2020 May. Available from https://www.networld2020.eu

[3] European Commission. *EU digital divide infographic* [online]. Available from https://digital-strategy.ec.europa.eu/en/library/eu-digital-divide-infographic

[4] 3GPP TR 38.811 v15.2.0, study on new radio (NR) to support non- terrestrial networks. Jan 2019.

[5] 3GPP ts 22.822 v16.0.0, study on using satellite access in 5G; stage 1. Jun 2018.

[6] 3GPP TR 23.737 v17.0.0, study on architecture aspects for using satellite access in 5G. Dec 2019.

[7] 3GPP TR 28.808 v0.4.0, study on management and orchestration aspects of integrated satellite components in a 5G network. Jan 2020.

[8] 3GPP TR 38.821 v16.0.0, solutions for NR to support non-terrestrial networks (NTN). Jan 2020.

[9] 3GPP SA WG2, update to fs_5gsat_arch: study on architecture aspects for using satellite access in 5G. Dec 2018.

[10] Guidotti A., Cioni S., Colavolpe G, *et al.* 'Architectures, standardisation, and procedures for 5G satellite communications: A survey'. *Computer Networks*. 2020, vol. 183, p. 107588.

[11] Handley M. 'Delay is not an option: low latency routing in space'. *Proceedings of the 17th ACM Workshop on Hot Topics in Networks*; Budapest, Hungary, 2018. pp. 85–91.

[12] Leyva-Mayorga I., Soret B., Roper M., *et al.* 'Leo small-satellite constellations for 5G and beyond-5G communications'. *IEEE Access: Practical Innovations, Open Solutions*. 2020, vol. 8, pp. 184955–64.

[13] 3GPP TS 38.410 v16.1.0, NG-RAN; NG general aspects and principles. Mar 2020.

[14] 3GPP TR 38.801 v14.0.0, study on new radio access technology: radio access architecture and interfaces. Apr 2017.

[15] ETSI EN 302 307-1. *Digital video broadcasting (DVB); second generation framing structure, channel coding and modulation systems for broadcasting, interactive services, news gathering and other broadband satellite applications; part 1: DVB-S2*; 2014.

[16] ETSI EN 302 307-2. *Digital video broadcasting (DVB); second generation framing structure, channel coding and modulation systems for broadcasting, interactive services, news gathering and other broadband satellite applications; part 2: DVB-S2 extensions*; 2009 May.

[17] ETSI EN 301 790. 'Digital video broadcasting (DVB); interaction channel for satellite distribution systems. 2015'. 2015.

[18] 3GPP TS 38.401 v16.1.0, NG-RAN; architecture description. Mar 2020.

[19] 3GPP TS 38.470 v16.1.0, "NG-RAN; F1 general aspects and principles. Mar 2020.

[20] 3GPP TR 38.809 v16.0.0, NR; background for integrated access and backhaul radio transmission and reception (release 16). Sep 2020.

[21] 3GPP TR 38.874 v16.0.0, NR; study on integrated access and backhaul. Dec 2018.
[22] Vanelli-Coralli A., Guidotti A., Foggi T., Colavolpe G., Montorsi G. '5G and beyond 5G non-terrestrial networks: trends and research challenges'. Bangalore, India, 2020. pp. 163–69.
[23] Kodheli O., Guidotti A., Vanelli-Coralli A. 'Integration of satellites in 5G through leo constellations'. *IEEE Global Communications Conference (GLOBECOM 2017)*; Singapore, 2017. pp. 1–6.
[24] Guidotti A., Vanelli-Coralli A., Conti M, *et al.* 'Architectures and key technical challenges for 5G systems incorporating satellites'. *IEEE Transactions on Vehicular Technology*. 2019, vol. 68(3), pp. 2624–39.
[25] ITU-T, FG-NET-2030. 'Network 2030 architecture framework'. 2020.
[26] '5G-PPP white paper, European vision for the 6G ecosystem. Jun 2020.
[27] 3GPP TR 22.261 v17.2.0, service requirements for the 5G system'. 2020.
[28] Sirotkin S. *5G radio access network architecture: the dark side of 5G*. John Wiley & Sons; 2021 Feb 23. Available from https://onlinelibrary.wiley.com/doi/book/10.1002/9781119550921
[29] de Cola T., Bisio I. 'QoS optimisation of eMBB services in converged 5G-satellite networks'. *IEEE Transactions on Vehicular Technology*. 2021, vol. 69(10), pp. 12098–110.
[30] Burleigh S.C., De Cola T., Morosi S., Jayousi S., Cianca E., Fuchs C. 'From connectivity to advanced Internet services: a comprehensive review of small satellites communications and networks'. *Wireless Communications and Mobile Computing*. 2021, vol. 2019, pp. 1–17.
[31] Di B., Song L., Li Y., Poor H.V. 'Ultra-dense LEO: integration of satellite access networks into 5G and beyond'. *IEEE Wireless Communications*. 2019, vol. 26(2), pp. 62–69.
[32] Rajendran S., Meert W., Lenders V., Pollin S. 'Unsupervised wireless spectrum anomaly detection with interpretable features'. *IEEE Transactions on Cognitive Communications and Networking*. 2019, vol. 5(3), pp. 637–47.
[33] Vázquez M.Á., Henarejos P., Gil J.C., Parparaldo I., Pérez-Neira A. 'Artificial intelligence for SATCOM operations'. *Proceedings of SpaceOps Operation 22-25*; Cape Town, South Africa, 2020.
[34] Vázquez M.Á., Henarejos P., Pérez-Neira A, *et al.* 'On the use of AI for satellite communications'. *IEEE SSC Newsletter*; 2019.
[35] Bengio Y., Lodi A. 'Antoine Prouvost: machine learning for combinatorial optimization: a methodological tour d'horizon corr abs/1811.06128'. 2018.

Chapter 4
Flat antenna arrays for non-geostationary orbit communications

Maria Carolina Viganó[1]

This chapter provides an overview of flat panel array antennas. This type of antenna is becoming more and more actual because its features are a good match to the needs of the new non-geostationary orbit (NGSO) constellations. Basic knowledge of all the parameters involved when designing an array is provided without details nor the theory needed to design one. Different architectures of phased arrays are analyzed, and pertinent examples of existing products are provided.

4.1 Connectivity, connectivity, and connectivity

The roadmap of Internet connectivity can be understood with a comparison to the evolution of the telephone. In 20 years, we have gone from a telephone tethered to the wall, to a cordless phone, then the mobile phone. Internet connectivity is following the same trajectory. We had Internet dial-up linked to a telephone line, then wireless that allowed an untethered connection in a limited area, and now, we are looking for our Internet connectivity to be available everywhere, i.e., on the ground, in the sky, at sea, and in the desert. Increasingly, services are designed with the expectation that people can be connected everywhere all the time. The demand for connectivity everywhere is being made with the conditions of high quality and low price.

Connectivity – fast, cheap, and everywhere – is considered an essential for many businesses and services to operate. Emerging technologies and resulting societal changes are driving the need to be connected. The global pandemic of 2020 and 2021 has caused a societal shift toward working from home. Technologies requiring continuous connectivity, like Internet of things (IoT), autonomous vehicles, and telemedicine, form the basis of many services and businesses.

[1]Viasat Antenna Systems, Lausanne, Switzerland, Special thanks to Dr. D. Llorens del Rio and Carlene Lyttle reviewing and supporting this work.

4.1.1 MEO-LEO-GEO-HEO

For terrestrial connectivity, developments in 5G and 6G are working toward meeting that demand. But, where terrestrial cannot reach – due to geographical or financial constraints – or cannot provide enough capacity, satellite connectivity is the missing link. Satellite Internet has moved, in the last 20 years, from niche markets, where customers have few choices, receive subpar service quality, and pay more, to a more mainstream telecommunications service. This move began with high-capacity satellites at geosynchronous Earth orbit (GEO). More recently, companies are looking toward lower orbits (Figure 4.1) with larger constellations of satellites as a way to provide connectivity.

Companies looking to provide Internet worldwide using lower orbit satellites are doing so with constellations varying in number of satellites. From few dozens of satellites, in the case of O3b's mPower medium Earth orbit (MEO), to SpaceX's Starlink constellation which has more than 1 600 satellites as of mid-2021 and plans to operate up to 30 000 small low Earth orbit satellites (LEO) in the near future.

Recently, also HEO (highly elliptical orbit) satellites are becoming of interest, especially when specific areas like the polar regions are to be covered.

The shift from a few GEO satellites to a large number of lower orbit constellations has a significant impact on the requirements for ground terminals, in particular for fixed ground terminals.

Ground terminals can be classified as fixed or mobile. Fixed terminals usually include larger terminals used for enterprise and back-haul applications and smaller fixed terminals servicing the residential broadband satellite Internet market. Mobile terminals are used where satellite Internet is required for mobility connectivity on land, sea, and in the air.

For a ground terminal, a GEO satellite looks like a fixed point in the sky. Operationally, this means the fixed terminal needs to be pointed just once.

This is not the case for a fixed terminal connecting to a non-GEO satellite. In this case, the satellite appears to be moving across the sky at different speeds depending on the satellite altitude. The ability to track a moving satellite is thus a requirement for a ground terminal that is part of a LEO satellite system.

Differences in the design of mobility terminals for NGSO or GSO (geostationary) systems are less substantial. Indeed antennas designed for connecting to a GEO

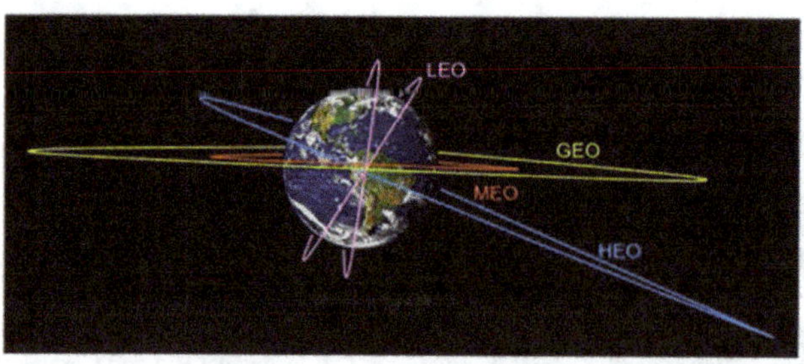

	distance from Earth
LEO	200-2000 km
MEO	2000-10000 km
GEO	35786 km
HEO	From 500 to 48000 km

Figure 4.1 Satellite orbits and their typical distance from Earth

satellite need to be able to steer the beam toward the satellite because the platform (aircraft, train, etc.) is moving. One of the few differences is the need for NGSO terminals to transition smoothly between satellites as they go in and out of the terminal field of view. This can be accomplished by fast switching or by creating more than one simultaneous beam.

The design parameters will be addressed in the following sections.

4.2 Design for NGSO antenna – specs

In this section, the main parameters useful for the antenna design will be discussed. Most of these requirements are common to any antenna application, while few of them like the handover architecture are prevalently addressed when referring to NGSO systems.

4.2.1 Frequency and frequency band

The first requirement that plays an important role in the antenna design is the radiating frequency. While electromagnetic theory is valid at all the different frequencies used for SATCOM, the implementation of each antenna architecture, especially for phased arrays, may be quite different. In the standard reflector technology, the main part changing with frequency is the reflector feed that needs to be scaled appropriately. Surface roughness of the reflector and accuracy of the manufacturing approach selected for the feed are the main limiting factors for high frequencies.

Similar considerations apply to phased array antennas, except that instead of a single feed there are multiple antenna elements to be manufactured and connected with properly designed beam forming networks (BFN).

If PCB (printed circuit board) technology is used for the array antenna, there are some parameters that need to be taken into consideration when the design is implemented. From vias dimension to clearance between metal details or finishing of the metal used, many of these are going to make design at high frequency more complicated than expected.

Another factor that complicates or sometimes even precludes the design of certain types of antenna is the width of the frequency band that needs to be targeted. For example, addressing a band as wide as the complete Ka band, with a single antenna, may limit the choice of architectures or may impact cost so substantially that separated antennas for different part of the band may be more suitable.

This is often, together with interference in full duplex systems, one of the main reasons to place the transmit and receive radiating functions in separated apertures for Ka band.

Figure 4.2 shows the frequencies used now for some of the broadband SATCOM GSO and NGSO user terminals. While most of the parameters and architectures analyzed here are generic, this chapter is by choice focused on the high frequency bands depicted in Figure 4.2, that correspond to the high-capacity systems developed in the last years.

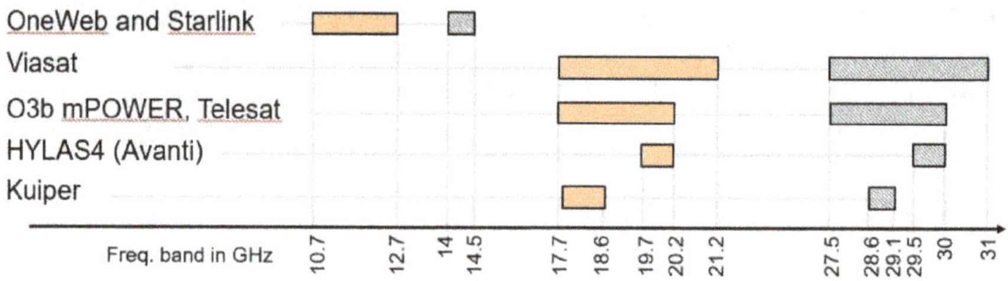

Figure 4.2 *Ku and Ka bands used by some of the current GSO and NGSO user terminals. In orange, the lower part of the band is used for reception, and in gray stripes, the upper part that is used for transmission.*

4.2.2 Instantaneous bandwidth

The instantaneous frequency is a requirement related to the radiating frequency but deserving of its own discussion. This parameter impacts the design of analog phased arrays (see sections 1.3.1 and 1.3.2) where phase shifters are used to modify the relative phase between the elements.

In the past, phased arrays that were designed to cover large bands did not require a large instantaneous bandwidth. This is, for example, the case of some phased arrays developed for radar applications. Nowadays, especially for communication applications, system trades for larger capacity result in larger link bandwidth. Many NGSO systems are designed today with large instantaneous bands as, for example, OneWeb and Telesat.

In a typical narrowband array, the phase shifters are set at the center of the frequency band of interest. The phases generated in this way by the phase shifters are not the correct ones for the edges of the band, which results in unwanted effects such as beam squint and intersymbol interference. These effects grow worse as the band grows wider, and they limit the instantaneous bandwidth of the array, especially for large arrays and large scan angles [1].

A way to mitigate the problem is to partition the array into smaller pieces with a reduced number of elements. These are often referred to as "subarrays" or "tiles"; by assigning a true time delay (TTD) control to each of them, the instantaneous bandwidth of the array can be improved.

This TTD unit can be implemented in an analog or digital way, even though the latter is surely more flexible and easier to implement.

4.2.3 Maximum scan angle

The maximum angle, measured from the antenna boresight, toward which the antenna needs to point is here referred to as "maximum scan angle." When considering terminals designed for GSO satellites, this parameter is dictated by the type of platform that will host the terminal and by the geographical location where the antenna will operate. For example, antennas deployed in northern or southern regions will need to be able to achieve larger scan angles than antennas designed for equatorial regions. This can be easily understood and visualized for a GSO system but cannot be generalized for NGSO systems. Depending on the application, NGSO satellite constellations can be designed to concentrate the coverage over specific areas by playing with the number of orbital inclinations and the number of satellites in each inclined orbit. In general, it is possible to conclude that the larger the number of satellites in a constellation the more reduced will be the required maximum scan angle on the user terminal.

For this reason, during the early days of an NGSO constellation, when the number of operational satellites is enough to provide a minimum level of service but still far from its complete number, companies have considered augmenting the scan angle of antennas by adding some tilting or motorized mechanism with the idea of removing it at a later stage when the constellation will be completed.

The maximum scan angle impacts greatly on the design of phased arrays since the interelement distance and the array lattice are chosen mainly based on this parameter. If a large distance between the elements is chosen, then the Grating Lobes (GL) will appear in the visible space when the main lobe is pointed away from boresight, making it impossible to comply with regulations as described in section 1.2.5. More details on how to design array lattices, choosing the right inter element distance, can be found in the classic antenna books as [2].

Designing and sizing the array for the maximum scan angle are also important because phased array performances degrade as the beam points further away from boresight. Mechanically steered antennas have similar performance wherever they point since the aperture is always the same relative to the direction of the satellite. In the case of flat array panels, the equivalent surface seen from the direction of the satellite is reduced with the scan angle in a $\cos(\vartheta)^q$ fashion (with q depending on the size of the chosen element). The more the beam is pointed away from boresight the more the gain drops, and the beam gets wider as depicted in Figure 4.3 [3].

4.2.4 Handover architecture

Mobility antennas connecting to GSO satellites may have to switch frequency and polarization often when crossing over different beams generated by the same satellite, but a handover to a different GSO satellite is a much rarer event. A small hitch in the connection can be possibly tolerated by the service if it only recurs after several hours. For NGSO systems, the handover may be happening as often as every few minutes, depending on the constellation altitude, the number of inclined orbits, or the number of satellites in each orbit. Hence, the handover needs to be seamless. This can be achieved in two ways, termed "break before make" (BBM) and "make before break" (MBB).

The first one relies on a rapid repointing of the antenna from the setting satellite toward the rising one. For this approach, phased arrays with fast repointing time are needed, together with quick network re-entry modem settings and/or caching techniques like the ones discussed in [4].

The second approach relies on the antenna being able to generate two beams (or on the use of two antennas), pointed to the setting and to the rising satellites,

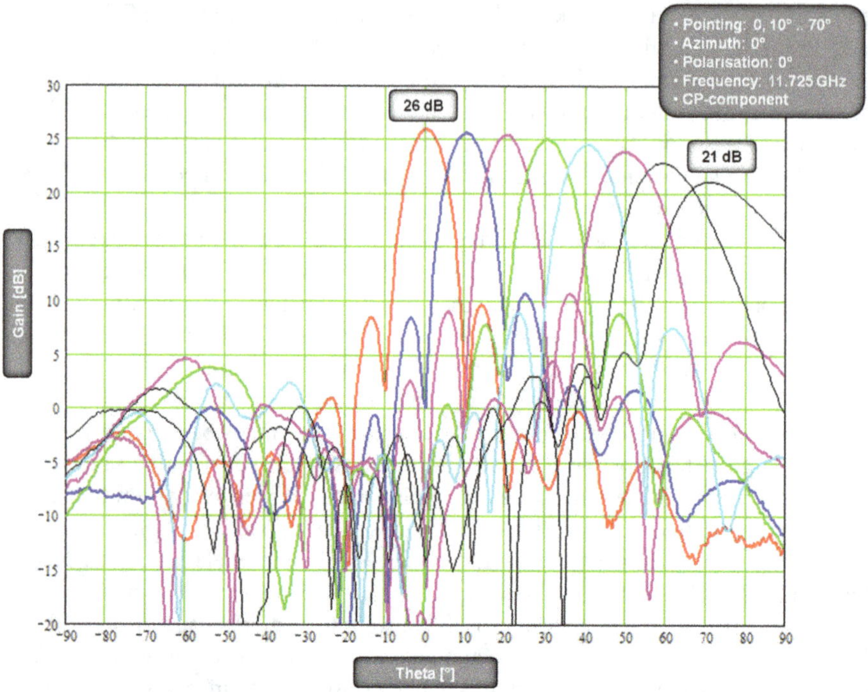

Figure 4.3 *Ku band phased array measurements for project NATALIA. Courtesy of Viasat Antenna Systems SA [3].*

respectively. This requires two independent modems, both active during the handover time. In this case, a quick network re-entry and the capability of fast beam repointing of the antenna are not necessary.

4.2.5 Mask compliance

Radio transmitters, including satellite user terminals, are required to comply with emission norms maintained by ITU (International Telecommunication Union) and other regional and national regulatory bodies.

The angular masks defined in these norms (Figure 4.4 [5]) limit the amount of energy that the terminals are allowed to radiate in directions away from their target satellite, with the goal of enabling the coexistence of terminals connected to different satellites while generating minimal interference. To comply with such masks, as the one in Figure 4.4, means to radiate a low quantity of energy in directions other than the one of the satellite the terminal wants to connect to. It is easy to understand that, as an example, having thousands of antennas with really high sidelobes all pointing in the same direction can cause large interference issues to a satellite that is exactly at that position up to an impossibility to operate.

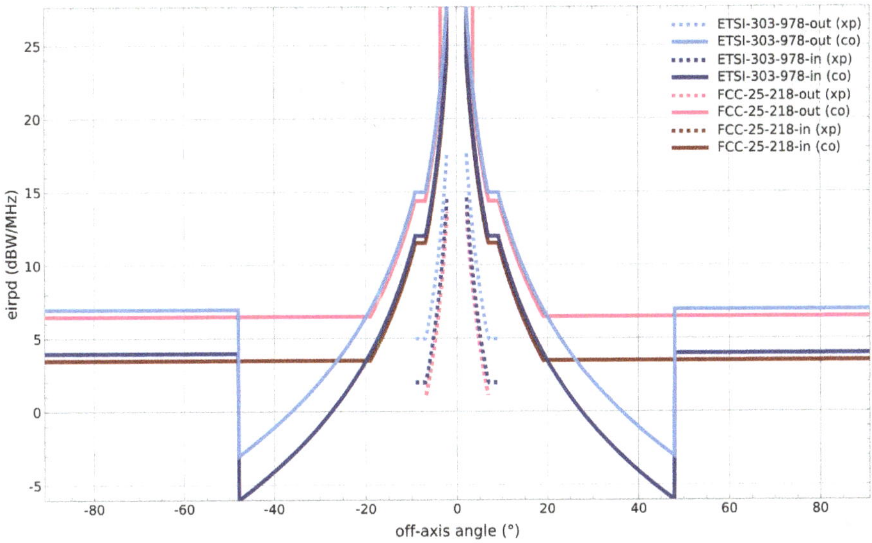

Figure 4.4 *ETSI (European Telecommunications Standards Institute) 303 978 and FCC (Federal Communications Commision) 25.218 for copolar and crosspolar allowed EIRPD inside and outside of the GEO arc*

However, these masks often date from an era when it could be safely assumed that every satellite terminal was a reflector of some kind, and as a result, they constrain the patterns of other types of antennas in ways that do not meaningfully help toward the stated goal of reducing the overall interference.

Compared to reflectors, a clear advantage of phased arrays is their capability to dynamically change and adapt their patterns. Phased arrays often have non-perfect circular apertures, which introduce a dependence on the azimuth for the pattern shape. This is true even for circular arrays when the beam is scanned off boresight.

The growing of NGSO constellations complicates the picture: the coexistence of GEO, LEO, MEO, and HEO satellites creates cases where interference cannot be avoided just following good design rules, since two satellites at different altitudes may appear at the same location for a ground terminal. Coordination and analyses, as the ones presented in [6], become necessary. EIRPD (equivalent isotropic radiated power density) masks become constellation specific, since allowable values vary depending on the many parameters that can be used to design a constellation. This presents an opportunity to include the characteristics of the antenna terminal in the design of the mask.

Attention needs to be placed in the antenna design also in reception; in this case, it is in the interest of the same terminal to avoid too much interference coming from other satellites. For this reason, regulatory bodies provide suggestions for pattern mask but compliancy is not required.

4.3 Types of flat panel array antennas

After a general overview of the main parameters used to design user antennas, this sections will focus on possible phased array architectures, their advantages, and their practical implementation for NGSO systems.

Phased array antennas can be classified in many ways; in this chapter, the division is based on the choice of the type of architecture: passive or active and analog or digital. These characteristics are used hereafter for describing more in detail the different array types. More attention is placed on analog architectures since nowadays more examples of this type are available in the market.

4.3.1 Analog passive phased arrays

Array antennas belonging to this category are characterized by distributed phase shifting function at element level, but a centralized amplification happening before (in the case of transmission or after in the case of reception) the beamforming is implemented.

The concept is depicted in Figure 4.5 for the transmit function case; an equivalent case can be outlined for the receive function case.

In this kind of architecture, the losses in the BFN have to be minimized in order to have good performances. This constitutes a limitation in the kind of technological implementation that can be used for the BFN. The most common way to implement

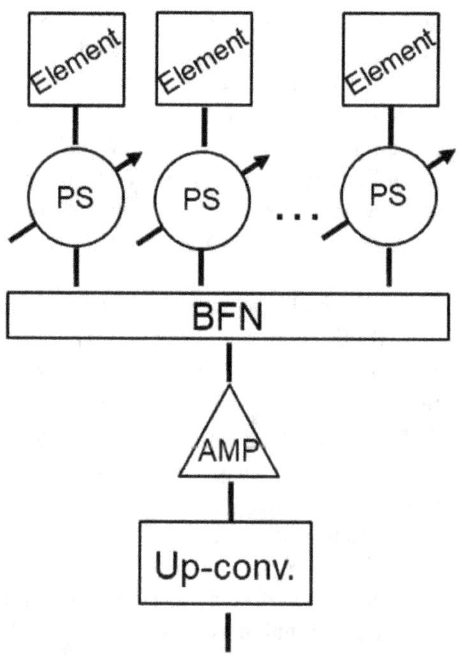

Figure 4.5 Analog passive phased array architecture example

a low loss BFN is to resort to waveguide technology. This technology exists since many years and usually results into a reliable but heavy and pricy solution.

Depending on the frequency band and the accuracy needed, the BFN can be manufactured with different materials and processes. The most common material used is metal, usually resulting into a heavy antenna structure. Stamping or molding metal, instead of machining, can result into a reduced quantity of material used and, for this reason, a lighter implementation. Metallized plastic can also be effectively used if the manufacturing process is well mastered [7, 8].

Solutions for low losses BFN not based on waveguide technology are possible. One example is the utilization of suspended stripline in PCB technology already in use since many years [9].

The fact of having only one amplifier for the complete array has also another important implication. It limits the type of taper that can be implemented over the array to only a phase taper. Having phase control at element level is usually sufficient to have good pattern shaping. Nevertheless, amplitude control, achieved for example in active arrays with adjustable gain amplifiers at element level, provides an extra degree of freedom. When designing for strict mask compliance, this additional degree of freedom may become necessary.

Regarding the phase shifting at the element level, this can be obtained with several techniques. One of these consists of using liquid crystals, like proposed in the products designed by Alcan or Kymeta (Figure 4.6). Both companies have reported a successful implementation for this new technology coming from the television display industry. By applying different voltages, it is possible to orient the molecules of the liquid crystal, changing in this way the dielectric properties of the medium. This principle is used to implement the required phase shifting.

Other ways to achieve phase shifting that can be mated with waveguide technology are reported in the literature; from the tunable substrate integrated waveguide phase shifter [10], to the use of ferrite materials [11] or the rotating metasurfaces proposed by Macquarie University [12]. Products developed by ThinKom Solutions

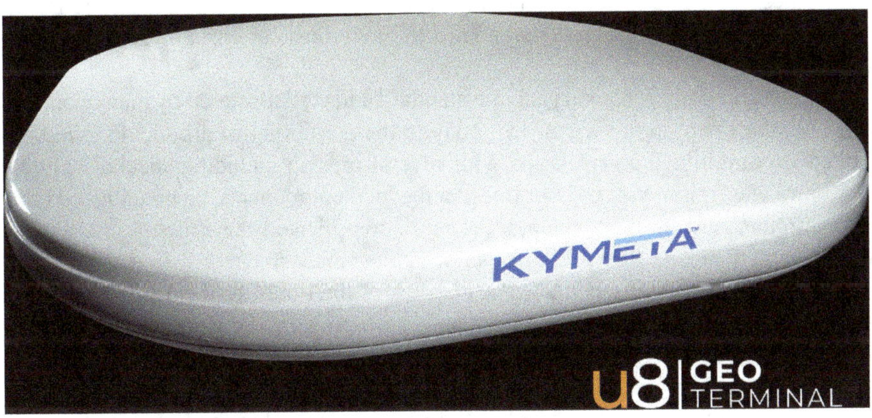

Figure 4.6 Kymeta™ u8 antenna

Figure 4.7 Thinkom ThinAir® KA2517 [13]

(Figure 4.7 [13]) are also based on a rotating platform (variable inclination continuous transverse stub technology) and have reliably been deployed on the market in both Ku and Ka band since many years.

The approach chosen to implement the phase shifting has an impact on the beam pointing speed. It is a parameter to keep in mind when designing an array for the mobility and/or for the NGSO networks.

An important advantage of this kind of configuration is usually a lower power consumption with respect to an active phased array or a digital one. This power saving comes from having a single, larger, usually more efficient amplifier instead of many distributed ones designed with cost and size constraints.

4.3.2 Analog active phased arrays

The category of analog active phased array is characterized by distributed analog amplitude and phase control at element level (Figure 4.8).

This architecture gives a large flexibility since it provides both phase and amplitude control at the element level.

The fact of having the amplification implemented really close to the element removes the constraint of having a quasi-lossless BFN. For this reason, in the active array category, PCBs are one of the most preferred options to implement the BFN, even though with respect to waveguide technology these are usually resulting into larger losses.

PCB technology is nowadays quite standard and well mastered by many companies. Advancement in this sector led today to the possibility of producing in a controlled way multilayered stack-ups with several presses including special features like blind vias, micro vias or laser ones, embedded components, or metal inserts for thermal management. Nevertheless, the more complicated the structure the higher the cost and the lower the yield that can be achieved.

High losses in PCB technology lead, most of the times, to the need for several stages of amplification. The number of stages depends on the gain of the amplifier used and the length of the BFN (since longer BFNs, connecting many elements, produce higher losses). Adding more stages of amplification is necessary in transmission to guarantee a certain level of the signal being fed to the last amplifier and then to the element or to the downconverter and modem in reception.

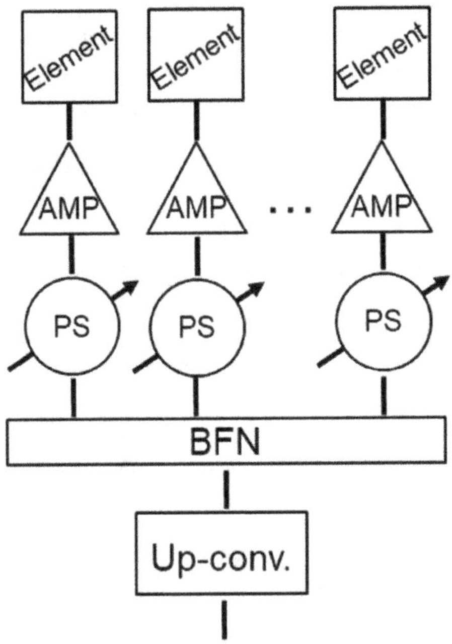

Figure 4.8 *Analog active array architecture example*

The phase shifting and the amplification functions are usually implemented in integrated circuits (IC) or monolithic microwave IC (MMIC).

The performances of this kind of architecture are mainly dictated by the active component technology chosen for the MMICs and on the interface of these MMICs with the radiating element.

Depending on the technology, indeed different noise figures (NF) in reception and RF (Radio Frequency) output power in transmission (as P1dB for the power at 1 dB compression point) can be achieved. Regarding low noise amplifiers, the NF can vary from <0.5 dB at high frequency for a GaAs (gallium arsenide) component to ~3 dB for less expensive components. In the same way, GaAs and GaN (gallium nitride) technology can usually provide more power output from the MMIC (easily up to a few watts) with respect to silicon-based MMICs, where numbers are more in the order of tens of dBm. In an active phased array, where thousands of elements are used, having large power amplifiers per element is not needed. For this reason, nowadays, silicon-based components are one of the preferred options for active phased arrays. The technological choice is also dictated by the market needs and the market size. For small volumes and special applications, where cost is less of an issue, GaAs, GaN, or hybrid solution involving even indium phosphide (InP) are preferred.

On top of the amplification function, these MMICs usually include also the phase shifting and in most of the cases some beamforming since in typical application one MMIC is serving more than one element [14].

Another difference with respect to passive arrays resides in the thermal structure that needs to be implemented. In passive arrays, a single amplifier is used, and the design effort on the thermal side mainly consists into removing the heat generated in a single place and spreading it in a passive or active manner.

Active phased arrays usually generate more heat than passive ones due to the efficiency of the small amplifiers. This heat is though already distributed across the aperture since one of these amplifiers is serving from one to eight antenna elements. The thermal difficulty in this kind of array is mainly to deal with large amounts of heat and to dissipate that in a passive or active way. Improvements in the MMIC amplifier efficiency allowed companies that were previously resorting to water cooling to move to more conventional fans or even passive cooling in some cases.

Several examples of this array architecture are present in the market. Most of them, like Viasat [15] and Ball ones [16], are based on a modular architecture that can be scaled up or down depending on the needs. This modularity choice was taken for two main reasons: being flexible to address markets with different performance needs and being able to assemble larger units where a single physical PCB would not be large enough to provide adequate performances for the application.

Some of these terminals have been designed mainly for mobility applications as Rockwell Collins Ku terminal or Viasat aero terminals (Figure 4.9). For this reason, they are already targeting a large scan angle and sometimes already include the possibility of creating two beams for a MBB.

In some other cases, like the Viasat MEO terminal (Figure 4.10), the design was aimed specifically to NGSO. The O3b constellation was considered, and for this reason, the array is designed to point the beam along the MEO arc and is capable of both BBM and MBB.

Figure 4.9 Viasat airborne solution for Ka band [15]

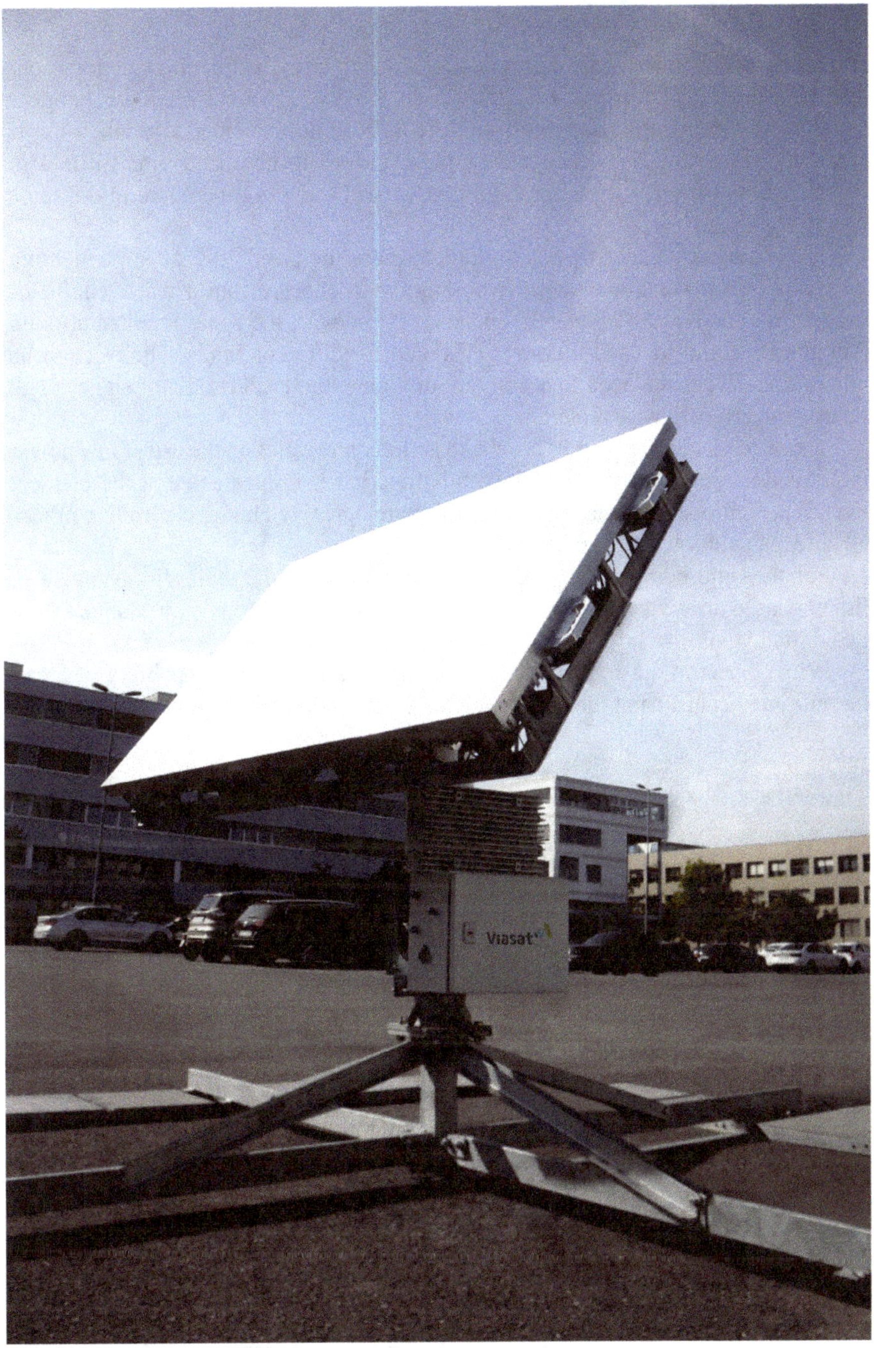

Figure 4.10 Viasat MEO Ka band terminal

4.3.3 Baseband/IF and digital BFN array

The baseband/IF (Intermediate Frequency) antenna array is the third category considered in this chapter. In this case, the up-conversion/down-conversion is happening close to the radiating elements, and the BFN is implemented at a much lower frequency (Figure 4.11). When the signal is moved to baseband often, it is also digitalized, and for this reason, digital phased arrays are grouped here in the same paragraph.

The main advantages of this architecture are the possibility of implementing the BFN with lower losses and the possibility to choose from a wider choice of components available at lower frequencies. Moreover, the same architecture and components could be used in arrays designed for different bands. The price to be paid in this case is the need to accurately distribute the local oscillator signal to all the up-converters/down-converters.

The success of an analog IF/baseband architecture relies on the use of low power up-converters/down-converters. Since this function is implemented at the element level, it is of utmost importance to have this parameter well under control if a practical antenna solution needs to be developed.

For the time being, the power consumption is also the main disadvantage of digital phased array; the development of low power components may change this in the coming years.

With respect to the analog version, the digital array has the capability of creating multiple beams using the overall aperture without degradation in performances.

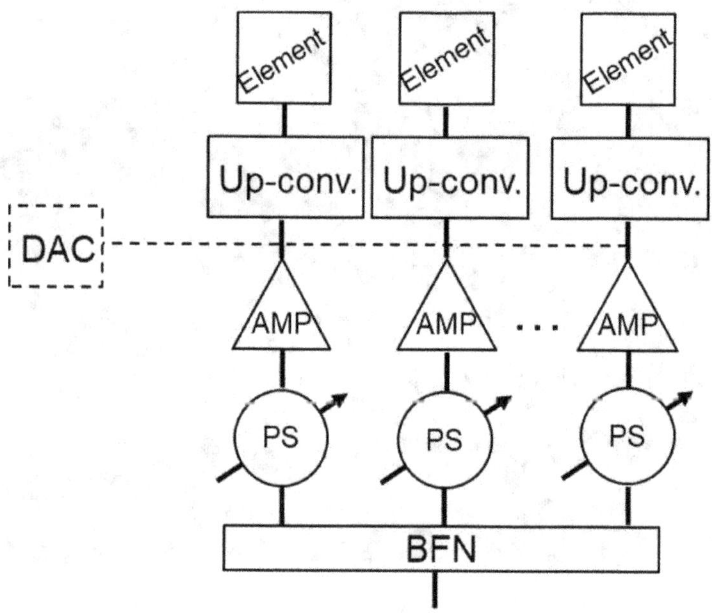

Figure 4.11 Baseband/IF array architecture example

Figure 4.12 Satixfy Ku digital phased array

The same can be achieved in the analog domain for a limited number of beams, by replicating the controls and implementing the same number of BFNs.

This is an achievable option when only few beams are needed, for example, in an antenna designed to support MBB for NGSO, but not practical when more than ten beams are to be active at the same time, as discussed nowadays for some NGSO ground stations.

Satixfy (Figure 4.12 [17]) is one of the first companies to propose commercially a digital BFN array for satellite IoT applications.

Another example of phased arrays belonging to this architecture category is the antenna developed by Hanwha Phasor (Figure 4.13). In this case, analog IQ (In-Phase and Quadrature) baseband signals coming from each antenna element are

Figure 4.13 Hanwha Phasor Ku band solution under test [18]

combined; thanks to a specifically developed ASIC (Application Specific Integrated Circuit) [18].

This approach results into a lower power consumption when compared to an array where the digitalization is also happening at element level.

4.4 Conclusions: are phased arrays finally there?

Phased array antennas are still today far from being widely spread in the market, raising sometimes the question if this technology would finally leave the Gartner's Hype Cycle stage of "peak of inflated expectations" and just end on a "trough of disillusionment."

We can cautiously say that, still according to the same scale, most of the phased array technologies described in the sections above are moving on the "slope of enlightenment" with customers starting to discern one from another. The differences and advantages of each are becoming more clear, and phased array manufacturers are now able to adapt the technology coming out with second and third generation products implementing market's feedbacks.

Being aligned with 5G and 6G technologies and following a sort of Moore's law for IC costing should enable phased arrays success and lead to a mainstream adoption.

While GSO mobility terminals could eventually survive a failure in achieving a drastic cost reduction, this is a fundamental need for NGSO constellations. In the past, most of NGSO companies focused on the satellites side only, somehow taking for granted that a proper user terminal would become available in time. The clock is ticking, and this, as of today, remains the field where more investments are needed to enable a successful NGSO business case.

References

[1] Haupt R.L. 'Factors that define the bandwidth of a phased array antenna' [online]. *IEEE International Symposium on Phased Array System & Technology (PAST)*; Waltham, MA, USA, 2019. Available from https://ieeexplore.ieee.org/xpl/mostRecentIssue.jsp?punumber=8982099

[2] Mailloux R.J. *Phased array antenna handbook*. 2nd edition. Boston, MA: Artch House; 2005.

[3] Baggen L., Vaccaro S., Padilla J., Sanchez R.T. 'A compact phased array for satcom application'. *IEEE International Symposium on Phased Array Systems and Technology, Boston*. 2013.

[4] Leng T., Xu Y., Cui G., Wang W. 'Caching-aware intelligent handover strategy for Leo satellite networks'. *Remote Sensing*. 2021, vol. 13(11), p. 2230.

[5] [ETSI 303 978]. Available from https://www.etsi.org/deliver/etsi_en/303900_303999/303978/02.01.02_60/en_303978v020102p.pdf

[6] *Presentation at ITU world radiocommunication seminar* [online]. Geneva. 2018. Available from https://www.itu.int/en/ITU-R/space/WRS18space/WRS18-EPFD-06122018.pdf

[7] Geterud E.G., Bergmark P., Yang J. 'Lightweight waveguide and antenna components using plating on plastics'. *7th European Conference on Antennas and Propagation (EuCAP)*; IEEE, 2013. pp. 1812–1815.
[8] Ghazali M.I.M., Park K.Y., Gjokaj V., Kaur A., Chahal P. '3D printed metalized plastic waveguides for microwave components'. *International Symposium on Microelectronics*. 2017, pp. 78–82.
[9] Yamashita E., Nakajima M., Atsuki K. 'Analysis method for generalized suspended striplines'. *IEEE Transactions on Microwave Theory and Techniques*. 2017, vol. 34(12), pp. 1457–63.
[10] Omam Z.R., Abdel-Wahab W.M., Raeesi A., *et al*. 'KA-band passive phased-array antenna with substrate integrated waveguide tunable phase shifter'. *IEEE Trans Antennas Propag*. 2017, vol. 68(8), pp. 6039–48.
[11] Cherepanov A.S., Guskov A.B., Yavon Y.P., Yufit G.A., Zaizev E.F. 'Innovative integrated ferrite phased array technologies for EHF radar and communication applications'. *International Symposium on Phased Array Systems and Technology*; Boston, MA,, 1996. pp. 74–77.
[12] Afzal M.U., Esselle K.P., Koli M.N.Y. 'A beam-steering solution with highly transmitting hybrid metasurfaces and circularly polarized high-gain radial-line slot array antennas'. *IEEE Transactions on Antennas and Propagation*. 2022, vol. 70(1), pp. 365–77.
[13] Available from https://www.thinkom.com/
[14] Rebeiz G.M. *UCSD, IEEE aerospace conference, plenary talk*; 8 March, IEEE, 2021.
[15] Available from https://www.viasat.com/phased-array/
[16] Available from https://www.ball.com/aerospace/markets-capabilities/capabilities/antennas
[17] Available from https://www.satixfy.com/#slide6
[18] Available from www.hanwha-phasor.com

Chapter 5

Low cost per bit for LEO satellite systems: radio-frequency impairments compensation

Bassel F. Beidas[1]

The relentless pursuit of continuous, global, broadband connectivity has recently intensified the interest in low-Earth orbit (LEO) satellite mega-constellations, which are capable of delivering massive throughput. This and the insatiable demand for low-cost, low-complexity user terminals conspire to cause analog radio-frequency (RF) components to exceed their tolerance limits. Advanced digital technological solutions are explored in this chapter to minimize the strong and frequency-selective in-phase/quadrature (I/Q) imbalance introduced by analog frequency-conversion circuits when signaling at multiple Giga (G)-Baud rates. Specifically, two characterization models of analog RF impairments are provided when frequency offset is present. Novel digital compensation algorithms with immunity to frequency offset are presented and categorized into equalization with image-rejection capability and image cancelation. Adaptive techniques are utilized to obtain compensation coefficients in an iterative manner using stacked construction, while the pursued coefficients are independent of frequency offsets. These methods are useful when using known data samples for initial factory calibration or in a decision-directed mode during field recalibration. Extensive computer simulations reveal that the proposed compensators provide lossless attenuation of the imbalance-induced, frequency-selective image in the presence of frequency offsets.

5.1 Introduction

The satellite scientific community has recently witnessed an eruption of research activity spurred by the ongoing deployment of LEO satellite systems which are envisaged to provide continuous, global, broadband connectivity. These emerging satellite mega-constellations, positioned at altitudes of less than 1,500 km, have the distinct advantages of significantly lower latency, smaller size, lower power consumption, and lower launch cost. LEO satellites are expected to become a strategic enabler of satellite-terrestrial

[1]Advanced Development Group, Hughes, USA

integrated networks [1]. The leading third-generation partnership project (3GPP) has already established new use cases for satellites in their fifth-generation (5G) standardization, designated as nonterrestrial networks [2].

Next-generation, mega-constellation LEO satellites are anticipated to deliver ultra-high capacity in the tens of Terabits per second. One of their strengths is their capability of guaranteeing broadband connectivity in rural and remote areas where service is currently nonexistent. In addition, LEO satellites are invaluable in providing backhaul links for mobile cellular networks in underserved regions which, because of infrastructure challenges, cannot be reached in a cost-effective manner by terrestrial-only networks. Moreover, due to their inherent resilience to natural disasters or wide-scale attacks, LEO satellites can offer an agile solution that ensures life-saving communication continuity during extreme emergency situations.

The efficiency of satellite systems needs to be optimized on many levels: payload mass efficiency through joint amplification of multiple frequency-compact carriers by a single high-power amplifier (HPA) [3, 4]; power efficiency by operating the HPA close to saturation; energy efficiency by employing adaptive coding and modulation (ACM) using capacity-approaching forward-error correction codes and high-order modulations. The spectral efficiency of satellite systems can be additionally increased by utilizing faster-than-Nyquist (FTN) signaling, for which advanced receivers are developed in References 5, 6 to realize the gains of FTN over nonlinear satellite links. Furthermore, aggressive frequency reuse in multibeam satellite systems can alleviate the tremendous spectrum scarcity. To tackle the resulting dominant co-channel interference (CCI) on the user link, signal processing solutions are developed in the form of precoding [7–9] and multiuser detection [10–12]. The performance of multiuser detectors is established for the forward link in References 13, 14 using a computationally efficient framework in terms of the theoretically achievable information rates. To mitigate spatiotemporal impairments of CCI under full frequency reuse, low-complexity receiver designs are explored in Reference 15 using code-division multiplexing for distinguishability and in Reference 16 utilizing an iterative divide-and-conquer paradigm for superior performance.

The inexorable quest for massive satellite throughput is motivating the transmission of broadband signals at symbol rates of multiple GBaud. Building low-cost, low-complexity user terminals is essential to the economic viability of next-generation systems. These two competing drivers conspire to cause the analog RF components to exceed their tolerance limits. In particular, analog mixers, antialiasing filters, and amplifications in quadrature frequency-conversion architectures induce mismatch between parallel I/Q arms [17, 18]. Also, there can exist direct-current (DC) offset due to local oscillator (LO) leakage. In addition, a frequency offset between converters in the transmitter and receiver is assumed to be present. When broadband signals are utilized, this mismatch in components creates strong image interference that is both frequency-independent and frequency-selective.

In an effort to maximize satellite transmission throughput while reducing the cost per bit, this chapter develops digital adaptive compensation methods intended to diminish analog RF impairments. This development takes into account the idiosyncratic features of satellite systems in the forward direction, namely from the gateway to the user terminals. First, complexity and cost are afforded at the gateway so that its components are well designed, and the I/Q imbalance at the transmitter can be neglected. Second, the gateway implements multicarrier data predistortion to mitigate nonlinear distortions resulting from the on-board HPA. Powerful multicarrier data predistortion algorithms using successive methods are developed in References 19–21, while inversion method with direct learning is provided in Reference 22. Third, driving down the cost of user terminals is of paramount importance as millions of them need to be employed, so quadrature frequency-conversion circuits are relied upon for their reconfigurability and high flexibility. Fourth, frequency-selective I/Q imbalance is expected to be strong, and with a large time span as the quadrature converters need to accommodate signaling with symbol rates of multiple GBaud. Fifth, adaptive methods that operate without *a priori* knowledge of RF impairments are required to accommodate variations from one receiver realization to another. Sixth, the proposed methods need to remain effective in the presence of unavoidable and large frequency offset resulting from satellite propagation. Seventh, the distortion introduced by the satellite multiplexing filters is reduced via fractionally spaced (FS) equalization, developed in Reference 23 to offer large gains with limited complexity. Eighth, a physical-layer signaling (PLS) code is reserved to aid in the process of field recalibration. Once it is detected in high signal-to-noise ratio (SNR) conditions, software is informed to retrain.

The open literature contains several promising techniques for digitally mitigating RF impairments on the receiver side in wireless terrestrial networks. Blind schemes are introduced in References 24, 25 but are only capable of alleviating frequency-independent I/Q imbalance. Another blind approach is developed in Reference 26 that is able to compensate for the more beneficial case of frequency-selective I/Q imbalance. Unfortunately, it requires that the received signal satisfies "properness" statistical property, which is hard to maintain in some applications. Training-based approach is developed in Reference 27 using adaptive interference cancelation and in Reference 28, 29 using adaptive signal separation, requiring access to a primary reference source that is based on a strong and independent image signal. This makes these approaches not utilizable when the signal and its image are co-located in frequency, a situation encountered in direct conversion to baseband. Also, all these methods do not address the effects of DC offset or frequency offset. For receivers employing orthogonal frequency-division multiplexing (OFDM), the technique in Reference 30 considers frequency offset with I/Q imbalance, but no DC offset, and is based on restoring known phase rotation between pilot symbols. There, the pilot signal has a rather restrictive structure: several identical adjacent symbols with additional phase rotation of $\pi/4$ in even-numbered symbols, a requirement that is not satisfied in satellite systems.

This chapter provides two characterization models, in the presence of frequency offset, of analog RF impairments including frequency-independent and frequency-selective I/Q imbalance and DC offset: postmixer and premixer models. Novel digital compensation algorithms, developed in Reference 31, that are robust to frequency offset are described and can be categorized into equalization with image-rejection capability and image cancelation. A development of equalization and cancelation in the context of two carriers through a single HPA using Volterra formulation is introduced in Reference 32. Adaptive techniques are then utilized to obtain compensation coefficients in an iterative manner using a stacked construction that is immune to frequency offset. Toward this end, an estimate of the frequency offset is intentionally injected in the reference signal so that the pursued coefficients would not depend on the frequency offset. This essentially delinks the coupled tasks of frequency estimation and RF impairments compensation. Optimization methods extended to the multicarrier scenario are contained in Reference 31. These methods are useful when using known data samples for initial factory calibration or in a decision-directed mode during field recalibration. Special consideration is given here to estimating the frequency offset by sending a test tone during initial factory calibration. For this, a minimum condition on the tone frequency is presented to guarantee that spectral peaks from the tone and its interfering image do not substantially overlap. Performance evaluations are conducted using extensive computer simulations and reveal that the proposed compensation algorithms provide lossless attenuation of the imbalance-induced image and remove DC offset.

The remainder of the chapter is organized as follows: section 5.2 provides analytical characterization of RF impairments encountered in quadrature frequency-conversion devices; section 5.3 describes algorithms for compensating RF impairments, along with an iterative adaptation; section 5.4 contains a frequency estimation technique that operates successfully in a channel with I/Q imbalance; section 5.5 contains numerical studies to evaluate the associated performances and illustrate various design concepts; finally, section 5.6 contains concluding remarks.

Notation: Lower-case underlined symbol $\underline{x}$ and upper-case bold symbol $\mathbf{X}$ denote vectors and matrices, respectively; round parentheses for signal $x(t)$ denote continuous-time, while square parentheses for signal $x[n]$ denote discrete time; superscripts $(\cdot)^*$, $(\cdot)^T$, and $(\cdot)^H$ denote conjugation, transposition, and Hermitian operations, respectively; $\mathrm{Re}\{\cdot\}$ denotes the real-part operation; and in-line asterisk '$*$' denotes convolution.

5.2 Characterization models of RF impairments

A general structure for I/Q down-conversion, with real-valued bandpass input $r(t) = \mathrm{Re}\{\tilde{r}(t) \cdot e^{j2\pi f_c t}\}$, is shown in Figure 5.1 and includes analog components needed to implement operations such as mixing, filtering, amplification, and analog-to-digital conversion (ADC). Analog components induce mismatch between parallel I/Q branches of the receiver tuner. In particular, mismatch in gain and phase between LO mixer arms,

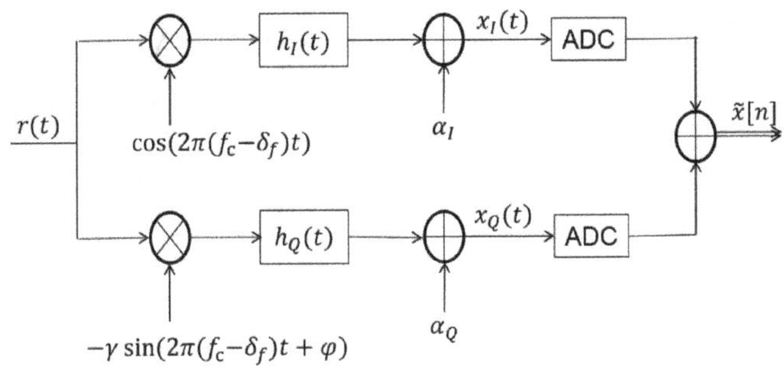

Figure 5.1 Quadrature frequency down-conversion with RF impairments

γ and φ, respectively, creates imbalance that is frequency-independent or constant over the signal bandwidth. Mismatch between antialiasing filters in the I/Q arms, $h_I(t)$ and $h_Q(t)$, causes imbalance that is frequency-selective. Also, there exists DC offset, α_I and α_Q, on each arm resulting from LO leakage. In addition, a frequency offset, δ_f, between the converters in the transmitter and receiver is assumed to be present to model any frequency instability in the oscillators as well as to incorporate frequency variation resulting from satellite propagation. The tuner model in Figure 5.1 produces sampled discrete-time version $\tilde{x}[n]$ at the ADC output.

Two analytical models are provided here that characterize the impact of RF impairments on the received baseband I/Q samples at the output of a wide-band tuner. They are referred to as postmixer or premixer models, depending on whether the imbalanced channel appears after or before the frequency off-set, respectively. Generalizing the modeling in References 26, 27 to include effects of frequency offset and DC offset, the down-converted complex-valued baseband signal at the ADC input, $\tilde{x}(t) = x_I(t) + jx_Q(t)$, can be mathematically expressed as

$$\tilde{x}(t) = \left[\tilde{r}(t) \cdot e^{j2\pi\delta_f t}\right] * g_1(t) + \left[\tilde{r}(t) \cdot e^{j2\pi\delta_f t}\right]^* * g_2(t) + \alpha, \tag{5.1}$$

where

$$g_1(t) = \tfrac{1}{2}\left[h_I(t) + \gamma e^{+j\varphi} \cdot h_Q(t)\right], \tag{5.2}$$

$$g_2(t) = \tfrac{1}{2}[h_I(t) - \gamma e^{+j\omega} \cdot h_Q(t)], \tag{5.3}$$

and $\alpha = \alpha_I + j\alpha_Q$. This postmixer analytical model is depicted in Figure 5.2(a).

The received signal without I/Q mismatch is $\left[\tilde{r}(t) \cdot e^{j2\pi\delta_f t}\right]$. As evident in (5.1)–(5.3), the I/Q imbalance creates a superposition of the signal with an interfering source, $\left[r(t) \cdot e^{j2\pi\delta_f t}\right]^*$. This interference is the signal's own image, as the conjugation in the time domain corresponds to frequency-domain reflection. The $g_1(t)$ and $g_2(t)$ are impulse responses associated with the signal and its image, respectively, and contain the joint impact of frequency-independent mismatch, from γ to φ, and frequency-selective mismatch, from antialiasing filters $h_I(t)$ to $h_Q(t)$.

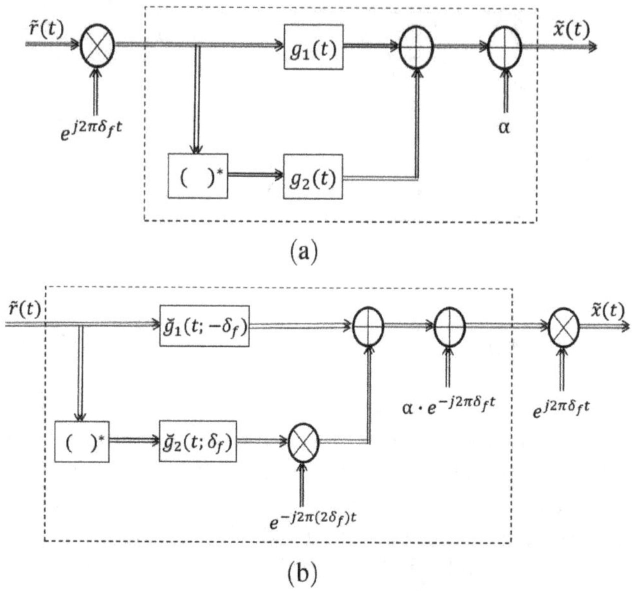

Figure 5.2 *Analytical baseband models of (a) postmixer and (b) premixer quadrature down-converters with RF impairments*

An equivalent premixer model, displayed in Figure 5.2(b), allows for the frequency offset to percolate through the imbalanced channel. For this, we use some basic properties of Fourier transforms dealing with convolutions and frequency shifting to generate a pre-mixer model as

$$\tilde{x}(t) = \left[\tilde{r}(t) * \breve{g}_1(t;-\delta_f) + \left(\tilde{r}^*(t) * \breve{g}_2(t;\delta_f)\right) \cdot e^{-j2\pi(2\delta_f)t} + \alpha \cdot e^{-j2\pi\delta_f t}\right] \cdot e^{j2\pi\delta_f t}, \tag{5.4}$$

where $\breve{g}_l(t;\delta_f) = g_l(t) \cdot e^{j2\pi\delta_f t}$ and $l = 1, 2$.

The rightmost exponential term in (5.4) represents standard linear phase rotation that can be corrected in the receiver using traditional methods. However, even if the frequency offset is perfectly corrected, the impact of frequency offset on the I/Q imbalance channel has three detrimental effects: (i) signal impulse response is phase-rotated by $-\delta_f t$ or shifted in the frequency domain as $G_1(f+\delta_f)$, (ii) image impulse response is phase-rotated by $\delta_f t$ or shifted in the frequency domain as $G_2(f-\delta_f)$, (iii) there exists an extraneous phase rotation at the rate of $(-2\delta_f)$ affecting only the image path.

An exemplary impulse response associated with the signal, $g_1(t) = g_{1,I}(t) + jg_{1,Q}(t)$, and its corresponding image, $g_2(t) = g_{2,I}(t) + jg_{2,Q}(t)$, is shown in Figure 5.3 for a system with symbol rate of 1 GBaud, generated by using passive filters for a wideband sixth-order Butterworth design with 3-dB cutoff frequency of 500 MHz. The image strength associated with these antialiasing filters alone is −20 dB relative to the signal, dominating receiver performance in the high SNR region.

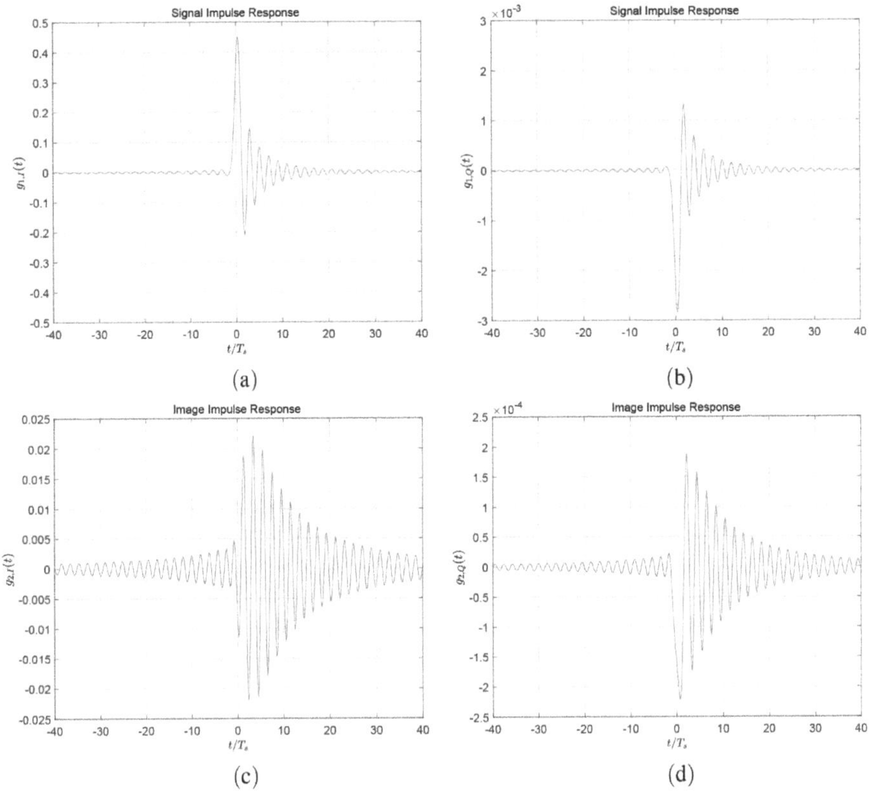

Figure 5.3 *Impulse response associated with wideband antialiasing filters: (a) in-phase signal $g_{1,I}(t)$, (b) quadrature signal $g_{1,Q}(t)$, (c) in-phase image $g_{2,I}(t)$, and (d) quadrature image $g_{2,Q}(t)$*

5.3 Digital compensation of RF impairments

We provide digital compensation techniques with robustness to frequency offset that are effective in attenuating the imbalance-induced image resulting from impairments in analog RF components. This is described in section 5.3.1 using equalization with image rejection and in section 5.3.2 for image cancelation. Stacked construction with immunity to frequency offset is presented in section 5.3.3. The computation of compensation parameters is made adaptive in section 5.3.4.

5.3.1 Equalization with image rejection

During normal operation, the premixer equalizer takes the form of suppression of distortion in the signal $\tilde{x}[n]$ and rejection of its image $\tilde{x}^*[n]$ simultaneously by applying coefficients $w_1[n]$ and $w_2[n]$ on each, respectively, followed by adding β to remove the DC offset. The resultant is then processed through a mixer to compensate for the frequency offset using an estimate, $\hat{\delta}_{f,\text{normal}}$, made during normal

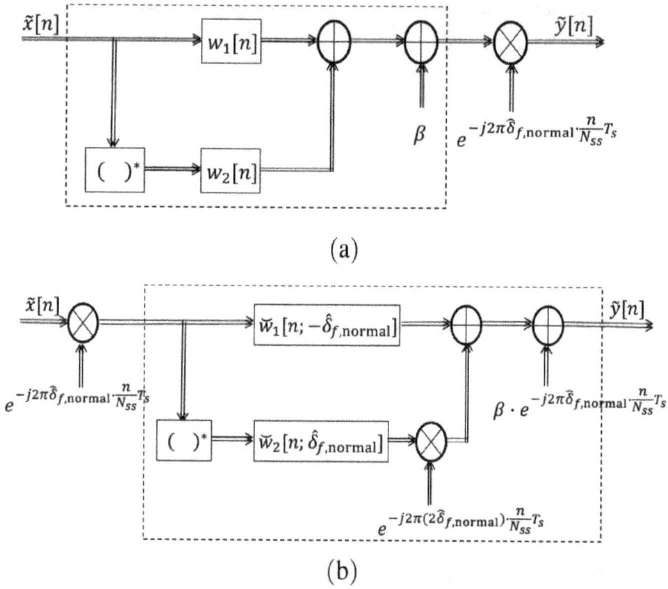

Figure 5.4 Structures for (a) premixer and (b) postmixer RF impairments equalizers with image rejection during normal mode

operation. This premixer equalizer with image rejection is depicted in Figure 5.4(a) and is expressed as

$$\tilde{y}[n] = \left(\tilde{x}[n] * w_1[n] + \tilde{x}^*[n] * w_2[n] + \beta\right) \cdot e^{-j2\pi\hat{\delta}_{f,\text{normal}} \cdot \frac{n}{N_{SS}} T_s} \tag{5.5}$$

where N_{ss} is the number of samples per symbol and T_s is the symbol duration.

An equivalent embodiment of postmixer equalization is presented in Figure 5.4(b) and is expressed as

$$\tilde{y}[n] = \left(\tilde{x}[n] \cdot e^{-j2\pi\hat{\delta}_{f,\text{normal}} \cdot \frac{n}{N_{SS}} T_s}\right) * \breve{w}_1[n; -\hat{\delta}_{f,\text{normal}}] + \left[\left(\tilde{x}[n] \cdot e^{-j2\pi\hat{\delta}_{f,\text{normal}} \cdot \frac{n}{N_{SS}} T_s}\right)^* * \breve{w}_2[n; \hat{\delta}_{f,\text{normal}}]\right] \cdot e^{-j2\pi(2\hat{\delta}_{f,\text{normal}}) \cdot \frac{n}{N_{SS}} T_s} + \beta \cdot e^{-j2\pi\hat{\delta}_{f,\text{normal}} \cdot \frac{n}{N_{SS}} T_s}, \tag{5.6}$$

where $\breve{w}_l[n; \hat{\delta}_{f,\text{normal}}] = w_l[n] \cdot e^{j2\pi\hat{\delta}_{f,\text{normal}} \cdot \frac{n}{N_{SS}} T_s}$ and $l = 1, 2$. This representation can be useful in some situations where frequency offset needs to be removed prior to RF impairments compensation, perhaps in the analog domain. Equation (5.6) describes the modifications needed to the compensation parameters, so they are effective in this case.

5.3.2 Image cancelation

During normal operation, the premixer canceler subtracts an estimate of the image from the received signal by applying cancelation coefficients $w[n]$ on the secondary

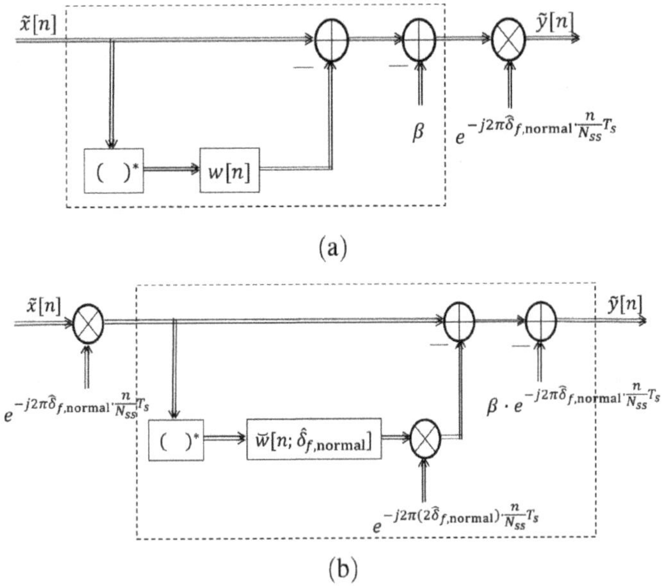

Figure 5.5 *Structures for (a) premixer and (b) postmixer RF impairments image cancelers during normal mode*

complex-conjugated path, followed by subtracting β to remove the DC offset. The resultant is then processed through a mixer to compensate for the frequency offset using an estimate, $\hat{\delta}_{f,\text{normal}}$, made during normal operation. This premixer image canceler is depicted in Figure 5.5(a) and is expressed as

$$\tilde{y}[n] = \left(\tilde{x}[n] - \tilde{x}^*[n] * w[n] - \beta\right) \cdot e^{-j2\pi\hat{\delta}_{f,\text{normal}} \cdot \frac{n}{N_{SS}} T_S} \tag{5.7}$$

The corresponding postmixer version, depicted in Figure 5.5(b), is expressed as

$$\begin{aligned}\tilde{y}[n] = &\left(\tilde{x}[n] \cdot e^{-j2\pi\hat{\delta}_{f,\text{normal}} \cdot \frac{n}{N_{SS}} T_S}\right) - \left[\left(\tilde{x}[n] \cdot e^{-j2\pi\hat{\delta}_{f,\text{normal}} \cdot \frac{n}{N_{SS}} T_S}\right)^* \right. \\ &\left. * \breve{w}[n; \hat{\delta}_{f,\text{normal}}]\right] \cdot e^{-j2\pi(2\hat{\delta}_{f,\text{normal}}) \cdot \frac{n}{N_{SS}} T_S} - \beta \cdot e^{-j2\pi\hat{\delta}_{f,\text{normal}} \cdot \frac{n}{N_{SS}} T_S}\end{aligned} \tag{5.8}$$

where $\breve{w}[n; \hat{\delta}_{f,\text{normal}}] = w[n] \cdot e^{j2\pi\hat{\delta}_{f,\text{normal}} \cdot \frac{n}{N_{SS}} T_S}$. Equation (5.8) illustrates the modifications needed for the cancelation coefficients when frequency offset is removed prior to RF impairments compensation.

A similar receiver compensation structure that uses filtering on the complex-conjugated path is used in Reference [26, Figure 3] but without taking frequency offset into account. There, the filtering coefficients are derived to restore the properness statistical property. Detailed performance comparison with this state-of-the-art method is made later in section 5.5.4.

5.3.3 Stacked construction with immunity to frequency offset

The above equalizer requires estimate of the RF impairment channel inverse, whereas the canceler needs estimate of the RF impairment channel model. We here introduce a stacked construction that is useful for this estimation, along with an added immunity to frequency offset. Toward this, we form vector $\underline{c}_s$ by stacking vectors relating to filter coefficients and DC-offset parameter as

$$\underline{c}_s = \begin{bmatrix} \underline{c}_1 \\ \underline{c}_2 \\ \beta \end{bmatrix}, \tag{5.9}$$

where $\underline{c}_1$ and $\underline{c}_2$ are coefficients for the signal and interfering image filters with memory span L_1 and L_2, respectively. The corresponding input vector $\underline{\tilde{u}}_s[n]$ to the compensator is comprised of samples from input $\tilde{u}[n]$ and its frequency image $\tilde{u}^*[n]$ and is expressed as

$$\underline{\tilde{u}}_s[n] = \begin{bmatrix} \underline{\tilde{u}}_1[n] \\ \underline{\tilde{u}}_2[n] \\ 1 \end{bmatrix}, \tag{5.10}$$

where

$$\underline{\tilde{u}}_1[n] = \begin{bmatrix} \tilde{u}[n], \tilde{u}[n-1], \cdots, \tilde{u}[n-L_1+1] \end{bmatrix}^T \tag{5.11}$$

and

$$\underline{\tilde{u}}_2[n] = \begin{bmatrix} \tilde{u}^*[n], \tilde{u}^*[n-1], \cdots, \tilde{u}^*[n-L_2+1] \end{bmatrix}^T \tag{5.12}$$

Using the stacked construction, the output of the compensator is mathematically expressed as the dot product between the coefficient vector $\underline{c}_s$ and the joint input vector $\underline{\tilde{u}}_s[n]$, containing both the signal and its image, or

$$\tilde{y}[n] = \underline{c}_s^T \cdot \underline{\tilde{u}}_s[n] \tag{5.13}$$

Note that the formulation for computing coefficients in (5.13) uses dot-product operation, while compensation, as in (5.5) or (5.7), uses convolution operation instead. The convolution coefficients can be simply extracted from vectors $\underline{c}_1$ and $\underline{c}_2$ by flipping their rows in the up/down direction.

One important aspect of the proposed estimation method is that the same training structure is utilized for RF channel inversion, used in equalization, and for RF channel modeling, used in cancelation. Another important aspect is the manner of addressing the frequency offset. The reference $d[n]$ contains intentional modification by a complex mixer that uses frequency offset estimate $\hat{\delta}_{f,\text{train}}$ made during training, so similar phase rotation as the incoming signal is experienced. This way, the pursued coefficients do not contain the frequency offset in them, essentially delinking the coupled tasks of frequency estimation and RF impairments compensation.

More specifically, to obtain the best inverse for equalization, the input signal and its reference are composed based on

$$\tilde{u}[n] = \tilde{x}[n], \tag{5.14}$$

$$\tilde{d}[n] = \tilde{z}[n] \cdot e^{j2\pi \hat{\delta}_{f,\text{train}} \cdot \frac{n}{N_{SS}} T_S}, \tag{5.15}$$

where $\tilde{z}[n]$ represents the baseband signal with no imbalance. In contrast, to obtain the best estimate of the RF channel model for cancelation, the input signal and its reference are composed based on

$$\tilde{u}[n] = \tilde{z}[n] \cdot e^{j2\pi \hat{\delta}_{f,\text{train}} \cdot \frac{n}{N_{SS}} T_S}, \tag{5.16}$$

$$\tilde{d}[n] = \tilde{x}[n] \tag{5.17}$$

If not addressed properly, frequency error creates a mismatch between the imbalanced channel for which it is trained versus that to which it is applied, thus harming performance. This is highlighted later using scattered plots in section 5.5.

Finally, for image cancelation, another computation is needed to obtain the coefficients $w[n]$ based on the best RF channel estimate found in $c_1[n]$ and $c_2[n]$ of $\underline{c}_s$, in (5.9), and can be expressed in the Fourier domain as [26]

$$W(f) = \frac{C_1(f)}{C_2^*(-f)} \tag{5.18}$$

The desired data $\tilde{z}[n]$ can be arbitrary but known set of samples used during initial factory calibration. For field recalibration, decision-directed mode can be used. In the latter, and in an effort not to interrupt normal traffic, one of the PLS codes is reserved. Once it is detected, data is captured into memory buffer, and software is informed to retrain. To ensure decisions are almost error-free, this process can be done using a codeword of robust quadrature phase-shift keying (QPSK) constellation under high-SNR conditions, ensuring performance above the forward-error correction (FEC) code threshold.

5.3.4 Adaptive computation of parameters

Techniques of adaptive solution to (5.13) are examined here without requiring *a priori* knowledge of RF impairments parameters. The first is based on least squares (LS). Given a block of N samples of the input $\tilde{u}[n]$ and reference $\tilde{d}[n]$, we form matrix $\mathbf{U_s}$, of size $N \times (L_1 + L_2 + 1)$, using stacked vectors as

$$\mathbf{U_s} = \left[\ \underline{\tilde{u}}[0], \underline{\tilde{u}}[1], \cdots, \underline{\tilde{u}}[N-1] \ \right]^T \tag{5.19}$$

with a corresponding reference vector $\underline{\tilde{d}}$, of size $N \times 1$, as

$$\underline{\tilde{d}} = \left[\ \tilde{d}[0], \tilde{d}[1], \cdots, \tilde{d}[N-1] \ \right]^T \tag{5.20}$$

The solution $\underline{c}_{s,\text{LS}}$ that minimizes the error $\left(\underline{\tilde{d}} - \mathbf{U_s} \cdot \underline{c}_s\right)$ in the LS sense is then

$$\underline{c}_{s,\mathrm{LS}} = \left(\mathbf{U}_s^H \mathbf{U}_s\right)^{-1} \mathbf{U}_s^H \underline{\tilde{d}} \tag{5.21}$$

To avoid matrix inversion in (5.21), stochastic gradient-based algorithms are provided that arrive iteratively at the solution without *a priori* knowledge of RF impairments parameters. Matrix inversion is computationally cumbersome and introduces performance instability. Two techniques can be utilized to adaptively compute the sought set of coefficients $\underline{c}$ that minimize error signal $\tilde{e}[n] = \tilde{d}[n] - \tilde{y}[n]$: least mean-squares (LMS) and recursive LS (RLS). The RLS is a preferred technique as it provides fast convergence and achieves performance that is somewhat independent of input signal statistics.

Using LMS criterion, the compensator coefficients are computed iteratively as

$$\underline{\tilde{c}}_{s,\ \mathrm{LMS}}[n+1] = \underline{\tilde{c}}_{s,\ \mathrm{LMS}}[n] + \mu \cdot \underline{\tilde{u}}_s[n] \cdot \tilde{e}^*[n] \tag{5.22}$$

where μ is a small positive number chosen to adjust adaptation speed.

Using RLS criterion, the compensator coefficients are computed iteratively as

$$\underline{\tilde{c}}_{s,\ \mathrm{RLS}}[n+1] = \underline{\tilde{c}}_{s,\ \mathrm{RLS}}[n] + \mu \cdot \underline{\tilde{k}}[n] \cdot \tilde{e}^*[n] \tag{5.23}$$

where

$$\underline{k}[n] = \frac{\lambda^{-1} \cdot \mathbf{P}[n-1] \cdot \underline{\tilde{u}}_s[n]}{1 + \lambda^{-1} \cdot \underline{\tilde{u}}_s^H[n] \cdot \mathbf{P}[n-1] \cdot \underline{\tilde{u}}_s[n]}, \tag{5.24}$$

$$\mathbf{P}[n] = \lambda^{-1} \cdot \mathbf{P}[n-1] - \lambda^{-1} \cdot \underline{k}[n] \cdot \underline{\tilde{u}}_s^H[n] \cdot \mathbf{P}[n-1], \tag{5.25}$$

and λ is the forgetting factor, $0 < \lambda \leq 1$. For initialization, we set $\mathbf{P}[0] = \epsilon^{-1} \cdot \mathbf{I}$ where $\mathbf{I}$ is the identity matrix of size $(L_1 + L_2 + 1) \times (L_1 + L_2 + 1)$ and ϵ is a small positive parameter selected to provide good performance.

5.4 Frequency offset estimation in the presence of RF impairments

During initial factory calibration and in the presence of RF impairments, an accurate estimate of frequency offset $\hat{\delta}_{f,\mathrm{train}}$ is needed for computing the compensator coefficients, as indicated in (5.15) or (5.16). The estimate of frequency error can be obtained by sending a test tone through the imbalanced channel.

When using an ideal receiver, free of RF impairments, determining the frequency of a noisy complex-valued sinusoid in additive white Gaussian noise (AWGN) is a classical estimation problem [33] that has been long investigated. A complex-valued discrete-time test tone with magnitude b and phase θ is described as

$$b \cdot e^{j\left(2\pi f_{\mathrm{tone}} \cdot \frac{n}{N_{SS}} T_S + \theta\right)};\ n = 0, 1, \cdots, N-1, \tag{5.26}$$

where f_{tone} is its spectral location. In Reference 33, the maximum-likelihood estimator is shown to be the frequency that maximizes the magnitude of the discrete Fourier transform (DFT), efficiently computed using fast Fourier transform (FFT). The optimal Cramer-Rao lower bound (CRLB) on its estimation error when the tone's other parameters, phase and amplitude, are unknown can be obtained by

$$\sigma_{\text{CRLB}} = \frac{1}{2\pi} \cdot \frac{N_{SS}}{T_S} \cdot \sqrt{\frac{6 \cdot N_{SS}}{N(N^2-1)}} \cdot \left(\frac{E_S}{N_0}\right)^{-1/2} \tag{5.27}$$

where E_s/N_0 is the per-symbol SNR.

However, when RF impairments are present, the received signal contains the test tone and its image. Specifically, substituting (5.26) into discrete-time (5.1) with algebraic manipulation yields

$$\begin{aligned}\tilde{x}_{\text{tone}}[n] &= b \cdot G_1\left(f_{\text{tone}} + \delta_{f,\text{train}}\right) \cdot e^{+j(2\pi(f_{\text{tone}}+\delta_{f,\text{train}}) \cdot \frac{n}{N_{SS}} T_S + \theta)} \\ &\quad + b \cdot G_2\left(-(f_{\text{tone}} + \delta_{f,\text{train}})\right) \cdot e^{-j(2\pi(f_{\text{tone}}+\delta_{f,\text{train}}) \cdot \frac{n}{N_{SS}} T_S + \theta)} + \alpha\end{aligned} \tag{5.28}$$

where $\delta_{f,\text{train}}$ is the frequency offset during the training process. From (5.28), it is clear that the impact of RF impairments includes a DC offset term, α, and that the received signal consists of two tones, desired tone at $(f_{\text{tone}} + \delta_{f,\text{train}})$ with unknown amplitude and phase resulting from $G_1\left(f_{\text{tone}} + \delta_{f,\text{train}}\right)$, and an image at $-(f_{\text{tone}} + \delta_{f,\text{train}})$ with unknown amplitude and phase resulting from $G_2\left(-(f_{\text{tone}} + \delta_{f,\text{train}})\right)$. The associated DFT of the test tone with RF impairments can be shown to be

$$\begin{aligned}\tilde{X}_{\text{tone}}(f) &= G_1\left(f_{\text{tone}} + \delta_{f,\text{train}}\right) \cdot e^{-j\pi(f-(f_{\text{tone}}+\delta_{f,\text{train}})) \cdot \frac{N-1}{N_{SS}} T_S} \\ &\quad \cdot \frac{\sin(\pi(f-(f_{\text{tone}}+\delta_{f,\text{train}})) \cdot \frac{N}{N_{SS}} T_S)}{N \sin(\pi(f-(f_{\text{tone}}+\delta_{f,\text{train}})) \cdot \frac{1}{N_{SS}} T_S)} \cdot b \cdot e^{j\theta} \\ &\quad + G_2\left(-(f_{\text{tone}} + \delta_{f,\text{train}})\right) \cdot e^{-j\pi(f+(f_{\text{tone}}+\delta_{f,\text{train}})) \cdot \frac{N-1}{N_{SS}} T_S} \\ &\quad \cdot \frac{\sin(\pi(f+(f_{\text{tone}}+\delta_{f,\text{train}})) \cdot \frac{N}{N_{SS}} T_S)}{N \sin(\pi(f+(f_{\text{tone}}+\delta_{f,\text{train}})) \cdot \frac{1}{N_{SS}} T_S)} \cdot b \cdot e^{-j\theta} \\ &\quad + (\alpha - \hat{\alpha}) \cdot e^{-j\pi f \cdot \frac{N-1}{N_{SS}} T_S} \cdot \frac{\sin(\pi f \cdot \frac{N}{N_{SS}} T_S)}{N \sin(\pi f \cdot \frac{1}{N_{SS}} T_S)}.\end{aligned} \tag{5.29}$$

In (5.4), $\hat{\alpha}$ is subtracted to remove the effect of DC offset before DFT computation and can be estimated as the mean of the tuner output using N samples, or

$$\hat{\alpha} = 1/N \cdot \sum_{n=0}^{N-1} \tilde{x}_{\text{tone}}[n] \tag{5.30}$$

Figure 5.6 provides an exemplary DFT output of a test tone and its image due to RF impairments.

Considering that in a well-designed set of antialiasing filters, $|G_1(f)|^2 \gg |G_2(f)|^2$, the DFT output in (5.4) achieves its global maximum at $(f_{\text{tone}} + \delta_{f,\text{train}})$, so it can be safely used to provide an estimate of $\delta_{f,\text{train}}$. However, when spectral peaks of the tone and its image overlap, the accuracy of the frequency estimate is expected to deteriorate as their superposition smears the desired peak and affects the frequency resolvability. To ensure that the spectral peaks from the tone and its image do not substantially overlap, a minimum condition on f_{tone}, denoted as $f_{\text{tone}}^{(\min)}$, is imposed (assuming $f_{\text{tone}} > 0$ with no loss in generality). Namely, to guarantee that spectral peaks from the tone and its image do not overlap up to the lth side-lobe, we choose

$$f_{\text{tone}} > f_{\text{tone}}^{(\min)} = l \cdot \frac{N_{SS}}{NT_S} + \delta_f^{(\max)}, \tag{5.31}$$

where $\delta_f^{(\max)}$ is the largest absolute value in the frequency-offset range and $l > 4$. In fact, as will be shown in section 5.5, this technique of estimating the frequency

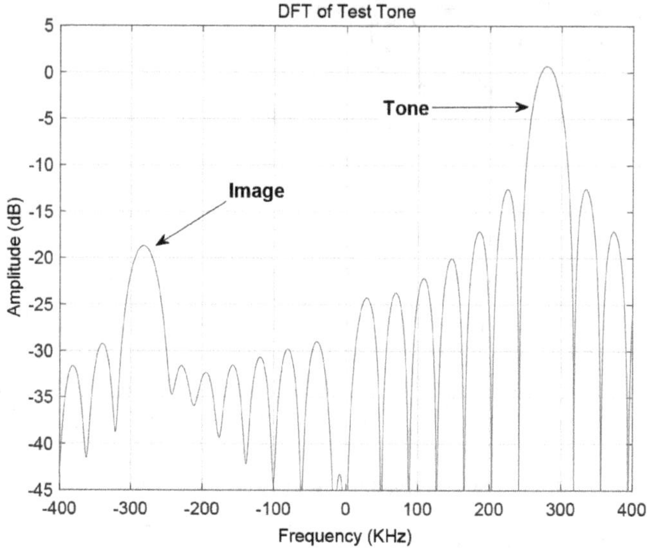

Figure 5.6 *Output of DFT when a test tone is passed through 1-GHz tuner with RF impairments*

parameter can approach the performance of the ideal receiver given that f_{tone} is large enough to ensure sufficient separation of the spectral peaks, resulting from the desired tone and its interfering image.

5.5 Numerical studies

Extensive computer simulations are implemented to demonstrate the effectiveness of the proposed methods of RF impairments compensation depicted in Figures 5.4 and 5.5. The transmit and receive filters are a pair of matching root-raised cosine (RRC) pulses with roll-off factor of 0.05. The constellation employed is amplitude phase-shift keying (APSK). On the receiver side, a wideband quadrature frequency-converter, or tuner, is implemented with bandwidth spanning 1 GHz, following the mathematical model in Figure 5.1. In particular, the antialiasing filters $h_I(t)$ and $h_Q(t)$ are designed using sixth-order Butterworth criterion with single-sided cut-off frequency of 500 MHz. The corresponding impulse responses of the frequency-selective IQ imbalance are as displayed in Figure 5.3 for the signal and its image. In addition, a mismatch in gain, γ, and phase, φ, of 15% and 10°, respectively, is used at the LO mixer. DC offset parameters α_I and α_Q are chosen as 0.05 and −0.05. The sampling rate at the ADCs outputs is 2 Giga samples per second. The composite is down-converted directly to DC or zero frequency. Frequency offset δ_f is considered of up to 500 kHz, during training mode, representative of initial factory calibration. However, the coefficients are applied during normal operation with a different frequency offset, selected as large as 4 MHz to account for frequency variation due to satellite propagation.

Unless otherwise stated, the coefficients of RF impairments compensators are obtained at the end of a training mode with 20,000 samples processed via the RLS iterative technique with forgetting factor $\lambda = 1$. The per-symbol SNR, E_s/N_0, is 25 dB when training during factory calibration. In addition, all results are reported at the output of a 35-tap traditional FS LMS equalizer, operating at two samples per symbol at its input, intended to eliminate inter-symbol interference (ISI) that is remaining from RF impairments compensators or to eliminate ISI that is not accounted for during training.

5.5.1 Frequency offset estimation

Figure 5.7 depicts the simulated performance of frequency offset estimation in terms of root mean-square error (MSE) when sending a test tone through a 1-GHz tuner with RF impairments versus the choice of tone frequency location. This is done during initial factory calibration in the presence of RF impairments, where the number of samples N of 20,000 or 50,000 are captured and stored in a memory buffer. The figure also includes the minimum values of f_{tone} with $l = 5$ in (5.30) and the optimal CRLB of (5.27) when no RF impairments are included. The estimation is based on either locating the peak at the FFT output or extracted as the slope of the LS line formed by phase measurements, after removing the estimated DC offset as explained in section 5.4. Results indicate that very good performance is achieved if minimum tone frequency value is set, approaching the ideal CRLB with no RF impairments.

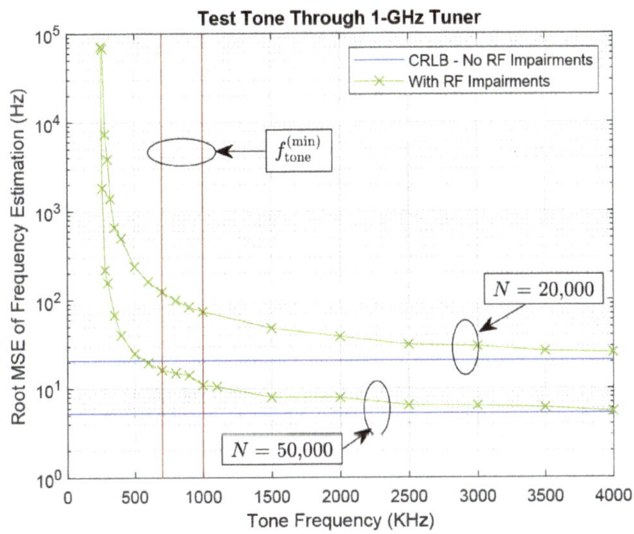

Figure 5.7 Performance of frequency offset estimation when using test tone through 1-GHz tuner with RF impairments

5.5.2 Noiseless scatter plots

Figure 5.8(a) presents the scatter plot at the FS equalizer output without the proposed RF impairments compensation and exhibits considerable clustering due to image interference, for which the FS equalizer cannot compensate. To emphasize the immunity of the stacked construction to frequency offset, we first illustrate in Figure 5.8(b) the amplitude response of compensation coefficients, $w_1[n]$ and $w_2[n]$, at the end of training, outlined in section 5.3.4, without injecting the frequency offset estimate in (5.15), but instead the frequency offset estimate is removed prior to training. The associated scatter plot is in Figure 5.8(b) showing very little improvement even for small values of frequency offset of 500 kHz during training. This confirms the detrimental effects of frequency offset mentioned in section 5.2 that creates mismatch between the imbalanced channel for which it is trained versus that to which it is applied. In contrast, the corresponding results when the frequency offset estimate is injected into the desired reference are in Figure 5.8(d),(e), illustrating that the proposed RF impairments compensator very effectively reduces the clustering with about 23 dB improvement in MSE, beyond the state-of-the-art FS equalizer.

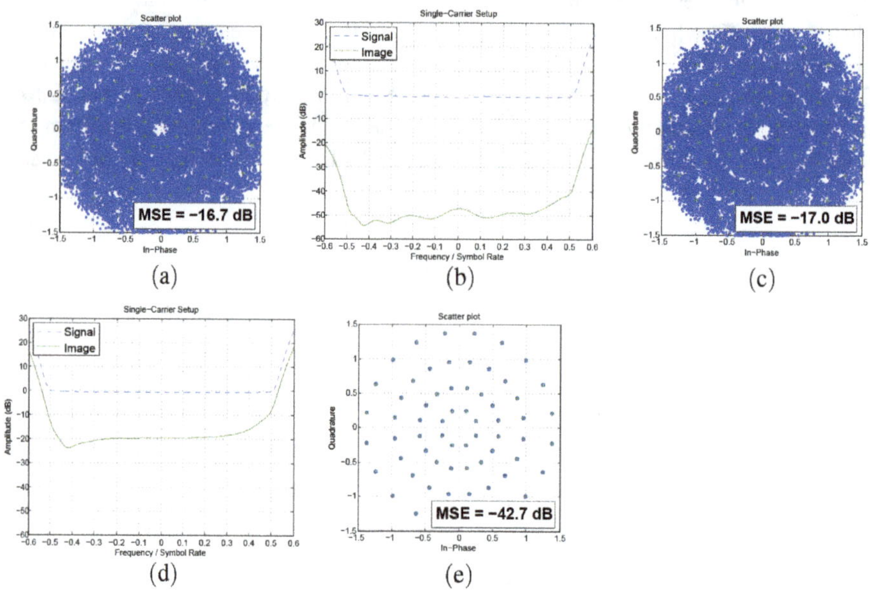

Figure 5.8 Case of single-carrier 64APSK signal through 1-GHz tuner with RF impairments: (a) noiseless scatter plot without RF impairments compensation; (b) amplitude response of equalization coefficients when frequency offset estimate is removed prior to training; (c) noiseless scatter plot with equalization when frequency offset estimate is removed prior to training; (d) and (e) are with the proposed equalization when frequency offset estimate is injected during training

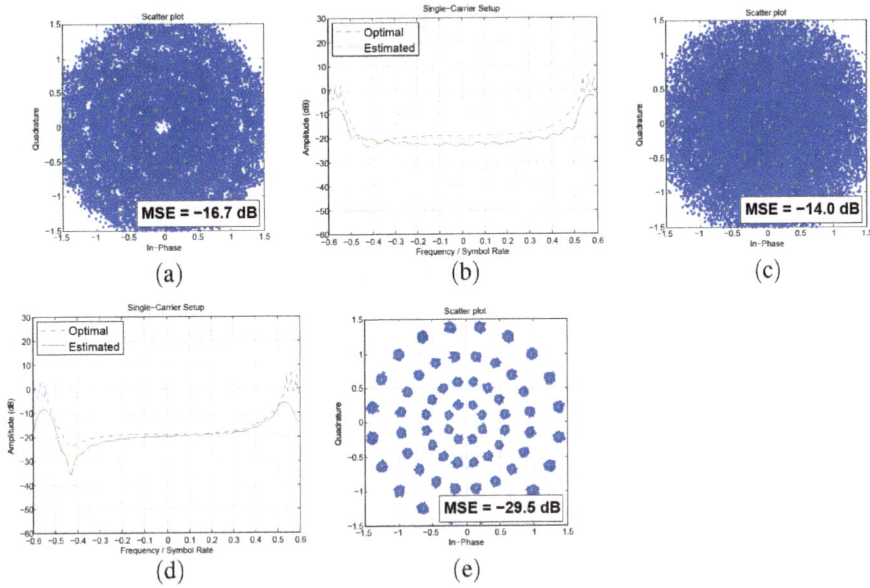

Figure 5.9 *Case of single-carrier 64APSK signal through 1-GHz tuner with RF impairments: (a) noiseless scatter plot without RF impairments compensation; (b) amplitude response of cancelation coefficients when frequency offset estimate is removed prior to training; (c) noiseless scatter plot with cancelation when frequency offset estimate is removed prior to training; (d) and (e) are with the proposed cancelation when frequency offset estimate is injected during training*

Furthermore, the excellent results associated with the proposed compensation are achieved despite using a frequency offset of 500 kHz for training and 4 MHz during application.

Figure 5.9 includes the counterpart results when image cancelation is used for compensation and displays similar pattern of results. One important note to make from Figure 5.9(e) is that cancelation is not as effective as equalization as it requires an additional step to generate cancelation coefficients based on estimated channel model coefficients as in (5.18). Figure 5.9(d) also includes the amplitude response of the optimal cancelation coefficients derived under perfectly known RF impairments channel. The finite accuracy of the estimation in this case limits the performance improvement even when using powerful RLS during training.

5.5.3 Transient behavior

The stochastic gradient-based methods are compared here in terms of their transient behavior including the choice of adaptation parameters such as μ and λ. Figure 5.10 shows the statistical average of the amplitude of error

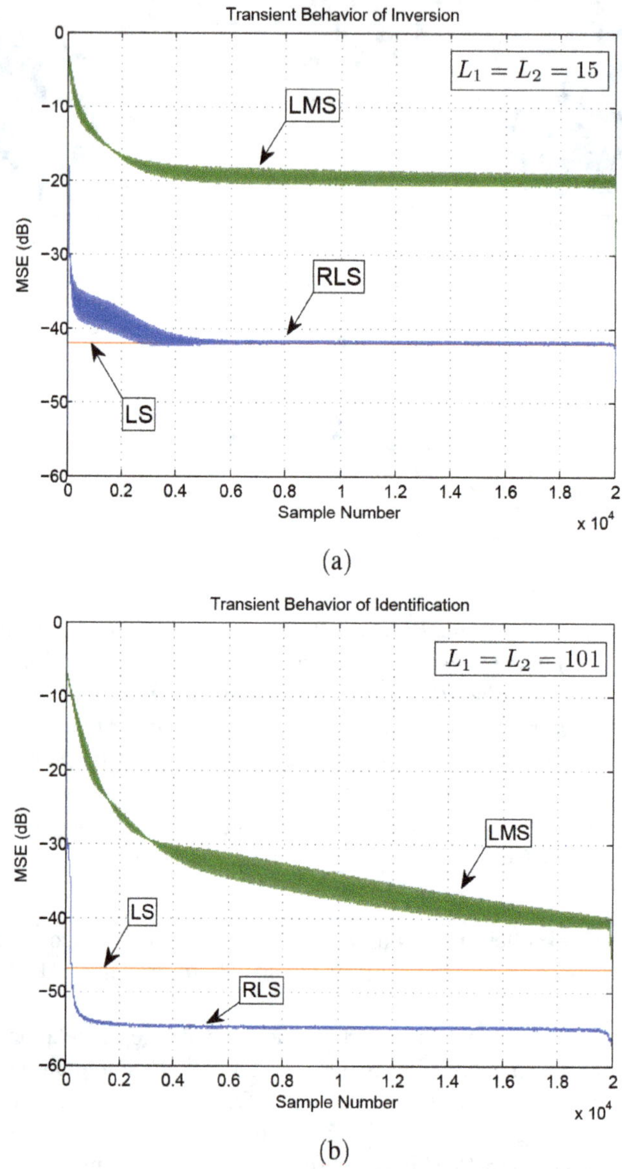

Figure 5.10 *Transient behavior for RLS and LMS when compared with LS for (a) inversion and (b) identification*

signal $\tilde{e}[n]$ as it evolves over the sample number n when using LMS at $\mu = 0.01$ and RLS at $\lambda = 0.99$. The first part of the figure relates to channel inversion, for equalization, when $L_1 = L_2 = 15$, whereas the second part of the figure relates to the channel identification, for cancelation, when $L_1 = L_2 = 101$. Both parts of the figure include results for the LS method, computed based on a data block of the same size, to provide theoretical support for convergence.

Based on Figure 5.10, we can remark that the RLS settles at a lower value of the error when compared with LMS. Also, the performance of LS method is achieved by RLS when providing channel inversion and is actually surpassed by RLS when providing channel identification. The latter may be attributed to the numerical instability of the matrix inversion step needed for LS when the number of parameters is large. Further, faster convergence of RLS over LMS is demonstrable.

5.5.4 Performance comparisons

More comparisons of the investigated compensation techniques are performed, including comparisons with a state-of-the-art technique that is effective at mitigating frequency-selective I/Q imbalance. In particular, a block moment-based receiver scheme from Reference 26 is chosen that applies compensation coefficients at the image path to restore the properness condition for the span of the compensator. The coefficients are calculated based on Reference [26, Equation (25)] with additional Newton-like iterations applied to improve their performance. Figure 5.11 reports these simulations in terms of MSE as it varies with respect to the number of compensation coefficients. Three curves are generated, one for a system employing equalizer with image rejection in (5.5), a second for image canceler in (5.7), and a third for a system implementing block moment-based scheme which outperforms its LMS counterpart. For

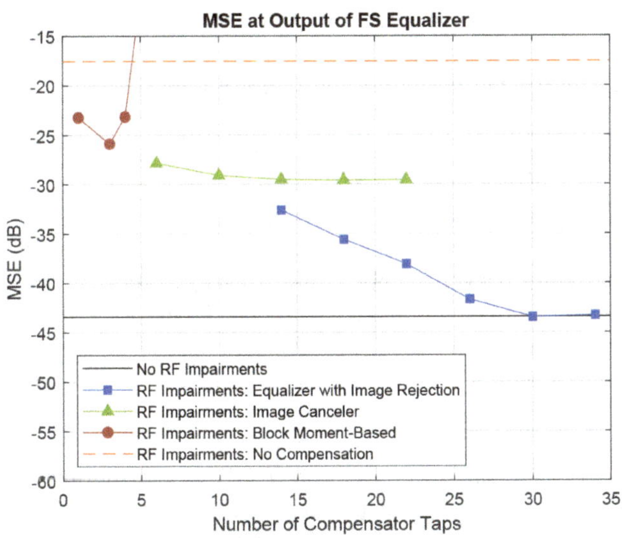

Figure 5.11 Performance comparisons versus the number of compensation taps for single-carrier 64APSK signal through 1-GHz tuner when using N=50,000

favorable conditions to the moment-based scheme, frequency variations and DC offset parameters are assumed not present. In addition, the figure includes the bounding performance of the scenarios when RF impairments are not compensated and when compensated perfectly.

As evident in Figure 5.11, the moment-based solution performs better than without compensation, offering about 8.4 dB improvement in MSE when the number of taps is three. However, no improvement is obtained when the memory span of the moment-based compensator exceeds four taps, even when using 15 Newton-like iterations and 50,000 samples during training. The state-of-the-art scheme relies on the properness assumption which can be weak in some applications, especially when using sample statistics for nulling the complementary autocorrelation function. This suggests that the moment-based compensation cannot cope with I/Q imbalance channels of large memory span encountered when transmitting signals in the GBaud range. In contrast, the system that implements image cancelation can tolerate large memory span providing 3.6 dB in additional benefit over the state-of-the-art when using 14 taps. Furthermore, the system that implements equalization with image rejection provides lossless compensation when using 15 taps per arm.

5.6 Conclusion

This chapter has provided two characterization models of analog RF impairments in the presence of frequency offset: postmixer and premixer models. Novel digital compensation algorithms with immunity to frequency offset have been presented, categorized into equalization with image-rejection capability and image cancelation. Adaptive techniques are utilized to obtain compensation coefficients in an iterative manner using stacked construction, while the pursued coefficients are independent of the frequency offset. These methods are useful when using known data samples for initial factory calibration or in a decision-directed mode during field recalibration. Special consideration is given to estimating the frequency offset by sending a test tone through the imbalanced channel during initial factory calibration. Extensive computer simulations have revealed that the proposed compensators provide lossless attenuation of the imbalance-induced image and remove DC offset, in the presence of frequency offsets.

The presented analysis and techniques, with their attractive performance, are beneficial to other important lines of research, such as when utilizing the extremely high frequency (EHF) band of the radio spectrum [34, 35]. Exploiting the large, commercially available bandwidth in the Q/V/E-band can substantially overcome spectrum limitations and reduce equipment size. Consequently, the application and experimental verification of digital solutions for suppressing frequency-selective image induced by quadrature frequency-conversion circuits, when operated in the EHF band, can aid the feasibility of LEO satellite systems in achieving ultra-high capacity.

References

[1] Sharma S.K., Chatzinotas S., Arapoglou P.-D. (eds.) *Satellite communications in the 5G era*. Telecommunications. London, UK: Institution of Engineering and Technology; 2018.

[2] 'Study on using satellite access in 5G (release 16)'. [Technical Specification Group Services and System Aspects] *Document 3GPP TR 22.822, 3rd Generation Partnership Project*. 2018.

[3] Beidas B.F. 'Intermodulation distortion in multicarrier satellite systems: analysis and turbo Volterra equalization'. *IEEE Transactions on Communications*. 2011, vol. 59(6), pp. 1580–90.

[4] Beidas B.F. 'Powerful nonlinear countermeasures for multicarrier satellites: progression to 5G' in *Satellite communications in the 5G era*. Telecommunications. London, UK: Institution of Engineering and Technology; 2018. pp. 209–47.

[5] Piemontese A., Modenini A., Colavolpe G., Alagha N.S. 'Improving the spectral efficiency of nonlinear satellite systems through time-frequency packing and advanced receiver processing'. *IEEE Transactions on Communications*. 2013, vol. 61(8), pp. 3404–12.

[6] Beidas B.F., Seshadri R.I., Eroz M., Lee L.-N. 'Faster-than-Nyquist signaling and optimized signal constellation for high spectral efficiency communications in nonlinear satellite systems'. *IEEE Military Communications Conference (MILCOM)*; Baltimore, MD, 2014. pp. 818–23.

[7] Chatzinotas S., Ottersten B., De Gaudenzi R. (eds.) *Cooperative and cognitive satellite systems*. UK: Academic Press; 2015.

[8] Vazquez M.A., Perez-Neira A., Christopoulos D, *et al*. 'Precoding in multibeam satellite communications: present and future challenges'. *IEEE Transactions on Wireless Communications*. 2016, vol. 23(6), pp. 88–95.

[9] Arapoglou P.-D., Ginesi A., Cioni S., *et al*. 'DVB-S2X-enabled precoding for high throughput satellite systems'. *International Journal of Satellite Communications and Networking*. 2016, vol. 34(3), pp. 439–55.

[10] Andrenacci S., Angelone M., Candreva E.A., *et al*. 'Physical layer performance of multi-user detection in broadband multi-beam systems based on DVB-S2'. *Proceedings of the 20th European Wireless (EW) Conference*; 2014. pp. 1–5.

[11] Caus M., Perez-Neira A.I., Angelone M., Ginesi A. 'An innovative interference mitigation approach for high throughput satellite systems'. *IEEE 16th International Workshop on Signal Processing Advances in Wireless Communications (SPAWC)*; Stockholm, Sweden, 2015. pp. 515–19.

[12] Cocco G., Angelone M., Pèrez-Neira A.I. 'Co-channel interference cancelation at the user terminal in multibeam satellite systems'. *International Journal of Satellite Communications and Networking*. 2017, vol. 35(1), pp. 45–65.

[13] Colavolpe G., Modenini A., Piemontese A., Ugolini A. 'Multiuser detection in multibeam satellite systems: theoretical analysis and practical schemes'. *IEEE Transactions on Communications*. 2017, vol. 65(2), pp. 945–55.

[14] Ugolini A., Colavolpe G., Angelone M., Vanelli-Coralli A., Ginesi A. 'Capacity of interference exploitation schemes in multibeam satellite systems'. *IEEE Transactions on Aerospace and Electronic Systems*. 2019, vol. 55(6), pp. 3230–45.

[15] De Gaudenzi R., Alagha N., Angelone M., Gallinaro G. 'Exploiting code division multiplexing with decentralized multiuser detection in the satellite multibeam forward link'. *International Journal of Satellite Communications and Networking*. 2018, vol. 36(3), pp. 239–76.

[16] Beidas B.F., Seshadri R.I. 'Superior iterative detection for co-channel interference in multibeam satellite systems'. *IEEE Transactions on Communications*. 2020, vol. 68(12), pp. 7660–71.

[17] Mirabbasi S., Martin K. 'Classical and modern receiver architectures'. *IEEE Communications Magazine*. 2000, vol. 38(11), pp. 132–39.

[18] Mak P.-I., U S.-P., Martins R. 'Transceiver architecture selection: review, state-of-the-art survey and case study'. *IEEE Circuits and Systems Magazine*. 2007, vol. 7(2), pp. 6–25.

[19] Beidas B.F., Seshadri R.I., Becker N. 'Multicarrier successive predistortion for nonlinear satellite systems'. *IEEE Transactions on Communications*. 2015, vol. 63(4), pp. 1373–82.

[20] Beidas B.F. 'Adaptive digital signal predistortion for nonlinear communication systems using successive methods'. *IEEE Transactions on Communications*. 2016, vol. 64(5), pp. 2166–75.

[21] Beidas B.F., Seshadri R.I. 'OFDM-like signaling for broadband satellite applications: analysis and advanced compensation'. *IEEE Transactions on Communications*. 2017, vol. 65(10), pp. 4433–45.

[22] Piazza R., Bhavani Shankar M.R., Ottersten B. 'Data predistortion for multicarrier satellite channels based on direct learning'. *IEEE Transactions on Signal Processing*. 2014, vol. 62(22), pp. 5868–80.

[23] Cioni S., Colavolpe G., Mignone V., *et al*. 'Transmission parameters optimization and receiver architectures for DVB-S2X systems'. *International Journal of Satellite Communications and Networking*. 2016, vol. 34(3), pp. 337–50.

[24] Valkama M., Renfors M., Koivunen V. 'Advanced methods for I/Q imbalance compensation in communication receivers'. *IEEE Transactions on Signal Processing*. 2001, vol. 49(10), pp. 2335–44.

[25] Fettweis G., Lohning M., Petrovic D, *et al*. 'Dirty RF: a new paradigm'. *Proceedings of the IEEE 16th International Symposium on Personal, Indoor and Mobile Radio Communications (PIMRC)*; 2005. pp. 2347–55.

[26] Anttila L., Valkama M., Renfors M. 'Circularity-based I/Q imbalance compensation in wideband direct-conversion receivers'. *IEEE Transactions on Vehicular Technology*. 2008, vol. 57(4), pp. 2099–113.

[27] Valkama M., Renfors M., Koivunen V. 'Compensation of frequency-selective I/Q imbalances in wideband receivers: models and algorithms'. *SPAWC-2001. Third IEEE Signal Processing Workshop on Signal Processing Advances in Wireless Communications*; Taiwan, China, 2001. pp. 42–45.

[28] Yu L., Snelgrove W.M. 'A novel adaptive mismatch cancellation system for quadrature IF radio receivers'. *IEEE Transactions on Circuits and Systems II*. 1999, vol. 46(6), pp. 789–801.

[29] Pun K.P., Franca J.E., Azeredo-Leme C., Chan C.F., Choy C.S. 'Correction of frequency-dependent I/Q mismatches in quadrature receivers'. *Electron Lett*. 2001, vol. 37(23), pp. 1415–17.

[30] Xing G., Shen M., Liu H. 'Frequency offset and I/Q imbalance compensation for direct-conversion receivers'. *IEEE Transactions on Wireless Communications*. 2005, vol. 4(2), pp. 673–80.

[31] Beidas B.F. 'Radio-frequency impairments compensation in ultra high-throughput satellite systems'. *IEEE Transactions on Communications*. 2019, vol. 67(9), pp. 6025–38.

[32] Beidas B.F., Seshadri R.I. 'Analysis and compensation for nonlinear interference of two high-order modulation carriers over satellite link'. *IEEE Transactions on Communications*. 2010, vol. 58(6), pp. 1824–33.

[33] Rife D., Boorstyn R. 'Single tone parameter estimation from discrete-time observations'. *IEEE Transactions on Information Theory*. 1974, vol. 20(5), pp. 591–98.

[34] Kourogiorgas C.I., Lyras N., Panagopoulos A.D, *et al*. 'Capacity statistics evaluation for next generation broadband MEO satellite systems'. *IEEE Transactions on Aerospace and Electronic Systems*. 2017, vol. 53(5), pp. 2344–58.

[35] Rossi T., De Sanctis M., Domenico S.D., Ruggieri M., Cianca E. 'Low-complexity blind equalization in Q/V band satellite links: an experimental assessment'. *IEEE Transactions on Aerospace and Electronic Systems*. 2021, vol. 57(6), pp. 4465–71.

Chapter 6
Flexibility/complexity trade-offs in payload design

Florian Vidal[1], Hervé Legay[1], George Goussetis[2], Ségolène Tubau[1], and Jean-Didier Gayrard[1]

6.1 Introduction

GEO satellite telecommunication networks offer either mobile or fixed services that take profit from the wide coverage offered by the satellite platforms [1]. These services may be multicast, broadcast, or unicast. Broadcast is used for radio or TV services, whereas multicast and unicast are more utilized for data services, mobile communications, and broadband internet and multimedia services [2]. Broadband services require high data rates per user terminal (UT) and consequently large frequency bands and high equivalent isotropic radiated power (EIRP). To meet this challenge, satellites implement antenna architectures based on multibeam coverages with high antenna directivities and frequency reuse to increase the available aggregated frequency bandwidth [3]. In a 5G context, other requirements such as wide coverage, latency, and reliability emerge as priorities. In this environment, LEO satellite megaconstellations bring promising advantages compared to GEO satellites:

- A constellation of non-GEO satellites may, in the case of a polar orbit, for example, offer a seamless quality of service even at the earth poles, which is not possible with GEO satellites because of low user elevations at these latitudes.
- The round-trip time latency for a GEO satellite is 500 ms because of the 36,000 km path between the satellite and the Earth, which is far from the 1 ms objective of 5G [4]. This requirement is essential, for example, for real-time health monitoring or finance transactions [4]. When considering a satellite at 1200 km altitude, this round-trip latency is reduced to 16 ms which makes them a more viable solution than GEO satellite for applications requiring low latency.

[1]Thales Alenia Space, Toulouse, France
[2]Heriot-Watt University, Edinburgh, UK

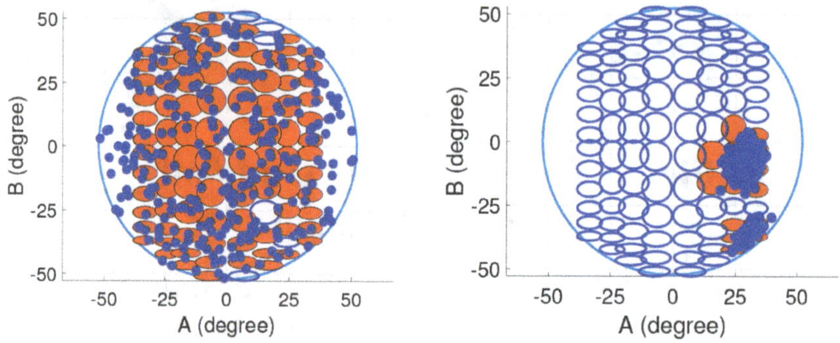

(a) Uniform user distribution (b) Non-uniform user distribution

Figure 6.1 Antenna FOV with different traffic distribution, the beams with traffic are colored, users are represented by dots

- The high number of satellites in megaconstellations reduces the criticality of losing a satellite (graceful service degradation). Moreover, spare satellites may be placed in orbit in case of failure at low cost compared to the overall system CAPEX.

The previous points underline that LEO megaconstellations seem to be a relevant solution to support 5G networks. LEO satellites experience a diversity of traffic distributions in their antenna field of views (FOV) as they orbit around the Earth. At the payload level, introducing flexibility is of interest to address user distributions that may range from uniform over all the FOV (Figure 6.1(a)) to very dense concentrations. The former would correspond, for example, to scenarios with hubs like cities (Figure 6.1(b)). In a uniform traffic scenario, a uniform sharing of resources among beams is optimal to meet the user demands. This solution can be provided by payloads with a static resource allocation associated and non-steerable multiple beam antennas. This approach was selected by OneWeb with 16 non-steerable, highly-elliptical beams [5]. However, in a non-uniform scenario, only a limited number of the beams are potentially active. In that case, flexibility in resource allocation is essential to meet the demand. A lack of flexibility would both waste resources in non-populated areas and be insufficient to match the throughputs required in densely populated areas. Satellite operators would then reduce the data rates proposed to their customers and lose potential earnings. Flexibility necessitates adaptive payloads including, for instance, inter-satellite links or reconfigurable antennas which come with added cost and complexity. This technological choice was followed by Telesat and SpaceX Starlink megaconstellations [5].

Depending on the implemented technologies and system architecture, flexibility at payload level enables the satellites to reallocate their resources in power, frequency, time, or coverage to adapt to these various scenarios:

- In the time domain, beam hopping (BH), whereby satellite resources are shared in the time domain, is increasingly being considered an attractive solution [6]. It consists of splitting time into slots and illuminating a set of beams at each time slot. Ferrite switches are a mature technology for the implementation of BH [7]. BH brings flexibility to initially non-flexible antenna solutions, such as antenna farms with focus optics and single feed per beam (SFPB).
- In the frequency domain, flexible frequency allocation between beams that best matches the capacity request and flexible channelization are already available with a payload digital core, such as the digital transparent processor (DTP) [8].
- In the power domain, the flexibility for power allocation is available thanks to multiple port amplifiers (MPA) [9] for passive multiple beam antennas and distributed amplification for active antennas. As the available power must be shared between an increased number of beams, the complexity of the MPA increases accordingly with possible drawbacks in terms of isolation and linearity.
- In the spatial domain, flexibility implemented with beam steering may be achieved with mechanically steerable antennas or phased arrays. The phased array solutions require a beam forming network (BFN) that can be done either digitally in the DTP [10], or in an analog way, for example, with phase shifters, butler matrices or quasi-optical beam formers (QOBF) [11]. Hybrid beamforming strategies combining analog and digital techniques can provide trade-offs between the complexity of the BFN and spatial flexibility [12]. Interference mitigation by means of beamforming such as the precoding approach is also an alternative to increase throughputs [13]. It however implies further complexity in channel estimation and beamforming coefficient computations.

There is thus a wide range of solutions to implement flexibility in satellite payloads. There is consequently a need for benchmarking these solutions in performances and in complexity.

In the following, a methodology to benchmark payload architectures taking into account the different domains of flexibility is presented. This methodology implements a resource allocation algorithm. The resource allocation algorithm must be able to leverage the flexibility capabilities of the considered payloads. A parameter to characterize the non-uniformity of user distributions is introduced. Thanks to this parameter, the ability of various LEO payload architectures to support different scenarios of user distributions that may be encountered along their orbits is assessed. To illustrate the benefits of this methodology, different LEO satellite payload architectures with different levels of flexibility and complexity are benchmarked. For brevity, the chapter focuses on the forward downlink throughput between the satellite and the UTs. It is expected that the study of the return link would not alter the presented methodology or conclusions. These architectures are first described in section 6.2. An approach to parameterize payload designs is presented in section 6.3. The non-uniformity parameter and the resource allocation algorithm used to compute throughputs are described in section 6.4; finally, results are presented in 6.5, including considerations on payload complexity.

6.2 Examples of LEO satellite payloads

A megaconstellation scenario with polar orbiting satellites at 1200 km altitude is assumed. The satellites operate in the Ka-band to access more frequency bandwidth which is key for broadband services. One particular issue with LEO satellites is the difference in attenuation experienced between the nadir and the edge of coverage due to increasing path losses. In this context, path losses reflect the Radio Frequency (RF) signal attenuation as it propagates in the atmosphere. They are caused by the expansion of the wave front in free space (free space losses, defined in (6.1) with d, the distance between the satellite and the UT and λ the wavelength) or by absorption (rain, clouds and gazes).

$$\text{FreeSpaceLoss} = \left(\frac{4\pi d}{\lambda}\right)^2 \tag{6.1}$$

Attenuations due to atmospheric gases may be estimated as discussed in Reference 14. For preliminary studies, it is sufficient to only consider free space losses as defined in (6.1). In this context, FOV is defined as the region where users have the satellite in line-of-sight with over 20° of elevation angle. Under this value, satellites may be masked by building or reliefs. Furthermore, the link budgets are poorer due to higher distances between the satellite and the UTs. The variation of free space loss as a function of the elevation angle is presented in Figure 6.2(a). In case of 20° of elevation for a user on the ground, the user is at a scan angle of 52° that corresponds to an extra path loss of 6.2 dB.

To provide a uniform EIRP over the coverage, more directive beams are needed to compensate for increased path losses at high elevations. One alternative is to use high power amplifiers (HPA) operating at different power levels. This strategy is, however, more expensive than having a single HPA design. The other alternative is to launch more satellites such that the scanning angle is reduced and, consequently, the path loss differences between the center and edge of FOV. This solution is also costly as the number of satellites to be manufactured increases.

Coverage where more directive beams are used at larger scan angles is referred to as isoflux. An example of isoflux coverage is schematically depicted in Figure 6.2(b). It ensures an almost constant power flux on the ground. It is noted that the projection of the beams of an isoflux coverage on the surface of the Earth results in comparable footprints. This implies that due to the Earth's curvature and the pointing angle, beams at the edge of the coverage have larger footprints than if they were pointed at nadir.

Given these comparable footprints, over multiple orbits, all beams would experience similar maximum traffic per beam. Moreover, coordination constraints with users of GEO satellite systems suggest steep beam slopes along the north-south axis for reducing interference; consequently, beams with elliptical shapes were specified. With elliptical beams, the operator of satellites in polar orbit can switch off a set of beams (to avoid interference from the ground to the GEO satellites) with less impact on the users in the north–south direction. Reference 15 shows that the users

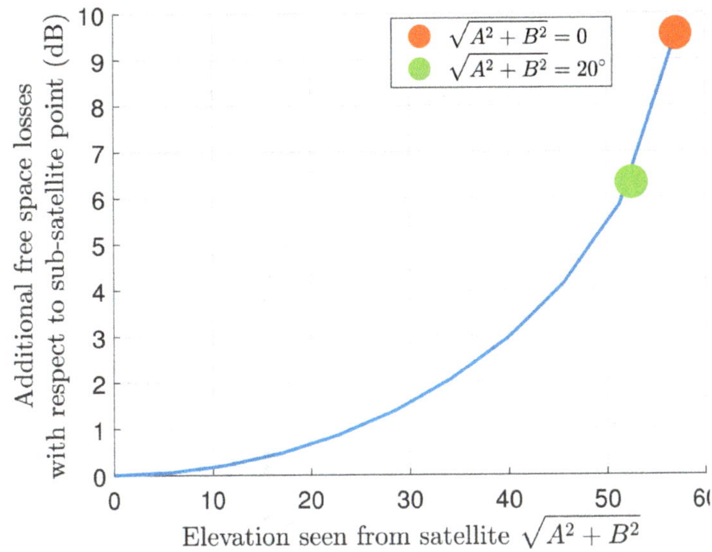

(a)

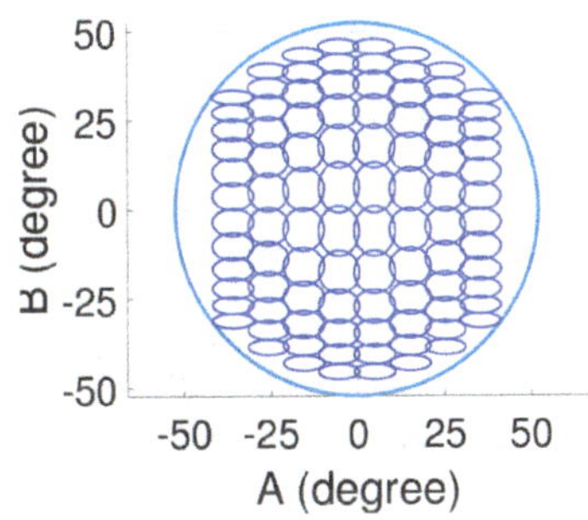

(b)

Figure 6.2 *(a) Free space loss for users at high elevations at 1200 km altitude; (b) multiple beam coverage with the isoflux characteristic formed by 96 beams along 8 columns. The ellipses represent the 3 dB contour of the beams*

under the switched off beams can be served by adjacent satellites either by applying a pitch to the adjacent satellites to tilt the beams or by expanding the coverage of the satellite.

In the remaining, coverages are represented in *AB* coordinates, which satisfy (6.2).

$$\theta = \sqrt{A^2 + B^2}$$
$$\phi = \arctan\left(A/B\right) \tag{6.2}$$

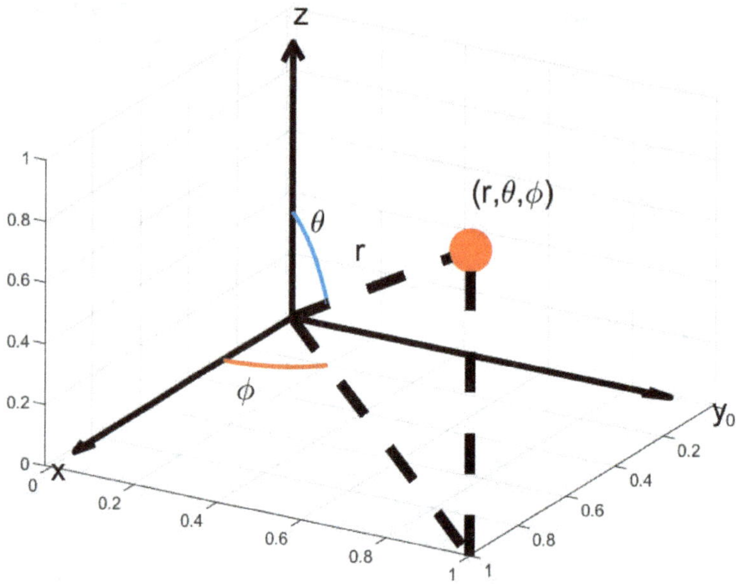

Figure 6.3 Spherical coordinates

In (6.2), θ and ϕ are the spherical coordinates in the satellite coordinate system; the Z axis in Figure 6.3 is pointing toward the Earth and the Y axis is in the direction of the satellite's velocity.

6.2.1 Multiple fixed beam coverage with static resource allocation

An antenna solution to comply with the isoflux requirement is the antenna concept shown in Figure 6.4(a) and described in Reference 16 based on QOBF and cylindrical reflectors. This antenna architecture produces several beams from the same radiating aperture and is therefore a compact solution for LEO satellite platforms. The accommodation constraint took into account 17 satellite platforms fitting into a

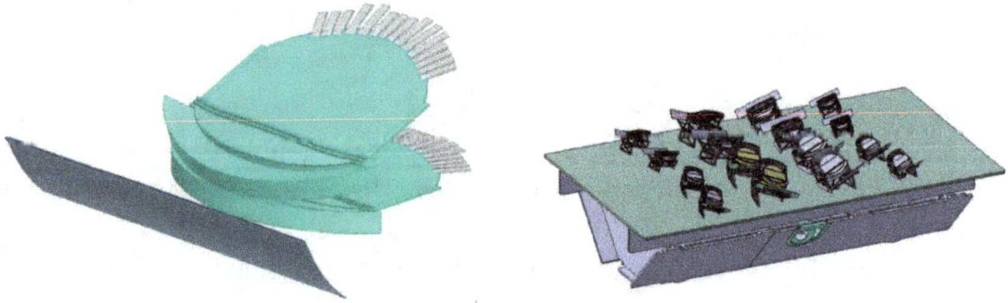

(a) (b)

Figure 6.4 (a) QOBF with cylindrical reflector and (b) accommodation of the whole antenna set up on a small satellite platform

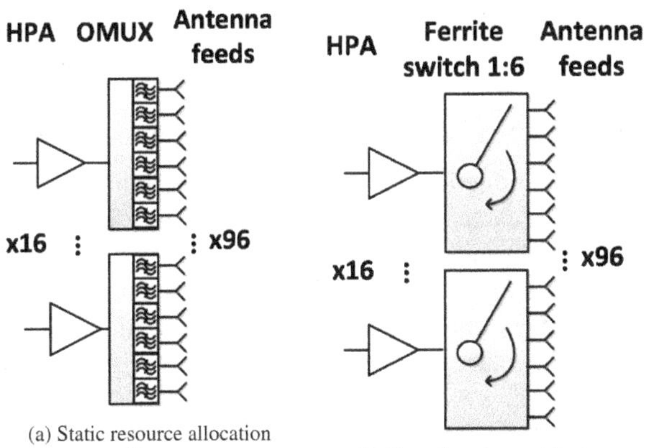

(a) Static resource allocation architecture

(b) Beam hopping architecture

Figure 6.5 Payload output section architectures

5-m diameter launcher fairing, which imposes a maximum volume of 1.5 m × 5 m × 0.5 m for the antenna farm [16]. Figure 6.4(b) shows the accommodation of the Tx and Rx antennas on a small platform suitable for allowing multiple satellites per launch. This solution developed during the CNES study 'Architectures d'antennes multifaisceaux pour constellations' proved to have equivalent radiating performance as conventional array of horn antennas but with the benefit of compactness [16].

This antenna may support a payload architecture with a fixed frequency plan in a SFPB configuration. SFPB implies that each beam is amplified by a single HPA. An issue with isoflux coverage is the irregular beam lattice, which is contrary to traditional GEO hexagonal four-color schemes [17]. A graph coloring algorithm based on the DSATUR algorithm [18] was used to allocate the frequency bands to each beam. This algorithm is efficient to provide frequency plans avoiding the reuse of frequency bands between two neighbor beams that would generate interference. A six-color reuse scheme is utilized to achieve acceptable interference mitigation. Even if more aggressive frequency reuse is targeted, the amplified bandwidth per beam is limited by the available power of the LEO satellite platform. Splitting the frequency band into more sub-bands would reduce interference, at the cost of less bandwidth per beam, hence less capacity.

The static resource allocation architecture in Figure 6.5(a) relies on travelling wave tube amplifiers (TWTAs). They were chosen because of the required power and of the operating band (Ka-band). Output multiplexers (OMUX) are represented but they do not play any role in the throughput calculations. This architecture does not have a DTP as it does not require on-board routing. A static mapping between forward uplink and downlink frequencies can be implemented. Adding a DTP would increase power consumption/dissipation and the complexity of the satellite platform to provide and dissipate this power. This architecture delivers fixed coverage and fixed resource allocation, which may be optimal for uniform distributions (Figure 6.1(a)) but may not meet the demands of scenarios where the distribution of users in the FOV is not uniform (Figure 6.1(b)). More flexibility is required in this case and other flexible solutions are proposed in the following. This payload architecture follows the same principles as OneWeb satellites payloads.

6.2.2 Multiple beam coverage with beam hopping capability

As discussed in the introduction, the time domain solutions employing BH [6] may offer flexibility. An example of implementation with ferrite switches at post-amplification level is illustrated in Figure 6.5(b). In this case, as many ferrite switches as high-power amplifiers are needed to allocate time slots. The integration of ferrite switches implies more mass, power and accommodation requirements; consequently, this architecture is more complex than the previous one. Insertion losses due to the switches are neglected as they are estimated to be under 0.2 dB [7]. They may, however, not be negligible in the case where several switch stages are implemented. In Figure 6.5(b), each TWTA is connected to a set of six possible beams, among which only one is illuminated during each time slot. In the considered architecture, 16 beams out of a total of 96 beams are active during each time slot. During each time slot, each beam illuminated transmits signal over the entire available forward downlink frequency band. The optimization algorithm for the time slot allocation to beams is presented in section 6.4.

6.2.3 Multiple beam coverage with steering capabilities along ones axis

The architecture considered in this section is a hybrid one. It uses a passive linear array of QOBF to form beams (multiple feeds per beam antenna) in columns of fixed coordinate A and steers the beams along the B axis thanks to an ABF (analog beam former). ABF can be based on Monolithic Microwave Integrated Circuit (MMIC) phase shifters. These beams are steered on a slot-by-slot basis. Flexibility is limited to steering beams along just one axis. However, this architecture is more flexible than the BH architecture because beams can be steered precisely over users positions, limiting losses in gain. This architecture also implements isoflux in one direction to compensate for the path losses previously described. Losses due to QOBF are very low [11]. They are consequently neglected for the following analysis. For comparison purposes, the time domain flexibility with BH is also implemented with this architecture. As with the previous architecture, 16 beams are illuminated per time slot, but now with a maximum of two beams per column since each ABF is connected to two ports of the processor. Consequently, this payload cannot reallocate power from one column to another. The HPA at the output of each ABF in Figure 6.6 corresponds to the distributed amplification along each steering column. Solid state power amplifiers (SSPA) are chosen for this payload architecture. The reasons are accommodation purposes and lower required power per HPA than in SFPB architectures.

The hybrid architecture presented in Figure 6.6 is a mix between analog phase shifting and quasi-optical technologies. Using a mix of analog/digital or quasi-optical/digital such as proposed in Reference 19 in a GEO use case, may also be of interest.

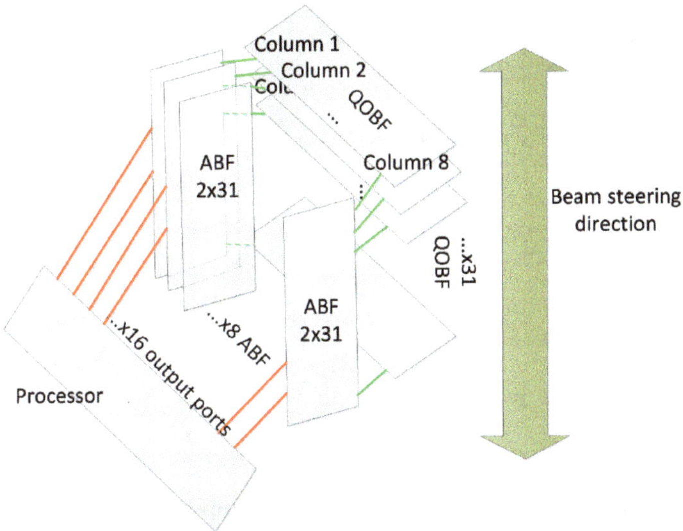

Figure 6.6 *Hybrid steering architecture with analog plus quasi-optical beamforming*

6.2.4 Other architectures not considered in the benchmark

Payload architectures may also implement a DTP to leverage the flexibility to allocate frequency bands in the beams formed for users as well as gateways. The DTP may also include digital beamforming (DBF). When considering a high number of beams and radiating elements, DBF is a preferable beamforming solution compared to a fully ABF. To give an idea of the complexity of an ABF, to form one thousand beams with one hundred radiating elements, an ABF would need one thousand 100-way RF power dividers and 100,000 (100 × 1000) analog phase-shifters. This solution is very challenging for packaging and design. Analog beamforming would be interesting for lower numbers of beams or radiating elements as the hardware solution would be less complex. Furthermore, this solution would dissipate and consume less power than the digital one.

Instead of an array of lenses as presented in Figure 6.6, an array of radiating elements such as horn antennas can be implemented. An example of circular direct radiating array (CDRA) lattice with 351 radiating elements is presented in Figure 6.7(a). This antenna (designed for a MEO use case) is active, which implies one HPA integrated with each radiating element. This technology enables power sharing between beams and benefits from a graceful degradation in case of the breakdown of HPAs. Another advantage of active antennas is that HPAs compensate for the losses in the repeater. A drawback of this architecture is the constraint of having one RF chain per radiating element. A RF chain is defined as a sequence of electronic components such as HPAs, frequency converters, Analogue to Digital Converters (ADCs) and Digital to Analogue Converters (DACs) and DTP ports. The more RF chains

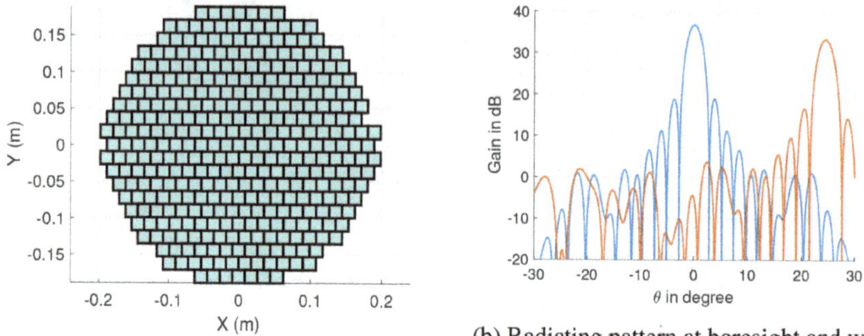

(a) Radiating elements layout for the CDRA.

(b) Radiating pattern at boresight and when the main beam is steered to 24.5°.

Figure 6.7 Presentation of the CDRA

there are, the more stringent are the power, mass and mechanical requirements. The CDRA with full DBF was compared to a payload with an architecture similar to what was presented in section 6.2.3 in Reference 20.

6.3 Payload models

Section 6.2 defined the payload architectures to benchmark. The current section proceeds with characterizing these payloads. These include the radiation patterns of the antennas, how the antenna is monitored: BH, beam steering and how power is distributed: SFPB, multiple feeds per beam.

6.3.1 Radiation patterns

For the architecture with static resource allocation and the BH architecture, the radiation patterns of QOBF are computed with a ray-tracing method [21]. The radiation of the antenna subsystem (doubly curved reflector illuminated by QOBF) can then be calculated with GRASP, a commercial software package for satellite antenna design and analysis. A minimum 10 dB taper of the reflector ensures lower side lobes to mitigate inter-beam interference and interference with GEO satellites.

For the steerable case, the simulated radiation patterns of the QOBF are combined into a phased array. Each column of beams represents one input port of a QOBF. The amplitude coefficients of the linear array are chosen according to the Taylor law to reduce the side lobe levels on the north-south axis. On the west-east axis, side lobes are mitigated with a tapering of QOBF apertures.

Alternatively, closed-form analytical expressions or measurements can be used to represent the antenna radiation patterns. ITU proposes radiation pattern models [22] that may help to compute capacities and interference check for non-geostationary orbit fixed-satellite service.

6.3.2 Antenna monitoring

Two antenna monitoring techniques are taken into account: BH and beam steering. In BH, beams are already formed and follow the lattice in Figure 6.2. BH consists of allocating the optimal time slots to the beams in order to minimize the interference. The time slot allocation must adhere to power sharing constraints (each amplifier is shared among a fixed set of beams). The architecture illustrated in Figure 6.6 relies on beam steering. In this case, the antenna is able to steer toward the optimal directions that are not pre-defined as in BH. To reduce the complexity of the optimization in the beam steering cases, the FOV of the antenna is sampled with possible pre-formed beam positions as illustrated in Figure 6.8. Subsequently, beams that serve users are kept and considered in the resource allocation problem to reduce the size of the optimization problem.

6.3.3 HPA

One of the constraints of the resource allocation algorithm is related to the power sharing among beams (constraint later specified in 6.4). Power sharing is formalized in binary matrices M of size $N_{\mathrm{HPA}} \times N_{\mathrm{b}}$, with N_{HPA} the number of HPAs which can be either single HPAs or banks of distributed HPAs, such as in the steering architecture or in the MPA case [9], and N_{b} the number of beams. $M_{h,b}$ equals 1, when power from HPA h can be allocated to beam b, 0 otherwise. For clarity, matrices are below represented in bold. In the BH architecture, the connection matrix links HPAs to the beams they can potentially illuminate. The connection matrix of the architecture with a static resource allocation and the BH architectures is defined in (6.3).

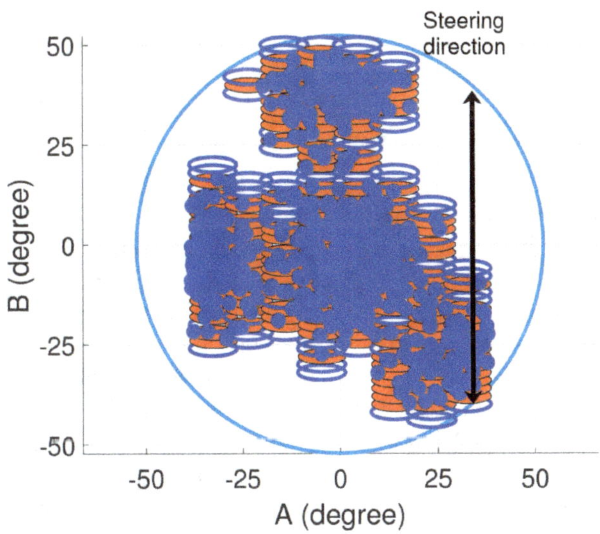

Figure 6.8 Directive beams are generated over users with isoflux implemented in the A direction

$$\mathbf{M} = \left.\overbrace{\begin{bmatrix} 1 & 1 & 1 & 1 & 1 & 1 & 0 & \cdots & 0 \\ 0 & 0 & 0 & 0 & 0 & 0 & 1 & \cdots & 0 \\ \cdots & \cdots & \cdots & \cdots & \cdots & \cdots & \cdots & \ddots & \cdots \\ 0 & 0 & 0 & 0 & 0 & 0 & 0 & 0 & 1 \end{bmatrix}}^{\text{6 beams linked to same switch/HPA}}\right\}\text{16 amplifiers} \tag{6.3}$$

For the steering architecture, each steering column has a distributed amplification fed by a pool of HPAs. This way, each column of beams as represented in Figure 6.8, share a common bank of power formed by the amplifiers at the output of each ABF (see Figure 6.6). It is to be noted that for this architecture the matrix depends on the position of the pre-formed beams. These positions are derived from the position of users in the FOV as illustrated in Figure 6.8. The connection matrix of the bent pipe and the steering architecture is defined in (6.4).

$$\mathbf{M} = \left.\overbrace{\begin{bmatrix} 1 & 1 & 1 & 1 & 1 & 1 & 0 & \cdots & 0 \\ 0 & 0 & 0 & 0 & 0 & 0 & 1 & \cdots & 0 \\ \cdots & \cdots & \cdots & \cdots & \cdots & \cdots & \cdots & \ddots & \cdots \\ 0 & 0 & 1 & 0 & 0 & 0 & 0 & 0 & 0 \end{bmatrix}}^{\text{Pre-formed beams}}\right\}\text{8 pools of HPAs} \tag{6.4}$$

Examples of connection matrices can be found in Reference 23 to express the constraints for a resource allocation problem. Constraints on power are further detailed in section 6.4.4.

6.4 Resource allocation

6.4.1 Generation of user distribution scenarios

This section presents a method to generate different profiles of UT distributions without known traffic models. As a minimum, an objective of 100 Mbps per household [24] was set as a target by the European Commission. This value is considered to be a realistic data rate demand per terminal. Scenarios of user distributions were generated by differently distributing 300 UTs with the aforementioned traffic demands. In this way, the target throughput per satellite would be 30 Gbps, which is in the range of current megaconstellation projects [5]. For link budget computations, all user antennas are assumed to be 60 cm diameter. That diameter corresponds to a fixed satellite service application with a figure of merit, $GT = 15$ dB/K. A method is proposed to generate user distributions that would cover several statistically possible profiles of distributions from uniform distribution to uneven ones. To generate a uniform scenario, AB coordinate pairs in the antenna FOV are randomly selected following a uniform law. To add non-uniformity, denser user demands must be generated in some areas. These areas of higher traffic density, in the remaining referred

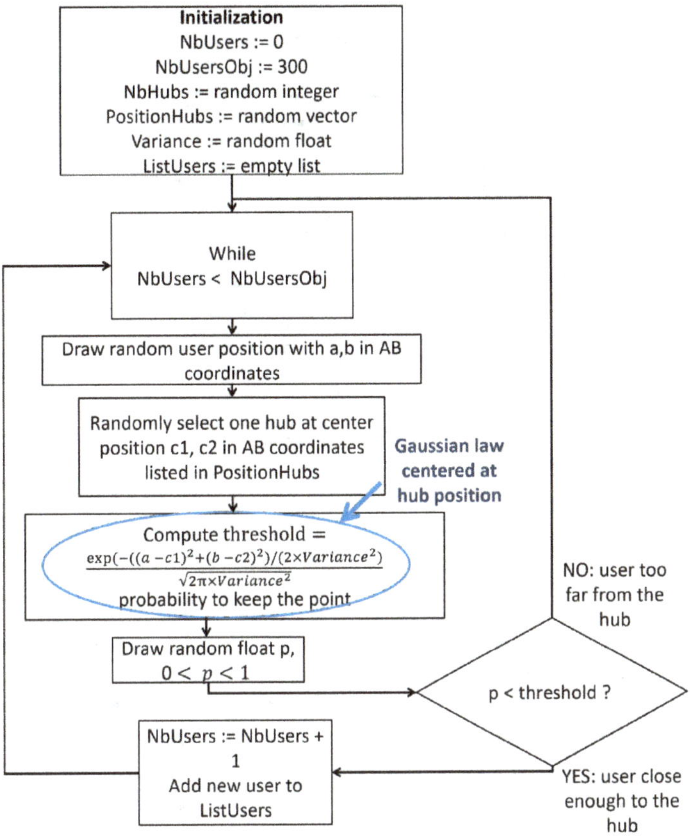

Figure 6.9 User distributions generation process

to as hubs, are centered around positions that are randomly chosen across the FOV. The users are drawn according to the uniform law and randomly kept following the Gaussian law centered at hub positions. The variance of this law and the number of hubs can be tuned to increase the non-uniformity of the distribution. There is no direct relationship between the non-uniformity indicator that will be presented in the next section. It is consequently important to cover all the traffic scenario profiles by randomly varying these last two parameters. The process of scenario generation is described in Figure 6.9.

6.4.2 Distribution characterization of traffic scenarios

Here is defined a parameter that captures in a single number how uniformly user traffic is spread across the satellite FOV. This user traffic can be either simulated (e.g. following the methodology presented in the previous section) or known (see section 6.5.2). We start by producing the user demand distribution over the coverage area. This is performed by partitioning the FOV into elliptical cells, each of whose

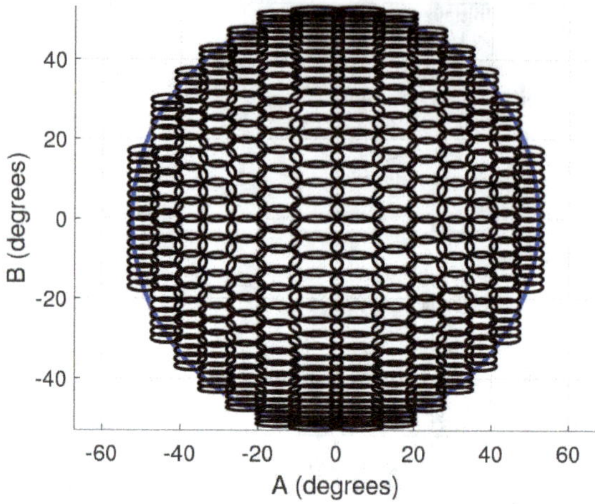

Figure 6.10 Elliptical iso Earth projection surface cells to compute ATD

surface area is commensurate to the dimension of the average beam footprint on Earth. An example of a cell lattice is given in Figure 6.10. Note that the number of cells is not related to the actual number of beams produced by the payload. Based on a traffic scenario, we can compute the aggregate traffic demand, ATD, per cell. The parameter to quantify non-uniformity, denoted μ, is defined in (6.5) as the standard deviation of the aggregated demands in each cell normalized by the mean of this parameter in these same cells.

$$\mu = \frac{\text{standard deviation of } ATD \text{ across all the cells in the coverage}}{\text{mean of } ATD \text{ across all the cells over coverage}} \tag{6.5}$$

Based on the above definitions, Figure 6.11 shows examples of user distributions with the corresponding value of the non-uniformity coefficient μ. The results in (6.5) highlight the relation between the non-uniformity parameter and the ability of the payload architectures to achieve high throughputs. High ATD leads to lower

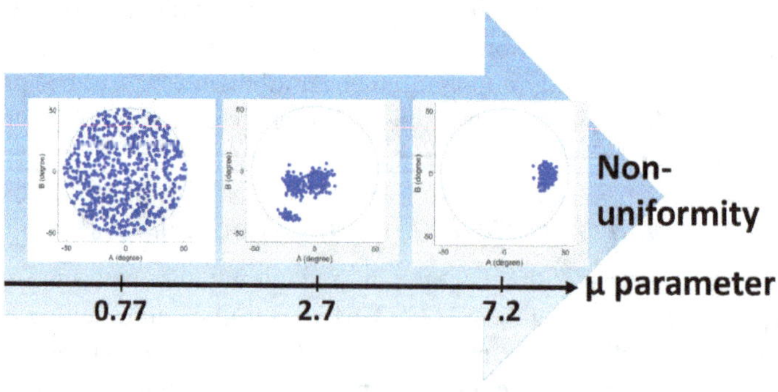

Figure 6.11 Evolution of users distribution with non-uniformity parameter

throughputs. This is due to the limited available frequency band and the impact of reusing this resource due to inter beam interference.

6.4.3 Resource allocation algorithms

The topic of resource allocation for satellite applications emerged as the satellites acquired the ability to flexibly allocate their resources in time, frequency, and power. Reference 25 focused on power optimization based on traffic demands per beam. By neglecting inter-beam interference, a closed form expression of the optimum power levels can be determined. Accounting interference, however, makes the power allocation problem NP-hard. To solve this issue, metaheuristics were proposed, such as genetic algorithms, simulated annealing (SA), or particle swarm optimization [26]. Time slot allocation in case of BH or frequency chunks allocation can also be optimized with these heuristics, such as in Reference 6, where a genetic algorithm is used to allocate time slots for a payload capable of BH or in Reference 27, where frequency plan and time slot allocation are optimized for various payload designs. Frequency flexibility is granted by the filtering and routing abilities of the DTP. Reference 23 utilizes the power and the frequency flexibility to meet the demand in all the beams. Metaheuristics may require lengthy computations, which can be detrimental for on-board implementation or if a large number of payload architectures need to be evaluated. Greedy algorithms also exist to provide, within short running times, relevant solutions. Greedy algorithms are here referring to one-shot algorithms. They are not iterative like the previously mentioned metaheuristics and focus on quickly finding a local optimum instead of looking for a global optimum through lengthier iterative optimization methods. They were investigated in Reference 28 for coverage with beams of different sizes. Greedy algorithms based on graph coloring were used in References 29, 30 and 31 in a space division multiple access (SDMA) scheme, where beams are steered to UTs' directions. An iterative method is proposed in Reference 32 to allocate carriers while minimizing inter-beam interference. Another approach to allocate resources is integer linear programming (ILP). It consists in solving a linear problem only with integer variables. The problem can be solved with a specific ILP solver, e.g. Gurobi [33]. This approach was compared to Reference 28 and concluded that ILP is better than greedy algorithms on small-sized problems. For large-sized problems, the gap with the upper bound increases, suggesting that the algorithm converges further and further from the optimum. Reference 29 also compares a greedy algorithm with ILP. The ILP algorithm surpasses the greedy one when the number of users increases. In a time/frequency resource allocation [34], also uses ILP. In the following, the choice was to focus on greedy and SA algorithms. The greedy algorithm provides quasi-immediate solutions (which can be nevertheless quite far from the optimum), which is interesting for quickly narrowing down the most promising payload architectures. SA is more likely to reach closer to the global optimum as it performs global research in the field of feasible solutions. This last algorithm was used instead of ILP as ILP requires a stringent mathematical description of the problem, which may not allow to test all the domains of flexibility of

Table 6.1 Some resource allocation algorithms found in the literature with the domain of flexibility they leverage and their optimization methods.

	Analytical	Metaheuristics	ILP	Greedy
[25]	√ Power			
[26]		√ Power		
[6]		√ Frequency		
[27]		√ Time/frequency		
[23]		√ Time/frequency		
[28]				√ Frequency
[30]				√ Frequency
[32]	√ Time/frequency			
[35]			√ Frequency	
[29]				√ Frequency
[34]			√ Time/frequency	

the payloads (frequency, power, and beamforming). Moreover, SA is a heuristic that can be relatively easily tuned with only a small set of parameters controlling the convergence of the algorithm and the domain of feasible solutions that can be evaluated. Table 6.1 recaps the mentioned algorithms. This list is not exhaustive.

6.4.4 Description of the considered resource allocation algorithm

In the case of the BH and steerable payload architectures, a resource allocation algorithm is required to allocate time slots to beams and allocate the time slots over the busy regions while avoiding interference. A SA algorithm is used to solve this NP-hard problem that is close to the frequency assignment problem [36]. SA brings advantageous features in terms of its simplicity and convergence; studies involving a given traffic scenario with different starting points have shown to converge to the same optimum solution, thereby providing confidence on the accuracy of the method. This type of algorithm was also used in Reference 23 to solve the resource allocation problem and is described by a flow graph in Figure 6.12.

The SA provides the optimum time slots for beams association by allocating time slots to beams with unserved demand while accounting for inter-beam interference. An initial solution is provided to the algorithm, then the algorithm explores the domain of feasible solutions with the capacity of new random solutions being estimated. During this exploration phase, the algorithm is more likely to accept solutions with poor throughputs. This probability of acceptance decreases with the number of iterations as the temperature T decreases. The decrease speed is monitored with the cooling factor α. In the end, the algorithm only accepts the modifications in the time slots allocation that improve the current configuration in local research. The considered parameters for the SA algorithm are given in Table 6.2.

The initial temperature was set to explore a wide range of solutions at the start of the iterations. The cooling factor and the number of iterations are jointly tuned to balance exploration and local research phases.

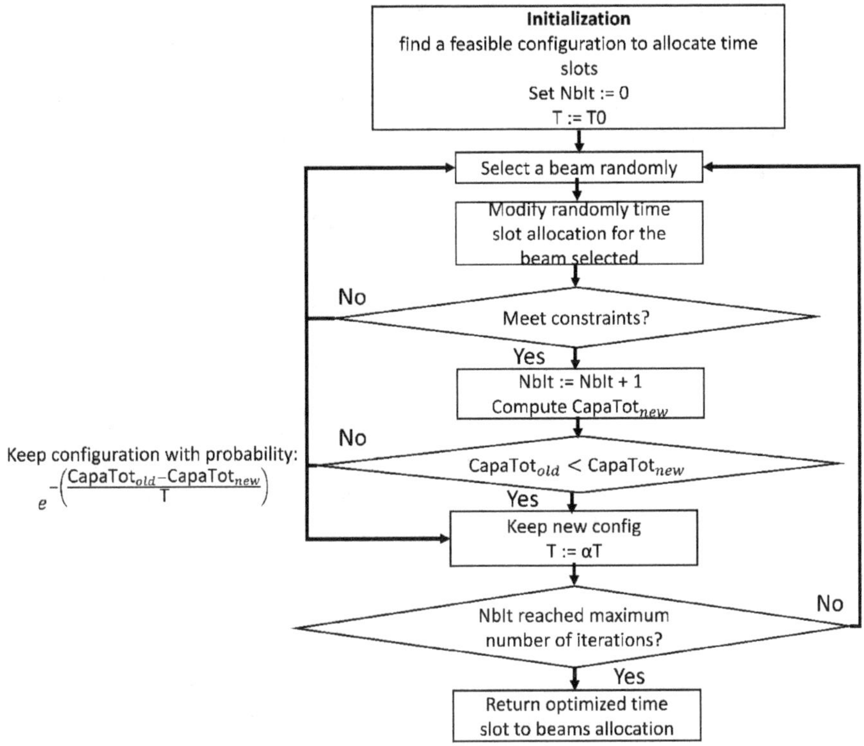

Figure 6.12 Simulated annealing algorithm

The allocation of a time slot to a beam is stored in the binary matrix $\boldsymbol{T}$, $\boldsymbol{T} \in \mathcal{T} \subset \mathbb{N}^{N_b \times N_t}$, where $T_{b,l}$ equals 0 if the beam b is off at time slot l and 1 if this beam is on during this time slot. $\mathcal{T}$ is the set of time slot matrices that adhere to the following payload constraints:

- For all the payloads, it is impossible to allocate more power to a set of beams than what is available with the HPA connected to these beams by the M matrix defined in section 6.3.3. This corresponds to the condition $\boldsymbol{M}(T_l \odot W) \leq P$, where T_l is the lth column of matrix T and $\odot$ the Hadamard product that multiplies element wise vectors and matrices. The vector P, $P \in \mathbb{R}^{N_{\text{HPA}} \times 1}$ contains the maximum power that can be allocated to beams for each amplifier or distributed amplification. N_{HPA} is the number of HPAs in the static and BH

Table 6.2 Simulated annealing algorithm parameters.

Initial temperature	Cooling factor α	Number of iterations
1000 K	0.9993	20,000

architectures or the number of steering columns which corresponds to N_{HPA} distributed amplification banks. The vector W, $W \in \mathbb{R}^{N_b \times 1}$ stores the power allocated per beam.

- For all payload architectures implementing BH, the number of beams illuminated at each time slot is assumed to be 16. That constraint can be formalized by: for all time slots l, $\sum_b T_{b,l} \leq 16$.
- In the case of SFPB architectures, the beam to amplifier connection matrix must be taken into account. It ensures that an amplifier does not amplify more than one beam within the same frequency band. This constraint is not taken into account in this optimization because at each time slot, each amplifier illuminates at most one beam over the entire frequency band available. However, in the case of a system with flexible frequency slot allocation, this constraint must be accounted for.
- For the hybrid architecture, due to beam forming constraints in the DTP, only two beams at maximum per column of fixed coordinate A can be formed. This constraint can be formalized by: for all time slot l, for all steering column C_i with i ranging from 1 to 8, $\sum_{b \in C_i} T_{b,l} \leq 2$.

The UTs' data rates are derived from the spectral efficiencies calculated using carrier-to-noise power ratios, interference levels and considering the DVB-S2 standard [37]. A design choice to have a constant power density (units W/Hz) per beam is followed to have comparable spectral efficiencies over the coverage. Given the isoflux requirements presented in (6.2), this approach guarantees an iso EIRP density over the coverage area to achieve a quasi-uniform spectral efficiency. Users are allocated to beams from which centers they are the closest. In (6.6), $(CN)_i^j$ represents the carrier-to-noise ratio of user j belonging to beam i.

$$\left(\frac{C}{N}\right)_i^j = P_i g_i^j \left(\frac{4\pi d_j}{\lambda}\right)^{-2} \left(\frac{G}{T}\right)_j \frac{1}{kB_i} \tag{6.6}$$

where P_i is the allocated power to beam i, d_j the distance between the satellite and user j, g_i^j the gain of beam i toward user j, $(GT)_j$ the antenna gain-to-noise-temperature of the UT, k the Boltzmann constant and B_i the frequency bandwidth allocated to beam i.

The carrier to interference power ratio of user j belonging to beam i is defined as:

$$\left(\frac{C}{I}\right)_i^j = \frac{P_i g_i^j}{\sum\limits_{i', i' \neq i} P_{i'} g_{i'}^j b_{i',i}} \tag{6.7}$$

where $b_{i',i}$ is the time ratio when beams i' and i are illuminated at the same time. If these two beams are simultaneously illuminated during all the time slots they are active, this coefficient is equal to 1. If they are never illuminated at the same time this coefficient is equal to 0.

For user resource access, a Time Division Multiple Access (TDMA) scheme with one carrier per beam is considered. When a beam is allocated at a certain time slot, the frequency band available to users is allocated in a greedy way such that users with the best link budgets are first served. Spectral efficiencies are computed according to DVB-S2 standard and SINR levels given in (6.8) utilizing adaptive coding and modulation (ACM) techniques.

$$\text{SINR}_i^j = \frac{1}{\left(\left(\frac{C}{N}\right)_i^j\right)^{-1} + \left(\left(\frac{C}{I}\right)_i^j\right)^{-1} + \left(\left(\frac{C}{I_{\text{intermod}}}\right)_i^j\right)^{-1}} \tag{6.8}$$

ACM adapts the waveform to maintain the communication link between the satellite and UTs depending on the link condition. Lower SINR signals can be addressed with more robust modcod at the cost of lower spectral efficiencies. On the opposite, higher spectral efficiencies can be reached in the case of high SINR with the use of complex symbol constellations. $(CI)_{\text{intermod}}$ are estimated from the modulation and output back-off operated at amplifiers level [38] and [39] provide methodologies to study intermodulation interference, respectively for TWTA and SSPA.

In the case of the steering architecture, beams are generated at user positions to simplify the resource allocation problem (see Figure 6.8). Radiation pattern models are described in section 6.3.1.

6.5 Results

6.5.1 Architectures performance

The total available bandwidth for the forward downlink is assumed to be 2.5 GHz. A total of 50 different traffic scenarios with varying aggregate demands are produced, following the methodology presented in section 6.4.1. All the payloads were designed to provide the same throughput on a uniform scenario with an aggregate demand of 30 Gbps. This demand corresponds to the targeted throughput per satellite of the megaconstellation Figure 6.13.

Figure 6.14 presents how each payload architecture performs on the generated 30 Gbps scenarios (only the geographical distribution of this demand varies). The matching ratio (MR) is defined as the ratio between the total throughput allocated by the payload over the aggregated user demand in the FOV of the satellite. For all the payload architectures, the MR tends to decrease with non-uniformity. For the static resource allocation architecture, the non-flexibility of the resource allocation quickly limits the MRs achieved. BH achieves better throughputs, reaching more than two times the throughput of the architecture with the static resource allocation for μ values higher than 2. The hybrid architecture performs best. Its benefits are particularly noticeable for μ between 1 and 7. For μ values over 7, corresponding to high traffic non-uniformity, the performance of the steering architecture converges toward that of BH. Payload architectures with different levels of complexity demonstrate different levels of flexibility. The most flexible solution, the beam steering architecture, have the best MRs over the highest range of μ parameters. However, it is also the most complex architecture as it will be further discussed in

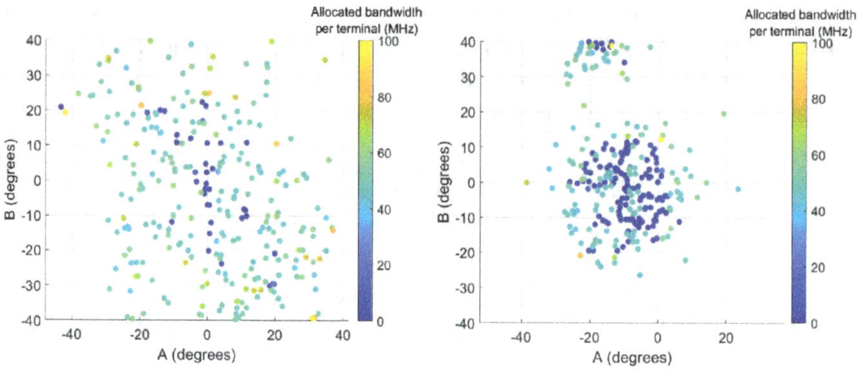

(a) $\mu = 1.3$ (b) $\mu = 2.4$

Figure 6.13 Impact of non-uniformity on the bandwidth allocation

section 6.5.3. The architecture with the static resource allocation is the least flexible with inferior MRs for most μ parameters, especially for the highest μ, but it is also the simplest. BH is an intermediate solution between these two architectures in terms of flexibility and complexity. The same conclusions can be drawn from

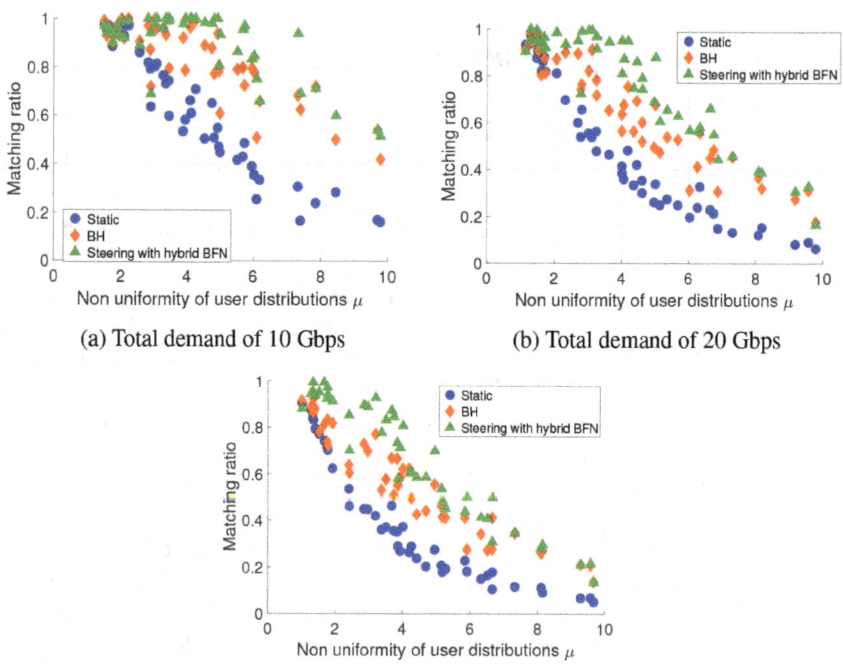

(a) Total demand of 10 Gbps (b) Total demand of 20 Gbps

(c) Total demand of 30 Gbps

Figure 6.14 Variation of matching ratio achieved by the payload architectures

scenarios with total aggregated demand of 10 Gbps (Figure 6.14(a)) and 20 Gbps (Figure 6.14(b)) with better MR achieved as the aggregated demands are lower in these scenarios. These trends have been confirmed for a scenario including 1000 users and a total demand of 20 Gbps (not shown here for brevity). The decrease of capacity with non-uniformity is due to the difficulty for the resource allocation algorithm to allocate bandwidth to users terminal while avoiding interference. Reusing frequencies for close UTs would increase interference and lead to lower capacities. This phenomenon is illustrated in Figure 6.13. The considered payload architecture was the one with the static resource allocation on two scenarios with an overall demand of 30 Gbps.

The μ parameter characterizes the different levels of non-uniformity over which a user distribution can range. However, all μ values may not be reached in realistic scenarios and possibly might occur with different probabilities. In the next section, an analysis employs the non-uniformity analysis results to select the most adapted payload architectures for specific worldwide user distributions.

6.5.2 Analysis on an orbit with a realistic traffic

In the case of LEO satellite constellations, estimating the throughput at each orbital position is a costly process as the resource allocation algorithm should be run over one or more complete orbits for each payload architecture considered. Machine learning, e.g. based on neural network, may be a used to estimate the capacity of a constellation [40]. An alternative analytical approach is proposed in this section to quickly benchmark payload architectures for specific traffic scenarios. A simple model of MR as a function of the aggregated demand and non-uniformity is proposed. It is derived from the numerical results presented in Figure 6.14. A statistical overall capacity achieved by each architecture over two polar orbits is deduced from the models. It permits one to determine the most adapted architecture for two different realistic scenarios of users' worldwide distributions; 1200 km polar orbits were chosen to achieve global coverage of Earth. The polar orbit is used by the OneWeb megaconstellations [5].

MR functions are modeled with one parameter μ_0 defined in (6.9). This function is chosen because as μ approaches 0 (uniformity), the MR converges to 1 (all the user demand is met). μ was squared to reflect the fact that above a certain non-uniformity, MR decreases more slowly. As μ_0 grows, the MR decreases more rapidly (Figure 6.15).

$$MR_{\text{model}}(\mu) = 1 - \exp\left(\frac{-\mu_0}{\mu^2}\right) \tag{6.9}$$

μ_0 is higher for the steering architecture, showing its greater flexibility with respect to the other solutions. BH is the second architecture with higher μ_0. The least flexible is the architecture with static allocated resources with lowest μ_0 parameter.

The μ_0 values are calculated with the least square method which consists of minimizing the sum of the squares of the residuals $r_i = MR_i - MR_{\text{model}}(\mu_i)$, with, $i = 1, ...n$, where n is the number of user distribution scenarios, here 50; μ_i represents

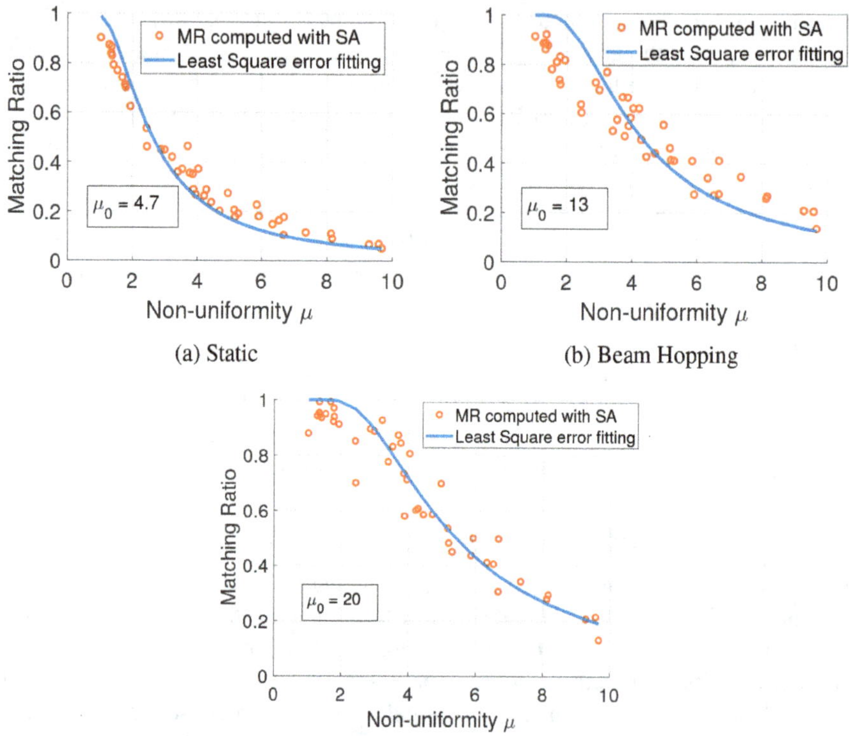

(a) Static

(b) Beam Hopping

(c) Beam Steering

Figure 6.15 Least square error interpolation results with the MR function model selected for (a) the static resource allocation; (b) BH; (c) beam steering architectures for a total demand per scenario of 30 Gbps

the non-uniformities of each generated user distributions, obtained with the method described in section 6.4.1. MR_i is the associated MR computed with SA and MR_{model} the function expressed in (6.9).

A payload benchmark study based on the non-uniformity of realistic traffic issued from a commercial forecast analysis is next presented. These data are derived from the ESA study next generation high data rate trunking systems [41] in the framework of an ARTES future program. Several other databases are available on the Internet and can be used to model aero, fixed satellite service (FSS) or maritime traffic depending on the targeted market [42]. The first traffic profile contains a mix of data for aero, maritime, and terrestrial markets exhibited in Figure 6.16(a). The other traffic profile is only based on the aero data of this study (see Figure 6.16(b)). The scenarios are adapted such that the maximum demand in the satellite FOV never exceeds 30 Gbps throughput for which the payloads were initially designed.

In Figure 6.17, the mixed demand presented in Figure 6.16(a) seen by the satellite orbiting is shown with the non-uniformity of the traffic in its FOV and the throughput allocated by different payloads.

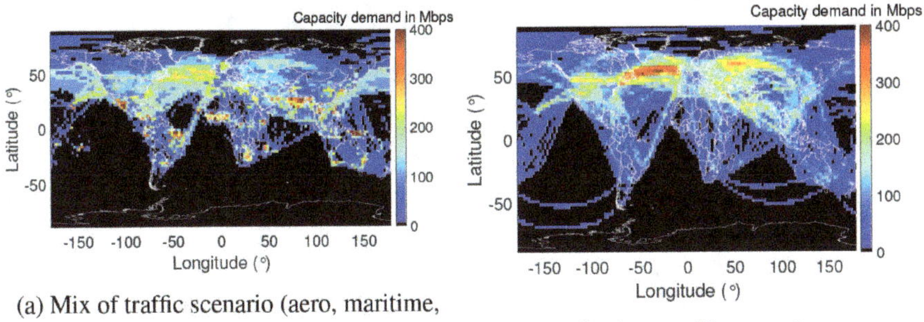

(a) Mix of traffic scenario (aero, maritime, terrestrial

(b) Aero traffic scenario

Figure 6.16 Various geographical capacity demand distributions

In the cases where the aggregate demand is less than 10 Gbps, it is supposed that the satellite completely fulfills the demand. Otherwise for each payload architecture and for the demand at each orbital position, the parameter μ_0 introduced in (6.9) is linearly interpolated from the μ_0 obtained for aggregate demands of 10, 20, and 30 Gbps. The gaps in the curves correspond to the flyovers of the south pole where there is no demand. Non-uniformity may reach high values; however, most of the time these values are associated with low aggregate demands so payloads architectures are able to meet the demand in these cases.

For the mixed scenarios, in the BH and in the steering cases, the throughputs follow the demand with MR near 1 along all the orbits. In Figure 6.17(a), the MR of the architecture with static resource allocation decreases to 0.7 at the 1600th minute when the satellite flies over New Guinea, where the backhaul and inland terrestrial demands are expected to be important. The high demands in these sparse islands surrounded by the Pacific Ocean induce highly non-uniform user distributions.

The average throughput and the minimum MR over the orbital positions are given in Table 6.3 for both traffic scenarios. The mixed scenario proved to have less uniform user distributions than the aero scenario. In the mixed scenario, the most non-uniform regions correspond to islands with land terminals demanding high data rates. The choice of the optimum payload may be driven by the maximum average throughput provided. BH would be chosen in this case as it performs similarly to the steering architecture but with less complexity in the payload. Another driver may be the worst-case MR over the coverage. In this case, the steering architecture is the best one with 99% of the total demand matched for all orbital positions. However, the choice for the payload architecture must also account for the complexity of each candidate solution. The next subsection proposes to add the complexity aspects to the system trade-off.

The study can be extended at the constellation level. In that case, the capacity computations must also take into account the presence of intersatellite links and GW / satellite links. An example of constellation level analysis is given in Reference 5.

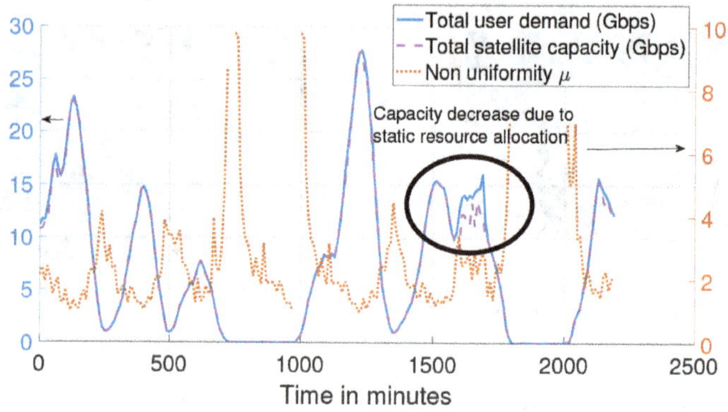

(a) Static resource allocation

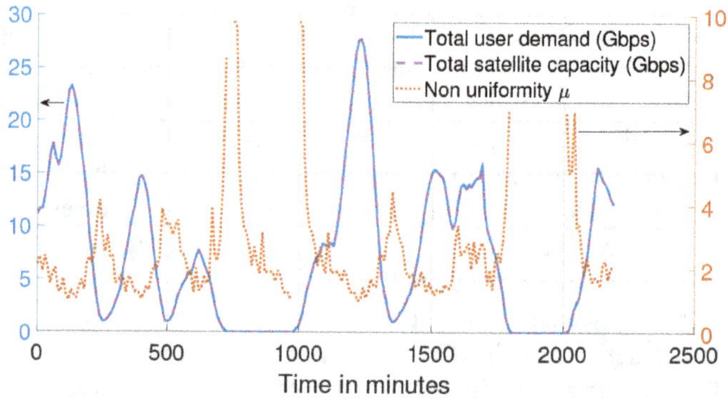

(b) Beam Hopping

Figure 6.17 For the mixed traffic scenario, variation of non-uniformity, total user demand and estimated capacity. The curves for the steering architecture were not displayed as the provided capacity is very close to the demand and is similar to the one with BH

Table 6.3 Payloads performances derived from the MR model shown in (6.9).

	Mixed traffic scenario		Aero traffic scenario	
Architectures	**Average throughput over orbits (Gbps)**	**Minimum MR**	**Average throughputs over orbits (Gbps)**	**Minimum MR**
Static	7.2	0.71	5.7	0.89
BH	7.4	0.97	5.8	0.99
Steering	7.4	0.99	5.8	0.99

6.5.3 Trade-off between satellite allocated capacity and payload complexity

The previous sections 6.5.1 and 6.5.2 compared the different payload architectures based on the capacity they could effectively achieve. In section 6.5.2, the results obtained from realistic scenarios present the advantages of BH and beam steering compared to the architecture with static resource allocation. These benefits were demonstrated in terms of average throughput and minimum MR over two 110 minutes orbits. To conclude the benchmark, this section examines the complexity of such flexible payload architectures. The estimation of the complexity focuses on the payload mass and DC power consumption. On the one hand, mass is related with accommodation constraints and launch costs. On the other hand, power is linked with constraints on thermal design for heat dissipation as well as the mass and cost of HPAs, solar panels and batteries. Mass and power therefore play then a crucial role in the assessment of the satellite complexity.

The estimation of mass takes into account the QOBFs (which have the role of beam formers and radiating elements) and reflectors for the payload with static resource allocation. In Reference 16, the mass of the QOBFs is modeled as a function of the number of formed beams. According to this reference, the configuration QOBFs plus reflectors implies 2.5 beams/kg. With recent developments mentioned in Reference 16 on QOBF with up to 54° scanning angle and new manufacturing techniques (including molded plastics and additive manufacturing), the mass can be lowered to achieve 5.1 beams/kg. For the BH architecture, ferrite switches are additionally taken into account. Commercial solutions are available and the mass is around 550 g per ferrite switch. For the beam steering architecture, the analog beam former mass was estimated to 3 kg. Mass estimations are detailed below in Table 6.4.

To a first approximation, the power consumed only accounts for the HPAs as they are expected to play a major role in power consumption. Their efficiency depends on the amplifier technology: typical values indicate that SSPAs have 35% of estimated power efficiency whereas TWTAs have an estimated 60% power efficiency. All architectures have a total of 500 W aggregated RF power. Even if these mass and power budgets are not exhaustive, they give an overview of the complexity of the payload subsystems that are benchmarked with the main components that contribute to the mass and power budgets.

Table 6.4 Mass and power budgets

	Static res. alloc.	**Beam hopping**	**Beam steering**
Mass ferrite switches	0	8.8 kg	0
Mass ABFs	0	0	3 kg
Mass QOBFs	19 kg	19 kg	49 kg
Total mass estimated	19 kg	28 kg	52 kg
DC power consumed by HPAs	830 W	830 W	1400 W

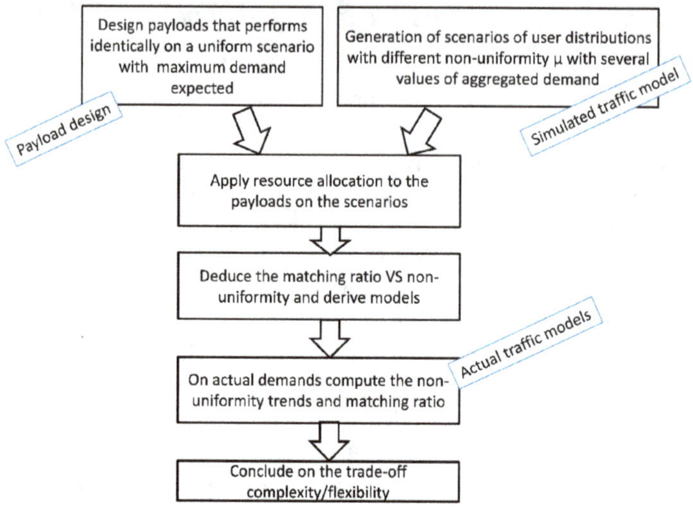

Figure 6.18 Benchmark method

Considering the earlier results and Table 6.4, BH stands as the best solution with average throughputs and minimum MR reaching closely the ones of the beam steering architecture in mixed and aeronautical scenarios. These performances are leveraged with 1.9 times less mass and 40% less power consumption than the beam steering architecture. Even if the payload with the static resource allocation is the less complex architecture with 30% less mass than BH, its lowest MR (70%) does not make it a reliable solution in the scenarios analyzed. In use cases with narrower FOVs and less expected demand, this configuration, similar to the OneWeb approach could have been considered. Other parameters when estimating the complexity of a payload solution are the manufacturing cost and TRL (technical readiness level) of components.

The diagram Figure 6.18 sums up the methodology presented in the chapter.

6.6 Conclusion

Before launching constellations of LEO satellites, a careful benchmark the flexibility/complexity of payloads and antenna subsystems shall be done. Orbiting satellites experience a diversity of user distributions. At the same time, low-cost systems are needed to limit the overall cost of launching a megaconstellation. A method has been applied to several multiple-beam antenna architectures in a megaconstellation use case. To benchmark the presented solutions, a method to compare the flexibility of antenna and payload solutions in servicing non-uniform user demand distributions was developed. The method was applied to three payload architectures. These payloads all involved an innovative quasi-optical beamformer as a primary radiator. The three architectures enable beam hopping, beam steering, as well as static resource allocation. A resource allocation algorithm was used to estimate the performances of the BH and beam steering payloads in each scenario. A measure of the non-uniformity of the user distribution is introduced to observe how non-uniformity affects the throughput of each payload solution. The choice for the most appropriate payload depends on the non-uniformity parameter and encountered demands during the satellite orbits and the complexity of each payload architecture. Complexities have in turn been estimated in regards of the mass and power required by each payload. In the considered user distribution scenario, the beam hopping and the beam steering architectures prove to be efficient solutions compared to the architecture with static resource allocation. However, with 1.9 times less mass and 40% less consumed power, the beam hopping architecture appears to be the most adapted payload design.

References

[1] Evans B., Werner M., Lutz E., *et al.* 'Integration of satellite and terrestrial systems in future multimedia communications'. *IEEE Wireless Communications.* 2005, vol. 12(5), pp. 72–80.

[2] Maral G., Bousquet M. *Satellite Communications Systems*. Hoboken, NJ: Wiley; 2005.

[3] Palacin B., Fonseca N.J.G., Romier M. 'Multibeam antennas for very high throughput satellites in Europe: technologies and trends'. 2017 11th European Conference on Antennas and Propagation (EUCAP); Paris, France, 2005.

[4] Agiwal M., Roy A., Saxena N. 'Next generation 5G wireless networks: a comprehensive survey'. *IEEE Communications Surveys & Tutorials*. 2005, vol. 18(3), pp. 1617–55.

[5] del Portillo I., Cameron B.G., Crawley E.F. 'A technical comparison of three low earth orbit satellite constellation systems to provide global broadband'. *Acta Astronautica*. 2005, vol. 159, pp. 123–35.

[6] Anzalchi J., Couchman A., Topping C, *et al.* 'Beam hopping in multi-beam broadband satellite systems'. 27th IET and AIAA International Communications Satellite Systems Conference (ICSSC 2009); Edinburgh, UK, 2010.

[7] Lejay B. 'High power ferrite switch matrix development for advanced telecom payload at Thales Alenia space' in *4th ESA workshop on advanced flexible telecom payloads*; 2019.

[8] Voisin P. 'Flexible payloads for telecommunication satellites – a Thales alenia space perspective' in 3rd ESA Flexible Payloads Workshop; 2016.

[9] Angeletti P., Lisi M. 'Multiport power amplifiers for flexible satellite antennas and payloads'. *Microwave Journal*. 2010.

[10] Bailleul P.K. 'A new era in elemental digital beamforming for spaceborne communications phased arrays'. *Proceedings of the IEEE*. 2016, vol. 104(3), pp. 623–32.

[11] Legay H., Ségolène T., Etienne G, *et al.* 'Multiple beam antenna based on a parallel plate waveguide continuous delay lens beamformer'. *Proc Int SympAntennas Propag.* 2016, pp. 118–19.

[12] Sohrabi F., Yu W. 'Hybrid digital and analog beamforming design for large-scale antenna arrays'. *IEEE Journal of Selected Topics in Signal Processing.* 2016, vol. 10(3), pp. 501–13.

[13] Devillers B., Perez-Neira A., Mosquera C. 'Joint linear precoding and beam-forming for the forward link of multi-beam broadband satellite systems'. IEEE Global Communications Conference (Globecom 2011); Houston, TX, 2011.

[14] ITU. *Attenuation by atmospheric gases and related effects*; 2019. pp. 676–12.

[15] Su Y., Liu Y., Zhou Y., Yuan J., Cao H., Shi J. 'Broadband leo satellite communications: architectures and key technologies'. *IEEE Wireless Communications.* 2019, vol. 26(2), pp. 55–61.

[16] Tubau S., Vidal F., Legay H. 'Novel multiple beam antenna farms for mega-constellations'. *ESA Antenna Workshop*; 2019.

[17] Fenech H., Amos S., Hirsch A., Soumpholphakdy V. 'VHTS systems: requirements and evolution'. 2017 11th European Conference on Antennas and Propagation (EUCAP); Paris, France, EUCAP, 1979.

[18] Brélaz D. 'New methods to color the vertices of a graph'. *Communications of the ACM.* 1979, vol. 22(4), pp. 251–56.

[19] Tugend V., Thain A. 'Hybrid beamforming with reduced grating lobes for satellite applications'. 12th European Conference on Antennas and Propagation (EuCAP 2018); London, UK, 1979.

[20] Vidal F., Legay H., Goussetis G., Strober T., Gayrard J.-D. 'Benchmark of MEO multibeam satellite adaptive antenna and payload architectures for broadband systems'. *2020 10th Advanced Satellite Multimedia Systems Conference and the 16th Signal Processing for Space Communications Workshop (ASMS/SPSC)*; 2020, Graz, Austria, 1979. Available from https://ieeexplore.ieee.org/xpl/mostRecentIssue.jsp?punumber=9268689

[21] Doucet F., Fonseca N.J.G., Girard E., Legay H., Sauleau R. 'Analytical model and study of continuous parallel plate waveguide lens-like multiple-beam antennas'. *IEEE Trans Antennas Propag.* 2018, vol. 66(9), pp. 4426–36.

[22] ITU. *Satellite antenna radiation patterns for non-geostationary orbit satellite antennas operating in the fixed-satellite service below 30 Ghz. recommendation ITU-R S1528*; 2001.

[23] Cocco G., de Cola T., Angelone M., Katona Z., Erl S. 'Radio resource management optimization of flexible satellite payloads for DVB-S2 systems'. *IEEE Transactions on Broadcasting.* 2018, vol. 64(2), pp. 266–80.

[24] 'European electronic communications code'. *Official Journal of the European Union.* 2018, p. 12.

[25] Choi J.P., Chan V.W.S. 'Optimum power and beam allocation based on traffic demands and channel conditions over satellite downlinks'. *IEEE Transactions on Wireless Communications.* 2005, vol. 4(6), pp. 2983–93.

[26] Aravanis A.I., Shankar M. R. B., Arapoglou P.-D., Danoy G., Cottis P.G., Ottersten B. 'Power allocation in multibeam satellite systems: A two-stage multi-objective optimization'. *IEEE Transactions on Wireless Communications*. 2015, vol. 14(6), pp. 3171–82.

[27] Alberti X., Cebrian J.M., Del Bianco A., *et al.* 'System capacity optimization in time and frequency for multibeam multi-media satellite systems'. 5th Advanced Satellite Multimedia Systems Conference and the 11th Signal Processing for Space Communications Workshop (ASMS/SPSC); Cagliari, Italy, 2010.

[28] Camino J.T., Stephane M., Christian A, *et al.* 'A greedy approach combined with graph coloring for non-uniform beam layouts under antenna constraints in multibeam satellite systems'. 7th Advanced Satellite Multimedia Systems Conference and the 13th Signal Processing for Space Communications Workshop; 2014.

[29] Houssin L., Artigues C., Corbel E. 'Frequency allocation problem in a SDMA satellite communication system'. *Proceedings 39th International Conference on Computers and Industrial Engineering (CIE39)*; Troyes, France, 2009. pp. 1611–16.

[30] Kiatmanaroj K., Artigues C., Houssin L., *et al.* 'Greedy algorithms for time-frequency allocation in a SDMA satellite communication system'. Proceedings 9th International Conference of Modeling, Optimization and Simulation; 2012.

[31] Kiatmanaroj K., Artigues C., Houssin L, *et al.* 'Frequency allocation in a SDMA satellite communication system with beam moving'. *Proceedings IEEE International Conference on Communications (ICC)*; 2012.

[32] Lei J., Castro M.A.V. 'Joint power and carrier allocation for the multi-beam satellite downlink with individual SINR constraints'. *IEEE International Conference on Communication (ICC)*; 2010.

[33] *Gurobi optimiser* [online]. 2020. Available from https://www.gurobi.com/products/gurobi-optimizer/

[34] Alouf S., Altman E., Galtier J, *et al.* 'Quasi-optimal bandwidth allocation for multi-spot MFTDMA satellites'. *Proceedings of IEEE Infocom 2005 Conference*; 2005.

[35] Camino J.T., Artigues C., Houssin L, *et al.* 'Mixed-integer linear programming for multibeam satellite systems design: application to the beam layout optimization'. *Annual IEEE Systems Conference (SysCon)*; 2016.

[36] Zhi-Quan L., Shuzhong Z. 'Dynamic spectrum management: complexity and duality'. *IEEE Journal of Selected Topics in Signal Processing*. 2008, vol. 2(1), pp. 57–73.

[37] ETSI. *Digital Video Broadcasting (DVB): Second Generation Framing Structure, Channel Coding and Modulation Systems for Broadcasting, Interactive Services, News Gathering and Other broadband Satellite Applications: Part 2: DVB-S2 Extensions (DVB-S2X). EN 302 307-2*. 2014.

[38] Aloisio M., Angeletti P., Casini E., Colzi E., D'Addio S., Oliva-Balague R. 'Accurate characterization of TWTA distortion in multicarrier operation

by means of a correlation-based method'. *IEEE Transactions on Electron Devices*. 2009, vol. 56(5), pp. 951–58.

[39] D'Addio S., Valenta V. 'Non-linearity assessment of SSPA-based active antennas by means of a time-domain correlation method' in *4th ESA Workshop on Advanced Flexible Telecom Payloads*; 2019.

[40] Kisseleff S., Shankar B., Spano D., Gayrard J.-D. 'A new optimization tool for mega-constellation design and its application to trunking systems'. Presented at Advances in Communications Satellite Systems. 37th International Communications Satellite Systems Conference (ICSSC-2019); 2019, Okinawa, Japan.

[41] *ESA study next generation high data rate trunking systems* [online]. Available from https://artes.esa.int/projects/next-generation-high-data-rate-trunking-systems

[42] Al-Hraishawi H., Lagunas E., Chatzinotas S. 'Traffic simulator for multi-beam satellite communication systems'. *10th Advanced Satellite Multimedia Systems Conference and the 16th Signal Processing for Space Communications Workshop (ASMS/SPSC)*; Graz, Austria, 2020. Available from https://ieeexplore.ieee.org/xpl/mostRecentIssue.jsp?punumber=9268689

Chapter 7

Novel multiple access for non-geostationary orbit communications

Xiaojuan Yan[1,2], Kang An[3], Zhiqiang Feng[2], and Qianfeng Zhang[2]

The task of multiple access strategy is to link users as flexible as possible with limited spectrum and power resources as far as possible. The design and implementation of appropriate multiple access scheme are of great importance for satellite networks since their on-board resource is severely limited while the service requirements of terminal devices are ever-increasing, especially in rural areas. In this case, novel multiple access schemes, such as non-orthogonal multiple access (NOMA) and enhanced ALOHA, have been introduced in non-geostationary orbit (NGSO) communication to further improve resource utilization efficiency in recent years. This chapter focuses on these two novel access schemes which are relevant for future satellite systems.

Section 7.1 introduces and compares three commonly used orthogonal multiple access (OMA) schemes. Section 7.2 is devoted to the performance analysis of a NOMA-based NGSO satellite system, where the key points of the NOMA scheme are specifically addressed. Section 7.3 focuses on the performance improvement brought by the enhanced ALOHA. Finally, the last section summarizes the content of this chapter.

7.1 Orthogonal multiple access

The OMA scheme is a strategy with which multiple terminals can connect to the same medium and transmit their respective carrier signals, without severe degradation in the performance of the communications system. For example, the beamforming technique combined with the use of precoding at the gateway/satellite can divide a satellite footprint into beams and also realize the space division multiple access (SDMA). Then, within a spot beam, many user signals access the same satellite transponder

[1]School of Information Science and Engineering, Southeast University, Nanjing, China
[2]Guangxi Key Laboratory of Ocean Engineering Equipment and Technology, Qinzhou, China
[3]Sixty-third Research Institute, National University of Defense Technology, Nanjing, China

in the user uplink, while in the user downlink, user signals are broadcasted to their corresponding receivers. With an OMA scheme, the transmission medium can be divided into specified "channels," i.e., time, frequency, and code, such that the signals $s_i(t)$ at time t and in channel i are orthogonal to signals in other channels, as

$$\int s_i(t)s_j(t) = 0 \text{ for } i \neq j \tag{7.1}$$

and thus, any signal out of the signal multiplexed can be extracted at the receiver.

This section briefly introduces the concept, advantages, and disadvantages of three OMA schemes commonly used in satellite communications as follows:

Frequency-division multiple access (FDMA): FDMA is a channel access technique in the data link layer. For FDMA, the available spectrum bandwidth is divided into sub-bands which are separated from each other by guard bands. Each sub-band can be accessed by modulating a radio carrier at the center frequency of the sub-band. At the receiver, the sub-bands are filtered out and the FDMA signals are reconstructed. In satellite networks, the FDMA receiver can be located either onboard the satellite (e.g., if a different multiplexing scheme is used in the feeder link) or at the ground station [1].

FDMA is a conventional multiplex scheme in satellite communications, whose advantages are:

- Mature technology, easy implementation, and cost effective.
- Not prone to the near-far problem due to the frequency filtering.
- No user synchronization required, no restrictions on the baseband signal, modulation methods, channel coding, carrier signal rate, and bandwidth.

whose disadvantages are:

- Low frequency utilization efficiency due to the required guard band between channels.
- Uplink power control is required to limit adjacent channel interference and the influence of non-linearities.
- Exact frequency control is required to limit adjacent channel interference.
- Power resources cannot be fully used if the transponder is not operated at saturation area to reduce intermodulation noise during multi-carrier operation.

Time-division multiple access (TDMA): TDMA is a channel access method used to facilitate channel access without interference. For TDMA, the available time resource is divided into frames which are again divided into time slots. These slots are assigned to different earth stations or terminals to transmit their information on the same frequency. To take account of the phase uncertainty within the time slot, guard time is used to avoid the overlap of traffic bursts. If a user's slot allocated arrives, its signal segment will be transmitted as a high rate burst fitting into the corresponding time slot [2]. To ensure that bursts arrived at the satellite do not overlay but are in sequence, every user should be managed with the common time reference to send their bursts.

TDMA technology is a mature and well- understood method, which offers the following advantages:

- No frequency guard band between channels, TDMA achieves higher throughput efficiency than power-limited FDMA when increasing channels. No need for precise narrow bandwidth filters, as is needed in FDMA.
- Facilitate the realization of comprehensive service, changing traffic demands, and interconnection with terrestrial digital communication equipment.
- Good for digital communications and satellite on-board processing.
- For pure TDMA, no intermodulation occurs, and thus, the transponder can be operated near maximum power output or saturation level. No back-off is needed. Uplink power control is not required.

which has the following drawbacks:

- Because of the high burst rate, TDMA terminals must provide high peak transmit power.
- Subject to multipath distortion because of its sensitivity to timing.
- Requires network-wide timing synchronization.
- Because of the high transmission rate, adaptive equalizers may be required in the terminals and onboard the satellite or in the ground station.

Code-division multiple access (CDMA): For CDMA, each user is assigned a unique digital code sequence, which is used to encode the user's digital data signal before modulation. Since the cross correlations between the code of the desired user and the codes of the other users are small, a certain receiver can decode the received signal and recover the original data, but other sender signals seem like noise with respect to the desired signal. The bandwidth of the code signal is chosen to be much larger than the bandwidth of the user's original data signal, the encoding process enlarges (spreads) the spectrum of the signal and is therefore also known as spread-spectrum modulation. All users transmit simultaneously at the same time and on the same carrier frequency.

Since no feedback of the frame structure is required, CDMA offers the following advantages:

- With the CDMA scheme, soft handovers are possible for the forward and return links.
- Signal spreading with user-specific codes improves privacy interference rejection capability.
- With a Rake receiver, CDMA signals transmitted simultaneously from more than one satellite can be easily detected and combined, and thus, optimum satellite diversity can be achieved with maximum ratio combining.

- Transmit signal seems buried in noise and has low power spectral density, then the low probability of intercept/detection can be obtained in certain frequency bands.
- Different from the TDMA, whose number of users is hard-limited, CDMA is soft-limited in users' number and graceful degradation in signal-to-noise ratio when the number of simultaneously transmitted user signals is exceeded.
- High frequency efficiency since user signals are distinguished by nearly orthogonal digital codes and the same frequency band can be used in all spot beams and satellites.

CDMA has the following drawbacks:

- Fast and exact power control is required to achieve equal power levels of the user signals received by satellite or ground station. Otherwise, strong user signals would suppress weaker signals (near-far problem).
- The transponder backoff is necessary because of the non-constant signal amplitude.
- High complexity of the CDMA receiver onboard the satellite brings by high chip rates, complex signal processing, and multiple Rake signal demodulators.

Figure 7.1 gives the principle of FDMA, TDMA, and CDMA. In addition to these three access schemes, the beamforming technique combined with the use of precoding at the gateway/satellite can divide a satellite footprint into beams and also realize the SDMA. Table 7.1 gives a comprehensive comparison of FDMA, TDMA, CDMA and SDMA from various perspectives.

Despite that the application of FDMA/TDMA/CDMA schemes can effectively avoid intra-beam interference and simplify signal detection, the fact that a single orthogonal resource block only can serve one user restricts further improvement of the spectrum efficiency and capacity for satellite networks. Suffering from low spectrum utilization efficiency and a limited number of users, OMA schemes cannot provide an enhanced performance at high resource efficiency to meet the explosive growth of traffic demand for future satellite networks. Therefore, new multiple access schemes, which can harmoniously integrate with OMA techniques in existing satellite architectures, should be taken into account.

7.2 NOMA-based NGSO satellite system

7.2.1 NOMA scheme

Recently, a novel multiple access scheme, referred to as non-orthogonal multiple access, which is abbreviated as NOMA has been proposed as a promising multiple access principle. The idea of NOMA is to serve multiple users in the same band/slot/code and abandon any attempt to provide orthogonal access to different users as in conventional OMA. Orthogonality naturally drops when the number of active users is higher than the number of degrees of freedom and "collisions" appear. One

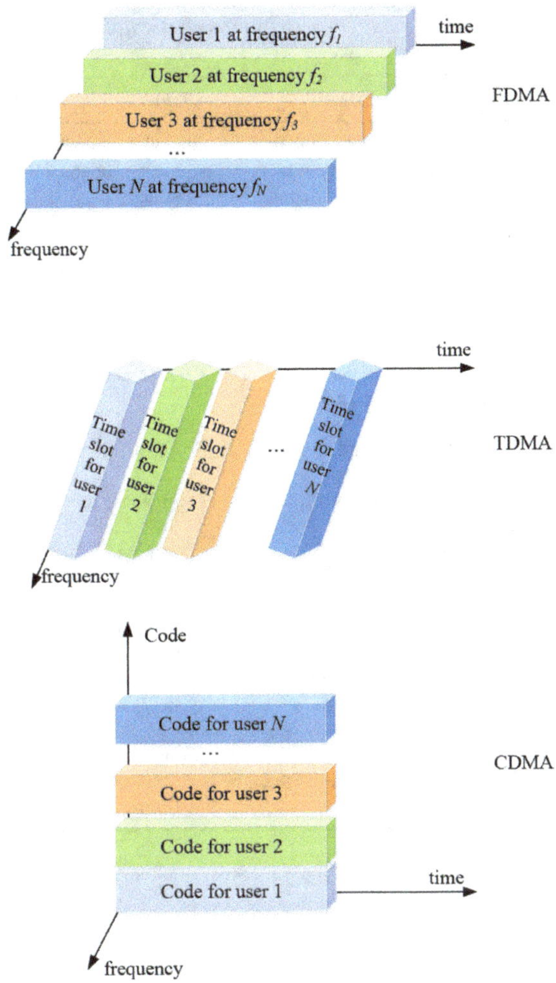

Figure 7.1 Principle of FDMA, TDMA, and CDMA

possible way of controlling collisions in NOMA is to share the same signal dimension among users and exploit power (power domain NOMA) versus code (code domain NOMA) domains, whose principles are illustrated in Figure 7.2.

As shown in Figure 7.2(a), the key idea of power domain NOMA is to superpose multiple signals in the power domain, and thus, the available resources, i.e., time and frequency can be shared among multiple users. At the receiver, advanced multiuser detection technique, such as the successive interference cancellation (SIC) which differentiates the users according to the assigned power levels, is used for multiuser detection. And thus, the NOMA scheme can provide an improved spectral efficiency at the cost of reasonable increased complexity [3].

Table 7.1 Multiple access schemes used in current satellite networks

Technique	FDMA	TDMA	CDMA	SDMA
Concept	Divide the frequency band into disjoint sub-bands	Divide the time into non-overlapping time slots	Spread the signal with orthogonal codes	Divide the space into sectors
Active terminals	All terminals active on their specified frequency	Terminals are active in their specified slot on same frequency	All terminals active on same frequency	Number of terminals as per beam depends on FDMA/TDMA/CDMA
Signal separation	Filtering in frequency	Synchronization in time	Code separation	Spatial separation using smart antennas
Handoff	Hard handoff	Hard handoff	Soft handoff	Hard and soft handoff
Advantages	Robust against varying channel conditions	Permits flexible rates	Optimal use of the available bandwidth	Increases system capacity and transmission quality
Disadvantages	Inflexible because available frequencies are fixed and guard bands are required	Requires guard space and synchronization problem	Complex receivers and requires power control to avoid near-far problem	Inflexible since network monitoring is required to avoid intracell handoffs
Current applications	Inmarsat, MSAT, MSS, and Mobilesat	Iridium, ICO, and FLTSATCOM	Globalstar, Odyssey, and Ellipso	All satellite systems

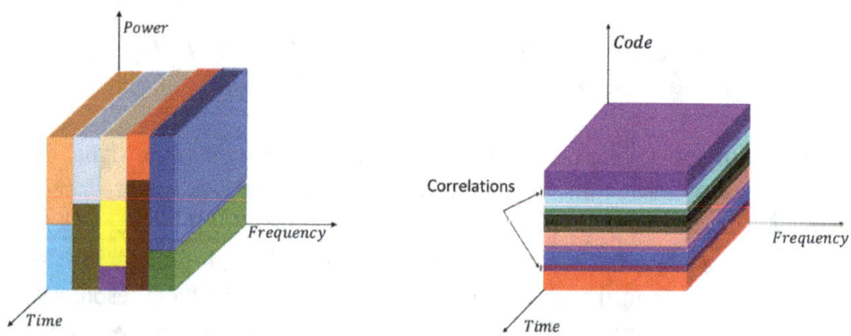

(*a*) power domain NOMA (*b*) code domain NOMA

Figure 7.2 Resource allocation in (a) power domain NOMA and (b) code domain NOMA

As shown in Figure 7.2(b), by overlapping parts of codes, code domain NOMA can provide service to users whose number is more than the number of available resources. Although non-orthogonal codes increase the probability of errors in detecting the active users at the receiver, advanced multiuser detection techniques, such as SIC, message passing algorithm, and minimum mean square error can be used to effectively recover the transmitted data [4]. The prominent code domain NOMA schemes, such as sparse code multiple access [5], pattern division multiple access [6], and multi-user shared access [7], all enable system overloading and flexible resource allocation by relaxing orthogonality requirements.

Since in recent literature, many of the NOMA schemes in satellite networks imply a power domain case, this chapter focuses on the application of the power domain NOMA scheme in satellite networks and calls power domain NOMA shortly as NOMA for simplicity.

7.2.2 The key points of the NOMA scheme

In this chapter, the NOMA scheme and its variations are introduced in various satellite networks, including integrated/hybrid and heterogeneous architectures. Particularly, in all cases, only two users are paired to form a NOMA group because of twofold: (1) interference and additional complexity introduced at the receiver will be aggrandized greatly as the number of users admitted in one NOMA group is greater than two. (2) According to the result in Reference 8, the number of users admitted in one NOMA group is limited by users' different demands, and the sum and ergodic rate were maximized when only two users are in a NOMA group. For convenience, a user with a good link condition is denoted as User p and the other one is denoted as User q in all cases.

As shown in Figure 7.3, prior to the superposing coding operation, the transmitter will select users to be served via a scheduling strategy, such as a random selection strategy. Then, in a downlink scenario, a linear superposition of multiple users'

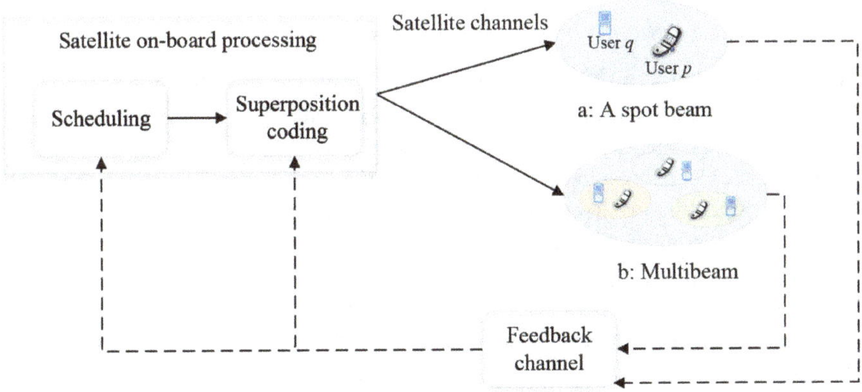

Figure 7.3 System model of NOMA-based conventional satellite networks: (a) spot beam and (b) multiple beams

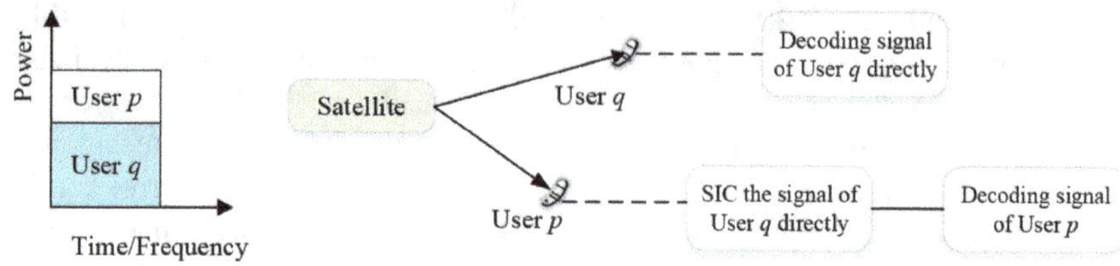

Figure 7.4 The principle of the NOMA in a downlink scenario

signals is broadcasted by allocating different power to each user. Indeed, the power is allocated based on users' feedback channel information and it is applied at each frame. On the receiver side, different detection strategies are executed by different users. As shown in Figure 7.4, User q, which may be closer to the edge of the beam and/or has lower antenna gain, is allocated with more transmission power to ensure that the intra-interference caused by User p is relatively low and can decode itself information directly. While User p, although experiences a relatively better channel condition, must adopt the SIC strategy to firstly decode and remove User q's signal, and then, decode its information since less power resource is allocated to it and its information is buried underneath.

In an uplink scenario, Users p and q simultaneously transmit to the satellite over the same time/radio block, with their maximum or controlled transmission power. Thus, with the NOMA scheme, a superposed signal is received at the satellite and the SIC strategy must be adopted to detect each user's signal. Opposite to the downlink environment, the signal of User p is decoded firstly and directly at the satellite, because its signal is stronger than that of User q. While the signal of User q must be decoded by applying the SIC strategy, which means that contribution from the signal of the User p has been already decoded and removed, the signal of User q still can be observed even if its channel condition is far worse than that of User p.

7.2.3 NOMA scheme in satellite networks

7.2.3.1 Conventional NOMA scheme in satellite networks

For the scenario depicted in Figure 7.3(a), it can be either one or two beams isolated from several spot beams, adopting 4/7-frequency reuse such that cooperation among beams is non-essential or adopting a full-frequency reuse strategy with one user in each spot beam. Since the first scenario is similar to a terrestrial cell, the strategies on user pairing and power allocation in satellite networks can refer to those made in terrestrial networks, such as the larger difference in channel gains, the larger superiority of the NOMA scheme achieved. However, it is worth noting that the link characteristics of satellite networks are different from those in terrestrial networks, i.e., non-negligible delay, as well as path loss, resulting from long distance and mobility from aeronautical or vehicular users all can cause significant errors on channel estimation, which will significantly degrade the estimation on users' channel state information (CSI) accuracy and further influence the user pairing processing.

Thus, from the transmission to reception, the link budget of User I can be modeled as

$$Q_I = L_I G_s(\varphi_I) G_I |g_I|^2 \tag{7.2}$$

- where L_I: free space loss of User I is $L_I = \left(\frac{c}{4\pi f_I h_I}\right)^2$ with c, f_I, and h_I being the light speed, the frequency, and the distance from User I to satellite, respectively. Due to the fact that NOMA users are served within the same frequency and spot beam coverage area, we assume $h_p = h_q = h$ and $L_p = L_q = L$ for simplicity.
- $G_s(\varphi_I)$: let φ_I, $\varphi_I = \arctan(d_I/h)$ with d_I denoting the distance from the beam center to User I, stand for the angle between User I and the beam center with respect to the satellite. Then, the beam gain of User I is [9]

$$G_s(\varphi_I) = G_{\max}\left(\frac{J_1(u_I)}{2u_I} + 36\frac{J_3(u_I)}{u_I^3}\right)^2 \tag{7.3}$$

where $G_{\max}$ is the maximum antenna gain, $J_n(\cdot)$ is the Bessel function of first kind and n-th order [10], and $u_I = 2.07123\sin\varphi_I/\sin\varphi_{I3\mathrm{dB}}$ with $\varphi_{I3\mathrm{dB}}$ being the one-sided half-power beamwidth, which can be written as $\varphi_{I3\mathrm{dB}} = \arctan(R/h)$.

- G_I: the antenna gain at User I. For simplicity, $G_p = G_q$ is considered here.
- $|g_I|^2$: the channel power gain of satellite link is assumed to follow a Shadowed-Rician fading model, which is mathematically tractable and has been widely applied in various fixed and mobile satellite services for a variety of frequency bands, such as the UHF-band, L-band, S-band, and Ka-band [11–13]. In this case, the probability density function of $|g_I|^2$ is given by [14]

$$f_{|g_I|^2}(x) = \alpha_I e^{-\beta_I x}{}_1F_1(m_I; 1; \delta_I x) \tag{7.4}$$

where $\alpha_I = \frac{(2b_I m_I)^{m_I}}{2b_I(2b_I m_I+\Omega_I)^{m_I}}$, $\delta_I = \frac{\Omega_I}{2b_I(2b_I m_I+\Omega_I)}$, and $\beta_I = \frac{1}{2b_I}$ with $2b_I$ and Ω_I being the average power of the multipath and the LoS components, respectively, $m_I(m_I > 0)$ denoting the Nakagami-m fading parameter, and ${}_1F_1(a; b; c)$ representing the confluent hypergeometric function [10, Eq. (9.100)].

Since the satellite channel, which is assumed in (7.4) to be block fading, requires regular estimation over a certain period of time [15]. Here, we assume that L unit energy training symbols, i.e., $E[|x_I|^2] = 1$, are transmitted from User I to the satellite in L time slots, the signal received at the satellite is

$$y_l = g_I x_l + n_l, \quad l = 1, 2, \ldots, L \tag{7.5}$$

where n_l is the noise at the satellite with zero mean and δ^2 variance. By using the maximum likelihood or least squares estimator [16], the estimate of g_I in (7.5) can be given in [17]

$$\hat{g}_I = \frac{1}{L}\sum_{l=1}^{L} y_l x_l^* = g_I + \frac{1}{L}\sum_{l=1}^{L} n_l x_l^* = g_I + e_I \tag{7.6}$$

where $\hat{g}_I$ and e_I are the estimated channel coefficient and estimated channel error with $e_I \sim \mathcal{CN}(0, \delta^2/L)$, respectively.

For the OMA scheme, such as a TDMA scheme commonly applied in satellite networks, a unit energy signal, x_I, is transmitted from the satellite to User I with a transmission power, P_s, in different time slots. The received signal at the User I is $y_I = \sqrt{P_s\Theta_I}\hat{g}_I x_I + n_I$ with $\Theta_I = L_I G_I G_s(\varphi_I)$ and n_I being the noise at the User I with zero mean and δ^2 variance. Thus, the signal-to-interference-plus-noise ratio (SINR) of User I is

$$\gamma_I^{\mathrm{T}} = \frac{P_s\Theta_I\left|\hat{g}_I\right|^2}{\delta^2} \tag{7.7}$$

With the NOMA scheme, the satellite can broadcast a superposed signal $x(x = \sqrt{\alpha P_s}x_p + \sqrt{(1-\alpha)P_s}x_q)$ to satellite users, the received signal at User $I(I = p, q)$ is

$$y_I = \sqrt{\Theta_I}\hat{g}_I x_I + n_I \tag{7.8}$$

where $\alpha(0 \le \alpha \le 1)$ denotes a fraction of the transmit power P_s allocated to User p. According to the principle of the NOMA scheme, the user with the worse channel quality decodes its information directly. Thus, the instantaneous end-to-end SINR of User q can be expressed as

$$\gamma_q^{\mathrm{N}} = \frac{(1-\alpha)P_s\Theta_q\left|\hat{g}_q\right|^2}{\alpha\Theta_q P_s\left|\hat{g}_q\right|^2 + \delta^2} = \frac{(1-\alpha)\gamma_q^{\mathrm{T}}}{\alpha\gamma_q^{\mathrm{T}} + 1} \tag{7.9}$$

where $\Theta_q = L_q G_q G_s(\varphi_q)$. Based on the criterion of SIC, user with good channel gain, User p, first decodes the information from User q. In this paper, the decoding SINR is

$$\gamma_{p\to q}^{\mathrm{N}} = \frac{(1-\alpha)\Theta_p P_s\left|\hat{g}_p\right|^2}{\alpha\Theta_p P_s\left|\hat{g}_p\right|^2 + \delta^2} \tag{7.10}$$

where $\Theta_p = L_p G_p G_s(\varphi_p)$. By comparing (7.9) with (7.10), we can find that $\gamma_{p\to q}^{\mathrm{N}}$ is larger than γ_q^{N} because of the assumption $\Theta_p > \Theta_q$, implying that the information of User q could be correctly decoded at User p. After subtracting the decoded information, User p decodes its information and the SINR of User p can be written as

$$\gamma_p^{\mathrm{N}} = \frac{\alpha P_s\Theta_p\left|\hat{g}_p\right|^2}{\delta^2} = \alpha\gamma_p^{\mathrm{T}} \tag{7.11}$$

The achieved rates of Users p and q with the NOMA scheme are

$$R_p^{\mathrm{N}} = \log(1 + \alpha\gamma_p^{\mathrm{T}}) \tag{7.12}$$

and

$$R_q^{\mathrm{N}} = \log(1 + (1-\alpha)\gamma_q^{\mathrm{T}}/(\alpha\gamma_q^{\mathrm{T}} + 1)) \tag{7.13}$$

While that with the TDMA scheme are

$$R_p^{\mathrm{T}} = 0.5\log(1+\gamma_p^{\mathrm{T}}) \tag{7.14}$$

and

$$R_q^{\mathrm{T}} = 0.5\log(1+\gamma_q^{\mathrm{T}}) \tag{7.15}$$

If assuming the capacity achieved by User p with the NOMA scheme is better than that with the TDMA scheme, we have

$$\log(1+\alpha\gamma_p^{\mathrm{T}}) \geq 0.5\log(1+\gamma_p^{\mathrm{T}}) \Rightarrow \alpha \geq 1/\left(\sqrt{1+\gamma_p^{\mathrm{T}}}+1\right) \tag{7.16}$$

Similarly, assuming the capacity achieved by User q with the NOMA scheme is better than that with the TDMA scheme, we get

$$\log(1+(1-\alpha)\,\gamma_q^{\mathrm{T}}/(\alpha\gamma_q^{\mathrm{T}}+1)) \geq 0.5\log(1+\gamma_q^{\mathrm{T}}) \Rightarrow \alpha \leq 1/(\sqrt{1+\gamma_q^{\mathrm{T}}}+1) \tag{7.17}$$

From (7.9) and (7.11), we find that an increasing α improves the capacity of User p, but simultaneously degrades the capacity of User q. From (7.16) and (7.17), we note that when $1/(\sqrt{1+\gamma_q^{\mathrm{T}}}+1) \leq \alpha \leq 1/(\sqrt{1+\gamma_q^{\mathrm{T}}}+1)$, the performance of Users p and q with the NOMA scheme outperform those with the TDMA scheme. Thus, we can draw a conclusion that when the value of α falls in the area $[1/(\sqrt{1+\gamma_q^{\mathrm{T}}}+1), 1/(\sqrt{1+\gamma_q^{\mathrm{T}}}+1)]$, the larger power allocated to the better channel link user (User p), i.e., α gets larger, the better the performance of User p is. At the same time, the performance of User q with the NOMA scheme is still better than that with the TDMA scheme due to the constraint of (7.17).

Figure 7.5 from Reference 18 presents the achievable system capacity for TDMA and NOMA schemes in the presence of imperfect CSI with Users p and q experiencing average shadowing and frequent heavy shadowing, respectively. Figure 7.6 from Reference 19 presents the achievable ergodic capacity between

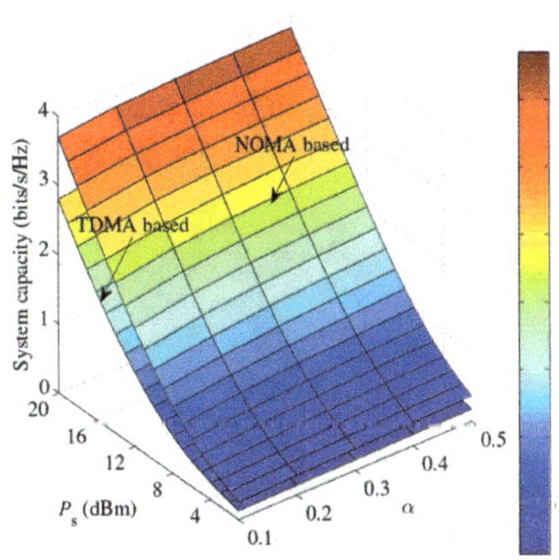

Figure 7.5 System capacity for two access schemes versus P_s and α, with the variance of estimated channel error being 0.5

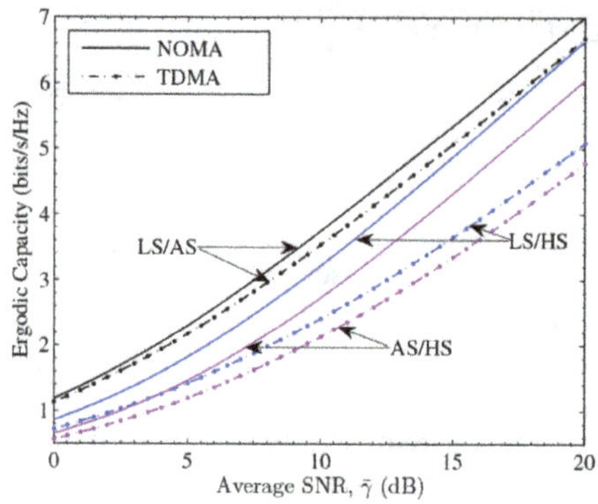

Figure 7.6 The ergodic capacity versus the average SNR $\bar{\gamma}$ for different multiple access schemes

the NOMA and the TDMA schemes for different fading severities of satellite links with perfect CSI, where $\alpha = 1/(\sqrt{1+\gamma_q^{\mathrm{T}}}+1)]$ and label (light shadowing (LS)/HS) denotes the link shadowing severity of User-p/User-q. As observed in these two figures, the performance achieved with the NOMA scheme is superior to that of the TDMA scheme.

As for the uplink NOMA scheme, users simultaneously transmit to the satellite over the same time/radio block, with their maximum or controlled transmission power. Thus, with the NOMA scheme, a superposed signal

$$y = \sum_{l=p,q} \sqrt{\Theta_l P_l}\hat{g}_l x_l + w \tag{7.18}$$

is received at the satellite and the SIC strategy is adopted to detect each signal. Especially, the signal of User p, which has a better channel link quality to the satellite, is decoded firstly and directly. While the signal of User q must be decoded by applying the SIC strategy, with perfect SIC assumed, the signal of User q is detected after subtracting the signal of User p from the received information. Thus, the achievable SINRs of Users p and q can be respectively derived as

$$\gamma_p^{\mathrm{N}} = \frac{\Theta_p P_p |\hat{g}_p|^2}{\Theta_q P_q |\hat{g}_q|^2 + \delta^2} = \frac{\gamma_p^{\mathrm{T}}}{\gamma_q^{\mathrm{T}} + 1} \tag{7.19}$$

and

$$\gamma_q^{\mathrm{N}} = \frac{\Theta_q P_q |\hat{g}_q|^2}{\delta^2} = \gamma_q^{\mathrm{T}} \tag{7.20}$$

Thus, with the NOMA scheme, the achievable sum rate is

$$R^{\mathrm{N}} = \log(1+\gamma_p^{\mathrm{N}}) + \log(1+\gamma_q^{\mathrm{N}}) = \log(1+\gamma_p^{\mathrm{T}}+\gamma_q^{\mathrm{T}}) \tag{7.21}$$

which is larger than that achieved in the same time slot with the TDMA scheme, given as

$$R^{\mathrm{T}} = 0.5\log(1+\gamma_p^{\mathrm{T}}) + 0.5\log(1+\gamma_q^{\mathrm{T}}) = 0.5\log(1+\gamma_p^{\mathrm{T}}+\gamma_q^{\mathrm{T}}+\gamma_p^{\mathrm{T}}\gamma_q^{\mathrm{T}}) \tag{7.22}$$

As for the full-frequency reuse scenario, if without beam cooperation, it is similar to the two-user interference channel. Thus, information intended to Users p or q can be divided into two parts, namely the common part, which can be decoded by these two users, and the private part, which only can be decoded by the intended user. To reduce the interbeam interference, Users p and q can jointly decode, regenerate, and cancel the common signals from their received signal and recover their corresponding private information. While with beam cooperation, the most suitable encode/decode strategy can be independently adopted within each pairing beams to improve the achievable rate regions.

For the scenario depicted in Figure 7.3(b), a multibeam satellite [20] is another scenario where the application of the NOMA scheme has been shown to be very beneficial [21, 22]. By serving users simultaneously in each spot beam with the same resource block, i.e., time/frequency resource and allocating different frequency resources to different beams in a cluster, the satellite system is overloaded at a relatively low complexity since only two users in each spot beam are admitted to form a NOMA group. To serve this type of communication, existing works have shown that the introduction of the NOMA scheme can further provide an enhanced performance gain over the OMA scheme. However, two key challenges for this case are how to effectively group users and design the power allocation factor for each beam, especially with the increase in beam numbers.

7.2.3.2 Conventional NOMA scheme in cognitive satellite terrestrial networks

With the rapid growth in satellite traffic, the licensed spectral resources appear to be insufficient and the problem of spectrum scarcity is becoming more and more obvious, which motivates the use of cognitive radio (CR) technology and the emerging cognitive satellite-terrestrial networks (CSTN) [23, 24]. Interestingly, we note that the motivation for applying the NOMA scheme is exactly the same as CR, i.e., increase the spectrum efficiency, but the solution provided by the NOMA scheme is to explore the power domain for multiple access. It is obvious that a further performance improvement can be observed if we integrate the NOMA scheme with CSTN. On the one hand, the use of the CR strategy can ensure spectrum sharing between a satellite network, termed as primary/cognitive network, and a terrestrial network which acts as a cognitive/primary network, and thus, the spectrum utilization efficiency of the overall system will be increased. On the other hand, the introduction of the NOMA scheme in CSTN ensures multiple users, access and further improves spectrum utilization without extra resource consumption.

However, in the CSTN, transmit power of the cognitive network must be constrained so that the communication of the primary user will not deteriorate. Moreover, interference from the shared network cannot be neglected, which means that channels of NOMA users suffer interference not only from the intra-group but also from co-channel interference. For example, although the achievable rates of

Users p and q still can be written as that given in (7.12) and (7.13), the effect of co-channel interference and limited transmit power must be taken into consideration in γ_p^T and γ_q^T. Thus, to serve this type of network, power allocation among NOMA users should be carefully designed to ensure user fairness. As shown in Reference 25, incorporating the NOMA scheme with the CSTN can provide an enhanced spectrum utilization efficiency compared to OMA, if power allocation coefficients are reasonably designed.

7.2.3.3 Conventional NOMA scheme in integrated satellite-terrestrial networks

Similarly to the CSTN, integrated satellite-terrestrial networks (ISTN) are also motivated by the fact that the utilization efficiency of spectrum resources is not sufficient [26]. But different from the CSTN, within which satellite and terrestrial networks may be managed by themselves the controller, satellite, and terrestrial networks in the ISTN are managed by a core network and use the same frequency band.

However, frequency reuse between satellite and terrestrial networks cannot always meet the rapid growing traffic requirement and user fairness. And thus, implementing the NOMA scheme in the ISTN is useful to support a further enhanced spectrum efficiency [27, 28]. Particularly, in the ISTN, the satellite network is complementary to terrestrial networks to extend coverage and improve reliability, which means that applying the NOMA scheme in cellular networks is more beneficial than that in satellite networks. By setting the rate achieved with the TDMA schemes as the QoS target for the far user, the Reference 27 has shown that a Max-Min user pairing scheme can achieve higher performance over a random algorithm. However, two key challenges, such as the co-channel interference caused by frequency reuse and signals from satellite and terrestrial transmitters in these heterogenous networks, need to be carefully addressed for the interference management purpose to achieve a good performance.

7.2.4 The application of cooperative NOMA scheme

In satellite communications, direct links between the satellite and terminals are sometimes deteriorated and even unavailable when terminals are deployed in masking effect areas or in spot beam edge. Under this condition, integrating the NOMA scheme with the relay technique can improve system performance in terms of reliability and capacity. As shown in Figure 7.7, four types of cooperative NOMA-based satellite networks are considered in this subsection and described as follows.

7.2.4.1 Cooperation among users

Under the circumstance of Figure 7.7(a), User p can act as a DF relay node and forward information to User q, since with the SIC strategy, the information of the user with bad channel quality is available to the user with good link condition [29]. In this case, during the first time phase, the satellite node broadcasts a superposing signal to

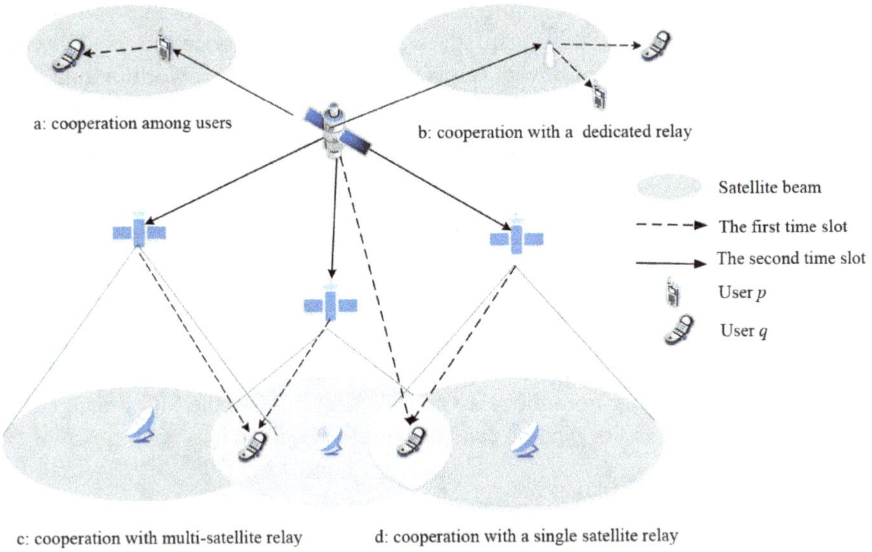

Figure 7.7 System model of cooperative NOMA-based satellite networks: (a) cooperation among users, (b) cooperation with a dedicated relay, (c) cooperation with multi-satellite relays, and (d) cooperation with a single satellite relay

User p, User p decodes the superposed signal and then forwards the corresponding information to User q in the second phase, the received SINR at User q is

$$\gamma_q = P_p Q_{pq} \left|h_{pq}\right|^2 / \delta_q^2 \tag{7.23}$$

where $Q_{pq} = G_p G_q \big/ d_{pq}^u$, P_p is the transmit power at User p, d_{pq}, h_{pq}, and δ_q^2 are the distance, the channel coefficient, and the variance of AWGN of User $p \rightarrow$User q, and u is the path loss exponent. Thus, by exploiting the prior information available in the NOMA scheme, the deployment of the relay node is unnecessary, which is very economic and beneficially for areas with low-density populations and/or areas without grid power supply.

Moreover, even when the link of satellite$\rightarrow$User q is available, the system performance achieved with this type of cooperation NOMA scheme is still superior to that with the TDMA scheme [30]. This is because although two time slots are needed in both NOMA and TDMA schemes, the reliability of User q can be further enhanced by cooperating among NOMA users.

7.2.4.2 Cooperation with a dedicated relay

If the direct links of Users p and q are all unavailable, as shown in Figure 7.7(b), a relay node with AF or DF protocol must be adopted to forward signals to users simultaneously. Since only two time slots are needed for multiple users to access this type of cooperative NOMA scheme, while four time slots are needed for these two users with the TDMA scheme. In this regard, the waiting time for service and reliability of users can be reduced and improved, respectively [31]. Here, it is worth noting that if an AF relay protocol is adopted, the superposed information received at the relay node is firstly amplified with a fixed or a variable gain factor $G = \sqrt{1/\left(P_s\Theta_{sr} + \delta_r^2\right)}$, with Θ_{sr} and δ_r^2 being the channel gain and the AWGN variance of the satellite→relay link, and then forwarded to NOMA users. In this case, the received signal at Users I $(I = p, q)$ is

$$y_I = G\sqrt{\Theta_{sr}}(\sqrt{\alpha P_s}x_p + \sqrt{(1-\alpha)P_s}x_q)\sqrt{P_r}h_{rI} + n_I \tag{7.24}$$

where $n_I^2 = P_r/\delta_d^2$, P_r is the transmit power at the relay node and h_{rI} is the channel coefficient of the link from the relay node to User I. With the SIC strategy, the achievable SINRs of Users p and q can be respectively given by

$$\gamma_q = \frac{(1-\alpha)\overline{\gamma}_{sr}\rho_{sr}\overline{\gamma}_{rd}\rho_{rq}}{\alpha\overline{\gamma}_{sr}\rho_{sr}\overline{\gamma}_{rd}\rho_{rq} + \overline{\gamma}_{rd}\rho_{rq} + \overline{\gamma}_{sr}\rho_{sr} + 1} \tag{7.25}$$

and

$$\gamma_p = \frac{\alpha\overline{\gamma}_{sr}\rho_{sr}\overline{\gamma}_{rd}\rho_{rp}}{\overline{\gamma}_{rd}\rho_{rp} + \overline{\gamma}_{sr}\rho_{sr} + 1} \tag{7.26}$$

where $\overline{\gamma}_{sr} = P_s/\delta_r^2$, $\overline{\gamma}_{rp} = P_r/\delta_d^2$, $\rho_{sr} = \Theta_{sr}$, $\rho_{rq} = |h_{rq}|^2$, and $\rho_{rp} = |h_{rp}|^2$.

While with the TDMA, the achievable SINRs of User I is

$$\gamma_I^T = \frac{\overline{\gamma}_{sr}\rho_{sr}\overline{\gamma}_{rd}\rho_{rI}}{\overline{\gamma}_{rd}\rho_{rI} + \overline{\gamma}_{sr}\rho_{sr} + 1} = \frac{\gamma_{sr}^T\gamma_{rI}^T}{\gamma_{sr}^T + \gamma_{sI}^T + 1} \tag{7.27}$$

where $\gamma_{sr}^T = \gamma_{sr}\rho_{sr}$ and $\gamma_{rI}^T = \gamma_{rI}\rho_{rI}$ are the SINRs of links from the satellite→relay node and relay node→User I, respectively.

Thus, the achievable rate with the NOMA scheme of Users p and q is

$$R_p^N = \log\left(1 + \frac{\alpha\gamma_{sr}^T\gamma_{rp}^T}{\gamma_{sr}^T + \gamma_{rp}^T + 1}\right) \tag{7.28}$$

and

$$R_q^N = \log\left(1 + \frac{(1-\alpha)\gamma_{sr}^T\gamma_{rq}^T}{\alpha\gamma_{sr}^T\gamma_{rq}^T + \gamma_{sr}^T + \gamma_{rq}^T + 1}\right) \tag{7.29}$$

The outage probability of two schemes with imperfect CSI is illustrated in Figure 7.8, in which we assume a LEO operates at the 1.6 GHz, the satellite→relay link experiences the frequent heavy shadowing, $\bar{\gamma}_{sr} = \bar{\gamma}_{rd}$, $\gamma_{thp} = -3$ dB, and $\gamma_{thq} = 3$ dB, h_{rI} follows a Nakagami-m fading with parameters $m_{rp} = 1$ and $m_{rq} = 0.5$. As observed, the performance achieved with the NOMA scheme is superior to that with the TDMA scheme by employing a suitable power allocation factor.

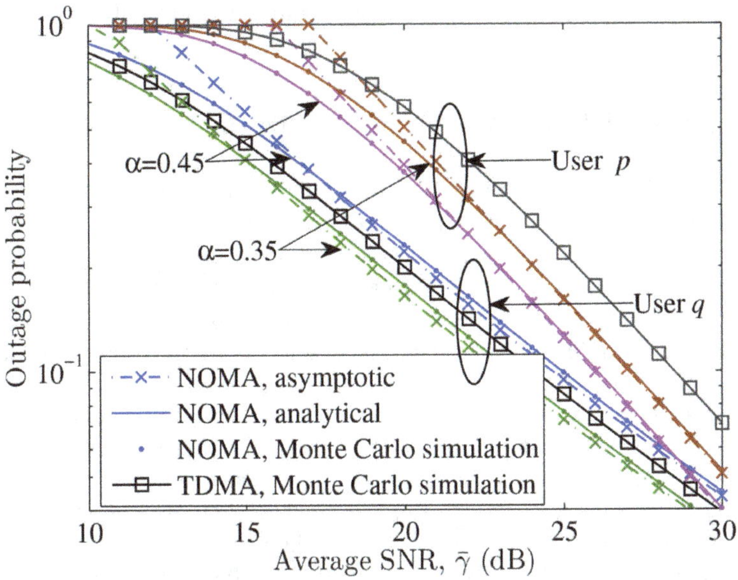

Figure 7.8 Outage probability versus $\bar{\gamma}$ for different power allocation factor α

While if the DF protocol is adopted, first, the relay node must decode the information intended to Users p and q with the SIC strategy, and then split the transmission power at the relay node to superpose those two signals and transmit to users.

7.2.4.3 Cooperation with multi-satellite relays

For the scenario considered in Figure 7.7(c), the direct link between the source satellite and the ground user is unavailable, satellites which have lower earth orbits act as relays to forward information [32]. During the first phase, signals for two time slots are superposed and transmitted from the source node, relays decode their corresponding signals by using the SIC strategy. During the second phase, relays forward different slot signals to the ground user, and SIC is adopted to combine its received signals. It is worth noting that the environment from the source node to relays is similar to that in the first case of Figure 7.3(a), in which power split is processed at the transmitter side. While the environment from relays to the ground user is similar to that of the NOMA uplink scenario, where multiple signals from the differed transmitter are received and SIC is applied at the receiver side.

7.2.4.4 Cooperation with a single satellite relay

If there is a poor direct link quality, as shown in Figure 7.7(d), a single satellite relay node can be applied to provide an enhanced source utilization efficiency. In this scenario, similar to the second case of Figure 7.7(a), the signal can be divided into two parts, i.e., source and relay parts. The relay part is transmitted with the help of the relay node, while the source part is transmitted directly from the source satellite.

At the ground user, signals from two different paths are combined and decoded by using the SIC scheme.

7.3 Enhanced ALOHA in satellite networks

For bursty traffic, random access (RA) is efficient for massive terminals to request resources frequently, especially for interactive satellite networks [33]. Both pure and slotted ALOHA schemes are the classical RA protocols which are suitable for satellite environment since they are all independent on the propagation delay.

Especially, pure ALOHA allows users to transmit packets whenever they are ready, while slotted ALOHA only permits a user to transmit its packet within the next upcoming time slot after the transmission has been initiated by the user. That means, although pure and slotted ALOHAs are RA schemes in the time domain, no time slots are defined in the pure ALOHA, while in the slotted ALOHA, time axis is divided into time slots with duration T_s corresponding to a packet duration plus a guard time accounting for synchronization errors [34]. In these two schemes, transmitters use the same carrier frequency and an acknowledgment will be returned by the receiver (i.e., the satellite or the ground station) to the transmitter, if a packet is successfully transmitted.

Although the required synchronization make slotted ALOHA have the ability to provide an enhanced throughput (almost twice) than that provided with the pure ALOHA scheme, the network throughput it can provide is still poor due to data collisions in the case of more than one terminal is transmitted in a time slot. In this case, no acknowledgment is returned and terminals must retransmit their packets after having waited for an additional random delay. Therefore, both pure and slotted ALOHA schemes are not practical to operate in the high load region in satellite communications due to the large latencies caused by large number of retransmissions and poor throughput.

In this section, we mainly review three recently proposed enhanced slotted ALOHA schemes, i.e., contention resolution diversity slotted ALOHA (CRDSA) [35], irregular repetition slotted ALOHA (IRSA) [36], and code slotted ALOHA (CSA) [37], suitable for satellite networks.

7.3.1 Contention resolution diversity slotted ALOHA

To solve packet collisions, CRDSA, which transmits multiple packet replicas (typically one packet repetition) in random slots of the same frame and uses the SIC strategy at the receiver side, is proposed to improve the probability of successful decoding [35].

As shown in Figure 7.9, an RA frame of the CRDSA consists of N ($N = 5$ in Figure 7.9) time slots. We assume the duration of each time slot equals the packet transmission time, and M ($M = 4$) users, which are synchronized to a common clock that determines the start of each slot, share the medium. The transmitter sends l ($l = 2$ set here) replicas of the same medium access (MAC) packet physically per RA frame. Although replicas are randomly put in these l slots, the payload can still

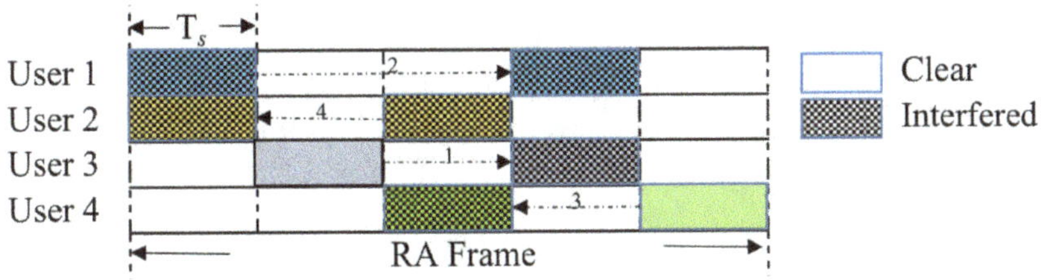

Figure 7.9 CRDSA frame structure and interference cancellation procedure

remove the interference contribution of a correctly decoded packet from all the N slots selected for transmission, due to the same preamble and the localizing information within the frame.

After the receiver stores an incoming MAC frame, the SIC iterative procedure starts from the clear bursts, i.e., the first packet of User 3 in Slot 2 in Figure 7.9. Once it is successfully decoded, its twin replica, namely the second packet of User 3 in Slot 4, can be removed from the frame. In this way, the second packet of User 1 in Slot 4 can be correctly decoded and its twin in Slot 1 is also removed from the frame. Similarly, the second packet of User 4 in Slot 5 can be correctly decoded, since it has not been subject to collisions and therefore its twin packet in Slot 3 can be also removed from the frame, releasing a replica of the packet of User 2 from the collision.

By adopting the SIC technique, the CRDSA can greatly extend the maximum load and further improve the achievable throughput and packet loss ratio. According to the conclusion made from Reference 35 that compared to the slotted ALOHA, the CRDSA improves 25 times with 2 replicas and 58 times with 3 replicas in throughput. However, unrecoverable collision still may arise from users' replicas transmitted in the same slots and/or load is too high.

7.3.2 Irregular repetition slotted ALOHA

Different from the CRDSA, which fixes the repetition rate, IRSA allows a user to transmit a random and non-constant replica in the frame. The specific and optimized packet repetition scheme probabilities can be derived by exploiting the bipartite graphs techniques as shown in Reference 36, which derives the mass probability distribution p_l for the probability that each burst is transmitted l times (l varies from burst to burst according to a given probability distribution) in each MAC frame with $1 \leq l \leq l_{max}$, where l_{max} is the maximum burst repetition number.

Especially, in Reference 36, the frame status is described as $G = (B, S, E)$ with set B consisting of the burst nodes, set S consisting of the slot nodes, and set E consisting of the edges used to connect burst nodes to slot nodes when bursts are transmitted in the corresponding slots. As shown in Figure 7.10 from Reference 36, edge is labeled as 1 if the corresponding burst replica has been revealed, otherwise, it is labeled as 0. The iterative interference cancellation (IC) process starts from slot within which the packet is clean, i.e., the second slot, as illustrated in

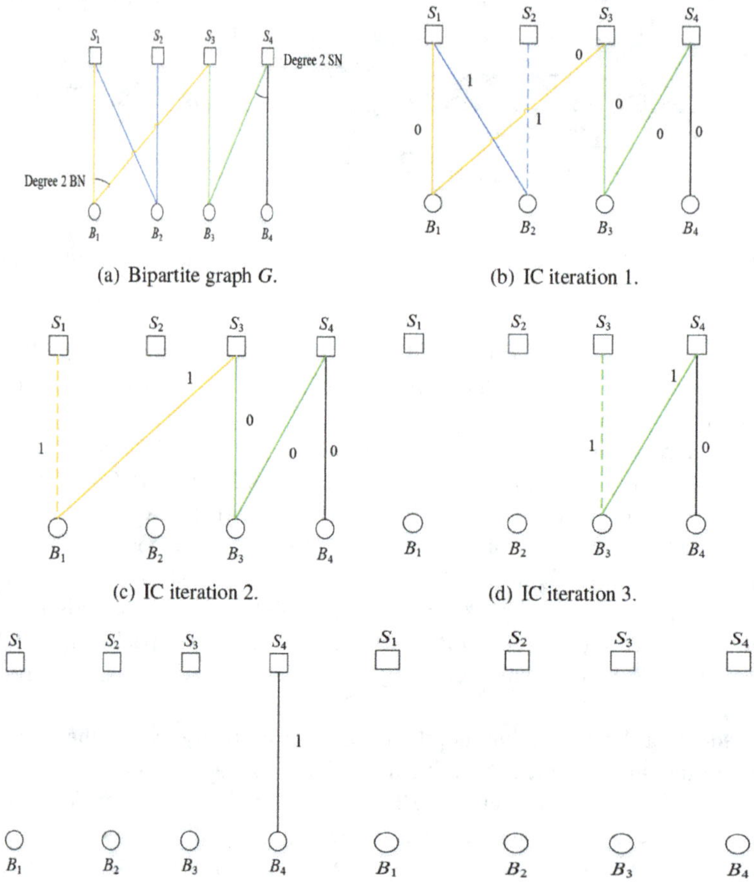

Figure 7.10 *Graph representation of the IC iterative process*

Figure 7.10(b). When a burst is recovered, the contribution of the corresponding burst and its replica can be revealed, and the iteration is carried out until all edges in the graph are revealed (as shown in Figure 7.10(b)–Figure 7.10(f)) or some remaining edges cannot be revealed. According to the conclusion made from Reference 38 that IRSA and CRDSA TCP almost achieved the same performance when they are adopted in satellite scenario. But IRSA is more complex than CRDSA because of the randomized and variable packet replicas' number [39]. Based on the work in Reference 40, which demonstrated that RA can benefit from NOMA scheme, the power domain was induced in IRSA strategy in Reference 41, thus, within a slot bursts may transmit with different transmission powers. This chapter formulated the density evolution analysis and optimized the degree distribution for different number of power levels. Simulation results reported from Reference 41 show that compared with IRSA scheme, the proposed IRSA-NOMA scheme has advantages in terms of a higher throughput with a given PLR, due to the power difference with the IRSA-NOMA scheme is larger than that with the common IRSA scheme.

7.3.3 Code slotted ALOHA

As opposed to the IRSA and CRDSA schemes only exploiting packet repeated and SIC in the framework of RA, with the CSA scheme, user packets can be encoded prior to transmission in the MAC frame. As shown in Reference 37, the burst of an active user in CSA is divided into k data segments prior to the transmission, all segments are the same length in bits. Then, these segments are encoded by the user via a packet-oriented linear block code generating n_h encoded segments. For each transmission, the (n_h, k) code is randomly selected from a finite code-book family C. These segments are then further encoded via a physical layer code before transmission over the multiple access channel. In CSA the physical layer protection is applied at the individual (packet) slice level instead of at the packet level [39].

On the receiver side, segments that are received in clean slices (i.e., segments not experiencing collisions) are first decoded at the physical layer. Once the clean signal is detected, its corresponding packet can be recovered and the information about the user is extracted, such as the code C_h adopted by the user and the positions of the other segments in the MAC frame are extracted. For each active user, the receiver becomes aware of, the maximum-a-posteriori erasure decoding of the code C_h is performed in order to recover as many encoded segments as possible for the user. This procedure is iterated until the maximum number of iterations is reached.

Simulated throughput results for CSA reported in Reference 37 show small superior to IRSA for rates lower than 1/3, while remarkably superior to IRSA for rates comprised between 1/3 and 1/2, even simple binary two-dimensional component codes are used. For rates larger than 1/2, the CSA protocol must rely on component codes with large enough dimensions and high enough coding rate. From the implementation point of view, CSA is a more complex RA scheme due to its associated signaling mechanism [39].

7.4 Summary

Multiple access is crucial for research and development for NGSO and future satellite communications. The multiple access in satellite needs to face up to a variety of challenges and requirements, such as high resource utilization efficiency and further increased access rate. This chapter begins with the three OMA schemes commonly used in satellite networks. Since the NOMA scheme can harmoniously integrate with OMA techniques in existing satellite architectures, the performance of several NOMA-based satellite models is analyzed. The formation and verification of this performance also verify the superiority of these access strategies used in satellite networks. Finally, this chapter reviews three enhanced ALOHA schemes (CRDSA, IRSA, and CSA) used in satellite networks and their principles, to show the efforts of scholars and researchers have done in novel multiple access for satellite networks.

References

[1] Lutz E, Werner M, Jahn A. *Satellite systems for personal and broadband communications*. Berlin, Heidelberg: Springer-Verlag; 2000. Available from http://link.springer.com/10.1007/978-3-642-59727-5

[2] Available from https://www.techopedia.com/definition/5089/time-division-multiple-access-tdma

[3] Saito Y., Kishiyama Y., Benjebbour A., Nakamura T., Li A., Higuchi K. 'Non-orthogonal multiple access (NOMA) for cellular future radio access'. *Proceedings of IEEE VTC'13*; Dresden, Germany, 2013.

[4] Alam M., Zhang Q. 'A survey: non-orthogonal multiple access with compressed sensing multiuser detection for MMTC'. *ArXiv: 1810.05422 [Eess. SP]*. n.d

[5] Nikopour H., Baligh H. 'Sparse code multiple access'. *IEEE 24th Annual International Symposium on Personal, Indoor and Mobile Radio Communications (PIMRC)*; London, IEEE, 2000. pp. 332–36.

[6] Alam M., Zhang Q. 'Performance study of SCMA codebook design'. *IEEE Wireless Communications and Networking Conference (WCNC)*; San Francisco, CA, IEEE, 2017. pp. 1–5.

[7] Alam M., Zhang Q. 'Designing optimum mother constellation and codebooks for SCMA'. *ICC 2017 - IEEE International Conference on Communications*; Paris, France, IEEE, 2017. pp. 1–6.

[8] Zeng M., Yadav A., Dobre O.A., Tsiropoulos G.I., Poor H.V. 'Capacity comparison between MIMO-NOMA and MIMO-OMA with multiple users in a cluster'. *IEEE Journal on Selected Areas in Communications*. 2017, vol. 35(10), pp. 2413–24.

[9] Cuevas E.G., Weerackody V. 'Technical characteristics and regulatory challenges of communications satellite earth stations on moving platforms'. *Johns Hopkins APL Technical Digest*. 2015, vol. 33(1), pp. 37–51.

[10] Gradshteyn I.S., Ryzhik I.M. *Table of Integrals, Series, and Products*. 7th ed. New York, NY: Academic Press; 2007.

[11] An K., Lin M., Ouyang J., Zhu W.-P. 'Secure transmission in cognitive satellite terrestrial networks'. *IEEE Journal on Selected Areas in Communications*. 2003, vol. 34(11), pp. 3025–37.

[12] An K., Lin M., Zhu W.-P., Huang Y., Zheng G. 'Outage performance of cognitive hybrid satellite–terrestrial networks with interference constraint'. *IEEE Transactions on Vehicular Technology*. 2003, vol. 65(11), pp. 9397–404.

[13] Guo K., An K., Zhang B. 'On the performance of the uplink satellite multiterrestrial relay networks with hardware impairments and interference'. *IEEE Systems Journal*. 2003, vol. 13(3), pp. 2297–308.

[14] Abdi A., Lau W.C., Alouini M., Kaveh M. 'A new simple model for land mobile satellite channels: first- and second-order statistics'. *IEEE Transactions on Wireless Communications*. 2003, vol. 2(3), pp. 519–28.

[15] Arti M.K. 'Two-way satellite relaying with estimated channel gains'. *IEEE Transactions on Communications*. 2003, vol. 64(7), pp. 2808–20.

[16] Zhang X., Gao Q., Gong C., Xu Z. 'User grouping and power allocation for NOMA visible light communication multi-cell networks'. *IEEE Communications Letters*. 2017, vol. 21(4), pp. 777–80.

[17] Arti M.K. 'Channel estimation and detection in hybrid satellite–terrestrial communication systems'. *IEEE Transactions on Vehicular Technology*. 2016, vol. 65(7), pp. 5764–71.

[18] Yan X., An K., Liang T., *et al.* 'The application of power-domain non-orthogonal multiple access in satellite communication networks'. *IEEE Access: Practical Innovations, Open Solutions*. 2019, vol. 7, pp. 63531–63539.

[19] Yan X., Xiao H., Wang C.-X., An K., Chronopoulos A.T., Zheng G. 'Performance analysis of NOMA-based land mobile satellite networks'. *IEEE Access: Practical Innovations, Open Solutions*. 2018, vol. 6, pp. 31327–39.

[20] Lu W., An K., Liang T. 'Robust beamforming design for sum secrecy rate maximization in multibeam satellite systems'. *IEEE Transactions on Aerospace and Electronic Systems*. 2019, vol. 55(3), pp. 1568–72.

[21] Caus M., Vázquez M., Pérez-Neira A. 'Noma and interference limited satellite scenarios'. *Proceedings of the IEEE Asilorma' 17*; Pacific Grove, CA, IEEE, 2017. pp. 1–5.

[22] Beigi N.A.K., Soleymani M.R. 'Interference management using cooperative noma in multi-beam satellite systems'. *IEEE International Conference on Communications (ICC 2018)*; Kansas City, MO, 2018.

[23] An K., Lin M., Ouyang J., Zhu W.-P. 'Secure transmission in cognitive satellite terrestrial networks'. *IEEE Journal on Selected Areas in Communications*. 2016, vol. 34(11), pp. 3025–37.

[24] An K., Liang T., Zheng G., Yan X., Li Y., Chatzinotas S. 'Performance limits of cognitive-uplink FSS and terrestrial FS for ka-band'. *IEEE Transactions on Aerospace and Electronic Systems*. 2019, vol. 55(5), pp. 2604–11.

[25] Yan X., Xiao H., Wang C.-X., An K. 'On the ergodic capacity of NOMA-based cognitive hybrid satellite terrestrial networks'. *IEEE/CIC International Conference on Communications in China (ICCC)*; Qingdao, China, IEEE, 2017. pp. 1–5.

[26] Ruan Y., Li Y., Wang C.-X., Zhang R. 'Energy efficient adaptive transmissions in integrated satellite-terrestrial networks with SER constraints'. *IEEE Transactions on Wireless Communications*. 2018, vol. 17(1), pp. 210–22.

[27] Zhu X., Jiang C., Kuang L., Ge N., Lu J. 'Non-orthogonal multiple access based integrated terrestrial-satellite networks'. *IEEE Journal on Selected Areas in Communications*. 2017, vol. 35(10), pp. 2253–67.

[28] Lin Z., Lin M., Wang J.-B., de Cola T., Wang J. 'Joint beamforming and power allocation for satellite-terrestrial integrated networks with non-orthogonal multiple access'. *IEEE Journal of Selected Topics in Signal Processing*. 2019, vol. 13(3), pp. 657–70.

[29] Ding Z., Peng M., Poor H.V. 'Cooperative non-orthogonal multiple access in 5G systems'. *IEEE Communications Letters*. 2015, vol. 19(8), pp. 1462–65.

[30] Yan X., Xiao H., Wang C.-X., An K. 'Outage performance of NOMA-based hybrid satellite-terrestrial relay networks'. *IEEE Wireless Communications Letters*. 2018, vol. 7(4), pp. 538–41.

[31] Yan X., Xiao H., An K., Zheng G., Tao W. 'Hybrid satellite terrestrial relay networks with cooperative non-orthogonal multiple access'. *IEEE Communications Letters*. 2018, vol. 22(5), pp. 978–81.

[32] Bai L., Zhu L., Zhang X., Zhang W., Yu Q. 'Multi-satellite relay transmission in 5G: concepts, techniques, and challenges'. *IEEE Network*. 2018, vol. 32(5), pp. 38–44.

[33] Clazzer F. 2017 Apr. 'Modern Random Access for Satellite Communications'. [PhD thesis]. Genova, University of Genoa

[34] Bai J., Ren G. 'Polarized MIMO slotted ALOHA random access scheme in satellite network'. *IEEE Access: Practical Innovations, Open Solutions*. 2017, vol. 5, pp. 26354–63.

[35] Lee M., Lee J.-K., Lee J.-J., Lim J. 'R-CRDSA: reservation-contention resolution diversity slotted ALOHA for satellite networks'. *IEEE Communications Letters*. 2012, vol. 16(10), pp. 1576–79.

[36] Liva G. 'Graph-based analysis and optimization of contention resolution diversity slotted ALOHA'. *IEEE Transactions on Communications*. 2011, vol. 59(2), pp. 477–487.

[37] Paolini E., Liva G., Chiani M. 'Coded slotted ALOHA: a graph-based method for uncoordinated multiple access'. *IEEE Transactions on Information Theory*. 2015, vol. 61(12), pp. 6815–6832.

[38] Celandroni N., Ferro E., Gotta A. 'RA and da satellite access schemes: a survey and some research results and challenges'. *International Journal of Communication Systems*. 2014, vol. 27(11), pp. 2670–2690.

[39] De Gaudenzi R., Del Rio Herrero O., Gallinaro G., Cioni S., Arapoglou P.-D. 'Random access schemes for satellite networks, from VSAT to M2M: a survey'. *International Journal of Satellite Communications and Networking*. 2018, vol. 36(1), pp. 66–107.

[40] Choi J. 'NOMA-based random access with multichannel ALOHA'. *IEEE Journal on Selected Areas in Communications*. 2017, vol. 35(12), pp. 2736–2743.

[41] Shao X., Sun Z., Yang M., Gu S., Guo Q. 'NOMA-based irregular repetition slotted ALOHA for satellite networks'. *IEEE Communications Letters*. 2019, vol. 23(4), pp. 624–627.

Chapter 8

Radio resource management for non-geostationary orbit

Xiaokai Zhang[1], Kang An[2], and Min Jia[3]

In this chapter, we introduce a potential game-based approach to implement collaborative user scheduling and power allocation in the uplink multi-beam non-geostationary satellite orbit (NGSO) system. First of all, a framework of the multi-beam uplink NGSO system is proposed, where full frequency reuse and co-channel interference among different spot beams are considered. To provide broadband NGSO service effectively, we formulate an initial optimization problem of maximizing uplink sum transmission capacity and then transform it into an interference mitigation problem to address the mathematical intractability. Specifically, a game-theoretic model is implemented to solve the transformed optimization problem, which is proved to be a potential game and existence of Nash equilibriums (NEs). Moreover, an iterative algorithm with low computational complexity, motivated by the finite improvement property, is designed to implement collaborative user scheduling and power allocation to the NE point. Finally, the simulation results prove the convergence and effectiveness of the proposed potential game-based approach.

8.1 Introduction

In recent years, the Internet over satellite has become a worldwide network based on standard communication protocols, helping to realize communication anything, anyone, anytime, and anyplace [1], which greatly facilitates our daily life and provides more intelligent services [2, 3]. The number and type of satellite devices have achieved tremendous growth. Thanks to improvements in the launch technologies as well as miniaturization, satellites could be effectively implemented for distributed control and automation in a smart grid, environmental monitoring, and emergency management scenarios [4–6]. The NGSO system offers truly global coverage, which contains tens of thousands of satellites and serves a huge number of devices, users,

[1]College of Communications and Engineering, Army Engineering University, Nanjing, China
[2]Sixty-third Research Institute, National University of Defense Technology, Nanjing, China
[3]Communication Research Center, School of Electronics and Information Engineering, Harbin Institute of Technology, Harbin, China

and/or objects [7–11]. There are some famous NGSO systems, such as OneWeb, Starlink, etc.

Radio resource allocation is one of the fundamental problems for the NGSO system. In such a system, there exists satellite-to-satellite links in space. Besides, one satellite generally serves a huge number of remote satellite-based Internet objects worldwide. Therefore, the spectrum efficiency of the satellite is a significant performance criterion [12]. The multi-beam satellite networks, oriented to increase the spectrum efficiency, have attracted enormous attention in the NGSO system [13–16], which allows us to reuse the available bandwidth sufficiently separated beams. However, the frequency reuse schemes cannot dramatically increase the channel capacity owing to the interval of interference protection. There exists considerable co-channel interference when the coverage area overlaps among the adjacent beams, particularly for the beam-edge users in the full frequency band reuse case [17]. Therefore, co-channel interference mitigation or interference alleviation is pivotal to further improve the performance of the whole system.

The crucial applications in the NGSO system are offering a wide range of backhaul links and data offloading solutions for huge number of user devices [18]. To satisfy the diverse traffic demands of all user devices, a multi-beam NGSO system must provide flexibility and efficient exploitation of the available resources for one satellite. Hence, proper and effective radio resource allocation in uplink is essential to increase the capacity of the whole multi-beam NGSO system [19]. However, most of the existing works about radio resource allocation mechanisms mainly focus on downlink scenarios [6, 20–23], which take into consideration the co-channel interference and the satellite power consumption in the downlink transmission as a priority. Besides, the user scheduling and power allocation are coupled with each other. But most radio resources only concentrate on one aspect [24], either channel allocation or power allocation, which achieves limited resource utilized improvement.

The existing papers for uplink scenarios mainly focus on a single beam satellite system or orthogonal frequency reuse [25], where there is no co-channel interference, the users can be orthogonally scheduled, and the power allocation problem can be solved by a convex optimization approach, such as a water-filling algorithm. Actually, fixed radio resources are allocated to each beam in most currently applied multi-beam NGSO systems [6, 26]. The fixed allocation scheme would inevitably result in the mismatch between traffic demand and allocated capacity [27, 28]. Also, to further improve the spectrum efficiency, the total bandwidth is reused by all beams, i.e. the frequency reuse factor is equal to 1 (worst-case scenario). There exists considerable co-channel interference, especially on the beam edge. As the user scheduling is discrete while the power strategy space is continuous, obtaining the maximum capacity becomes the non-convex non-deterministic polynomial-time hardness (NP)-hard problem [29]. To solve the non-convex radio resource allocation problem in uplink multi-beam satellite communication, some heuristic algorithms have been proposed [29]. However, heuristic algorithms cannot guarantee that the globally optimal solution is found, their performance is heavily affected by the optimization parameters. Besides, the computational complexity is substantial to find nearly optimal solutions, especially for the onboard processing satellite systems with limited computing resources.

Game theory shows clear interactional decision-making progress, which has been extensively investigated in wireless radio resource allocation [30]. Recently, game-theoretic approaches have been applied in satellite networks, which are more adaptable to users joining and leaving the systems and are robust for network dynamics [30]. Nevertheless, the state-of-the-art works only concentrate on downlink power allocation from a game-theoretic perspective. Unlike static resource allocation in Reference 24, the authors in Reference 31 propose a dynamic Stackelberg game model to maximize cost-efficiency which describes the relationship between the satellite system's profit, interference pricing, and user's power allocation. Moreover, the essential discussions and proofs for the pricing rationality are provided in Reference 32. However, to the best of our knowledge, there is no prior work focusing on resource allocation in the uplink NGSO system from a game theory perspective. Therefore, a collaborative user scheduling and power allocation scheme is of urgent interest to improve the spectrum efficiency of the uplink multi-beam NGSO system through the game-theoretic view.

In Reference 5, the author proposed to allocate more radio resources to the bottleneck user in poor channel conditions to satisfy the Quality-of-Service (QoS) requirement, which leads to inefficient resource utilization for the overall system. Therefore, we consider that the users prefer to participate in resource allocation as long as there is existence of transmission demand and leave the system when the information transmission has ended. In this way, more information will be transmitted for non-real-time services over a long period, thereby improving resource utilization. The designed approach should be able to meet the needs of equipment to join and leave the systems and be robust for network dynamics.

In this chapter, we propose a potential game-based approach to implement uplink collaborative user scheduling and power allocation in a multi-beam NGSO system. The main contributions of our work are summarized as follows:

- First of all, a framework for a multi-beam uplink NGSO system is proposed, where full frequency reuse and co-channel interference among different spot beams are considered. To obtain the maximum uplink sum capacity of the multi-beam satellite, we transform the no-concave NP-hard optimization problem into minimizing the sum of the co-channel interference mitigation problem, where two optimization problems have the same monotonicity under strategy profile change.
- To address the mathematical intractability, we then propose a game-theoretic model to solve the minimum sum of interference mitigation problem, and we prove the proposed game is a potential game and the existence of NE, where the NE is either the global or local optimal solutions.
- Subsequently, an iterative algorithm, motivated by the finite improvement property, is designed to implement collaborative user scheduling and power allocation to obtain an NE point. The complexity of the proposed algorithm can be solved by polynomial time and the convergence of the proposed algorithm

is proved to be guaranteed. The convergence and effectiveness of the potential game-based approach are verified via simulations. Besides, we analyze the fairness of the proposed algorithm by using Jain's fair index (JFI).

The rest of this chapter is organized as follows. Section 8.2 introduces the multi-beam satellite uplink system model and reformulates the resource allocation problem. In section 8.3, a game-theoretic approach is proposed and the existence of NE is analyzed. In addition, an iterative algorithm is designed to implement collaborative power allocation and user scheduling in section 8.4. Section 8.5 shows simulation results. Finally, conclusions are drawn in section 8.6.

8.2 Problem statement

8.2.1 System model

For the multi-beam NGSO system, the satellite illuminates the vast coverage area with spot beams, which is illustrated in Figure 8.1. In this model, a frequency reuse factor $R_f = 1$ is considered to improve the spectrum efficiency, which means all channels are reused in one spot beam. The set $\mathscr{L}$ is denoted as L spot beams. Defining the user set as $\mathscr{N}$, the subset of the users, located at the $l-th$ spot beam coverage area, is presented by $\mathscr{N}_l$, where $\mathscr{N}_l \subseteq \mathscr{N}$ and $\cup \mathscr{N}_l = \mathscr{N}$. One user in $l-th$ spot beam is denoted by n_l, where $n_l \in \mathscr{N}_l$. We assume that all users are equipped with a very small aperture terminal. The users in the same spot beam utilize the frequency division multiple access (FDMA) to avoid inter-beam interference [5]. We consider that the total available bandwidth is B_{tot}, including K subchannels and denoted by $\Omega = \{1, 2, \cdots, K\}$. It is worth noting that each subchannel is assigned to one user in one spot beam, but one uplink user may occupy several subchannels, for example, using carrier aggregation technology.

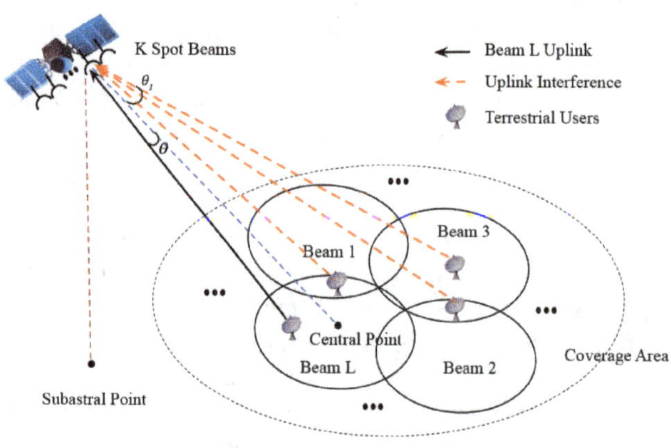

Figure 8.1 L spot beams uplink system model

The transmitted symbol from user n_i to the spot beam i on subchannel k is denoted by $x_{n_i,k}$. Therefore, the received signal in the *l-th* spot beam on the sub-channel k can be expressed as

$$y = \sum_{i=1}^{L} \sqrt{P_{n_i,k} G_{n_i,l,k}} x_{n_i,k} + n_0 \tag{8.1}$$

where $P_{n_l,k}$ is the transmitted power of user n_l in subchannel k, $\sqrt{P_{n_l,k} G_{n_l,l}} x_{n_l,k}$ denotes the signal coming from user n_l in subchannel k, $\sum_{i=1,i\neq l}^{L} \sqrt{P_{n_i,k} G_{n_i,l}} x_{n_l,k}$ is the sum of the interference from other beams coverage users on $k-th$ subchannel, including side lobes interference, and n_0 is the additional white Gaussian noise with the power of σ^2. We consider that the large scale of fading of the different sub-channels is homogenization, and the small scale of fading is heterogeneous among different users and subchannels. Moreover, $G_{n_i,l,k}$ denotes the channel gain, which responds from the $i-th$ beam coverage user n_i to spot beam l onboard receiver on $k-th$ subchannel. $G_{n_i,l,k}$ includes free-space path loss and antenna gain and the small scale of fading, etc. $G_{n_i,l,k}$ is given by Li *et al.* [31]

$$G_{n_i,l,k} = \frac{g_{n_i}(\varepsilon)\, g_l(\theta) f_\varepsilon \left|h_{n_i,l,k}\right|^2}{(4\pi d/\lambda)^2} \tag{8.2}$$

where ε denotes the elevation angle from n_i to the satellite, d is the straight-line distance from n_i to the satellite, λ denotes the wavelength, f_ε denotes the other loss of n_i at the direction ε, and $h_{n_i,l,k}$ denotes the shadowing and channel fading of the satellite link.

The $l-th$ beam satellite antenna gain g_l is given by Lu *et al.* [13]

$$g_l\left(\theta_{n_i,l}\right)[\text{dBi}] = G_{s,\max}\left(\frac{J_1\left(u_{n_i,l}\right)}{2u_{n_i,l}} + 36\frac{J_3\left(u_{n_i,l}\right)}{u_{n_i,l}^3}\right)^2 \tag{8.3}$$

where $u_{n_i,l} = 2.07123 \sin\left(\theta_{n_i,l}\right) / \sin\left(\theta_{3dB}\right)$, J_m is the first kind of Bessel's function of order m, $G_{s,\max}$ is the maximum antenna gain of the satellite, θ_{3dB} is the half-power beamwidth angle, and $\theta_{n_i,l}$ represents the angular position of the user n_i from the *l-th* beam, which determines by the coverage area center point, subastral point, and the satellite orbital altitude.

As illustrated in Figure 8.2, the oblique projection in 3D space is taken into account. θ describes the deviation angle between user and beam center that can be expressed as [31]

$$\theta = \arccos\left(\frac{d^2 + d_{sc}^2 - 2R^2\left(1 - \cos\frac{d_{cu}}{R}\right)}{2d_{sc}d_{su}}\right) \tag{8.4}$$

where d_{sc} denotes the straight-line distance from the beam center to the satellite as shown in Figure 8.2, d_{pu} is the on-Earth distance between the user and the subastral point, d_{cu} is the Earth surface distance from the user to the beam center, d_{pu} represents the Earth surface distance between the user and the subastral point, h denotes

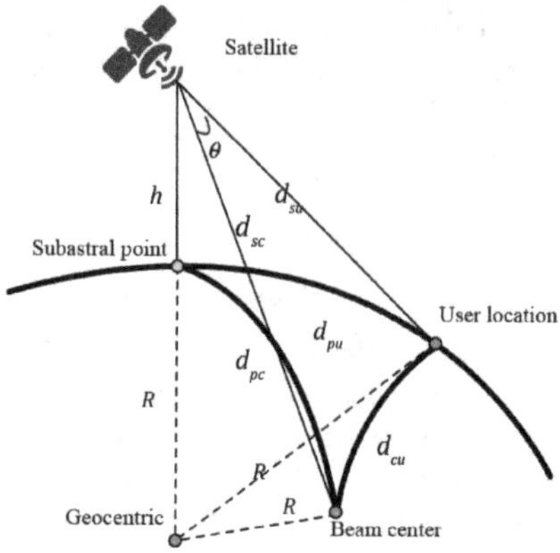

Figure 8.2 Orientation angle with oblique projector in 3D space

the height of the satellite, and R means radius of the Earth. d_{su} and d_{sc} can be calculated by the Pythagorean theorem within h, d_{pu}, and d_{pc} [31].

We consider that the users are equipped with a small parabolic antenna. According to ITU-R S.465 [25], the antenna pattern of the user is represented by

$$g_{n_i}(\varepsilon)\,[\mathrm{dBi}] = \begin{cases} G_{t,\max}, & 0^o < \varepsilon \le 1^o \\ 32 - 25\log\varepsilon, & 1^o < \varepsilon \le 48^o \\ -10, & 48^o < \varepsilon \le 180^o \end{cases} \tag{8.5}$$

where $G_{t,\max}$ denotes the maximal terrestrial antenna gain of the main lobe. The off-boresight angle ε of the user can be calculated by $\varepsilon = \arccos(\cos(\phi)\cos(\varphi))$, where ϕ denotes the horizon offset angle of the beam center, and φ denotes the vertical offset angle of the beam center.

Dissimilar to the conventional channel allocation schemes with the same fading through the entire bandwidth [33], we consider each subchannel follows independently identically distribution fading. The small scale of satellite channel fading is assumed to follow a shadowed Rician fading model [33–35], which is mathematically tractable and has been widely applied in various fixed and mobile satellite services for a variety of frequency bands, such as the Ultra-High Frequency (UHF)-band, L-band, S-band, and Ka-band. The probability density function of $|h|^2$ is shown as [35]

$$f_{|h|^2}(x) = \alpha\exp(-\beta x)\,{}_1F_1(m, 1, \delta x) \tag{8.6}$$

where ${}_1F_1(\cdot,\cdot,\cdot)$ denotes the confluent hypergeometric function and $\alpha = \frac{(2bm)^m}{2b(2bm+\zeta)^m}$, $\delta = \frac{\zeta}{2b(2bm+\zeta)}$, and $\beta = \frac{1}{2b}$. Besides, $2b$ is the average power of the scatter component, ζ denotes the average power of the line-of-sight component, and m is the Nakagami fading parameter. The small scale of fading follow quasi-static block fading.

8.2.2 Interference mitigation problem formulation

A scheduling vector $\mathbf{S}_k = [n_1, n_2, \cdots, n_l, \cdots, n_L]$ is utilized to represent users simultaneously scheduled across all spot beams at the $k-th$ subchannel, where $[\mathbf{S}_k]_l = n_l$. Thus, the scheduling strategy space is given by $\Lambda = \{\mathbf{S}|n_l \in \mathcal{N}_l, \forall l = 1, 2, \cdots, L\}$, where $\mathbf{S} = [\mathbf{S}_1, \cdots \mathbf{S}_k, \cdots \mathbf{S}_K]$. A transmit power vector contains the transmit power values of each scheduled user on $k-th$ subchannel, $\mathbf{P}_k = [P_{n_1,k}, \cdots, P_{n_l,k}, \cdots, P_{n_L,k}]$, where $[\mathbf{P}_k]_l = P_{n_l,k}$. If the $[\mathbf{S}_k]_l$ is null, the corresponding power allocation is zero. Furthermore, the maximum transmitting power for n_l of all subchannels is given by $p_{n_l}^{\max}$. The power strategy space is $\Xi = \left\{\mathbf{P}|0 \le \sum_\Omega p_{n_l,k} \le p_{n_l}^{\max}, \forall l = 1, 2, \cdots, L\right\}$, where $\mathbf{P} = [\mathbf{P}_1, \cdots \mathbf{P}_k, \cdots \mathbf{P}_K]$.

The Signal to Interference and Noise Ratio (SINR), experienced by the user n_l in spot beam l at the $k-th$ subchannel, is given by

$$SINR_{n_l,k} = \frac{P_{n_l,k} G_{n_l,l,k}}{\sum_{i=1,i\neq l}^{L} P_{n_i,k} G_{n_i,l,k} + \sigma^2} \tag{8.7}$$

For the NGSO system, there exists a large amount of non-real-time services, such as data offloading. If we directly implement the approaches toward fixed QoS requirements for the non-real-time services, more resources will be allocated to these bottleneck users in the poor channel condition, which is inefficient in resource utilization. Therefore, we consider that the user would participate in the resource allocation as long as there is existence of transmission demand, and leave the system when the information transmission has ended. In this way, more information will be transmitted for non-real-time services over a long period, thereby improving resource utilization. The designed approach can satisfy the requirements of equipment for joining and leaving the systems and be robust for network dynamics. To facilitate the above assumption, one of the conventional utilities for radio resource allocation is to maximize the sum transmission capacity.

The corresponding weight sum capacity of beam l coverage users is given by the following Shannon's formula

$$C_l(\mathbf{S}, \mathbf{P}) = \sum_\Omega \alpha_{n_l,k} \log_2 \left(1 + SINR_{n_l,k}\right) \tag{8.8}$$

where $\alpha_{n_l,k}$ is a weight coefficient, which denotes the priority of different users. In a multi-beam satellite communication system, a different service type of user owns the unique capacity requirement, specific task, and service priority. Therefore, to obtain the maximum weighted capacity of the whole system, the optimization problem can be formalized as

$$\arg\max_{\mathbf{S}\in\Lambda,\mathbf{P}\in\Xi} \sum_{l=1}^{L} C_l(\mathbf{S}, \mathbf{P}) \tag{8.9a}$$

$$s.t. \quad 0 \le \sum_\Omega p_{n_l,k} \le p_{n_l}^{\max} \tag{8.9b}$$

The optimization of (8.9a) is a non-convex NP-hard problem. Besides, the power allocation and user scheduling are coupled with each other. The set of user scheduling strategy space is discrete while the power strategy space is continuous, which cannot be solved directly.

From another perspective, the main reason hindering the increase of the system sum capacity is co-channel interference. Hence, interference mitigation or interference alleviation would increase the sum capacity [36]. Furthermore, we consider maximizing the negative weight sum of the inverse SINR as the performance metric of interference mitigation. The optimization problem is transferred as

$$\arg\max_{S\in\Lambda, P\in\Xi} \sum\nolimits_{l=1}^{L} \sum\nolimits_{\Omega} -w_{n_l,k} / \mathrm{SINR}_{n_l,k} \tag{8.10a}$$

$$s.t. \quad 0 \le \sum\nolimits_{\Omega} p_{n_l,k} \le p_{n_l}^{\max} \tag{8.10b}$$

where the coefficient $w_{n_l,k}$ denotes the positive weighted coefficient of different users, which is the same as the corresponding weight coefficient $\alpha_{n_l,k}$. The reasons for transforming the sum transmission capacity optimization problem (8.9a) to the proposed optimization problem (8.10a) exist two folds. The first is from the perspective of function monotonicity. The function of (8.9a) and (8.10a) have the same monotonicity with the sum weighted SINR, i.e. $\sum_{\Omega} \sum_{l=1}^{L} \alpha_{n_l,k} \mathrm{SINR}_{n_l,k}$, which is an increasing function of the $\mathrm{SINR}_{n_l,k}$. Therefore, the optimization problem (8.9a) and (8.10a) have the same monotonicity under strategy profile changes. Another reason is from the perspective of the physical meaning. The minimization of sum inverse SINR means reducing the co-channel interference, i.e. interference mitigation or interference alleviation, which can increase the sum transmission capacity.

However, the optimization problem (8.10a) is still an NP-hard problem. The burden of the central station requires advanced algorithms with reduced complexity in handling such a large scale of coupled discrete and continuous variables. In order to solve the problem (8.10a) in an efficient manner, we propose the game-theoretic approach to obtain the optimal solutions.

8.3 A game-theoretic approach for radio resource allocation

8.3.1 Game model

To reduce the calculation complexity for the proposed interference mitigation problem (8.10a), a game-theoretic approach is formulated to construct an analytical paradigm, which aims to address the mathematical intractability.

The game is denoted by $G = \lfloor \mathcal{L}, \{\mathcal{A}_l\}_{l\in\mathcal{L}}, \{u_l\}_{l\in\mathcal{L}} \rfloor$. $\mathcal{L} = \{1, 2, \cdots, L\}$ represents the set of game players. One subchannel only schedules one user at each spot beam, which means no intra-beam interference. Therefore, the game players can be all users in one spot beam, which could set a virtual agent as a player during the game. $\mathcal{A}_l$ is the actions of player l, and u_l is the utility function of player l, which requires careful design to obtain better results. In order to avoid the "tragedy of the commons" [13], co-channel interference is required to be considered as the penalty. Hence, the utility function of u_l is designed as (8.13) on the next page, where $A_l \in \mathcal{A}_l$

is the joint strategy of player l, $A_l = \mathbf{S_k} \times \mathbf{P_k}$, and a strategy profile of all the players excluding l is denoted by $A_{-l} \in \mathscr{A}_1 \times \cdots \mathscr{A}_{l-1} \times \mathscr{A}_{l+1} \times \mathscr{A}_L$, where $\times$ denotes the Cartesian product.

8.3.2 The existence of NE

To find stable and optimal or sub-optimal solutions for interactional game processing, we have to find out whether the equilibrium of the proposed game is existing. In addition, the relationship between the proposed game and the optimization problem (8.10a) requires to be clarified.

Definition 1: a profile $A^* = \left(A_1^*, \cdots, A_L^*\right)$ is a pure strategy NE point of a *game* if and only if no player can improve its utility by deviating unilaterally [37], i.e.

$$u_l\left(A_l^*, A_{-l}^*\right) \geq u_l\left(A'_l, A_{-l}^*\right), \forall A'_l \in \mathscr{A}_l \backslash \left\{A_l^*\right\} \tag{8.11}$$

Theorem 1: the proposed G is a potential game. And, the pure strategy NE point is the global or local optimal solution of the potential function.

Proof: we address the potential function for the game G as

$$F\left(A_l, A_{-l}\right) = -\sum_{\Omega} \sum_{l=1}^{L} w_{n_l,k} / \mathrm{SINR}_{n_l,k} \tag{8.12}$$

that is, the objection of the optimization problem (8.10a). The utility function of l-th user is designed as (8.13).

$$u_l\left(A_l, A_{-l}\right) = -\sum_{\Omega}\left(w_{n_l,k} \frac{\sum\limits_{i=1, i\neq l}^{L} P_{n_i,k} G_{n_i,l,k} + \sigma^2}{P_{n_l,k} G_{n_l,l,k}} + \sum_{i=1, i\neq l}^{L} \frac{w_{n_i,k} P_{n_l,k} G_{n_l,l,k}}{P_{n_i,k} G_{n_i,i}} \right) \tag{8.13}$$

$$\begin{aligned}
F\left(A_l, A_{-l}\right) &= -\sum_{\Omega} \sum_{i=1}^{L} w_{n_l,k} \frac{\sum\limits_{i=1, i\neq l}^{L} P_{n_i,k} G_{n_i,l,k} + \sigma^2}{P_{n_l,k} G_{n_l,l,k}} \\
&= -\sum_{\Omega} w_{n_l,k} \frac{\sum\limits_{i=1, i\neq l}^{L} P_{n_i,k} G_{n_i,l,k} + \sigma^2}{P_{n_l,k} G_{n_l,l,k}} - \sum_{\Omega} \sum_{j=1, j\neq l}^{L} w_{n_j,k} \frac{\sum\limits_{i=1, i\neq j}^{L} P_{n_i,k} G_{n_j,i,k} + \sigma^2}{P_{n_j,k} G_{n_j,j,k}} \\
&= -\sum_{\Omega} w_{n_l,k} \frac{\sum\limits_{i=1, i\neq l}^{L} P_{n_i,k} G_{n_i,l,k} + \sigma^2}{P_{n_l,k} G_{n_l,l,k}} - \sum_{\Omega} \sum_{j=1, j\neq l}^{L} w_{n_j,k} \frac{P_{n_l,k} G_{n_j,l,k}}{P_{n_j,k} G_{n_j,j,k}} \\
&\quad \underbrace{- \sum_{\Omega} \sum_{j=1, j\neq l}^{L} w_{n_j,k} \frac{\sum\limits_{i=1, i\neq j, i\neq l}^{L} P_{n_i,k} G_{n_j,i,k} + \sigma^2}{P_{n_j,k} G_{n_j,j,k}}}_{v(A_{-l})} \\
&= u_l\left(A_l, A_{-l}\right) - v\left(A_{-l}\right)
\end{aligned} \tag{8.14}$$

It can be seen that the utility function for a user is made of two parts: the weight inverse-SINR of the user at it's receive node and the weight interference caused by the user to all the other users in the same system. Hence, a user benefits by reducing the interference caused to the other users in addition to reducing the interference at its receiver. In this way, each user's utility function incorporates a measure of the influence of its actions on the other users. Therefore, the utility function avoids the "tragedy of the commons". Besides, the potential function can be derived as (8.14), where $v\left(A_{-l}\right)$ is a non-contributing term for the player l. Thus, if player l (any player) turns its strategy from A_l to A'_l, we get

$$F\left(A'_l, A_{-l}\right) = u_l\left(A'_l, A_{-l}\right) - v\left(A_{-l}\right) \tag{8.15}$$

Then according to formulas (8.13) and (8.14), we obtain

$$F\left(A'_l, A_{-l}\right) - F\left(A_l, A_{-l}\right) = u_l\left(A'_l, A_{-l}\right) - u_l\left(A_l, A_{-l}\right) \tag{8.16}$$

Formula (8.16) indicates that the change of the $F\left(A_l, A_{-l}\right)$ is equal to the change of the personal utility function caused by any unilateral deviation of the player, which means that the proposed G is a potential game by the definition of Reference 37. According to the inherent metric of potential games, there exists at least one pure NE point, and all NE points are the global or local optimal solution of the potential function $F\left(A_l, A_{-l}\right)$ [38]. Therefore, the NE of the proposed potential game-based approach is the solution to the optimization problem of (8.10a). Therefore, Theorem 1 is proved.

If the $F\left(A_l, A_{-l}\right)$ is a measure of global network performance, the proposed game provides a framework, where the total utility of the whole network benefit so that each player obtains maximum own interests.

Theorem 2: the total number of subchannels is no less than the total number of users, i.e. $K \geq |\mathcal{N}|$, the global optimal potential function solution is by the best response orthogonal user scheduling and implementing the maximum transmission power.

Proof: if the number of subchannels is no less than the number of users with $K \geq |\mathcal{N}|$, the orthogonal user scheduling can be conducted with no interference, the potential function is given by

$$F\left(A_l^o, A_{-l}^o\right) = -\sum\nolimits_{\Omega} \sum_{i=1}^{L} \frac{w_{n_l,k}\sigma^2}{p_{n_l} G_{n_l,l,k}} \tag{8.17}$$

Now we assume that user scheduling is non-orthogonal, i.e. there exist at least two users n_q and n_m applying the same subchannel k. The (8.17) on the last page indicates that the orthogonal user scheduling is superior to the non-orthogonal one for potential function $F\left(A_l, A_{-l}\right)$. In addition, the orthogonal user scheduling potential function $F\left(A_l, A_{-l}\right)$, i.e. (8.16), increases monotonically with the transmission power. Therefore, the maximum transmission power of the user is implemented. The user scheduling strategy applies the best response by the $w_{n_l,k}$ and $G_{n_l,l,k}$ would obtain the global optimal potential function. Therefore, Theorem 2 is proved.

$$\begin{aligned} F\left(A_l^n, A_{-l}^n\right) = -\sum\nolimits_{\Omega} \Bigg[\sum_{l=1, l\neq q, l\neq m}^{L} \frac{w_{n_l,k}\sigma^2}{p_{n_l} G_{n_l,l,k}} + w_{n_q,k} \frac{p_{n_m} G_{n_m,m,k} + \sigma^2}{p_{n_q} G_{n_q,q,k}} \\ + w_{n_m,k} \frac{p_{n_q} G_{n_q,q,k} + \sigma^2}{p_{n_m} G_{n_m,m,k}} \Bigg] < -\sum\nolimits_{\Omega} \sum_{l=1}^{L} \frac{w_{n_l,k}\sigma^2}{p_{n_l} G_{n_l,l,k}} = F\left(A_l^o, A_{-l}^o\right) \end{aligned} \tag{8.18}$$

Theorem 2 can be seen as a special case of the proposed potential game-based approach. Due to theoretical proof of the effectiveness of the proposed approach for the scenario, the total number of subchannels is no less than the total number of users. We mainly investigate the scenario that the total number of subchannels is less than the total number of users in the following work, which is more practical in the real system.

8.4 NE iterative algorithm and implementation

In this section, an iterative algorithm, executed before the start of each transmission interval, is proposed to solve the optimization problem (8.10a). The collaborative user scheduling and power allocation are updated through iteration for each game player.

8.4.1 The proposed iterative algorithm

Since a couple of the discrete user schedules and continuous power allocation, it is challenging to achieve a 2D strategy, user scheduling and power allocation, for each iteration at the same time in an efficient way. Therefore, a decomposed iterative process is stated in Algorithm 1, where the maximization utility function of each player can contribute to improving the global network performance in light of the potential game's inherent property. Owing to the discrete size of the user scheduling strategy space, the selection of the best user scheduling strategy is the best response. Besides, to obtain the optimal power allocation strategy, the partial derivative of the utility function for $P_{n_l,k}$ equalize to zero. We get

$$\frac{\partial [u_l]_k}{\partial P_{n_l,k}} = \frac{\sum_{i=1,i\neq l}^{L} P_{n_i,k} G_{n_i,l,k} + \sigma^2}{P_{n_l,k}^2 G_{n_l,l,k} \Big/ w_{n_l,k}} - \sum_{i=1,i\neq l}^{L} \frac{w_{n_i,k} G_{n_l,l}}{P_{n_i,k} G_{n_i,l,k}} \tag{8.19}$$

So, the optimal power strategy is

$$P_{n_l,k}^* = \sqrt{\frac{w_{n_l,k}\left(\sum_{i=1,i\neq l}^{L} P_{n_i,k} G_{n_i,l,k} + \sigma^2\right)}{G_{n_l,l,k} \sum_{i=1,i\neq l}^{L} \frac{w_{n_i,k} G_{n_l,i,k}}{P_{n_i,k} G_{n_i,i,k}}}} \tag{8.20}$$

No decrease in system utility is guaranteed for each step of different subchannels in Algorithm 1. To be noticed, each iteration for each subchannel has to consider the power constraint for each user. The remaining power on $k-th$ subchannel of user n_l is

$$P'_{n_l,k} = P_{n_l}^{\max} - \sum_{k' \in \Omega, k' \neq k} P_{n_l,k'}^{t-1} \tag{8.21}$$

The condition for the end of the loop in Algorithm 1 is $A^t \approx A^{t-1}$, which means the scheduling strategy of A^{t-1} is almost same as A^t and system performance improvement for each step is less than 0.001.

According to Algorithm 1, the computation complexity of the proposed approach is on the scale of $O(N)$. Therefore, the execution time can be acceptable, even in the onboard processing satellite. The essence of the proposed potential game-based approach is to transform the NP-hard problem into a distributed individual utility improvement problem to address the mathematical intractability. Besides, the best response for each iteration guarantees that the proposed NE is an acceptable result.

8.4.2 The practical implementation

Moreover, the implementation of the algorithm can be executed in the ground gateway for a transparent forward satellite or onboard processing satellite, where the game player is all users in one spot beam, which can be virtualized during the calculation. The channel state information, users' location, and user transmission strategy profiles can be stored in the satellite database, which can be updated frequently. The convergent user scheduling and power allocation strategy profiles can be broadcasted to the users to implement in the uplink.

Algorithm 1 Collaborative user scheduling and power allocation iterative algorithm

Initialization:

Set t=1, a random user scheduling strategy profile is implemented, and assign a random initial power to the corresponding user in each subchannel.

Iteration Process:

repeat

t++;

for subchannel $k = 1$ to K **do**

1. User Scheduling: Keep the power strategy profile unchanged in the last iteration, the user scheduling strategies apply the best response according to the utility function for each channel, i.e.

$$[\mathbf{S_k}]_l^t = \underset{[\mathbf{S_k}]_l \in N_l, \mathbf{P}=\mathbf{P}^{t-1}}{\arg\max} \; u_{l,k}\left([\mathbf{S_k}]_l, [\mathbf{S_k}]_{-l}^{t-1}\right).$$

where $u_{l,k}$ is the utility function for n_l on subchannl k, denoting by

$$u_{l,k}(A_l, A_{-l}) = -\frac{\sum_{i=1,i\neq l}^{L} P_{n_i,k} G_{n_i,l,k} + \sigma^2}{P_{n_l,k} G_{n_l,l,k} / w_{n_l,k}} - \sum_{i=1,i\neq l}^{L} \frac{w_{n_i,k} P_{n_l,k} G_{n_l,l,k}}{P_{n_i,k} G_{n_i,i}}.$$

Specially, the power $P_{n_l,k}$ is $\min(P'_{n_l,k}, P_{n_l,k}^{t-1})$ for $l \in (1, \cdots, L)$, where $P'_{n_l,k}$ is defined by equation (1.20).

2. Power allocation: Keep the user scheduling strategy profile at the last step **User Scheduling**, updating its power strategy profile in each subchannel according to

$$P_{n_l,k}^{*} = \underset{P_{n_l,k} \in \Xi, \mathbf{S}=\mathbf{S}^t}{\arg\max} \; u_{l,k}\left(P_{n_l,k}, \mathbf{P}_{-n_l,k}^{t-1}\right),$$

$$P_{n_l,k}^{t} \leftarrow P_{n_l,k}^{t-1} + \delta\left[\min\left(P_{n_l,k}^{*}, P'_{n_l,k}\right) - P_{n_l,k}^{t-1}\right],$$

where $\delta > 0$ is iteration step size, $P_{n_l}^{*}$ is defined by (1.19).

end for

until $A^t \approx A^{t-1}$

Besides, the coefficient $w_{n_l,k}$ defines the weight of the scheduled user in the $l-th$ beam on subchannel k. For simplicity, we consider the weight coefficient is

$$w_{n_l,k} = \frac{G_{n_l,l,k}}{\sum_{i=1}^{L} G_{n_i,l,k}} \tag{8.22}$$

which can be seen as a trade-off between the fairness and sum capacity of the proposed system.

Remark 1: The proposed potential game-based user scheduling and power allocation approach can also be implemented in FDMA or orthogonal frequency division multiplexing based multi-beam satellite uplink scenario. For the Digital video broadcasting Second Generation Satellite (DVB-S2) standard, the uplink scenario is the forward link (i.e. the link from the ground station to the actuators) [39]. If there exist multiple ground stations in different satellite beam coverage areas with multiple sub-channels, the proposed approach can be directly implemented. For Digital video broadcasting Second Generation Return Channel Satellite (DVB-RCS2) standard [40], the proposed can be implemented for different return links of users in one time slot, where the requirement of the users is sum transmission capacity.

8.4.3 Convergence analysis and significance

For potential games, the improvement in an individual's utility leads to the improvement of network performance until stability. According to the finite improvement property [30, 37], every maximal improvement path must terminate in an NE point, where any pure strategy NE point maximizes the potential function either globally or locally [38]. In the proposed algorithm, the user scheduling and power allocation strategy updates are decomposed into two steps, which can ensure that the system utility of each step will not be reduced. Therefore, it ensures the convergence to the NE. Besides, due to the best response, the reached NE is acceptable.

8.5 Simulation results

In this section, simulations are provided to characterize the effectiveness of the proposed approach and evaluate the performance of the key system parameters. Specifically, we consider four spot beams scenarios, and the beam radius R is 300 kilometers as shown in Figure 8.3. Ten users are the random uniform distribution in each spot beam coverage area. To simplify the calculation, we consider the sub-astral point is (0,0) in Figure 8.3. $\theta_{n_i,l}$ can be obtained by (8.4), and the small scale of fading is considered as infrequent light shadowing [34, 35]. The under-loaded (the number of subchannels is more than the number of users) and equally-loaded (the number of subchannels is equal to the number of users) situations have been proved theoretically by Theorem 2. Therefore, we consider the over-loaded situation during the simulations to prove the effectiveness of the proposed approach. Specific simulation parameters are listed in Table 8.1, where the shadowing and fading parameter of the h in satellite uplink is given as $\left[b = 0.158, p = 19.4, \zeta = 1.29\right]$ [41, 42].

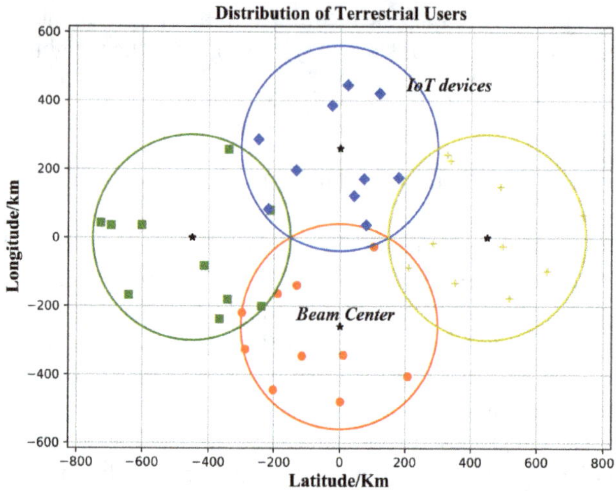

Figure 8.3 The users are uniform distributed in spot beam coverage area

To begin with, four scenarios are considered to verify the efficiency of the proposed algorithm in Figure 8.4. The dissimilar of the four scenarios is the maximal transmission power for users. Figure 8.4 depicts that the negative sum of inverse SINR is gradually increasing step by step, which means the potential function achieves the maximal improvement by the best response. According to the finite improvement property, each iteration converges to NE. Besides, the NE is an acceptable result for global performance. Therefore, Figure 8.4 demonstrates the correction of the proposed game-theoretic approach and the feasibility of the iterative

Table 8.1 Simulation parameters

Parameters	Values
Orbit	NGSO
System noise temperature	350 K
Maximum satellite antenna gain$G_{s,max}$	15dBi
User antenna diameter	0.9 m
3 dB angle of satellite antenna	0.5^o
Maximum user antenna gain$G_{t,max}$	7.38dBi
Off-boresight angleε	0°
Center frequency	4GHz
Satellite height	2000Km
The radius of the Earth	6400Km
Other loss f_ε	−10dB
Iteration step	0.01
The number of subchannels	24

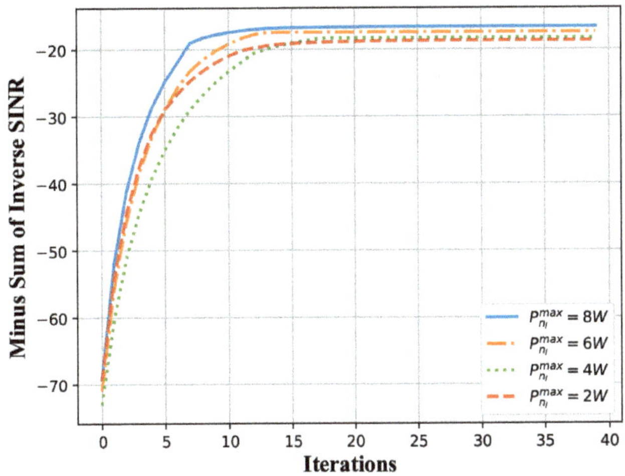

Figure 8.4 The negative sum of inverse SINR increasing by iterations in four scenarios versus different maximum transmission power

algorithm. Besides, the NE for the negative sum of inverse SINR is ascendant with the increase of the transmitted power.

The potential function achieves interference mitigation, which would directly improve the sum capacity of the whole system. Figure 8.5 shows the corresponding sum transmission capacity of the system during the iteration. Figures 8.4 and 8.5 have the same trend in different maximum transmission power, which imply

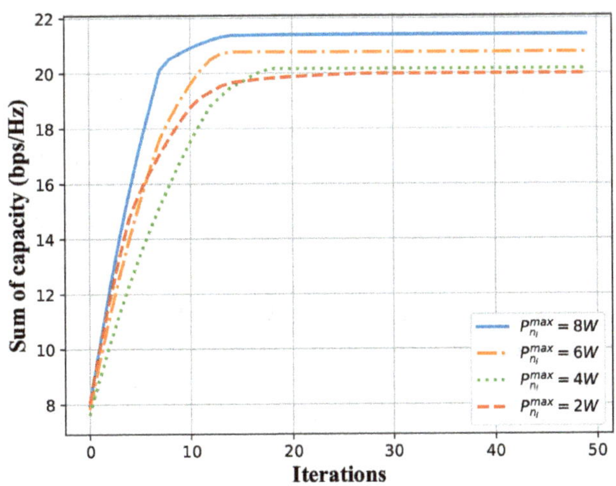

Figure 8.5 The sum capacity increasing by iterations in four scenarios versus different maximum transmission power

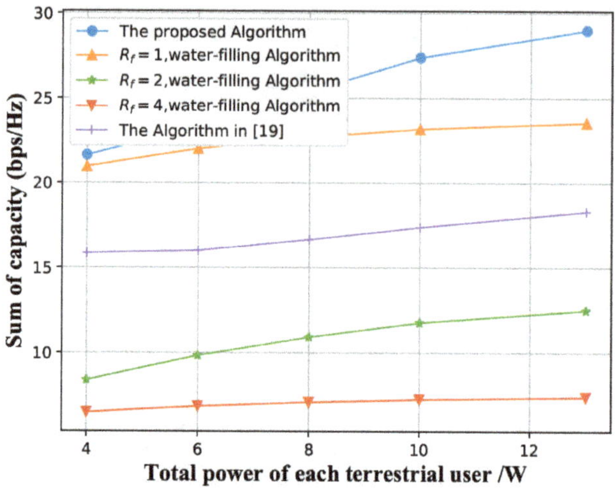

Figure 8.6 The average sum capacity caparison of different algorithm versus user maximum power constraint

the equivalent optimization problem of (8.9a) and (8.10a), i.e. the negative sum of inverse SINR and the sum of capacity are equal in performance metric.

To further prove the effectiveness of the proposed algorithm, we adopt the system sum-capacity for comparison. Figure 8.6 exhibits the sum capacity of the proposed approach and four conventional algorithms versus different maximal transmission power of users. With the help of spectrum reuse technology, co-channel interference could be mitigated. Therefore, multicolor frequency reuse is considered for comparison, where the spectrum reuse factors are one, two, and four. We consider that the power allocation coefficient is obtained through the water-filling algorithm, and user scheduling applies the best response. Besides, the total number of subchannels is evenly distributed into each spot beam for multi-color reuse. The water-filling algorithm can be seen as optimal in each spot without considering the co-channel interference among the adjacent beams. Besides, the proposed iterative algorithm in Reference 19, where the required SINR of each user is set to be the same. Apparently, the proposed interference mitigation approach achieves a supreme performance improvement in sum capacity compared with conventional approaches. Moreover, $R_f = 1$ is superior to $R_f = 2$ and $R_f = 4$ by implementing the water-filling power allocation algorithm, which indicates the higher frequency reuse would increase the spectrum efficiency.

Furthermore, fairness is another system-level performance metric. Here, we analyze the fairness by using JFI, which is defined by Gong *et al.* [21]

$$J = \frac{\left(\sum_{i=1}^{N} R_i\right)^2}{N \sum_{i=1}^{N} (R_i)^2} \tag{8.23}$$

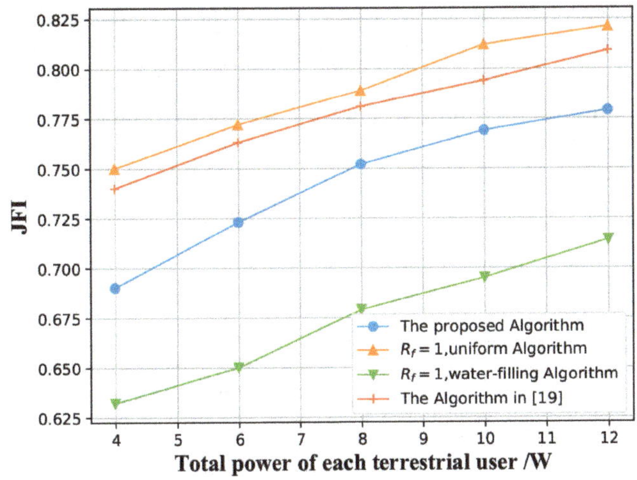

Figure 8.7 The average JFI of different algorithm versus user maximum power constraint

where R_i denotes the transmission rate of the $i-th$ user. Therefore, the range of JFI is $[1/N, 1]$ for N users. Figure 8.7 shows the JFI of the proposed approach and three conventional approaches versus the different maximal transmission power of users. Two conventional power allocation strategies are the water-filling algorithm and uniform allocation algorithm, where the corresponding user scheduling strategies apply the best response and the $R_f = 1$. Another conventional approach is the algorithm proposed in Reference 19, where the request SINR of each user is set to be the same. We implement the total transmission rate in each channel for scheduled users to obtain the JFI. Figure 8.7 describes that the uniform power allocation achieves higher fairness than the water-filling allocation. The algorithm in Reference 19 is based on the SINR request of each user, whose optimization problem has minimized the gap between the required SINR and SINR produced by resource allocation. In this paper, we set the required SINR of each user as the same for comparison. Therefore, the resource allocation tends to meet the required SINR goal. This is the reason the algorithm in Reference 19 achieves better fairness performance. If the required SINR of each user is different, the fairness would be totally dropped.

The proposed potential game-based approach achieves the general performance in terms of the fairness aspect based on the preset weight coefficient. However, the proposed potential game scheme can obtain a larger system sum capacity as shown in Figure 8.6. Achieving maximum sum capacity and fairness are contradictory to the inconsistent channel gain of each user. Therefore, there is a trade-off between fairness and system sum capacity. The proposed potential game-based algorithm can be adjusted by the weighted coefficient $w_{n_i,k}$ of the utility function of each user or potential function to balance the fairness and the sum capacity according to the requirement of the system. In addition, the water-filling algorithm is based on the Lagrangian algorithm. Due to the existence of co-channel interference, the water-filling algorithm is not the optimal solution. The algorithm in Reference 19 is based on the search algorithm, which cannot guarantee convergence and the solution is not optimal. The proposed potential game-based algorithm solves the problem by the best response, and convergence can be guaranteed. Besides, the potential function (negative of the inverse SINR) of the proposed algorithm is interference mitigation or interference alleviation, which has a clear physical meaning. The interference mitigation would increase the sum transmission capacity.

8.6 Conclusion

In this chapter, we proposed a potential game-based approach to implement collaborative user scheduling and power allocation in an uplink multi-beam NGSO system. A framework of the multi-beam uplink NGSO system was proposed, where full frequency reuse and co-channel interference among different spot beams were considered. To provide broadband NGSO service effectively, we formulated an initial optimization problem of maximizing uplink sum capacity and then transformed it into a sum co-channel interference mitigation minimization problem to address the mathematical intractability. Specifically, a game-theoretic model was adopted to solve the transformed optimization problem, which was proved to be a potential game and existence of NEs. Moreover, an iterative algorithm with low computational complexity, motivated by the finite improvement property, was designed to implement collaborative user scheduling and power allocation to the NE point. Finally, the simulation results proved the convergence and effectiveness of the proposed potential game-based approach.

Recently, there are exist a lot of game-based approaches to solve the radio resource allocation approach. As a subfield of economics and business management, auction theory has been introduced to provide an interdisciplinary technology for radio resource allocation in wireless systems, which can be implemented in asymmetric and incomplete information scenarios [43–45].

References

[1] Adu Ansere J., Han G., Wang H., Choi C., Wu C. 'A reliable energy efficient dynamic spectrum sensing for cognitive radio IoT networks'. *IEEE Internet of Things Journal*, vol. 6(4), pp. 6748–59. n.d.

[2] Zhang L., Liang Y.C. 'Joint spectrum sensing and packet error rate optimization in cognitive IoT'. *IEEE Internet of Things Journal*, vol. 6(5), pp. 7816–27. n.d.

[3] Li T., Yuan J., Torlak M. 'Network throughput optimization for random access narrowband cognitive radio internet of things (NB-cr-iot)'. *IEEE Internet of Things Journal*. 2018, vol. 5(3), pp. 1436–48.

[4] De Sanctis M., Cianca E., Araniti G., Bisio I., Prasad R. 'Satellite communications supporting internet of remote things'. *IEEE Internet of Things Journal*. 2016, vol. 3(1), pp. 113–23.

[5] Jia M., Zhang X., Gu X., Guo Q., Li Y., Lin P. 'Interbeam interference constrained resource allocation for shared spectrum multibeam satellite communication systems'. *IEEE Internet of Things Journal*, vol. 6(4), pp. 6052–59. n.d.

[6] Hu D., He L., Wu J. 'A novel forward-link multiplexed scheme in satellite-based internet of things'. *IEEE Internet of Things Journal*. 2018, vol. 5(2), pp. 1265–74.

[7] Jiao J., Sun Y., Wu S., Wang Y., Zhang Q. 'Network utility maximization resource allocation for NOMA in satellite-based internet of things'. *IEEE Internet of Things Journal*. 2020, vol. 7(4), pp. 3230–42.

[8] Li D., Wu S., Wang Y., Jiao J., Zhang Q. 'Age-optimal HARQ design for freshness-critical satellite-IoT systems'. *IEEE Internet of Things Journal*, vol. 7(3), pp. 2066–76. n.d.

[9] Ekpo S.C. 'Parametric system engineering analysis of capability-based small satellite missions'. *IEEE Systems Journal*. 2019, vol. 13(3), pp. 3546–55.

[10] Roumeliotis A.J., Kourogiorgas C.I., Panagopoulos A.D. 'Dynamic capacity allocation in smart gateway high throughput satellite systems using matching theory'. *IEEE Systems Journal*. 2019, vol. 13(2), pp. 2001–09.

[11] Zhang X., Guo D., An K., Zheng G., Chatzinotas S., Zhang B. 'Auction-based multichannel cooperative spectrum sharing in hybrid satellite-terrestrial IoT networks'. *IEEE Internet of Things Journal*. 2021, vol. 8(8), pp. 7009–23.

[12] An K., Liang T., Zheng G., Yan X., Li Y., Chatzinotas S. 'Performance limits of cognitive-uplink FSS and terrestrial FS for ka-band'. *IEEE Transactions on Aerospace and Electronic Systems*. 2019, vol. 55(5), pp. 2604–11.

[13] Lu W., An K., Liang T. 'Robust beamforming design for sum secrecy rate maximization in multibeam satellite systems'. *IEEE Transactions on Aerospace and Electronic Systems*. 2019, vol. 55(3), pp. 1568–72.

[14] Qi C., Wang X. 'Precoding design for energy efficiency of multibeam satellite communications'. *IEEE Communications Letters*. 2018, vol. 22(9), pp. 1826–29.

[15] Joroughi V., Vazquez M.A., Perez-Neira A.I. 'Generalized multicast multibeam precoding for satellite communications'. *IEEE Transactions on Wireless Communications*. 2017, vol. 16(2), pp. 952–66.

[16] Ugolini A., Colavolpe G., Angelone M., Vanelli-Coralli A., Ginesi A. 'Capacity of interference exploitation schemes in multibeam satellite systems'. *IEEE Transactions on Aerospace and Electronic Systems*. 2018, vol. 55(6), pp. 3230–45.

[17] Christopoulos D., Chatzinotas S., Ottersten B. 'Multicast multigroup precoding and user scheduling for frame-based satellite communications'. *IEEE Transactions on Wireless Communications*. 2018, vol. 14(9), pp. 4695–707.

[18] Sharma S.K., Chatzinotas S., Arapoglou P.-D. *Satellite communications in the 5G era*. Vol. 79; 2018 Jul 13. Available from https://digital-library.theiet.org/content/books/te/pbte079e

[19] Barcelo-Llado J.E., Vazquez-Castro M.A. Jiang Lei Hjorungnes A. 'Presented at GLOBECOM 2009 - 2009 IEEE global telecommunications conference; honolulu, HI, 2009'.

[20] Qi F., Guangxia L., Shaodong F., Qian G. 'Optimum power allocation based on traffic demand for multi-beam satellite communication systems'. *Proceedings of ICCT*; Jinan, China, 2011. pp. 873–76.

[21] Gong S., Shen H., Zhao K., *et al.* 'Toward optimized network capacity in emerging integrated terrestrial-satellite networks'. *IEEE Transactions on Aerospace and Electronic Systems*. 2020, vol. 56(1), pp. 263–75.

[22] Jiao J., Sun Y., Wu S., Wang Y., Zhang Q. 'Network utility maximization resource allocation for NOMA in satellite-based internet of things'. *IEEE Internet of Things Journal*. 2020, vol. 7(4), pp. 3230–42.

[23] Choi J.P., Chan V.W.S. 'Optimum power and beam allocation based on traffic demands and channel conditions over satellite downlinks'. *IEEE Transactions on Wireless Communications*. 2005, vol. 4(6), pp. 2983–93.

[24] Ge C., Wang N., Selinis I, *et al.* 'QoE-assured live streaming via satellite backhaul in 5G networks'. *IEEE Transactions on Broadcasting*. 2019, vol. 65(2), pp. 381–91.

[25] Digital Video Broadcasting *Interaction Channel for Satellite Distribution Systems*. 2005, vol. 4.1.

[26] An K., Liang T., Yan X., Li Y., Qiao X. 'Power allocation in land mobile satellite systems: an energy-efficient perspective'. *IEEE Communications Letters*. 2016, vol. 22(7), pp. 1374–77.

[27] Cocco G., de Cola T., Angelone M., Katona Z., Erl S. 'Radio resource management optimization of flexible satellite payloads for DVB-S2 systems'. *IEEE Transactions on Broadcasting*. 2018, vol. 64(2), pp. 266–80.

[28] Roumeliotis A.J., Kourogiorgas C.I., Panagopoulos A.D. 'Optimal dynamic capacity allocation for high throughput satellite communications systems'. *IEEE Wireless Communications Letters*. 2019, vol. 8(2), pp. 596–99.

[29] Brown N., Arguello B., Nozick L., Xu N. 'A heuristic approach to satellite range scheduling with bounds using Lagrangian relaxation'. *IEEE Systems Journal*. 2018, vol. 12(4), pp. 3828–36.

[30] Cai Y., Zheng J., Xu Y, *et al.* 'A joint game-theoretic interference coordination approach in uplink multi-cell OFDMA networks'. *Wireless Personal Communication*. 2014, vol. 80(7), pp. 1203–15.

[31] Li F., Lam K.Y., Liu X., Wang J., Zhao K., Wang L. 'Joint pricing and power allocation for multibeam satellite systems with dynamic game model'. *IEEE Transactions on Vehicular Technology*. 2015, vol. 67(3), pp. 2398–408.

[32] Li F., Lam K.Y., Chen H.H., Zhao N. 'Spectral efficiency enhancement in satellite mobile communications: A game-theoretical approach'. *IEEE Wireless Communications*. 2015, vol. 27(1), pp. 200–05.

[33] Zhang X., Zhang B., An K. 'Outage performance of NOMA-based cognitive hybrid satellite-terrestrial overlay networks by amplify-and-forward protocols'. *IEEE Access: Practical Innovations, Open Solutions*. 2015, vol. 7, pp. 85372–81.

[34] Yan X., Xiao H., An K., Zheng G., Chatzinotas S. 'Ergodic capacity of NOMA-based uplink satellite networks with randomly deployed users'. *IEEE Systems Journal*, vol. 14(3), pp. 3343–50. n.d.

[35] Zhang X., An K., Zhang B., Chen Z., Yan Y., Guo D. 'Vickrey auction-based secondary relay selection in cognitive hybrid satellite-terrestrial overlay networks with non-orthogonal multiple access'. *IEEE Wireless Communications Letters*. 2020, vol. 9(5), pp. 628–32.

[36] Menon R., Mackenzie A., Buehrer R., Reed J. 'Interference avoidance in networks with distributed receivers'. *IEEE Transactions on Communications*. 2009, vol. 57(10), pp. 3078–91.

[37] Yamamoto K. 'A comprehensive survey of potential game approaches to wireless networks'. *IEICE Transactions on Communications*. 2015, vol. E98.B(9), pp. 1804–23.

[38] Monderer D., Shapley L.S. 'Potential games'. *Games and Economic Behavior*. 2015, vol. 14(1), pp. 124–43.

[39] ETSI EN 302-307 vl.l.l 'Digital video broadcasting (DVB); second generation framing structure, channel coding and modulation systems for broadcasting, interactive services, news gathering and other broadband satellite applications'. 2004.

[40] Digital video broadcasting 'Part 2: lower layers for satellite standard'. *Second Generation DVB Interactive Satellite System*. 2012.

[41] Zhang C., Jiang C., Kuang L., Jin J., He Y., Han Z. 'Spatial spectrum sharing for satellite and terrestrial communication networks'. *IEEE Transactions on Aerospace and Electronic Systems*. 2019, vol. 55(3), pp. 1075–89.

[42] Rossi T., De Sanctis M., Maggio F., Ruggieri M., Hibberd C., Togni C. 'Smart gateway diversity optimization for EHF satellite networks'. *IEEE Transactions on Aerospace and Electronic Systems*. 2020, vol. 56(1), pp. 130–41.

[43] Al-Tous H., Barhumi I. 'Resource allocation for multiple-user AF-OFDMA systems using the auction framework'. *IEEE Transactions on Wireless Communications*. 2015, vol. 14(5), pp. 2377–93.

[44] Krishna V. *Auction theory*; 2009.

[45] Zhang X., Guo D., An K., Zheng G., Chatzinotas S., Zhang B. 'Auction-based multichannel cooperative spectrum sharing in hybrid satellite-terrestrial IoT networks'. *IEEE Internet of Things Journal*. 2021, vol. 8(8), pp. 7009–23.

Chapter 9

Inter-satellite links for non-geostationary orbit systems

Ramón Martínez Rodríguez-Osorio[1] *and*
Miguel Alejandro Salas Natera[1]

9.1 Introduction

The first generation of non-geostationary orbit (NGSO) communication satellite constellations developed in the 1990s experienced high levels of planning and deployment, from fully operational examples, such as Iridium or Orbcomm, to cancelled constellations such as Teledesic or Skybridge [1]. During the 1990s, although significant technological advances were achieved, non-successful concepts had in common the financial failures due to the required time to set up the system to provide the specified communication service.

By inter-satellite links (ISLs), we understand the technologies that make communication between spacecrafts possible. According to the International Telecommunication Union (ITU) [2], the inter-satellite service (ISS) is defined as "A radiocommunication service providing links between artificial satellites."

The new setup of massive NGSO constellations providing broadband communication services will imply the use of ISLs between spacecrafts as a mandatory feature. In fact, ISL has been identified as one of the features that will outperform the operation of massive NGSO constellations [3]. ISL improves the system performance in different aspects: latency reduction, reduction of the ground segment, security, larger service area and others, which are explained in section 9.3.

In order to show these benefits of ISL, Figure 9.1 compares a simple communication architecture with and without ISL. In case no ISL is available, the satellite network must successively relay the signal in a ground station (gateway) until it reaches a satellite in view of the destination. In contrast, the use of ISL allows the transfer of the signal through the constellation without the need to bounce the signal up and down between source and destination.

In NGSO constellations, where satellites are allocated in different orbital planes, two types of ISL can be distinguished: intra-plane ISL, which are communication

[1]Information Processing and Telecommunications Center, Universidad Politécnica de Madrid, Madrid, Spain

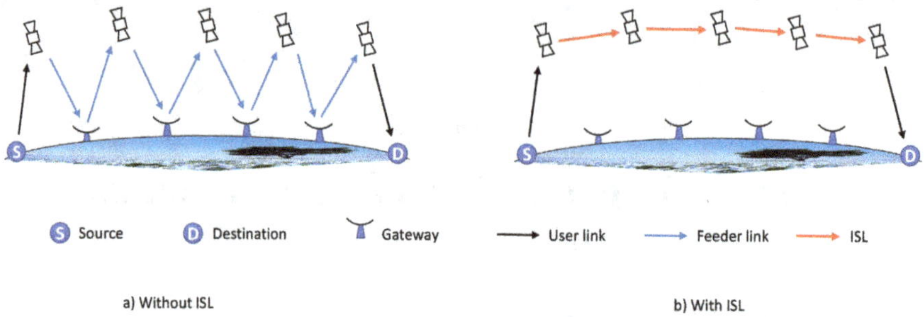

Figure 9.1 *No ISL vs ISL communication architecture (adapted from [4])*

links between satellites of the same orbital plane, and inter-plane ISL, which communicates between satellites allocated in adjacent orbital planes.

From an implementation perspective, two technologies are identified to set up ISL. The first group are presented by ISL in radiofrequency (RF), where onboard radio terminals are used to establish the ISL. The second group are optical ISL, where a communication link based on a non-guided laser beam is used between the two satellites. In both cases, an onboard processing payload is commonly used to facilitate the routing and avoid the degradation of the link budget figures in links passing through a large number of hops. Enabling technologies for ISLs will be covered in section 9.5.

9.1.1 Communication architectures using ISL

Different communication architectures can be implemented using ISL. These architectures are described below with additional examples:

- Cross-links: ISL is established between the satellites of a constellation in order to avoid or reduce the number of contacts with the ground. End-to-end communication is conveyed through the satellites in the constellation using a predefined routing strategy that aims at reducing latency or maximizing the data rate in the whole constellation. Cross-link architecture is used by Iridium, a pioneer constellation formed by 66 satellites in 6 orbital planes to provide voice and data services with a global coverage. Both, first (Block 1) and second generation (Iridium NEXT) satellites are equipped with four inter-satellite ISLs each with forward and aft of the reference satellite in the same orbital plane and with two Iridium NEXT (or Block 1) satellites that are in each of the two adjacent orbital planes. The ISL is operated in the 22.18–22.38 GHz band with beams that do not intersect the Earth's surface to avoid interference. Each of the eight ISS transponders has a necessary bandwidth of 21.3 MHz and uses horizontal polarization to transmit and receive [5, 6].

  Cross-links allow the direct communication between user terminals without any interaction with a terrestrial network. This is the case of LeoSat [7], a Low Earth Orbit (LEO) constellation of 108 interconnected satellites, where end-to-end communication between any two sites on the ground is feasible without requiring any gateway infrastructure. Each LeoSat satellite is equipped with four optical ISLs.

- Low Earth Orbit (LEO) to Geostationary Orbit (GEO): an ISL can be used to relay data from a LEO spacecraft to a ground station through a GEO satellite used as a relay. This architecture is used by Earth Observation (EO) satellites in order to access critical data with a minimum latency avoiding the use of an increasing number of received ground stations. Instead of direct transmission from the LEO satellite to the ground station with intermittent contact time of 10 minutes on average, this architecture allows the download of high-resolution images and observations from EO satellites increasing the available contact times with the ground station. Some examples can be cited: the first is ARTEMIS (Advanced Relay and Technology Mission Satellite) mission: in this ESA's mission, an optical relay link between SPOT-4 and the ARTEMIS geostationary satellite to transmit high-resolution images taken from SPOT-4 was demonstrated in 2001 [8]. The second is the optical link between Sentinel-1A and Alphasat, where a throughput of 1.8 Gbit/s was achieved in 2014 using a laser communication terminal (LCT) operating at 1,064 nm wavelength [9].
- Another example is represented by the payloads defined to implement the European Data Relay System (EDRS), which is equipped with ISL. EDRS uses relay satellites in geostationary orbits to relay information between fixed ground stations and spacecrafts and other terminals in NGSOs to increase the available time to transmit and receive data thus reducing latency [10]. Thanks to the use of ISL, EDRS provides real-time access to EO data, so the key services that benefit are Earth-observation missions supporting real-time critical services (monitoring of fires, floods, etc.), government and security services, rescue teams to operate in disaster areas, or relief forces operating in remote areas without communication support. The EDRS space segment is composed of two geostationary payloads, EDRS-A (9 degrees E) and EDRS-C (31 degrees E). EDRS-A includes an optical inter-satellite link (1.8 Gbit/s) and a Ka-band inter-satellite link (300 Mbit/s). EDRS-C consists of an optical ISL (1.8 Gbit/s).
- Formation-flying missions: several spatially distributed satellites with capacity for autonomous interaction and cooperation with one another in order to maintain the desired formation [11]. The performance of the formation-flying mission depends on the accuracy of maintaining the particular geometry of the formation, which depends on how accurately the relative positions and attitudes between spacecraft are known to all the satellites. It requires the implementation of a communication network for the exchange of data and control information using ISLs [12]. Some examples of formation-flying missions can be categorized in technology demonstration (e.g., PROBA-3, an ESA mission to demonstrate metrology and actuation techniques in formation-flying missions), Earth science (e.g., Flock-1, a Planet Labs Inc. mission to provide 3-5 m resolution Earth images), planetary science (e.g., NASA GRAIL's mission, to obtain high-quality gravitational field maps of the Moon to determine its interior structure), and astrophysics (e.g., OLFAR, a mission led by Delft University

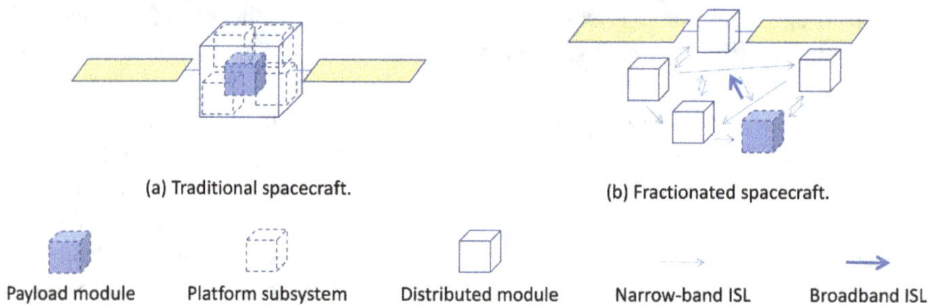

Figure 9.2 Use of ISL in a fractionated spacecraft mission (adapted from [14]

of Technology to produce sky maps by collecting cosmic signals at ultra-low wavelengths regimes, in the 0.3–30 MHz band) [11].

- Fractionated spacecrafts: one single satellite can be decomposed into diverse modules that communicate using wireless links and carry out the different operations of the mission in a distributed and coordinated topology [13, 14]. ISL is a critical enabling technology to decouple mission functions into separate spacecrafts, with the additional complexity of having a mobile and varying topology and reliable links to guarantee the autonomy of the spacecrafts being robust to faults. As shown in Figure 9.2, in a fractionated spacecraft mission ISL allows the information sharing and interaction between modules with links of variable capacity, multi-hop communications, concerns about the energy consumption, and distributed operation, making networking aspects complex [15].

However, ISLs have additional applications beyond communications missions. An example is Hera, a European Space Agency (ESA) mission that will be launched in October 2024 to investigate binary asteroid systems (in this case, Didymos and Dimorphos) [16]. In Hera, ISLs are used as an instrument to support different scientific objectives, such as the measurements of the gravity field of Dimorphos along with the dynamic properties of the asteroid [17].

9.1.2 Scope and contents of the chapter

In this chapter, we will mainly focus on the first cross-link architecture in broadband LEO or NGSO constellations. ISL in fractionated spacecrafts largely depends on the mission scope and concept of operations, and an appropriate set of requirements is not available at the time of writing the chapter. A similar situation occurs in scientific or planetary missions, where ISL requirements are defined by the particular instrument or mission objectives, and thus cannot be generalized.

The objective of this chapter is to present the use of ISL in broadband NGSO. Section 8.2 presents the status of ISL in current NGSO constellations. Section 9.3 presents the achievements that appear as a result of using ISL. Section 9.4 discusses the technical and operational challenges that the use of ISL entails. Next, section 9.5 describes the enabling technologies required for the implementation of ISLs onboard. A case study for the design of a RF antenna for ISL in broadband NGSO is presented in section 9.6, and finally section 9.7 summarizes the conclusions of the chapter.

9.2 Status of ISL in current NGSO constellations

The interest of ISL in present NGSO constellations for broadband services is clear. Examples of NGSO constellations that have considered the use of ISL are listed below:

- Telesat LEO [18] satellites incorporate optical intra-plane and cross-plane inter-satellite links [19] to have a fully meshed satellite network (Telesat Lightspeed Network);
- Starlink [20] described the use of optical ISL in the first filing in order to implement seamless network management and provide service continuity for traffic management when interference mitigation techniques are applied to ensure compliance with other systems [21]. However, the first Starlink satellites did not include laser crosslinks that have been added recently [22];
- OneWeb [23] did not consider ISL in the first versions of the system, although they did not discard the use of optical ISLs in subsequent deployments.

On the other hand, Amazon's Kuiper constellation does not consider the use of ISL in the original Federal Communications Commission (FCC) filing [24]. However, recent job postings from Amazon reveal an interest in optical ISL for the constellation [25].

However, ISLs are not only used in satellite constellations used for broadband communications. Spire [26], a constellation to build a global satellite network to distribute heterogeneous data with proper or hosted payloads (science, observation, weather, flight and maritime data) and formed by 3U CubeSats incorporates ISLs in the form of RF links in S band to meet data latency requirements [27]. Spire collects data-sensitive information such as Automatic Dependent Surveillance – Broadcast (ADS-B) position messages for aircraft that must be received by customers with the minimum latency. Spire is also considering the use of optical ISL in upcoming satellites.

9.3 Achievements acquired with ISL

ISLs bring a set of advantages to the performance of NGSO networks:

- Reduce the end-to-end latency: LEO constellations straightforwardly provide a reduction in the end-to-end latency compared to GEO or MEO systems due to the lower range paths the signal has to cover. The use of ISL produces an additional reduction in the latency as the signal can be routed through the satellite constellation without using long-range terrestrial links. Moreover, it must be considered that the propagation delay in the terrestrial link is affected by the relative permittivity of the transmission line: considering a relative permittivity of 1.46 (typical for a submarine optical fibre cable), the latency is increased over 20% relative to a radio link. However, the latency reduction must be accompanied by proper routing strategies to avoid unnecessary paths between communication sides. Thus, data with time-sensitive requirements are benefited from ISL. In Reference [28], it is proven that lower latency than any optical fibre terrestrial network can be achieved using ISL for distances above 3,000 km.

- Extend service area: in the absence of ISL, it is only possible to establish a connection if there exists a visible path between the user terminal, satellite, and gateway, both fulfilling the minimum elevation requirements imposed by the link budget. Thus, either a large number of gateways spread worldwide is required or the service area is significantly reduced. Without ISL in the constellation, customers located in maritime areas would not be able to access the system.
- Increase system capacity: with the use of ISL, on the one hand, more devices can access the network as the service area is increased, and on the other hand, the satellite becomes a node with additional communication paths to the user and feeder links.
- Reduce the number of sites for the gateways: considering gateways as the points on Earth where access to the Internet is carried out, the use of ISL allows the routing of the signal through the constellation until a gateway is reachable from another satellite.
- Decrease backhaul cost: as an indirect benefit for the satellite operator coming from the lower number of ground stations, the backhaul infrastructure is reduced. Thus, the cost of the backhaul lines interconnecting the ground segment centres is lowered.
- Security: the use of ISLs facilitate the end-to-end connectivity between user terminals without the need to contact intermediate ground stations. Point-to-point and point-to-multipoint communications can be set up. Thus, private data is not carried through any external network or gateway infrastructure.
- Improve satellite control: ISLs can be used to improve the positioning accuracy and orbit determination of navigation satellites [29, 30]. Precise orbit determination (POD) of LEO satellites is carried out using measurements from ground stations and onboard systems [31]. Although the precision of the ranging reaches centimetre level, additional accuracy could be achieved by using autonomous operations by the satellites. In order to reduce the dependency on ground measurements, the use of ISLs for range and range-rate measurements between satellites has been shown as a method to improve POD. In the case of constellations devoted to navigation services, the use of ISL improves the reliability of the positioning, and ISLs will be introduced in the next generation of global navigation satellites [29].
- Avoidance of interference to Geosynchronous (GSO) arc: NGSO constellations must avoid interference to GEO satellites when operating in lower latitudes close to the Equatorial plane. Relay links between northern and southern

hemispheres can be unfeasible if stations are close to the Equator [4]; a possible solution is routing the signal between the two hemispheres using ISLs.

- New sectors: the latency reduction and security achieved with ISL through NGSO constellations can outperform terrestrial and GEO satellite networks. Security is achieved if end-to-end links are established between NGSO terminals without using intermediate ground stations operated by third parties. Thus, sectors such as trading and financial providing new services and applications with very low latency and stringent security requirements will identify NGSO communications as key infrastructure.
- Pushing the integration of satellites in 5G and beyond networks: 5G is the current evolution of cellular networks that will transport a large amount of data at higher rates, connecting a large number of devices with minimal latency [32]. 5G will support services that exceed the capabilities of 3G and 4G, such as Machine-to-Machine (M2M) for industry automation, augmented reality, or 3D video, amongst others. Part of the cellular networks conforming to 5G is based on Non-Terrestrial Networks (NTN), which include any network involving non-terrestrial flying and orbiting objects, such as satellite communication networks in GEO, Medium Earth Orbit (MEO), LEO and Highly Elliptical Orbits (HEO), and Unmanned Aircraft System (UAS) that includes high altitude platform systems (HAPS) [33]. The use of NTN may contribute to fostering the roll out of 5G networks thanks to the increased coverage to underserved areas without terrestrial infrastructure, supporting critical missions when terrestrial networks are not operable due to natural disasters. Therefore, NTN improves the resilience, reliability and service continuity for users in moving platforms and M2M/IoT (Internet of Things) devices, and enables the scalability of 5G networks offering efficient multicast and broadcast resources to the network edge or the end user [34]. These enhancements provided by NTN are also envisioned for 6G systems [35]. The use of ISLs is considered in 3GPP in different architectures with NTN networks; moreover, ISLs are considered as a transport link in the Next-Generation Radio Access Network (NG-RAN) architecture when the gNB is on the satellite [36]. Figure 9.3 shows a typical scenario of a non-terrestrial network using a satellite or UAS platform with a regenerative payload. The NTN platform generates a service area using a set of beams under its field of view. User equipment (UE) terminals are served by the NTN platform in the service area. The feeder link represents the radio link between a gateway and the NTN platform. An ISL is included to represent the feeder link of a satellite in a constellation without contact with the gateway.
- Improvement in system synchronization and related aspects: in Global Satellite Navigation Systems (GNSS), the use of ISL range measurements can be used to estimate hardware delays by comparing clock offsets of ISL measurements with L-band measurements obtained before the time synchronization. The measurement of ISL outperforms time synchronization compared to L-band links with ground stations because of their wider coverage compared to the regional network and high precision [37]. Moreover, ISLs can be used to support

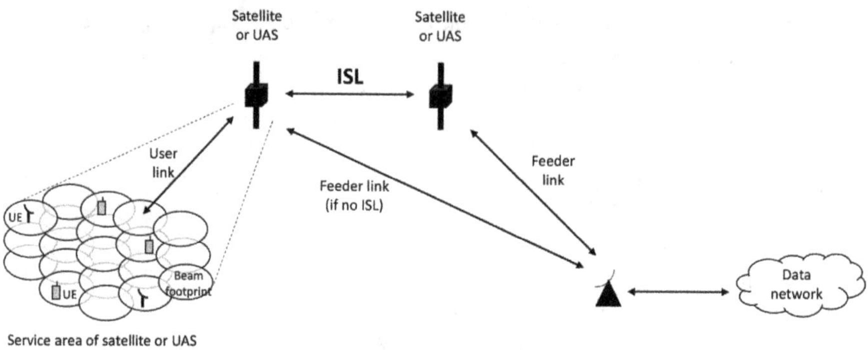

Figure 9.3 Non-terrestrial network typical scenario based on regenerative payload [36]

autonomous satellite navigation by using data from the rest of satellites in the constellation [38].

However, it is important to note that the use of ISL is not mandatory for other satellite constellation systems providing communication services with less stringent time requirements. This is the case for IoT and M2M applications where access to information in real time is not required and store-and-forward architectures are preferred: the satellite stores onboard the information sent from the ground terminal and forward the packets to the gateway once in contact.

9.4 Technical and operational challenges

The technical and operational challenges that accompany the use of ISLs in satellite constellations are listed in Table 9.1. These challenges can be divided into those affecting the satellite platform, the communication architecture and the system operation.

Next, the challenges associated with the use of ISL are explained:

- Challenges associated to the satellite platform [39]. The introduction of a new communication payload for ISL must comply with the requirements of the satellite bus. From a system engineering perspective, mass and power budgets are affected, meaning that the platform shall be designed jointly with the ISL payload. The impact on the platform is very sensitive if the use of small or even nanosatellites is required [13]. At least, the following platform subsystems are affected by ISL:
    - Attitude Determination and Control Subsystem (ADCS): first, depending on the beam size of the ISL terminal, the accuracy of the pointing system for link acquisition and tracking to maintain the link while moving is affected,

Table 9.1 Technical and operational challenges driven by the use of ISLs

Part of the system affected by the challenge	**Specific item affected**
Satellite platform	Attitude Determination and Control Subsystem (ADCS)
	Electrical Power Subsystem (EPS)
	Thermal Control Subsystem (TCS)
	Structure
	Onboard data handling (OBDH)
Communication architecture	Connectivity scheme
	Ground segment
	Frequency bands
	Traffic engineering
	Routing strategy
	Constellations with diverse topologies
System operation	Angular scanning of the ISL antenna
	Re-entry
	Other system aspects

being significantly more stringent in the case of using optical ISL. Second, the number, geometry and active ISL links will impose diverse and independent requirements on each ISL terminal.

- Electrical Power Subsystem (EPS): the introduction of additional communication links might require additional electrical power, thus needing larger solar panels and extra batteries to support the communication during eclipse periods. Moreover, the Power Conditioning and Distribution Unit (PCDU), in charge of delivering the required electrical power at different voltage levels to sensors and subsystems, must be redesigned to meet new requirements coming from the operation of the ISL.
- Thermal Control Subsystem (TCS): modifications in the TCS are required to maintain a stable temperature of the ISL payloads and transfer the dissipated power to a radiator [40].
- Structure: proper allocation of the ISL terminal shall be established, considering thermal control aspects and allowance of a wide angular range mostly for contact windows in inter-plane satellite links. Aspects such as the thermal dissipation of the ISL payload, terminal volume, size and mass shall be taken into account. One of the major differences between the previous ISL terminals and those required for current and future NGSO constellations comes from the different satellite buses. ISL payloads must adapt to new satellite platforms based on planar geometries [41] or future constellations using nanosatellites.
- On-board data handling (OBDH) and Telemetry, Tracking and Command (TT&C): new commands and telemetries will be introduced to monitor and command the ISL payload from ground control.

- Challenges associated to the communication architecture
  - Connectivity scheme: ISLs can be classified in intra-plane and inter-plane links. Intra-plane are ISLs with advanced and backward satellites located in the same orbital plane and can be considered as fixed links in circular orbits.
  - In the case of inter-plane links, depending on the constellation geometry, the ISL will face diverse challenges, due to the relative velocity between satellites, Doppler shift, tracking control problems, and link budget. As an example, let us consider Walker Star and Walker Delta constellations. In the former case, near the poles, satellites are much closer than in the Equator due to the quasi-polar inclination used. Moreover, large relative velocities between satellites in neighbouring planes will appear, making the ISL difficult in terms of tracking rate, Doppler shift and link re-establishment with another satellite when the orbital planes cross [42]. It can be seen in Figure 9.4, where in planes plotted in blue, satellites move in opposite directions. On the other hand, in Walker Delta constellations inclination can be modified to overlap satellite coverage areas depending on the service area requirements.
  - In Walker Delta constellations, where low and medium inclination planes with lower number of satellites per plane are allocated, the use of ISL is complex in terms of pointing, tracking, and also linking budget, as power requirement can be severe if the range between satellites is large. In the case of Walker Star, satellites move regularly in the orbital planes making easier the network design.

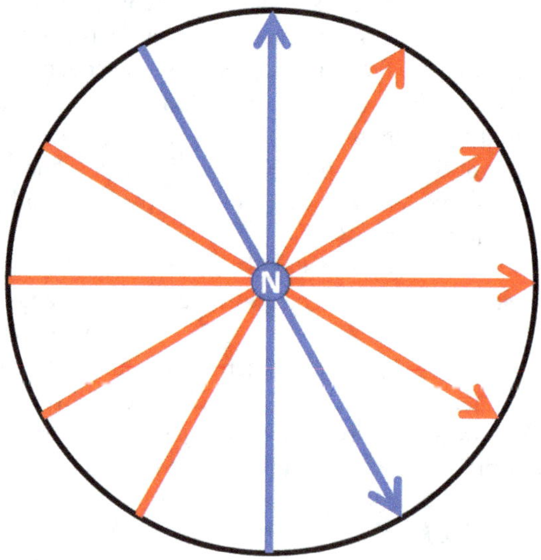

Figure 9.4 Polar view of a Walker Star constellation formed by 6 orbital planes (adapted from [42])

- An analysis of inter-plane connectivity for Walker Star constellation can be found in Reference [43], where aspects such as limitations of the small satellite platform or the multiple access methods are evaluated. In this reference, a simple approach for constellation design is provided considering transmission power and antenna gain as outputs for having full inter-plane connectivity.
- Ground segment: the use of ISL affects ground segment architecture in several aspects [44]. First, it is expected that ISLs will reduce the number of ground stations, but such reduction depends on the number of satellites, constellation parameters, link capacities, ground user demand, and routing strategy. Second, the allocation of gateways depends on the availability of a terrestrial broadband convection as well as topographic, meteorological, and economic factors. Finally, the computational complexity of the ground segment is increased with ISL, as the calculation of the best network configuration will require additional hardware and software resources. As a conclusion, an important idea is the fact that both ground and space segments shall be designed jointly.
- Frequency bands: for the ISS, ITU-R establishes the frequency band described in Table 9.2. Some of the bands have restrictions (e.g., only for GEO satellites) and others have limitations in the transmit power spectral density, and some must be with other services (e.g., Earth exploration-satellite service), so special attention must be paid for the selection and use of a particular frequency range.
- Traffic engineering: in a highly dynamic environment, advanced traffic engineering optimization for dynamically changing traffic loads in time and space and network topology must be implemented in order to efficiently route differentiated traffic demands over various nodes and links supporting differentiated services [45].
- Routing strategy: one of the most challenging aspects in NGSO constellation is the generation of the best inter-satellite route to connect any pair of network nodes. Two main options are available for the design of the routing strategy: ground-based (centralized) or autonomous (distributed). In the former case, the routing is decided by the ground segment and then the

Table 9.2 Frequency bands assigned to inter-satellite service by ITU [2]

Frequency bands			
22.55–23.55 GHz	56.9–57 GHz	66–71 GHz	174.5–174.8 GHz
24.45–24.75 GHz	57–58.2 GHz	116–119.98 GHz	174.8–182 GHz
25.25–27.5 GHz	59–59.3 GHz	119.98–122.25 GHz	185–190 GHz
32.3–33 GHz	59.3–64 GHz	122.25–123 GHz	191.8–200 GH
54.25–55.78 GHz	64–65 GHz	130–134 GHz	
55.78–56.9 GHz	65–66 GHz	167–174.5 GHz	

information of the updated routing tables is broadcasted to the satellites. In this case, an optimal solution can be found using the ground computation resources, with the impairment of having very frequent updates that requires a large ground station network to ensure availability of a satellite-to-ground connection [46]. Moreover, this centralized approach scales inefficiently with the number of satellites in the constellation, being O(N2) the computational complexity of the Dijkstra's shortest path algorithm [47]. On the other hand, in a distributed approach each node (satellite) of the network decides its paths independently typically using a link to exchange information between neighbour nodes. The distributed approach exploits the predictability of the LEO constellation topology and increases the resilience of the routing as it is realized in each node with independence of the rest of the network [46]. In contrast, a large effort to optimize the algorithm implementation on-board must be carried out to deal with the constraints of energy, mass and processing capabilities.

   - Constellations with diverse topologies: in contrast to Walker-based systems, current constellations include satellite in planes with different inclinations and altitudes, e.g. combining LEO and Very Low Earth Orbit (VLEO). These new deployments introduce another challenge to ISL as satellites must be capable of producing intra-plane and inter-plane ISLs, but also ISLs between satellites positioned at different altitudes.

- Challenges associated to the system operation
   - Angular scanning of the ISL antenna: depending on the field of view of the antenna, the number of visible satellites will be different. Thus, the routing strategy of the ISL in the constellation will be driven by the exploration range of the ISL antenna beam. In Reference 48, digital beam-steering and beam-switching with a Butler matrix are compared for inter-plane links in Walker Star constellations. Relative polar angles between satellites of 100±10 degrees are considered, which represents a challenge for ISL antenna in terms of angular scanning. Results show that in order to maximize data rate performance it is convenient to increase the number of antenna elements and reduce the matching period.
   - Limitations on re-entry: selection of adequate materials for the ISL payload in terms of complete burn up during re-entry is also required. A 1-to-10,000 probability of human casualty due to an impact of a satellite part after re-entry is usually accepted as a threshold by space agencies and nation states [49]. As an example, the silicon-carbide components whose melting point is 2730°C [28] used in the mirrors of optical ISL do not satisfy that requirement and have to be replaced by another material [4].
   - Other system aspects: operation of the ISL during constellation deployment or the update of the routing strategy by ISL are system operation aspects that shall be considered in the design of the constellation.

9.5 Enabling technologies for ISL

9.5.1 Introduction

Depending on the mission and communication requirements of the link, the ISL can be implemented with optical or RF technology. There are, as discussed in the previous sections, many in-orbit validated systems, experiments, and demonstrations that have been deployed to help in the implementation of the most suitable technologies for ISL terminals.

ISL can be implemented using RF (mainly in Ka and E bands) or optical communications technology. Both technologies present advantages and drawbacks and the decision to select one of them must be carried out under a trade-off considering diverse aspects like attitude control accuracy, power consumption, size and mass, or technology maturity.

Besides, for optical ISL terminals, enabling technologies are not only about terminal sub-systems but also about the attitude determination and control system onboard the platform that determines the feasibility of the connection between transmitter and receiver. In addition, aspects related to the platform such as the vibration is critical and shall be taken into account in the design process.

9.5.2 Radiofrequency

For RF implementation at Ka and E bands, the enabling technologies focus on the power amplifiers (PAs) and the antenna aperture to enable beam steering.

9.5.2.1 Power amplifiers

PAs are the first critical component in ISL RF terminals.

PA technologies at microwave frequencies typically use GaN technology for better efficiency and higher power, but it is not usual to find GaN devices that work in V or E bands (40–75 GHz and 60–90 GHz, respectively). Some experimental GaN devices show promising results but have not been commercialized yet, and it is challenging to find commercially available GaN devices above 40 GHz. Also, although SiGe and CMOS devices can work at higher frequencies, their power levels are low and combined solutions with many elements are needed to reach the required Effective Isotropic Radiated Power (EIRP) of the ISL.

Thus, while phased array and active antenna technology have proved effective at increasing the gain and EIRP of antennas, as well as providing beam-steering capabilities, as the frequency increases the half-wavelength dimension becomes smaller and they become difficult to fabricate. Increasing the number of elements also increases power consumption, so using high-power GaAs devices with fewer elements using subarrays architectures has become the best solution for E band links.

Efforts have been done in silicon-based millimetre-wave systems. At E band, several applications such as automotive radars [50–52], image sensing [53, 54] and short range and high data rate communications [55] can be implemented. In this sense, the main evolution of E and W band (75–100 GHz) amplifier is driven by the

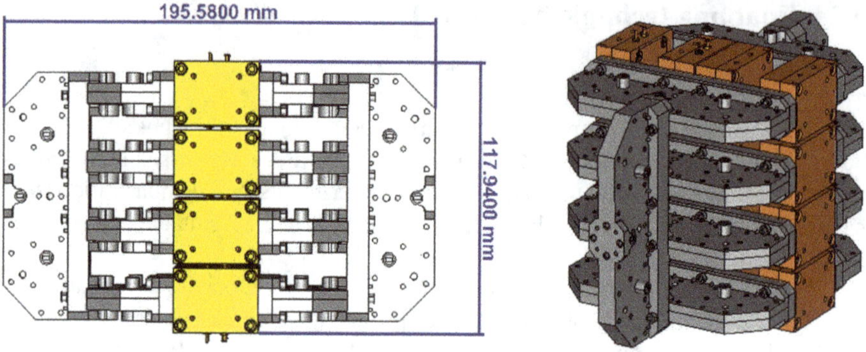

Figure 9.5 E band amplifiers combined for 5 W of RF linear power (~ 33 W of DC power). [ERAVANT authorized use of image]

automotive radar industry. Thus, the frequency range of Commercial-Off-The-Shelf (COTS) PAs is likely from 70 to 86 GHz while ISL frequencies at the E band are located between 60 and 70 GHz. Some transceivers using 65-nm CMOS technology in D-band (110–170 GHz) have been reported in the literature [56] for applications in WPAN and IoT.

There are some solutions in previous frequency bands like those proposed by MACOM, Analog Devices and ERAVANT. In this line, MACOM offers three PAs with a maximum output power of 23.4, 24, and 26 dBm [57]. Furthermore, Analog Devices proposes a PA for applications from 71 GHz to 76 GHz with 24 dBm of maximum output power [58]. ERAVANT (formerly SAGE Millimeter) provides E band power amplifiers with an output power of 24 dBm [59].

To date, no single silicon-based power amplifier working above 60 GHz has been reported with output power higher than 27.3 dBm. Figure 9.5 shows one solution to provide 5 W with E band amplifiers based on ERAVANT solution [60]. This solution needs 33 W of DC power, so it has an efficiency of 15%. The very low efficiency of the PAs at the E band is one of the major challenges to be solved if ISL for broadband or high-capacity satellite nodes is intended to be implemented with RF technology.

From the literature, Wagner and Rebeiz [61] present a set of wideband, fully integrated Eb and PAs designed in 0.12-µm silicon-germanium (SiGe) BiCMOS and working in the range of 60–75 GHz. The single, four-way, and eight-way combined PAs achieve saturated output powers of 16, 19.5 and 24 dBm with peak power-added efficiencies of 18%, 11% and 12%, respectively. Lin and Rebeiz [62] present a fully integrated 16-way power-combining amplifier for 67–92-GHz applications in an advanced 90-nm silicon germanium technology. As explained by the authors, the power combining amplifiers achieves a small-signal gain of 19.3 dB at 74 GHz, and output power of 25.3–27.3 dBm at 68–88 GHz with a maximum power added efficiency of 12.4%.

Regardless of the push for integration, silicon technologies are limited to generating large output powers at millimeter wave frequencies due to the breakdown voltage and transistor size and are often outperformed by other technologies such as GaAs, GaN and InP. Moreover, current technology must be adapted to the operation in space environment conditions.

9.5.2.2 Antenna apertures

Being the antenna aperture of one of the most critical part of one RF ISL terminal, it is important to highlight that future communication networks will use the satellite as a node. In this sense, the relative position and orientation between satellites changes over time and the antenna aperture needs to be reconfigured or the beam steered. Thus, the capacity of the antenna to steer its beam is an important technological challenge at millimetre wave frequencies that need to be solved if antenna technologyis going to be part of the future ISL terminals.

Regarding the size and power capacity of the new satellite platform intended for satellite constellation, low profile and high-power efficiency are two key specifications of the future ISL terminals. For this, the discussion here will be focused on three steering concepts: phased array, dielectric lens and frequency selective surfaces with switched-beams of shifted apertures, and reconfigurable antenna materials.

- Phased Arrays [63]: These antennas utilise constructive combining of incident signals at the individual antennas which are phase shifted before combining. Phased arrays can be considered as the most conventional method of electrical beam steering, and a wide spectrum of phase shifting techniques have been developed to suit different applications and frequency ranges.
- Dielectric lens and frequency selective surfaces with switched beams or shifted apertures: On higher frequencies, mainly upward of Ka band, radio waves can be redirected in similar fashion as visible light using dielectric lenses or frequency selective surfaces [50]. For beam steering, the aperture is moved in front of the antenna element or several antenna elements can be switched to change the beam. The main advantage of these concepts is that they allow the design of a fixed antenna and thus allow a simpler mechanical structure for the antenna system.
- Reconfigurable antenna materials [64]: In this case, the beam steering is achieved by using functional materials such as ferrites, ferroelectrics, Barium Strontium Titanate capacitors, filters, and phase shifters in thin- or thick-film technology and the Microwave Liquid Crystal technology beyond optics. With the evolution of LC mixtures in the early 2000-decade, new concepts with appropriate biasing schemes have enabled the design numerous high-performance microwave components and devices in the RF domain.

The presented reconfigurable antennas are based on different technologies but the most important is the layer that contains the reconfigurable RF components such as RF-MEMS, PIN-diodes and varactors, the photo conductors for optical switches as well as lighter and more precise positioners [65, 66]. Besides, the future in the development of steerable antennas seems to be in line with the development of new

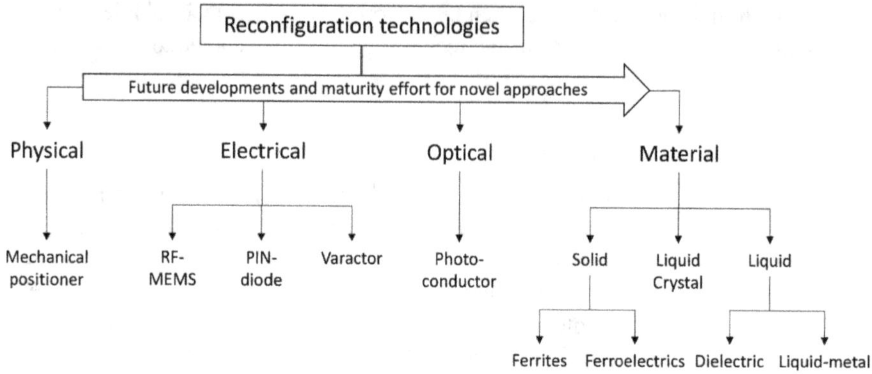

Figure 9.6 Proposed technology roadmap for reconfigurable antenna apertures in ISL terminals

reconfigurable and low-loss materials [67]. However, there are significant challenges related to the manufacturing techniques and the novel concepts and design that will propose novel RF reconfigurable components [68].

In the near future, researchers and engineers involved in circuit and system design for new communication platforms such as future satellite constellations in MEO/LEO need to follow up the direction of technology evolution that other sectors of industry trace. Figure 9.6 presents a categorization of reconfiguration technologies where the arrow represents the direction of efforts on technology development in order to move up in frequency and to increase communication capacities.

The Ka band ISL terminals need to be miniaturized since the main applications are for a low rate or low capacity such as IoT communication services [69]. These applications are related to small platforms that provide low power and require low volume and low mass, but the performance must be guaranteed for a lifetime above 3 years. The challenge for the implementation of the LEO to LEO ISL at the Ka band is the fact that when low power is available, the antenna size results large for small satellite platforms making integration unfeasible. Furthermore, large antennas require Antenna Pointing Mechanism (APM) that results in being bulky and heavy as can be observed in Figure 9.7.

For navigation satellites, there is the need for high gain and low side lobe antennas for long-lasting ISLs. These antennas consist of the RF parts, the structure, deployment and pointing mechanisms, and thermal and control subsystems. Figure 9.7 shows an ISL reflector antenna for navigation satellites excluding thermal control sub-system [70].

In order to achieve faster beam switching time and avoiding the use of APM systems that has several impairments such as power consumption, volume, weight, and the need for other complementary sub-systems, there is the option of using Phased Array Antennas (PAAs) at Ka band [70]. However, PAA also has technological impairments such as reduced scanning angle, phase centre stability, [71] and delay stability. Furthermore, depending on the operation frequency, the implemetation

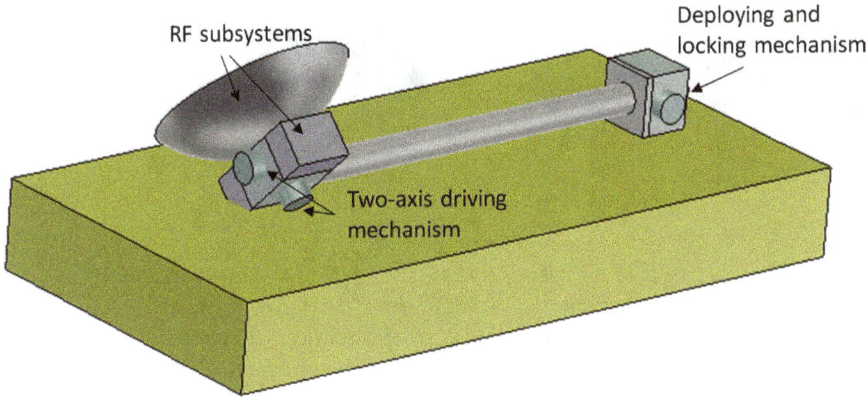

Figure 9.7 ISL reflector antenna for navigation satellites (adapted from [70])

of the PAA can result in a complex calibration system offering lower reliability. Finally, PAA requires exhaustive measurement for its verification and calibration [72]. The calibration process can be off-line, on-site, and online depending on the antenna implementation and accuracy requirement. It is assumed that when high gain PAA is required, a more complex calibration system and algorithm have to be implemented [71].

In the satellite communications business, during the last decades there is a huge need for high-speed radio links capable of achieving system capacities above the Tbps. In this scenario, from the RF perspective, the antenna and amplifier technologies play a vital role to achieve better efficiency of the channel. Thus, gigabit communication is possible in millimetre-wave technology [73] and so the E band is the best suitable band in the near future for millimetre-wave technology.

For broadband services, there are ISL antenna proposals based on E band technologies. In this case, antenna size results smaller at the time that less power is needed for the range between satellites providing higher capacity links. However, at E band frequencies the amplifier and electronically steered antenna technologies are not mature yet. Thus, E band RF components need further effort for development of new solutions that comply with the ISL requirement. Furthermore, aperture antenna technologies need to improve the steering and commuting solutions in terms of complexity, power consumption, volume, and cost.

Next, four aperture technologies from the state of the art that aim in the solution of E band needs of steerable beam antennas are introduced. These technologies are phased array, APM, Fresnel zone lenses, and magnetic surface or bull-eye antenna concept.

For large onboard phased array based on waveguide technology, SWISSto12 [74] proposes a modular approach for the realization of such large arrays. The modules are very light (5 times lighter compared to conventionally manufactured antennas), require minimal assembly, and simplify notably the integration in the spacecraft. Additional advantages of this technology are excellent RF performance,

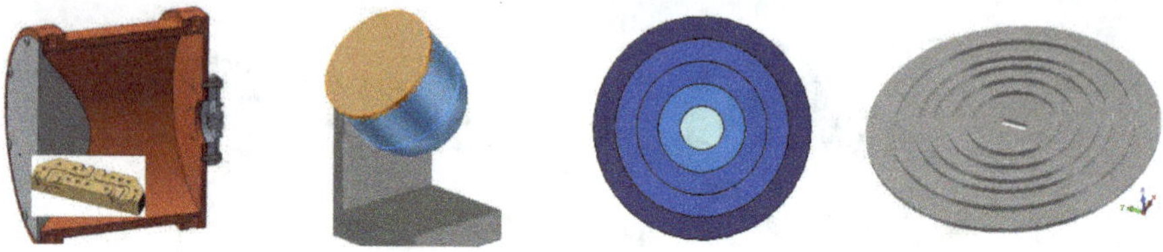

(a) (b) (c) (d)

Figure 9.8 Example of antenna technologies for RF ISL terminals. (a) 3D printed Eband antenna and diplexer [ANTERAL authorized use of image]; (b) Antenna over 2 axis positioner; (c) Switchable beam Fresnel zone antenna; (d) Bull eye antenna for small satellites

especially very low insertion loss due to the limited number of interfaces; clusters enhancing the mechanical performance of the payload, freedom in the array shape; etc. Designs are available from X- to Q-band [74].

Anteral [75] is presenting also a novel solution for E band antenna systems [76, 77] that achieve high gain antenna design with a low profile, weight and cost. This design based on 3D printing technologies has a realized gain of 40 dB being a suitable solution for in-plane RF ISL terminals for LEO constellations (Figure 9.8a).

Figure 9.8b shows an example of an antenna over a positioner like the X-band high-gain horn antenna from Surrey Satellite Technology Ltd. (SSTL) presented in Reference 78. The antenna can be mechanically steered toward the ground station while the satellite is moving. This antenna can radiate either right- or left-handed circularly polarized signals by modifying the feed position. It operates at X-band but the aperture can be implemented to operate at the E band. The antenna can achieve a wide scanning range, is robust, and has a low cost. Planar antennas are attractive for small satellites as they can be easily integrated with the satellite body.

Regarding the Fresnel zone lens antenna, one way to implement a multi-dielectric lens is to use solid rings of different materials with different ε_r as presented in Figure 9.8c. In Reference [79], a 3D printed-based model designed for 60 GHz with high gain and fed by a patch is presented. However, another approach to the solution of having different dielectric constants in the lens is possible, such as the one proposed in References [80–82], which consists of starting from a sheet of the same material and performing perforations in it in order to change its dielectric constant. It should be noted that this design technique avoids the problem of the commented assembly and is therefore particularly attractive for the design of Fresnel Zone Plates (FZP).

Authors in Reference [83] present a new approach to designing V-band Bull's eye antennas so as to produce multiple beams which are either fixed or discretely steerable (Figure 9.8d). Bull's eye antennas comprise concentric rings around a sub-wavelength aperture. Beam deflection is accomplished by adjusting the effective spacing of the rings, which they explain in terms of the coupling angle to free space and surface waves.

9.5.3 Optical

By optical ISLs we mean the application of Optical Wireless Communications (OWC) to unguided free space conditions, a technology named Free Space Optics (FSO). In FSO, an optical signal is transmitted through an unobstructed path to an optical receiver. In the case of ISLs, the medium is vacuum and does not include atmosphere. A FSO transmitter consists of a laser whose intensity is modified by an electrical signal modulated by the input data. The modulated laser beam is directed through a telescope to the receiver. After the signal crosses the unguided medium, it is received in the receiver telescope, where a photodetector converts the optical signal to an electrical signal that is demodulated to extract the information [84].

As reported in Reference 85, the first LEO-to-LEO optical ISL was successfully carried out in 2008. It consisted of communication between two large EO satellites: TerraSAR-X and NFIRE (Near-Field Infrared Experiment). Onboard optical terminals needed less than 25 seconds on average to lock on and establish a 5.6 Gbit/s link at distances between 3,700 and 4,700 km with a maximum range rate of 8,500 m/s [86].

Regarding tracking in optical ISL, the terminal has to detect the laser beam through the acquisition, tracking, and pointing (ATP) system that controls the optical antenna. It is important to emphasize that the laser beam divergence is very narrow, so a high precision ATP has to be installed onboard under additional miniaturization requirements [87].

Following, we present in Table 9.3 a comparison of nowadays optical ISL terminals developed. The power and capacity are related to the satellite market and so is the link distance. An important information about these terminals is their maturity. For the case of Mynaric CONDOR terminals, there are the Mk2 and the Mk3 for up to 5,000 km and more than 7,500 km, respectively [88]. The CONDOR Mk3 has a hyper hemispherical steering range and uses two wavelengths to separate transmission and reception (1553/1536 nm) and it is adaptable to flat panel satellite platforms [89].

Besides, General Atomics [90] presents the Optical Communication Terminal with a data rate up to 5 Gbps and for an estimated maximum range of 5,000 km. Furthermore, General Atomics is planning to make a validation experiment of the 40 Gbps terminal onboard a 12U satellite at the end of 2021.

Finally, one of the most mature providers of optical communication terminals is TESAT, offering different options for GEO-GEO, LEO-GEO, LEO-LEO and LEO-Ground station markets [91]. The ISL terminal options from TESAT have 1.8 Gbps and 10 Gbps, although the 100 Mbps CubeLCT terminal for CubeSat to GS is expandable for ISL [92]. The PIXL-1 is a technology demonstration mission that incorporates a CubeLCT terminal [93] in a 3U CubeSat to demonstrate the operation of the optical payload described in Reference [92]. TESAT ConLCT [94] (Figure 9.9) showed its performance with the coding and framing according to the SDA standard, data rate, wavelength compatibility, tracking tone, waveform capture, and continuous data transfer in a terrestrial test prior to its implementation for ISL in the SDA Tranche 0 constellation [95].

Table 9.3 Optical ISL terminals (adapted from [84])

Company	Terminal	Capacity [Gbps]	Link distance [km]	Power consumption (P_C) / Transmit power (P_{TX})	Mass [kg]/Volume [cm³]	Satellite market	Maturity
Mynaric	CONDOR Mk3	10 (variable)	8000	P_C = 150 W P_{TX} =3 W	OSA (Optical System Assembly): 35.1 x 21.0 x 17.0 Dual EB (Electronics Box): 16.1 x 33.6 x 25.5 Quad EB: 16.1 x 33.6 x 37.26	LEO-LEO	Acquisition, Tracking, and Pointing (ATP) availability 7+ years in LEO
TESAT	LCT 135	1.8	80000	P_C =150 W (acq.) P_C =120 W (comm) P_{TX} =2.2 W	53/ 60x60x70	GEO-GEO (backbone)	TRL9 15 years
TESAT	SmartLCT	1.8	45000	P_C = 130 W (comm.) P_{TX} = 5 W	30/ 35x35x30	LEO-GEO data relay	TRL6 10 years
TESAT	ConLCT	10	6000	86 W	15 (2 subunits)	LEO-LEO	--
General Atomics	1550 nm LCT	40	2500	—	NA/ 20x20x30 cm³	LEO-LEO	DEMO in 2021 of two 12U CubeSats
TESAT	CubeLCT	0.1	--	P_C =8 W P_{TX}=100 mW	0.36 / 9 x 9.5 x 3.5 (~ 0,3U)	CubeSat-GS*	PIXL-1 mission

*It is a Cubesat to GS optical terminal, expandable for optical ISL

Figure 9.9 TESAT ConLCToptical terminal (@TESAT) [TESAT authorized use of image]

9.5.4 Comparison

A comparison between RF and optical technology for implementing ISL can be realized using the following items [84, 85]:

- Bandwidth: thanks to the increase in the carrier frequency, usable bandwidth is in the order of 100 THz for optical links, which is five orders of magnitude relative to a typical RF carrier used in ISLs.
- Frequency allocation: in RF, frequency use is regulated and assigned by ITU. In congested bands, coordination is required and interference with adjacent carriers may appear due to the frequency reuse. On the contrary, optical systems are free from spectrum licensing, which reduces set-up time.
- Beamwidth-Beam divergence: optical beam suffers from beam divergence in the link, which causes that part of the transmit beam not to be captured by the receiver, implying a loss that increases with the length of the link. Similar to the antenna beamwidth in the antenna pattern, the optical beam divergence is proportional to λ/D_R, λ being the wavelength and D_R being the aperture diameter. As the wavelength used in optical links (typical 1,550 nm) is significantly lower than in RF (1 mm at 30 GHz), the intensity of the optical signal at the receiver is higher for the same transmit power.
- High directivity: the advantage of the directivity for an optical terminal is given by the ratio of antenna gain as

$$\frac{Gain_{opt}}{Gain_{radio}} \sim \frac{4\pi/\theta^2_{div(opt)}}{4\pi/\theta^2_{width(radio)}} \tag{9.1}$$

where $\theta_{div(opt)}$ and $\theta_{width(radio)}$ are the beam divergence of the optical terminal and the antenna beamwidth of the RF terminal, respectively. As the optical wavelength is much smaller than that of RF, the optical directivity is significantly higher.

- Power and mass requirements: thanks to the higher directivity of the beam in optical links, less power and aperture size are required compared to RF.
- Doppler shift: deviation in the frequency received relative to the transmit frequency depends on the range rate between the two ends of the communication. Doppler shift, thus, depends on the constellation orbits followed by the two satellites [96]. In ISLs the Doppler shift is especially in inter-plane links with upward and downward satellites and when the orbital altitudes are different. For example, the Doppler shift is around ±7.5 GHz in an ISL between LEO and GEO satellites. For a 2 GHz radio signal, the Doppler shift is reduced to ±140 KHz [85]. Thus, in order to avoid loss of data due to frequency mis-synchronization, mitigation techniques like using optical phase-locked loop (OPLL) or cooperative frequency tuning between transmitter and receiver are required.
- Platform effects and tracking: one of the features of optical transmitters is the narrow beam which can be on the order of 100 μrad [97]. Thus, complex and high accuracy pointing and tracking mechanisms are required on board to avoid pointing losses.
- Security: optical communication cannot be intercepted due to the narrowness of the laser beam.
- Interference: due to the narrower laser beams, optical links do not generate interference to other systems and are hard to jam or eavesdrop.
- Noise: in RF receivers, the noise is mainly due to the thermal noise and excess noise in amplifier, while noise in optical links comes from the variations in the photon flux being detected. Noise in the optical detector can be shot noise and excess noise factor from the APD detector, while thermal noise (Johnson noise) appears in the electronics preamplifiers 39. Another noise source in optical links is the background noise from the Sun and other sources, which can be controlled by reducing the receiver optical bandwidth [85].

Table 9.4 summarizes the comparison between RF and optical technologies for ISL.

Finally, Table 9.5 summarizes the most significant figures to compare the performance and main characteristics of RF and optical ISL terminals considering link budget and technology analysis. In this sense, within a similar case study, two proposals for RF ISL Terminals at 33.65 GHz and 70 GHz are compared to the TESAT ConLCT (TESAT's Constellation Laser Communication Terminal) [94]. For this, the range of RF scenario is set to 5,200 km while the specification for TESAT is 6,000 km, the antenna aperture diameter is 0.55 m and the transmit power is 15 W for the RF proposals.

The ISL distance is set to 5,200 km since at Ka band it is the larger range for which the throughput of 2 Gbps with the limitations of power and bandwidth

Table 9.4 Trade-off analysis of RF vs Optical for ISL

Item	RF	Optical
Frequency band	Regulated use by ITU Some ISL bands still far from technology maturity	Not regulated by ITU
Bandwidth	Limited depending in the frequency band	Large
Directivity	High (depending on aperture electrical size)	Significantly higher for the use of lower wavelengths
Beamwidth/Beam divergence	Narrow	Extremely narrow
Pointing mechanism	Mechanical or electronic	Mechanical or beam steering mirrors
Pointing requirements	Less stringent	Highly accurate
Power	Higher transmission power to compensate path losses	Lower transmit power due to lower beam size
Mass	Higher	Reduced (smaller aperture)
Interference	Typ. wider beam than laser Performance degradation due to neighbouring ISL interference	High directivity Beam spread 1,000 times smaller than RF
Doppler shift	Depends on carrier frequency	Large Requires complex mitigation techniques
Security	RF signal can be detected and interfered	Ultra-reliable communications
Platform vibration	Negligible effect	Very sensitive

detected. Furthermore, the power is set regarding the available power limited to 88 W at 70 GHz due to the low power added efficiency (PAE) of maximum 17% that matches well with the total optical terminal power required. It must be highlighted

Table 9.5 Performance analysis of RF and optical ISL for a LEO-to-LEO link

LEO to LEO link	RF		Optical (TESAT ConLCT terminal)
Frequency	33.65 GHz (Ka)	70 GHz (E band)	--
Bandwidth	0.7 GHz	1 GHz	--
ISL distance	5200 km	5200 km	6000 km
Throughput	2 Gbps	4.1 Gbps	10 Gbps
Transmit Power	15 W	15 W	10 W
Antenna diameter	0.55 m (15.3x)	0.55 m (15.3x)	3.6 cm
Beamwidth/Beam divergence	1.09 deg	0.53 deg	$<x10^{-3}$
Estimated mass	55 kg (3.7x)	48 kg (3x)	15 kg

that in RF scenarios, the DC power presented is only for driving the PAs. Besides, there is no COTS (Commercial-Off-The-Shelf) PA that provides 15 W of RF power. Thus, complex power combining networks and their implementation need to be developed.

With respect to the antenna diameter, it is noticed that a larger antenna and so a small beamwidth are needed to avoid high interferences. In the case of Ka band, the beamwidth doubles than the E band. Besides, optical beam is significantly narrower so a more accurate attitude control and mechanical pointing, acquisition and tracking system is required.

In conclusion, in order to decide on RF or optical technologies for ISL a trade-off analysis is required. Today, ISL technology efforts are being pushed by the deployment of NGSO constellations for broadband services, with Telesat and Starlink fostering optical ISL. However, although optical technology is mature, an important effort shall be carried out to miniaturize the optical terminals that have to be integrated into small satellite platforms. The use of RF ISL is limited by the low power efficiency of actual high PAs, which drastically reduces the capacity of the links and compromises the use of electrical power on board.

9.6 Case study for the design of an RF ISL antenna

9.6.1 Presentation of the case study

In this section, a representative case study to define the requirements of a RFISL in terms of link budget requirements and antenna selection is presented. The scope of the case study is the design of a terminal for ISL for a NGSO constellation formed by small satellites, with a special focus on the antenna used for the ISL. Both intra-plane and inter-plane links are considered in the analysis.

The case study analysis is based on a Walker-Star constellation with 648 satellites distributed in 18 planes with 36 satellites each. Satellites describe circular orbits at an altitude of 650 km with an inclination of 86.4 degrees. This constellation has been selected as it allows the analysis of inter-plane ISLs when satellites move in the same direction and in opposite directions. Figure 9.10 shows the 3D view of the scenario, where satellites in the same orbital plane have been plotted in the same colour.

9.6.2 Link budget analysis

As an initial hypothesis to establish the link budget, we assume that satellites establish either intra-plane with the previous or fore satellites or inter-satellite link ISLs with the nearest satellites in the adjacent planes. In both cases, the ISL range depends on the orbital altitude. For intra-plane links, the range depends on the number of satellites per plane, whereas for inter-plane satellite links the number of planes shall be considered.

We first evaluate the variation of free space propagation losses in intra-plane and inter-satellite link ISLs with the number of satellites and the number of

Figure 9.10 3D view of the constellation used for the analysis of the case study

orbital planes, respectively. Free space propagation losses L_{fs} in dB are calculated as

$$L_{fs} = 10log_{10}\left(\frac{4\pi d_{ISL}}{\lambda}\right)^2 \qquad (9.2)$$

where d_{ISL} represents the inter-satellite link distance and λ is the wavelength of the carrier frequency.

Figure 9.11 represents the ISL range and free space losses in intra-plane ISLs at an altitude of 650 km for two representative frequencies of ISL: 33.65 and 67.5 GHz. Variation with the number of satellites in the orbital plane is shown. A constellation with 36 satellites per plane has a range in the intra-plane links of 2,440 km which leads to propagation losses of 190.7 and 196.8 dB for the two frequencies, respectively. Due to the geometrical configuration, range and losses decrease when the number of satellites per plane increases, with the disadvantage of requiring more hops for a given end-to-end routing.

Figure 9.12 represents the range and free space propagation losses between satellites in adjacent planes assuming the two satellites are located in the Equatorial plane and as a function of the number of polar planes of the constellation. For 18 orbital planes, the distance between satellites in the equatorial plane at an altitude of 650 km is 2,440 km, which implies losses of 190.7 and 196.8 dB for 32.65 and 67.5 GHz, respectively. Similar to the behaviour in intra-plane links, increasing the number of planes, the link range is reduced and thus the propagation losses.

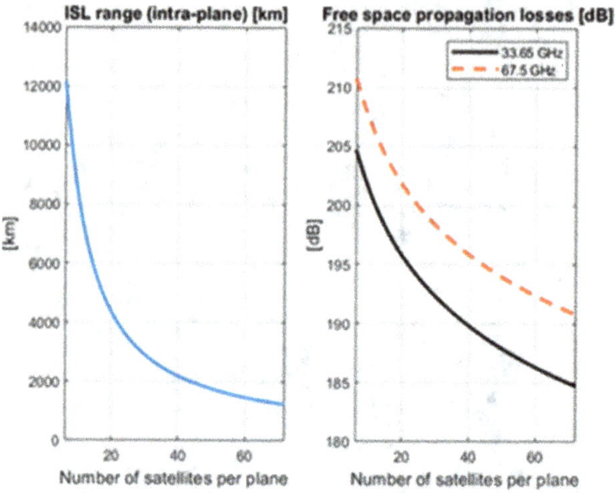

Figure 9.11 ISL range and free space propagation losses for intra-satellite links for an orbital altitude of 650 km

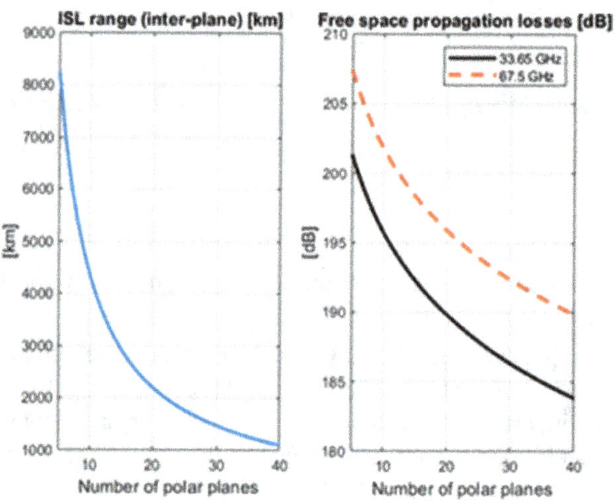

Figure 9.12 ISL range and free space propagation losses for inter-satellite links for an orbital altitude of 650 km (satellites located in the Equatorial plane)

Variations of the relative positions of satellites in adjacent planes the reference constellation can be seen in Figure 9.13.

Next, an analysis of the EIRP and G/T requirements is presented. Any radio link, and in particular, an ISL link can be sized according to the next expression:

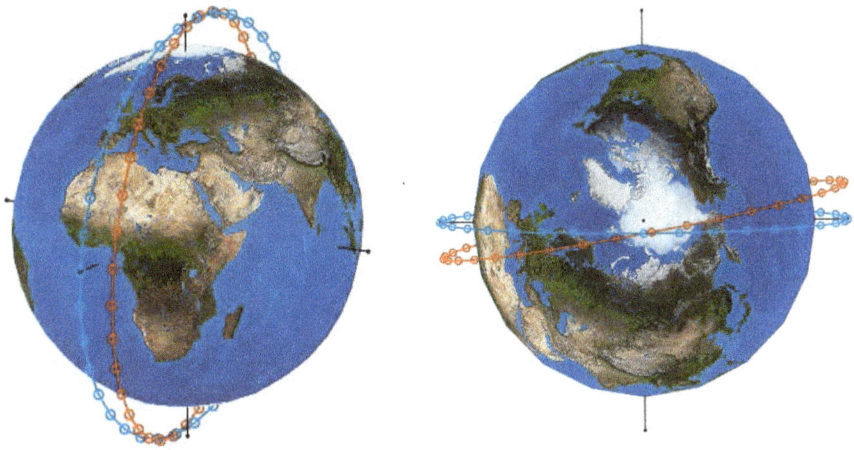

Figure 9.13 Detail of the relative positions of satellites in adjacent planes: front view (left) and zenital view (right)

$$\frac{C}{N} = EIRP + \frac{G}{T} - L_{fs} - L_{add} - K - B \qquad (9.3)$$

where

- C/N represents the received carrier power to noise power ratio (in dB)
- $EIRP$ is the figure of merit of the transmitting satellite terminal (in dBW)
- G/T is the figure of merit of the receiving satellite terminal (in dB/K)
- L_{fs} represents the free space propagation losses (in dB)
- L_{add} represents the additional losses in the link, e.g. pointing losses (in dB)
- K is the Boltzmann constant (−228,6 dB (W/Hz/K))
- B is the noise bandwidth (dB (Hz))

In order to evaluate the link requirements, some of the trade-offs that arises from the variations of system parameters are analysed. Figure 9.14 shows the EIRP and G/T trade-off of the ISL payload for an ISL range of 2,000 km, 33.65 and 67.5 GHz as carrier frequencies and a noise bandwidth of 25 MHz. The trade-off is presented for C/N requirements of 10, 15 and 20 dB. Pointing losses of 0.5 dB are considered at each of the two ends of the link.

Let us take as reference an EIRP of 41dBW in the transmit side and a G/T of 9 dB/K in the receive side, to achieve a C/N of 15 dB for the 33.65 GHz link. Considering a total noise temperature of 180 K, the required antenna gain in transmit and receive is 31.5 dBi and a transmit power of 10 W, which represents a −3 dB-beamwidth of 5.6 degrees. We have a trade-off between transmit power and beamwidth against the platform requirements of power and attitude control to minimize pointing losses, and the optimum EIRP and G/T requirements has to be selected according to the platform.

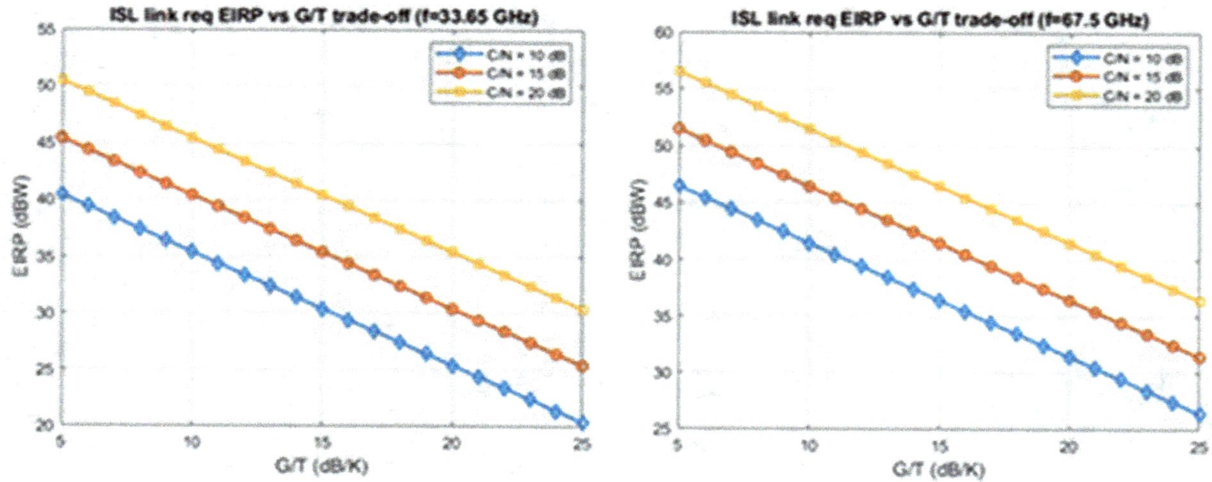

Figure 9.14 EIRP and G/T trade-off for an ISL of 2000 km at a frequency of 33.65 (left) and 67.5 GHz (right)

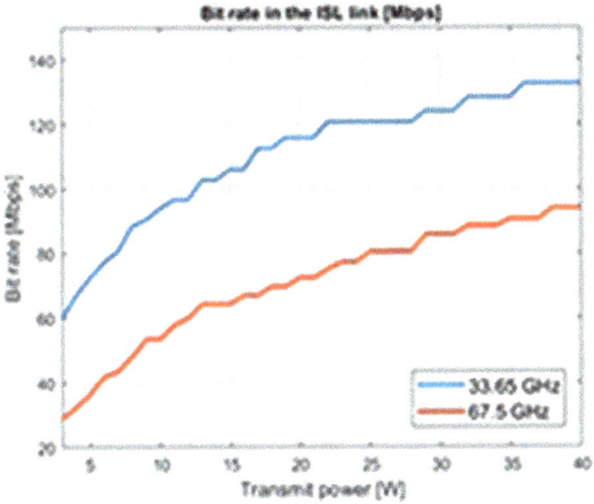

Figure 9.15 *Transmit power requirements for different bit rates in the ISL link*

The trade-off between transmit power and antenna gain can be analysed. Figure 9.15 shows the transmit power requirements for different bit rates using the DVB-S2X transmission system [98] for an antenna gain of 30 dBi and bandwidth of 25 MHz for 33.65 and 67.5 GHz. The higher the transmit power, the most efficient modulation and coding scheme can be used, increasing the bit rate in the link. The curve shows a staircase similar to the quantification of the symbol energy-to-noise power spectral density requirements of the transmission schemes in DVB-S2X. As expected, significant reduction of the achievable rate is appreciated when the frequency of the link is raised from 33.65 to 67.5 GHz.

9.6.3 Analysis of angular scanning requirements

The required scanning range to steer the antenna beam and maintain the communication between two satellites in adjacent planes, for ascending and opposite trajectories starting in the Equator to the pole, is presented.

We present elevation, azimuth, and range variation during an orbital period from a satellite to the neighbour in an adjacent plane when both move in the same direction (Figure 9.16) and when both satellites are moving in opposite directions (Figure 9.17). In both scenarios, the reference Walker-Star constellation with 18 planes and 36 satellites per plane, inclination 86.4 degrees and altitude 650 km is considered.

There are many differences between the two scenarios. First is the time during which both satellites are visible. In the first scenario, both satellites are visible all the time as they move in the same direction in the whole orbit, whereas in the second, the satellites are only visible when they are near the Equatorial plane. In the second scenario, satellites are only visible when the satellites are closer than 4,000 km, which represents 16.5% of the orbital period.

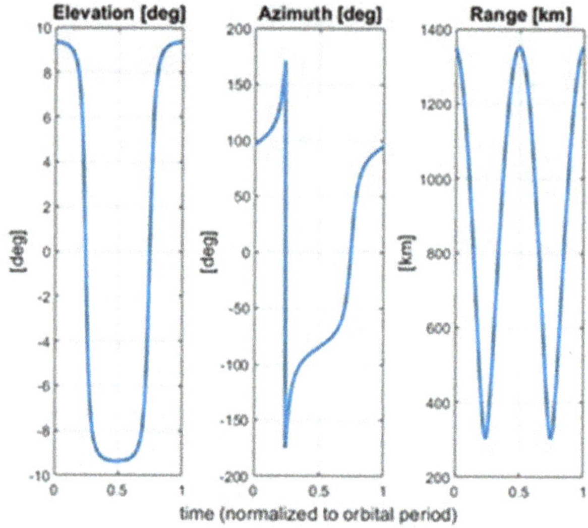

Figure 9.16 *Elevation, azimuth and range for an inter-plane ISL during an orbital period (satellites in the same direction)*

Second, variation in the link angle is also noticeable. In the first scenario, satellites get closer to the poles (300 km) and have a maximum range of 1,300 km close to the Equator. Close to the poles, satellite trajectories cross each other and change

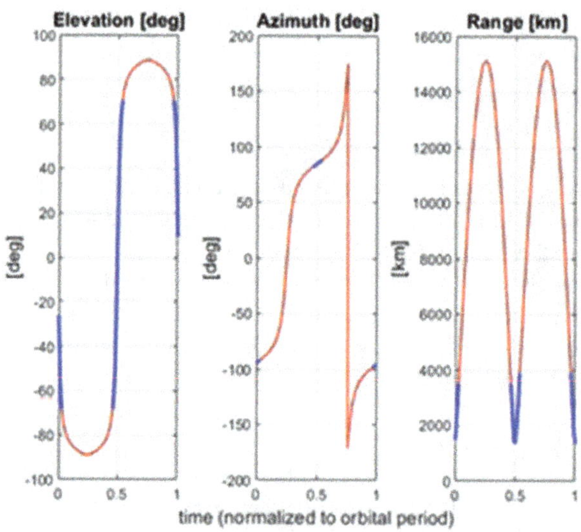

Figure 9.17 *Elevation, azimuth and range for an inter-plane ISL during an orbital period (satellites in opposite directions). Blue (thick): satellites visible; red (thin): non-visible satellites*

their relative position before and after the crossing. In the second scenario, satellites are visible for a short period of time (in the simulation of Figure 9.17, satellites are visible in the Equatorial plane) with a range variation between 1,500 and 4,000 km. Satellites move in opposite directions make the view angle changes rapidly while they are visible.

Third, due to the larger difference in the relative velocity of the two satellites in the second scenario, the acquisition and tracking procedures of the terminals (mainly in elevation) must be faster during the contact times. When the satellites orbit in the same direction, the angular range is concentrated below 10 degrees in elevation. When spacecrafts move in opposite directions, the elevation range reaches 70 degrees during the reduced contact time.

9.6.4 Requirements of the ISL antenna and candidate technologies

The following mission aspects and constraints will drive the selection of the ISL payload and antenna technology [13, 50]:

- Imposed by the mission:
  - Exploration: due to the relative movement of the satellites in inter-plane links, the antenna shall be able to steer the beam in a wide angular range that depends on the number of orbital planes in the constellation and the number of satellites per plane. Antenna aperture and beamwidth must comply with pointing and tracking requirements of the ISL.
  - Knowledge of the constellation status: antenna control unit must have information on the status of neighbour satellites to confirm the availability of a potential link.
  - Space environment: materials of the ISL payload must be compliant with the conditions imposed by the space environment (thermal, avoid out-gassing, radiation, effect of temperature, etc.).
- Imposed by the platform:
  - Launcher conditions: the ISL payload must be compliant with launcher vibrations and the conditions imposed by the space environment.
  - Deployment and allocation of satellites in the launcher: in multi-satellite launches, ISL antennas should minimize the impact on the satellite allocation to ensure the number of satellites per launch is not affected.
  - Power constraints: due to the limited available electrical power on-board, the antenna and RF design shall provide the minimum losses.
  - Mass: ISL payload will have to comply with the mass budget imposed by the platform.
- Imposed by the communications:
  - Link budget: antenna gains, transmit power and noise figures at the receiver shall be selected according to the variable range of the ISL.

- Duplex method: in general, ISLs will be used in transmit and receive modes. The best duplex method shall be defined according to the communications service. The selection of Frequency Division Duplex (FDD) or Time Division Duplex (TDD) will affect the antenna design.
- Frequency band: ISL link frequency shall be selected according to ITU-R and considering the state of the art of the technology of RF components.
- Polarization: circular polarization with a low axial ratio shall be used to minimize the polarization loss factor that linear polarization could produce in the ISL in case polarization axis is not properly aligned

A preliminary list of system-level requirements applicable to the antenna design:

- Antenna shall be directive with an EIRP constrained by the required ISL transmission rate and constellation geometry.
- Losses in the antenna shall be reduced. Due to the limitations of electrical power on-board, losses contribute to reducing the effective transmit power. Thus, the higher the losses, the lower the range or the lower the throughput available in the link.
- ISL throughput shall be as large as possible. This will lead to the use of millimetre wave frequencies where the available bandwidth is larger than in lower frequency bands.
- Beam scanning is required to maintain communication along the orbit while satellites in adjacent planes are visible.
- Antenna shall be lightweight.
- Integration of the ISL payload with the platform shall be easy.

From the analysis carried out previously in section 8.6 and the list of requirements, it is clear that a directive antenna must be used for the ISL. Mandatory requirements are low losses, beam-steering capability and minimal impact on the platform. The following technologies are candidates for the ISL antenna:

- Reflector antennas: characterized by high directivity, low losses, and low axial ratio, reflector antennas will require a mechanical subsystem named APM to steer the beam or a feed cluster. Impact on the platform is largely due to volume, mass, and the addition of the APM, which can be a limiting factor for small or planar satellites.
- Antenna arrays with electronic beam-steering: very powerful in terms of beam-steering and beamforming capabilities, antenna arrays shall be manufactured with low loss materials to avoid degradation of the radiation efficiency. Antenna arrays can be designed modularly (radiation surface, feed network, control subsystem), can be reconfigurable and can be attached to the satellite surface without significant effect on the satellite structure [12], although the final design would be bulky. As well, analogue or digital phase

shifting can be used, and even a hybrid configuration can be designed. Apart from the materials used in the feeding network of the radiating array, insertion losses of the phase shifters can be a limiting factor for the ISL specially in millimetre wave frequencies.

- Dielectric lens antenna: it is formed by a cluster of active antenna elements that individually illuminates a dielectric lens used as an aperture through a selection switch to synthesize a directive beam [99]. The scanning of this antenna operates like a switched beam antenna array, where only one of the elements is activated simultaneously by a switching network controlled by the on-board computer [100]. This technology is applicable in millimetre wave frequencies due to its low losses and low mass, easy manufacturing, and large angular exploration range.

Figure 9.18 shows a comparison of the three candidate technologies for the ISL antenna in terms of power requirements, mass, integration in the platform, losses, reconfigurability and calibration. Due to the complexity and number of components of the antenna array, it is important to emphasize that antenna arrays with electronic beam-steering require intensive calibration procedures before launch and in operation, which makes the integration and control of the antenna array complex [72, 101]. The most balanced technology is the dielectric lens antenna which outperforms the antenna arrays and reflector antennas in two critical parameters for a small satellite: mass and integration with the platform. A more exhaustive comparison amongst antenna technologies can be found in Reference [99].

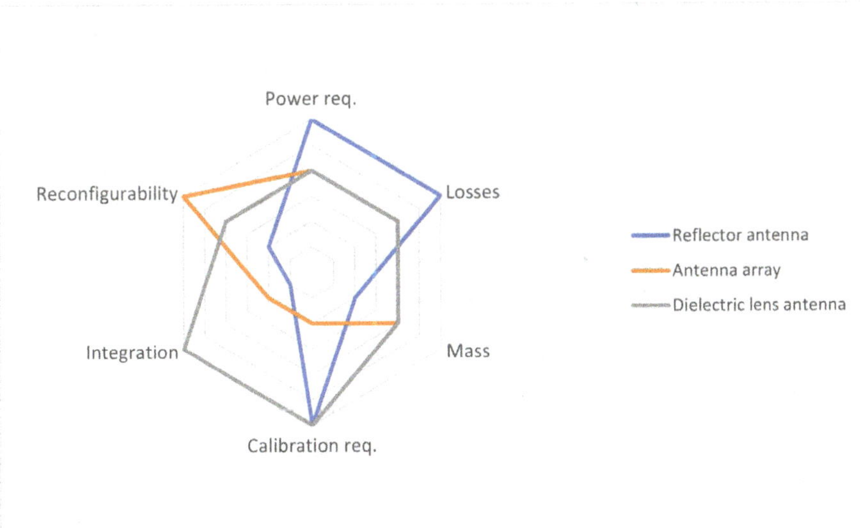

Figure 9.18 Comparison of antenna technologies for ISL

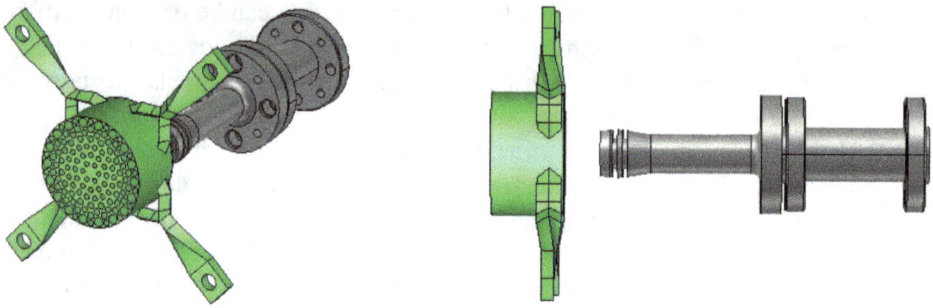

Figure 9.19 Elements of the prototype ISL antenna: front view (left); side view (right)

9.6.5 ISL antenna technology selection and antenna design

From the above discussion, the dielectric lens fed by a cluster of antenna elements is selected to design, manufacture, test, and measure an inter-satellite link antenna in Eband (70–85 GHz). Although this frequency band partially covers the band assigned to ISL (66–71 GHz), the prototype antenna presented here shall be considered as a technology demonstrator whose performance can be extended to the ISL band thanks to its broadband behaviour. Thus, presented results, mass, and volume figures can be extended to the 66 to 71 GHz band.

In a dielectric lens antenna fed by a cluster, beam scanning is produced illuminating the lens with a feed located out-of-the symmetry axis (broadside) of the antenna. In general, beams with a large scanning angle will experience a reduction in the gain relative to the broadside beam. Here, instead of using a feed cluster, we present a proof-of-concept of a single feed illuminating the lens in different positions to demonstrate the generation of scanning (steering) beams. The model of the prototype is shown in Figure 9.19, where the lens is plotted in green and the feed in grey. As an antenna feed, a horn is selected.

Figure 9.20 shows the synthesized beam by the antenna at 16 degrees relative to broadside obtained in simulation. Sidelobe level (SLL) is −20 dB and the back radiation is minimal thanks to the proper design of the lens.

Figure 9.21 shows the setup formed by lens, feed (horn), and mechanical interface to modify the relative position between lens and horn relative to the symmetry axis, and transition to a standard rectangular waveguide WR-12. The setup is finished with a mechanical interface with the instrumentation to be used for verification and testing in e.g. anechoic chamber.

Using the mechanical set-up of Figure 9.21, the antenna can be configured to steer the beam in diverse directions relative to broadside. Results in Figure 9.22 verifies the operation of the antenna in copolar and cross-polar patterns. Steering the beam toward 0, −8, −16, −24 and −31 degrees relative to broadside requires separation of 0, 2, 4, 6 and 8 mm between the feed and the symmetry axis of the system. SLL gets worse when the steering angle is increased, appearing a side lobe with a level of −15 dB at 45 degrees when the main beam is steered to −24 degrees.

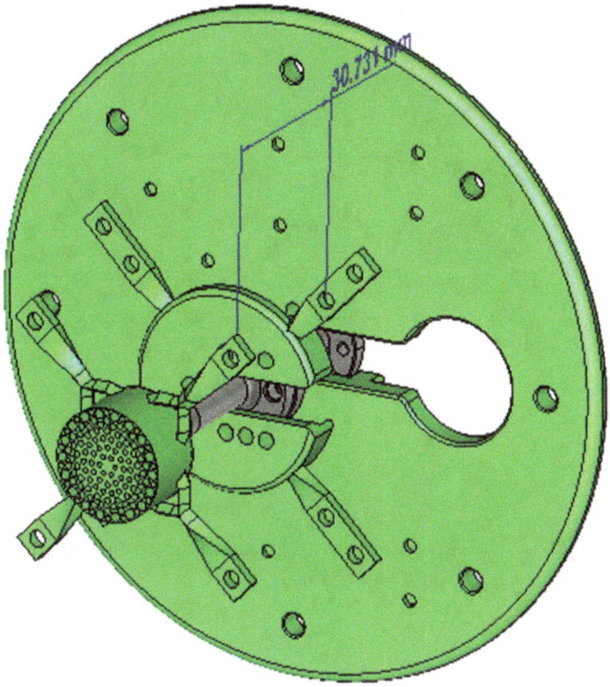

Figure 9.20 *Antenna beam (simulated) of a single feed and lens set up with a beam steered at 16 degrees (directivity of 22.6 dBi)*

Regarding Cross-Polar Discrimination (XPD), its value is larger than 32 dB in the five steering angles.

Finally, the simulated S_{11} of the complete setup is presented in Figure 9.23. In the whole frequency band, the S_{11} value is below −15 dB, which can be considered an acceptable figure for the demonstrator that shall be optimized in the final design.

9.6.6 Antenna prototype manufacturing and measurements

Figure 9.24 shows the measurement setup of lens, horn and mechanical support for the beam scanning. The dielectric lens has been fabricated using 3D printing with a grey resin with validated dielectric constant of 2.5. The horn has been machined in aluminium alloy AL7075, which provides steel-like strength and surface accuracy below 0.15 μm after the manufacturing process. The horn antenna mass is only 4 g, being the total mass of the final antenna given by the number of feeds, switching network, and support structure. The final lens design will be fabricated with 3D printing using Polyether Ether Ketone (PEEK). PEEK is a thermoplastic with excellent properties for space missions: lightweight, strength, temperature resistance, and stability. PEEK has been proposed as material for the 3D printing of CubeSats [102] and nanosatellite [103] structures. As well, thanks to its conductive properties, it

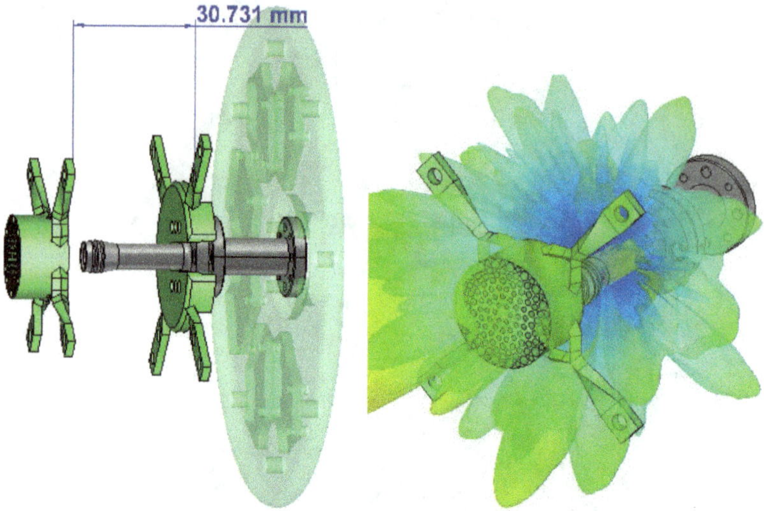

Figure 9.21 *Mechanical model of a single feed and lens, and relative position to synthesize a beam steered at 16 degrees*

has been used to manufacture the surface of reflector antenna and validated in space conditions [104].

In Figure 9.25, the complete ISL antenna setup prior to measurements for concept validation is presented. The measurement setup includes an additional transition from rectangular waveguide WR12 to rectangular waveguide WR10 for the connection to the instrumentation.

Figure 9.26 shows the set-up in a anechoic chamber, with a detail of the ISL antenna as antenna under test (AUT) in the positioner. Additional absorbing material is used around the antenna in the setup to mitigate reflections and obtain antenna patterns without multipath contribution.

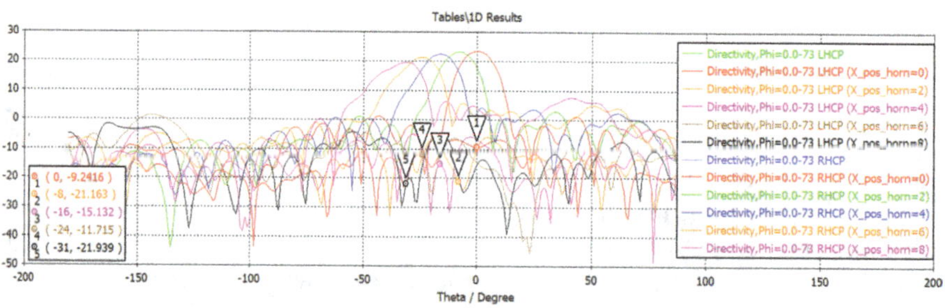

Figure 9.22 *Copolar and cross-polar antenna patterns simulated at 73 GHz of a single feed plus lens setup for different scanning angles (0, −8, −16, −24 and −31 degrees relative to broadside)*

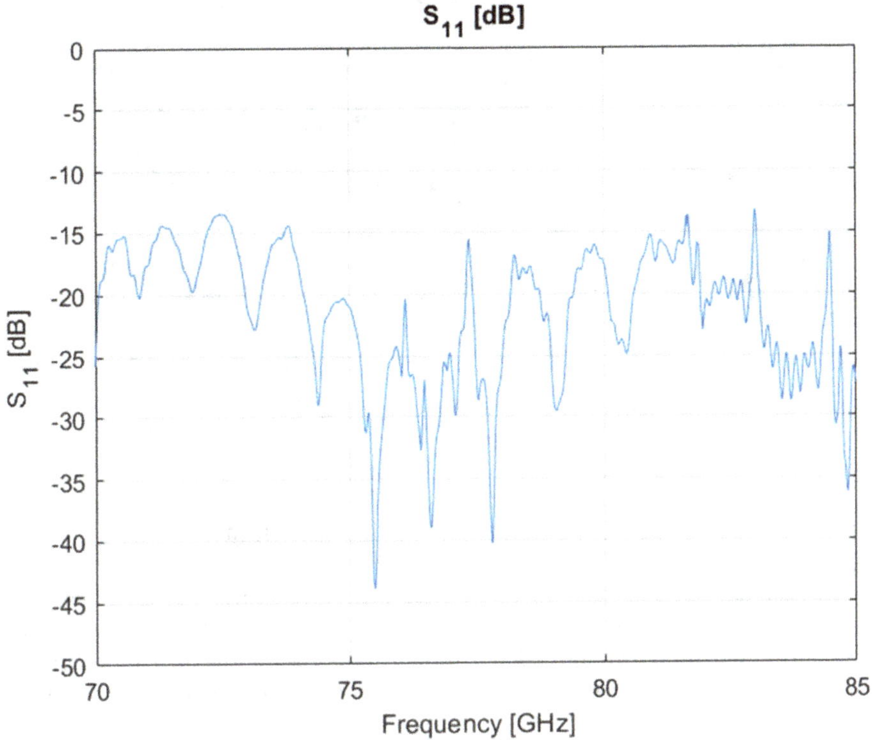

Figure 9.23 S_{11} *of the lens, horn plus transition from circular waveguide to rectangular WR12*

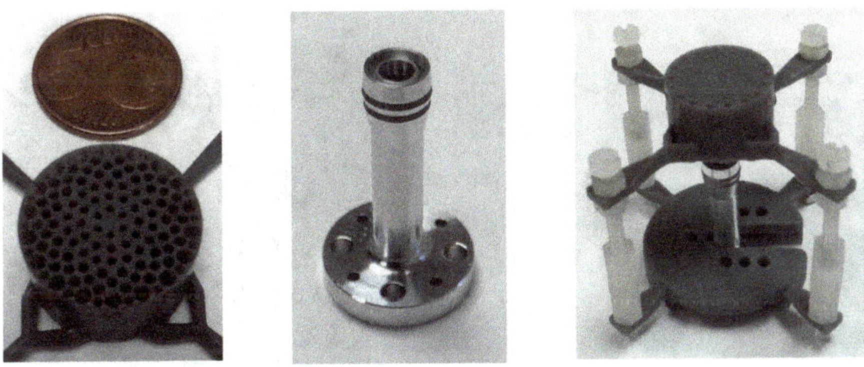

Figure 9.24 *Manufactured lens (left), horn (centre) and support (right) to modify the relative position between lens and hornto validate the beam scanning*

Figure 9.25 ISL antenna setup of lens, horn, waveguide transitions, and support to modify the relative position between lens and horn

Before validating the ISL antenna, radiation measurements of the feed (horn) are carried out. This intermediate validation will confirm the correct machining of the feed and the correct design of the circular waveguide to WR12 transition. Measurements in Figure 9.27 shows the radiation pattern of the horn (E-plane) in the range from 70 to 85 GHz realized from the flange. It is clear from the measurements that no spurious modes are generated in the transition or by machining imperfections in the manufacturing process.

In order to validate the simulations of the complete ISL antenna in the design phase, Figures 9.28 and 9.29 present the antenna patterns at broadside and at two scanning angles, respectively. From the measurements in broadside, measured antenna pattern and sidelobe level matches simulation results. Measurements of antenna patterns out of the broadside direction are used to validate the displacement of the horn to achieve the required scanning. Results confirm the relationship

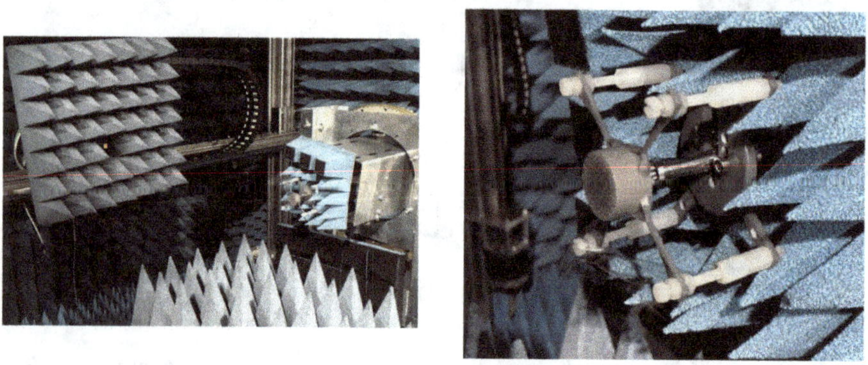

Figure 9.26 Measurement set-up for the ISL antenna in anechoic chamber: complete setup (left), antenna under test (right)

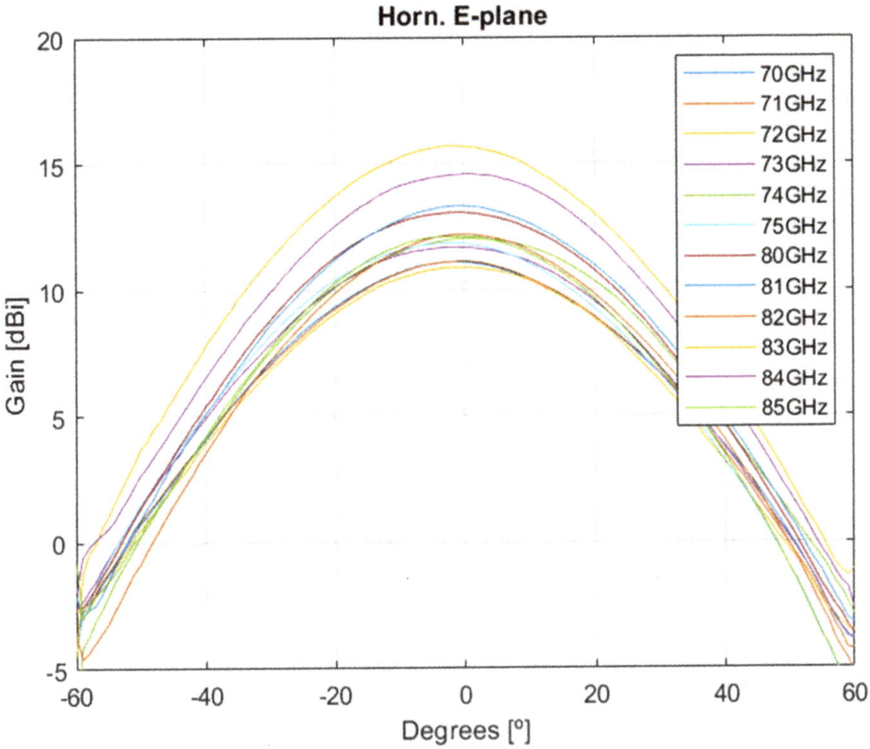

Figure 9.27 *Measured radiation pattern of the horn (E-plane)*

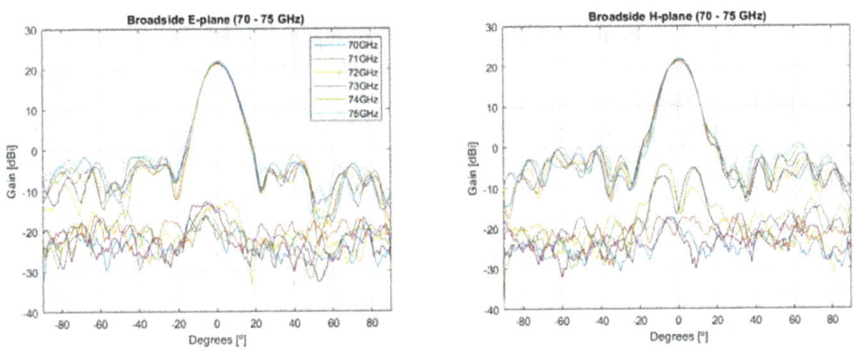

Figure 9.28 *Measured antenna patterns of the ISL antenna in anechoic chamberat 70–75 GHz at broadside: E-plane (left) and H-plane (right)*

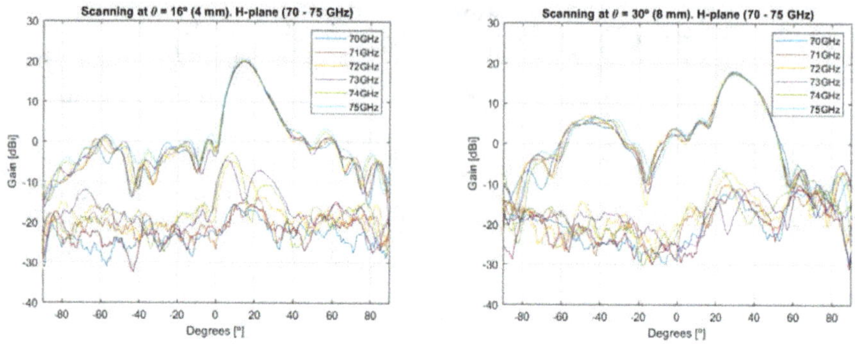

Figure 9.29 Measured antenna patterns of the ISL antenna in anechoic chamberat 70–75 GHzat scanning angles of 16 (left) and 31 (right) degrees(H-plane)

between the scanning at 16 and 30 degrees moving the feed off-axis 4 and 8 mm, respectively, as expected from the simulations.

9.7 Conclusions

The success of new NGSO constellations relies on a number of financial and technological issues, one of them being the use of ISLs between the satellites of the constellations. Thanks to ISL, the performance of the NGSO system is improved in terms of the number of ground station sites, end-to-end latency, capacity, service area and security. Moreover, the use of space systems for 5G is fostered by implementing ISL as a distant gateway station that can be reached from satellites located out of its field of view. In fact, most of the current NGSO system for broadband access makes use of ISL as an intrinsic part of the system. In addition, ISL facilitates diverse communication architectures and can be used in other-than-communication missions, such as data collection space systems, fractionated spacecrafts or to support scientific missions.

However, the introduction of ISLs in NGSO satellites has a large effect on the platform subsystems, communication architecture, and system operation. As shown in section 8.4, a number of complex technical and operational challenges arise in the presence of ISL.

ISL in broadband constellations can be RF or optical. Each of these technologies has pros and cons, so the selection of one of them must consider systems aspects like mission concept, platform requirements, communication link needs and technology maturity. In the case of RF ISLs, the roadmap should be focused on the design of PAs with higher efficiency in the ISL bands as well as advancing in antenna aperture with beam steering capabilities. For optical ISLs, as the links are free from limiting atmospheric effects, one critical aspect is the ATP subsystem to align the laser beam in the presence of vibrations of the platform.

With both technologies, lower mass and power requirements for the ISL terminal are required.

Finally, the chapter presents a case study where requirements for the design of a radio ISL in an NGSO constellation formed by small satellites. The case study concludes with the presentation of prototype of an ISL antenna operating in E band which meets the requirements of mass, losses, beam scanning, and integration with the platform. The measurement results of the proof of concept reinforces the idea that integration of ISL in NGSO constellations, even with small satellites, thanks to the evolution of RF and optical technology achieved since the first NGSO systems were proposed in the 1990s.

References

[1] Butash T., Garland P., Evans B. 'Non-geostationary satellite orbit communications satellite constellations history'. *International Journal of Satellite Communications and Networking*. 2021, vol. 39(1), pp. 1–5.

[2] International Telecommunication Union (ITU) 'Section IV. radio stations and systems, article 1.22, definition: inter-satellite service/inter-satellite radiocommunication service' in *ITU radio regulations*. ITU; 2016.

[3] Kodheli O., Lagunas E., Maturo N, *et al*. 'Satellite communications in the new space era: a survey and future challenges'. *IEEE Communications Surveys & Tutorials*. 2021, vol. 23(1), pp. 70–109.

[4] Hauri Y., Bhattacherjee D., Grossmann M., Singla A. '"Internet from space" without inter-satellite links'. *HotNets '20* [online]; Virtual Event, New York, NY, 2020. pp. 4–6. Available from https://dl.acm.org/doi/proceedings/10.1145/3422604

[5] Iridium Constellation LLC *Iridium next engineering statement, appendix 1*. 2013. Available from https://fcc.report/IBFS/SAT-MOD-20131227-00148/1031348.pdf [Accessed 30 Sep 2021].

[6] Pratt S.R., Raines R.A., Fossa C.E., Temple M.A. 'An operational and performance overview of the IRIDIUM low earth orbit satellite system'. *IEEE Communications Surveys & Tutorials*. 1999, vol. 2(2), pp. 2–10.

[7] LEOSAT Enterprises L. *LeoSat - satellite communication redefined.* Available from https://www.leosat.com [Accessed 26 Sep 2021].

[8] European Space Agency*Perfect images transmitted via a laser link between Artemis and SPOT 4*. 2001. Available from https://www.esa.int/Applications/Navigation/Perfect_images_transmitted_via_a_laser_link_between_Artemis_and_SPOT_4 [Accessed 26 Sep 2021].

[9] Benzi E., Troendle D.C., Shurmer I., James M., Lutzer M., Kuhlmann S. 'Optical inter-satellite communication: the Alphasat and Sentinel-1A in-orbit experience'. *14th AIAA International Conference on Space Operations (SPACEOPS 2016)*; Daejeon, South Korea, 2016.

[10] European Space Agency *European Data Relay Satellite system (EDRS) overview*. Available from https://artes.esa.int/european-data-relay-satellite-system-edrs-overview [Accessed 26 Sep 2021].

[11] Bandyopadhyay S., Subramanian G.P., Foust R., Morgan D., Chung S.-J., Hadaegh F. 'A review of impending small satellite formation flying missions'. *53rd AIAA Aerospace Sciences Meeting* [online]; Kissimmee, FL, Reston, VA, 2015. Available from https://arc.aiaa.org/doi/book/10.2514/MASM15

[12] Martinez Rodriguez-Osorio R., Fuey Ramírez E. 'A hands-on education project: antenna design for inter-cubesat communications [education column]'. [Education Column] *IEEE Antennas and Propagation Magazine*. 2012, vol. 54(5), pp. 211–24.

[13] Radhakrishnan R., Edmonson W.W., Afghah F., Martinez Rodriguez-Osorio R., Pinto F., Burleigh S.C. 'Survey of inter-satellite communication for small satellite systems: physical layer to network layer view'. *IEEE Communications Surveys & Tutorials*. 2016, vol. 18(4), pp. 2442–73.

[14] Poghosyan A., Lluch I., Matevosyan H, *et al. Unified classification for distributed satellite systems 2016* [online]. Available from http://golkar.scripts.mit.edu/fss/wp-content/uploads/2016/10/Unified-Classification-for-DSS_Sept-19-2016.pdf)

[15] Chi L., Sun F., Liu Z., Lin C. 'Overview of fractionated spacecraft technology'. Presented at 2020 International Conference on Computer Engineering and Application (ICCEA); Guangzhou, China, 2020. Available from https://ieeexplore.ieee.org/xpl/mostRecentIssue.jsp?punumber=9102208

[16] European Space Agency. *Hera ESA's planetary defence mission* [online]. Available from https://www.esa.int/Safety_Security/Hera [Accessed 30 Sep 2021].

[17] Michel P., Küppers M., Topputo F., Karatekin Hera Ö. *The ESA Hera mission: planetary defense and science return*. Available from https://www.unoosa.org/documents/pdf/smpag/PDC2021_Presentation_Day1/IAA-PDC-21-01-MASTERFILE.pdf

[18] Telesat *Telesat lightspeed*. 2021. Available from https://www.telesat.com/leo-satellites [Accessed 27 Sep 2021].

[19] Pachler N., del Portillo I., Crawley E.F., Cameron B.G. 'An updated comparison of four low earth orbit satellite constellation systems to provide global broadband'. *IEEE International Conference on Communications Workshops (ICC Workshops)*; Montreal, QC, 2021. pp. 1–7.

[20] Starlink 2021. Available from https://www.starlink.com [Accessed 27 Sep 2021].

[21] *Space exploration holdings, LLC. spacex non-geostationary satellite system. attachment A. technical information to supplement schedule S [online]*. 2016. Available from https://licensing.fcc.gov/myibfs/download.do?attachment_key=1158350 [Accessed 2021].

[22] Space.com *SpaceX paused Starlink launches to give its internet satellites lasers [online]*. 2021. Available from https://www.space.com/spacex-starlink-internet-satellites-lasers-launch [Accessed 30 Sep 2021].

[23] WorldVu Satellites Limited *OneWeb non-geostationary satellite system (LEO). phase 2: amended modification to authorized system* [Attachment B.

Technical Information to Supplement Schedule S - FCC]. 2021. Available from https://fcc.report/IBFS/SAT-MPL-20210112-00007/3493913 [Accessed 20 Sep 2021].

[24] Kuiper Systems LLC *Application of kuiper systems LLC for authority to launch and operate a non-geostationary satellite orbit system in ka-band frequencies* [online]. 2021. Available from https://fcc.report/IBFS/SAT-LOA-20190704-00057/1773885 [Accessed 7 Oct 2021].

[25] Amazon.com Services LLC *job position: optical test engineer*. Available from https://www.amazon.jobs/en/jobs/1514907/optical-test-engineer [Accessed 20 Sep 2021].

[26] Spire Global *Spire: global data and analytics [online]*. Available from https://spire.com [Accessed 20 Sep 2021].

[27] Brown C. *Using inter-satellite links to reduce data latency collected by the spire constellation*. San Luis Obispo, CA. 2021. Available from http://mstl.atl.calpoly.edu/~workshop/archive/2021/presentations/Day%201/Constellation/Conor%20Brown.pdf

[28] Handley M.J. 'Delay is not an option: low latency routing in space'. *Proceedings of the 17thACM Workshop on Hot Topics in Networks (HotNets '18)*; Redmond, WA, New York, NY: ACM, 2018. pp. 85–91.

[29] Kur T., Liwosz T., Kalarus M. 'The application of inter-satellite links connectivity schemes in various satellite navigation systems for orbit and clock corrections determination: simulation study'. *Acta Geodaetica et Geophysica*. 2021, vol. 56(1), pp. 1–28.

[30] *European space agency. optical inter-satellite links are best tech for galileo* [online]. 2019. Available from fromhttps://www.esa.int/Enabling_Support/Space_Engineering_Technology/Shaping_the_Future/Optical_inter-satellite_links_are_best_tech_for_Galileo

[31] Li X., Jiang Z., Ma F., Lv H., Yuan Y., Li X. 'LEO precise orbit determination with inter-satellite links'. *Remote Sensing*. 2021, vol. 11(18), p. 2117.

[32] *international telecommunication union. 5G-fifth generation of mobile technologies [online]*. 2021. Available from https://www.itu.int/en/mediacentre/backgrounders/Pages/5G-fifth-generation-of-mobile-technologies.aspx

[33] Lin X., Rommer S., Euler S., Yavuz E.A., Karlsson R.S. '5G from space: an overview of 3GPP non-terrestrial networks'. *IEEE Communications Standards Magazine*. 2021, vol. 5(4), pp. 147–53.

[34] 3rd Generation Partnership Project (GPP).) *TR 38.811 v15.2.0: study on new radio (NR) to support non-terrestrial networks (release 15)* [online]. Available from https://www.3gpp.org

[35] Slalmi A., Chaibi H., Chehri A., Saadane R., Jeon G. 'Toward 6G: understanding network requirements and key performance indicators'. *Transactions on Emerging Telecommunications Technologies*. 2021, vol. 32(3), pp. 1–5.

[36] *3rd generation partnership project (3GPP).'TR 38.821 v16.0.0: solutions for NR to support non-terrestrial networks (NTN)* [Release16 [online]. Available from https://www.3gpp.org

[37] Pan J., Hu X., Zhou S, *et al.* 'Time synchronization of new-generation BDS satellites using inter-satellite link measurements'. *Advances in Space Research.* 2018, vol. 61(1), pp. 145–53.

[38] Ananda M.P., Bernstein H., Cunningham K.E., Feess W.A., Stroud E.G. 'Global positioning system (GPS) autonomous navigation'. *IEEE Symposium on Position Location and Navigation. A Decade of Excellence in the Navigation Sciences*; Las Vegas, NV, IEEE, 1990. pp. 497–508.

[39] Panahi A., Kazemi A.A., Kress B.C., Kazemi A.A., Chan E.Y. 'Optical laser cross-link in space-based systems used for satellite communications'. *SPIE Defense, Security, and Sensing*; Orlando, Florida, WA: SPIE, 2010.

[40] Baister G.C., Dreicher T., Grond E.R., Gallmann L., Thieme B., Kudielka K. 'The OPTEL terminal development programme'. Presented at Enabling Technologies for Future Optical Crosslink Applications. Collection of the AIAA Space 2003 Conference and Exposition Technical Papers; Long Beach, CA, 2003.

[41] Arya M. 2016. 'Packaging and Deployment of Large Planar Spacecraft Structures'. [PhD Thesis]. California Institute of Technology

[42] Wood L. 'Appendix A: satellite constellation design for network interconnection using non-geo satellites ' in fun' in HuY. M.G., E F. (eds.). *Service efficient network interconnection via satellite.* Wiley; 2002. pp. 215–30.

[43] Leyva-Mayorga I., Soret B., Popovski P. 'Inter-plane inter-satellite connectivity in dense LEO constellations'. *IEEE Transactions on Wireless Communications.* 2021, vol. 20(6), pp. 3430–43.

[44] Chen Q., Yang L., Liu X., Guo J., Wu S., Chen X. 'Multiple gateway placement in large-scale constellation networks with inter-satellite links'. *International Journal of Satellite Communications and Networking.* 2021, vol. 39(1), pp. 47–64.

[45] Ravishankar C., Gopal R., BenAmmar N., Zakaria G., Huang X. 'Next-generation global satellite system with mega-constellations'. *International Journal of Satellite Communications and Networking.* 2021, vol. 39(1), pp. 6–28.

[46] Soret B., Smith D. 'Autonomous routing for LEO satellite constellations with minimum use of inter-plane links'. *ICC 2019 - 2019 IEEE International Conference on Communications (ICC)*; Shanghai, China, Shanghai, China, 2019. pp. 1–6.

[47] Madkour A., Aref W.G., Rehman F.U., Rahman M.A., Basalamah S. *A survey of shortest-path algorithms.* 2017 May 4. Available from https://arxiv.org/pdf/1705.02044.pdf

[48] Leyva Mayorga I., Roeper M., Matthiesen B., Dekorsy A., Popovski P., Soret B. 'Inter plane inter satellite connectivity in LEO constellations: beam switching vs'. *Beam Steering'. Proc. IEEE Global Commun. Conf*; GLOBECOM, 2021.

[49] *European space agency. reentry and collision avoidance* [online]. 2017. Available from https://www.esa.int/Safety_Security/Space_Debris/Reentry_and_collision_avoidance [Accessed 21 Sep 2021].

[50] Tiainen A. 2017. 'Inter-satellite Link. Antennas Review and The Near Future'. [Master Thesis]. Luleå University of Technology
[51] Park J., Ryu H., Ha K.-W., Kim J.-G., Baek D. '76–81-Ghz CMOS transmitter with a phase-locked-loop-based multichirp modulator for automotive radar'. *IEEE Transactions on Microwave Theory and Techniques*. 2015, vol. 63(4), pp. 1399–408.
[52] Oh J., Jang J., Kim C.-Y., Hong S. 'A W-band 4-GHz bandwidth phase-modulated pulse compression radar transmitter in 65-nm CMOS'. *IEEE Transactions on Microwave Theory and Techniques*. 2015, vol. 63(8), pp. 2609–18.
[53] Uzunkol M., Gurbuz O.D., Golcuk F., Rebeiz G.M. 'A 0.32 THz SiGe 4x4 imaging array using high-efficiency on-chip antennas'. *IEEE Journal of Solid-State Circuits*. 2013, vol. 48(9), pp. 2056–66.
[54] Sarkas I., Hasch J., Balteanu A., Voinigescu S.P. 'A fundamental frequency 120-GHz SiGe BiCMOS distance sensor with integrated antenna'. *IEEE Transactions on Microwave Theory and Techniques*. 2012, vol. 60(3), pp. 795–812.
[55] Yan Y., Zihir S., Lin H., Shin W., Rebeiz G., Inac O. 'A 155 GHz 20 Gbit/s QPSK transceiver in 45 nm CMOS'. *IEEE Radio Frequency Integrated Circuits Symposium*; Tampa, FL, IEEE, 2014. pp. 365–69.
[56] Suh B., Lee H., Kim S., Jeon S. 'A D-band multiplier-based OOK transceiver with supplementary transistor modeling in 65-nm bulk CMOS technology'. *IEEE Access: Practical Innovations, Open Solutions*. 2019, vol. 7, pp. 7783–93.
[57] *Macom [online]*. Available from https://www.macom.com/[Accessed 1 October 2021]
[58] *Analog devices [online].available from*. Available from https://www.analog.com/
[59] *Eravant*. Available from https://www.eravant.com/ [Accessed 1 October 2021]
[60] ERAVANT *E-band power ampifier 67 to 76 ghz, 35 db gain, +34 dbm p1db, in-line WR-12 waveguide [online]*. Available from https://www.eravant.com/67-to-76-ghz-35-db-gain-34-dbm-p1db-in-line-wr-12-waveguide-e-band-power-amplifier
[61] Wagner E., Rebeiz G.M. 'Single and power-combined linear E-band power amplifiers in 0.12-μm SiGe with 19-dBm average power 1-GBaud 64-QAM modulated waveforms'. *IEEE Transactions on Microwave Theory and Techniques*. 2019, vol. 67(4), pp. 1531–43.
[62] Lin H.-C., Rebeiz G.M. 'A 70–80-GHz SiGe amplifier with peak output power of 27.3 dBm'. *IEEE Transactions on Microwave Theory and Techniques*. 2016, vol. 64(7), pp. 2039–49.
[63] Mailloux R.J. (ed.) *Phased array antenna handbook*. Third Edition. Norwood, MA: Wiley; 2017.
[64] Mohammod A. *Reconfigurable antenna design and analysis*. Boston, MA: Artech House; 2021.

[65] Motovilova E., Huang S.Y. 'A review on reconfigurable liquid dielectric antennas'. *Materials (Basel, Switzerland)*. 2020, vol. 13(8), pp. 1–28.

[66] Christodolou C., Tawk Y., Lane S., Erwin S. 'Reconfigurable antennas for wireless and space applications'. *Proceedings of the IEEE*. 2012, vol. 100(7), pp. 2250–61.

[67] Jakoby R., Gaebler A., Weickhmann C. 'Microwave liquid crystal enabling technology for electronically steerable antennas in SATCOM and 5G millimeter-wave systems'. *Crystals*. 2020, vol. 10(6), p. 514.

[68] Ferrari P., Jakoby R., Karabey O.H., Rehder G.P., Maune H. 'Reconfigurable circuits and technologies for smart millimeter-wave systems' in *Millimeter-wave Systems*. Cambridge: Cambridge University Press; 2022 May 26.

[69] Cuzzola M. 'KA-band LEO-LEO ISL for small satellites'. *5th IAA Conference*. 2020, Rome.

[70] Rui Y., Wenjun G., Chunbang W., HongbinL. 'Inter-satellite link antenna' in *Technologies for Spacecraft Antenna Engineering Design*. Springer; 2020. pp. 103–35.

[71] Salas-Natera M.A., Rodriguez-Osorio R.M., de Haro Ariet L., Sierra-Perez M. 'Novel reception and transmission calibration technique for active antenna array based on phase center estimation'. *IEEE Trans Antennas Propag*. 2017, vol. 65(10), pp. 5511–22.

[72] Salas-Natera M.A., Martínez Rodríguez-Osorio R., de Haro-Ariet L. 'Procedure for measurement'. *'Characterization and Calibration of Active Antenna Arrays'. IEEE Transactions on Instrumentation and Measurement*. 2012, vol. 62(11), pp. 377–91.

[73] Sai Sandeep B., Vijayendra S., Vineeth Reddy B. 'High gain antenna design for inter satellite link at millimeter range'. *International Journal of Scientific & Technology Research*. 2020, vol. 9(1), pp. 1946–48.

[74] Swissto12 *Electronically steered antenna array for LEO [online]*. 2021. Available from https://swissto12.com/products/electronically-steered-antenna-arrays/ [Accessed 1 Oct 2021].

[75] Anteral *Innovative antennas. Passives and radar technologies* [online]. Available from https://anteral.com [Accessed 1 Oct 2021].

[76] Melero Frago M., Calero Fernández I., Arregui Padilla I, *et al*. 'Nuevas antenas y diplexores de bajo coste para radioenlaces 5G'. *XXXVI Simposium Nacional de la Unión Científica Internacional de Radio*; Vigo, Spain, 2021.

[77] Teberio F., Calero I., Arregui I., Martín-Iglesias P., Teniente J., Gómez-Laso M.A. 'High-yield waveguide diplexer for low-cost E-band 5G'. *Point-to-Point Radio Links'. European Microwave Week*. 2022.in press.

[78] SSTL [online]. Available from https://www.sstl.co.uk [Accessed 1 Oct 2021].

[79] Chi P.-L., Pao C.-H., Huang M.-H., Yang T. 'High-gain patch-fed 3D-printing Fresnel zone plate lens antenna for 60-GHz communications'. *IEEE International Symposium on Antennas and Propagation & USNC/URSI National Radio Science Meeting*; Boston, MA, 2018.

[80] Petosa A., Ittipiboon, A. 'A Fresnel lens designed using a perforated dielectric'. *9th International Symposium on Antenna Technology and Applied Electromagnetics*; 2002.

[81] Petosa A., Ittipiboon A. 'Design and performance of a perforated dielectric Fresnel lens'. *IEE Proceedings-Microwaves Antennas and Propagation*. 2003, vol. 150(5), pp. 309–14.

[82] Kadri I., Petosa A., Roy L. 'Ka-band Fresnel lens antenna fed with an active linear microstrip patch array'. *IEEE Trans Antennas Propag*. 2016, vol. 53(12), pp. 4175–78.

[83] Vourch C.J., Drysdale T.D. 'V-band bull's eye antenna for multiple discretely steerable beams'. *IET Microwaves, Antennas & Propagation*. 2016, vol. 10(3), pp. 318–25.

[84] Chaudhry A.U., Yanikomeroglu H. 'Free space optics for next-generation satellite networks'. *IEEE Consumer Electronics Magazine*. 2021, vol. 10(6), pp. 21–31.

[85] Kaushal H., Kaddoum G. 'Optical communication in space: challenges and mitigation techniques'. *IEEE Communications Surveys & Tutorials*. 2009, vol. 19(1), pp. 57–96.

[86] Smutny B., Hemmati H., Kaempfner H, *et al*. '5.6 Gbps optical intersatellite communication link'. *SPIE LASE: Lasers and Applications in Science and Engineering*; San Jose, CA, 2009.

[87] Li Q., Guo H., Xu S, *et al*. 'TRC-based high-precision spot position detection in inter-satellite laser communication'. *Sensors (Basel, Switzerland)*. 2020, vol. 20(19), p. 19.

[88] Mynaric.*Space terminal products* [online]. 2021. Available from https://mynaric.com/products/space/

[89] Mynaric. *CONDOR MK3. optical communications terminal for intersatellite operations [online]*. 2021. Available from https://mynaric.com/products/space/condor-mk3/ [Accessed 1 Oct 2021].

[90] *General atomics. optical communications terminal* [online]. 2020. Available from https://www.ga.com/space-systems/space-technologies

[91] *TESAT. Products* [online] 2021. Available from https://www.tesat.de/products [Accessed 1 Oct 2021].

[92] Rödiger B., Menninger C., Fuchs C, *et al*. 'High data-rate optical communication payload for cubesats'. *SPIE Optics + Photonics conference Digital Forum. SPIE Optical Engineering + Applications*; San Diego, CA: SPIE, 2020. pp. 1–13.

[93] TESAT *CubeLCT terminal datasheet* [online]. 2019. Available from https://www.tesat.de/images/tesat/products/CubeLCT_Data-Sheet.pdf

[94] TESAT *ConLCT terminal* [online]. 2021. Available from https://www.tesat.de/products

[95] TESAT TESAT ' S ConLCT modem successfully demonstrates sda laser communication standard. 2021. Available from https://www.tesat.de/news/press/921-tesat-s-conlct-modem-successfully-demonstrates-sda-laser-communication-standard

[96] Yang Q., Tan L., Ma J. 'Doppler characterization of laser inter-satellite links for optical LEO satellite constellations'. *Optics Communications*. 2009, vol. 282(17), pp. 3547–52.

[97] Song T., Wang Q., Wu M.-W., Kam P.-Y. 'Performance of laser inter-satellite links with dynamic beam waist adjustment'. *Optics Express*. 2016, vol. 24(11), pp. 11950–60.

[98] *European telecommunications standard institute (ETSI). digital video broadcasting (DVB); second generation framing structure, channel coding and modulation systems for broadcasting, interactive services, news gathering and other broadband satellite applications; part 2: DVB-S2 extensions (DVB-S2X)* [online]. 2021. Available from https://www.etsi.org/deliver/etsi_en/302300_302399/30230702/01.03.01_60/en_30230702v010301p.pdf

[99] Hong W., Hiang Z.H., Yu C., Zhou J., Chen P., Yu Z. 'Multibeam antenna technologies for 5G'. *IEEE Transactions on Antennas and Wireless Communications*. 2017, vol. 65(12), pp. 6231–49.in press.

[100] Imbert M., Romeu J., JofreL. *GHz WPAN applications. IEEE Antennas and Propagation Society International Symposium (APSURSI).*; Orlando, FL, 2013.

[101] Salas-Natera M.A., Martínez Rodríguez-Osorio R. 'Analytical evaluation of uncertainty on active antenna arrays'. *IEEE Transactions on Aerospace and Electronic Systems*. 2012, vol. 48(3), pp. 1903–1913.

[102] *European Space Agency (ESA). 3D printing cubesat bodies for cheaper, faster missions*. 2017. Available from https://www.esa.int/Enabling_Support/Space_Engineering_Technology/3D_printing_CubeSat_bodies_for_cheaper_faster_missions [Accessed 6 Oct 2021].

[103] Rinaldi M., Cecchini F., Pigliaru L., Ghidini T., Lumaca F., Nanni F. 'Additive manufacturing of polyether ether ketone (PEEK) for space applications: A nanosat polymeric structure'. *Polymers*. 2020, vol. 13(1), pp. 1–16.

[104] Kalra S., Munjal B.S., Singh V.R., Mahajan M., Bhattacharya B. 'Investigations on the suitability of peek material under space environment conditions and its application in a parabolic space antenna'. *Advances in Space Research*. 2019, vol. 63(12), pp. 4039–45.

Chapter 10

Non-geostationary orbit constellation design for global connectivity

Israel Leyva-Mayorga[1], *Beatriz Soret*[2,3], *Bho Matthiesen*[4,5], *Maik Röper*[4], *Dirk Wübben*[4], *Armin Dekorsy*[4], *and Petar Popovski*[1,5]

10.1 Introduction

Providing global connectivity is not possible with terrestrial infrastructure alone. This is due to a multitude of factors; the most important of which are geographical conditions and economic reasons. So far, it seems that every new mobile wireless generation has ambitions to connect sparsely populated areas, but terrestrial options have not proven to be cost-effective. While providing radio access merely necessitates the deployment of a base station (BS) or access point in the area of interest, connecting this infrastructure to the core network and, hence, to the Internet through backhaul and, possibly, fronthaul links is much more challenging. A clear use case is providing global connectivity to vessels in the open ocean, where deploying BSs and the necessary backhaul links (i.e., sea cables) to the many possible routes is not feasible.

In contrast, geostationary orbit (GEO) satellites have been used to provide global communication coverage for several decades, for instance, for TV broadcasting or maritime connectivity. In addition, Global Positioning System (GPS) is an example of a widespread medium Earth orbit (MEO) service. Even though GEO satellites are able to provide global service availability in underserved and disconnected areas [1], they are not efficient on their own as a competitive global connectivity solution. Due to the high altitude of the orbit, GEO satellites suffer from a long propagation delay and a high-signal attenuation. The latter aspect is specially

[1]Department of Electronic Systems, Aalborg University, Aalborg, Denmark
[2]Telecommunications Research Institute (TELMA), University of Malaga, Malaga, Spain
[3]Department of Electronic Systems, Aalborg University, Aalborg, Denmark
[4]Gauss-Olbers Center, c/o University of Bremen, Dept. of Communications Engineering, Germany
[5]Department of Communications Engineering, University of Bremen, U Bremen Excellence Chair, Germany

problematic when devices with energy and size restrictions attempt to communicate in the uplink, which are some of the defining characteristics of Internet of things (IoT) devices [2, 3].

Non-geostationary orbit (NGSO) satellite constellations represent a cornerstone in the NewSpace paradigm and thus have become one of the hottest topics for the industry, academia, but also for national space agencies and regulators. For instance, numerous companies worldwide, including Starlink, OneWeb, Kepler, SPUTNIX and Amazon have started or will soon start to deploy their own NGSO constellations [4], which aim to provide either broadband [5, 6] or IoT services [2]. One of the major drivers for such a high interest on NGSO constellations is that, with an appropriate design, they are capable of providing global coverage and connectivity. While global connectivity can also be provided by a small set of GEO satellites, NGSO constellations present three main advantages over terrestrial and GEO satellite communications:

1. **Short propagation delay:** Electromagnetic waves propagate faster in the vacuum than in optic fibre, which has typical refraction index of 1.44 to 1.5 [7]. Moreover, NGSO satellites are deployed at much lower altitudes than GEO satellites, which reduce the one-way ground-to-satellite (G2S) propagation delays to a few milliseconds. As a consequence, the end-to-end (E2E) latency with NGSO satellites over long distances may be competitive and even lower than that of terrestrial networks [7].
2. **Global connectivity:** NGSO satellites can provide coverage in remote areas where terrestrial infrastructure is not available. Furthermore, if appropriate functionalities are implemented, the data could be routed E2E by the satellites themselves.
3. **Feasible uplink communication from small devices:** Due to the relatively low altitude of deployment and the signals propagating mainly through free-space, it is feasible for small devices to communicate directly with low Earth orbit (LEO) satellites. This has led companies and organisations to aim for integrated space and terrestrial infrastructures using low-power wide-area network (LPWAN) technologies such as LoRaWAN and Narrowband IoT (NB-IoT) [8, 9].

Based on these advantages, some of the main use cases for NGSO constellations include:

1. **Backhauling:** Inter-satellite communication can be used to transmit the data towards the Earth, even when the source and destination are not within the coverage of the same satellite [10].
2. **Offloading:** NGSO constellations can serve as additional infrastructure in urban hot-spots where the capacity of the terrestrial network is temporarily exceeded, e.g., during sport and cultural events.
3. **Resilience:** Satellites in NGSO can serve as failback backhaul network for terrestrial BSs in case the primary backhaul fails, e.g., due to natural disasters.

4. **Edge computing and Artificial Intelligence AI as-a-service (AIaaS):** IoT devices have limited processing capabilities and limited energy supply (i.e., batteries). Therefore, NGSO satellites can be used as edge computing nodes [11] to reduce the computational load at the IoT devices. Furthermore, the satellites can gather data from several devices and locations along their orbit and use it, along with their computational capabilities, to provide AIaaS to devices where the data and/or processing capabilities are insufficient for AI [12].
5. **Earth observation:** NGSO satellites can be used as moving sensing devices that capture data, e.g., in the form of images or video, of physical phenomena at the Earth's surface or within its atmosphere. To obtain a sufficient resolution, LEOs are the preferred choice in most Earth observation satellite missions. Furthermore, sun-synchronous orbits, i.e., an orbit where the satellite maintains a constant angle towards the sun when viewed from Earth, often have favourable properties for Earth observation tasks.

For the use cases mentioned above, and many more, global connectivity is essential, as it allows to fully exploit the benefits of NGSO constellations. Specifically, it would allow the constellation to deliver the data generated on the ground, by aerial vehicles, or by the satellites themselves to the destination without heavily relying on additional (e.g., terrestrial) infrastructure.

Nevertheless, there are several key performance indicators (KPIs) that should be considered to determine whether an NGSO constellation design is appropriate for the target application. These include but are not restricted to

- **Service availability:** The fraction of the time in which the ground terminal is able to communicate with the constellation [8]. Through this chapter, we will assess this KPI based on the coverage of the constellation in different locations, including remote (e.g., polar) regions.
- **Transport capacity:** The maximum amount of data that can be transmitted by the constellation, E2E, per time unit.
- **Throughput:** Data rate experienced by the users.
- **Scalability:** Maximum number of devices supported by the constellation per unit area.
- **Inter-satellite connectivity:** The ability to achieve inter-satellite communication. Oftentimes it is assessed by the number of satellites with active connections [13] or by the number of satellites within the communication range [14].
- **Latency and reliability:** Probability that the data can be transmitted to the destination within a given time t.
- **Energy efficiency:** Since IoT devices and satellites are usually powered by batteries, minimising the energy required for communication is essential.

Several of these KPIs were considered by Del Portillo [6] to compare the OneWeb, Starlink (outdated configuration with $h > 1000$ km) and Telesat constellations.

Other KPIs have been defined for satellite constellations. For example, Soret *et al.* [10] emphasised the relevance of timing metrics beyond the packet delay, such as the age of information and its by-products, for some satellite tracking or remote sensing applications.

10.2 NGSO constellation design

Satellite constellations are groups of satellites organised in orbital planes. The N_{op} satellites in one orbital plane follow the same orbital trajectory, one after the other, and are usually uniformly spaced around the orbit. Furthermore, an orbital shell is a group of P orbital planes in a constellation that are deployed at approximately the same altitude; some orbital shells may implement minor variations of few kilometers called orbital separation. To maximise the coverage for communications, the organisation of satellites in one orbital shell usually belongs to one of the two basic types: Walker star and Walker delta (also called Rosette) [2, 5, 15, 16]. Satellite constellation design may include one or more orbital shells.

Walker star orbital shells consist of nearly polar orbits, with typical inclinations of $\delta \approx 90°$, which are evenly spaced within 180°. As such, the angle between neighbouring orbital planes is $180/P$.

Walker delta orbital shells, on the other hand, typically consist of inclined orbits, with typical inclinations of $\delta < 60°$, which are evenly spaced within 360°. As such, the angle between neighbouring orbital planes is $360/P$.

Due to the use of inclined orbits, Walker delta orbital shells do not provide coverage in polar regions or in the northernmost countries such as Greenland. However, this allows to keep the satellites within the areas where most of the population resides and, hence, where data traffic is generated and consumed.

Since both Walker star and delta geometries provide distinct advantages and disadvantages, some companies, such as SpaceX, have considered a mixed design consisting of multiple orbital shells. Specifically, the design of the Starlink constellation considered a Walker delta orbital shell at around 550 km and at around 1100 km. However, the Federal Communications Commission (FCC) granted permission to SpaceX to modify the constellation geometry and to lower the 2814 satellites in the 1100 km orbital shell to an altitude between 540 and 570 km [17]. Walker star, Walker delta, and an example of a mixed geometry, are illustrated in Figure 10.1. In addition, Table 10.1 shows the design parameters of some relevant NGSO constellations. The values in this table were obtained from the companies web pages, related papers [6, 14] and FCC filings and some of them have not yet been approved.*

Beyond the technical aspects, the dramatic increase in the number of objects put into orbit around the Earth due to the deployment of NGSO constellations has raised concerns on their long-term sustainability. Naturally, the more satellites orbit the Earth, the higher the risk of collision. Hence, measures to minimise the collision risk have been explored and should be adopted in commercial

*Updates on ongoing launches can be found at the New Space webpage https://www.newspace.im/

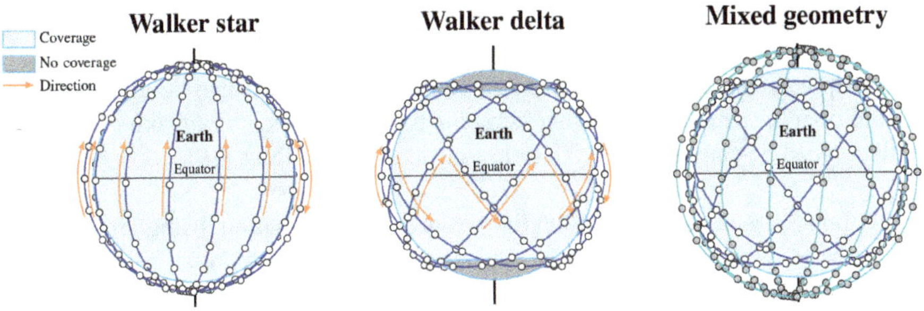

Figure 10.1 *Diagram of Walker star, Walker delta (Rosette) and mixed constellation geometries*

constellations [18]. In particular, deploying the orbital planes at slightly different altitudes, with differences of less than 4 km, greatly reduces the collisions caused by failed satellites: a scenario that cannot be avoided. However, this introduces slight asymmetries in the constellations that complicate several technical aspects; these will be further described in section 10.4.2. Another example of slight asymmetries in the constellations is that the satellites between neighbouring orbital planes may be shifted across the orbit so that the satellites in one orbital plane are rotated by a relatively small angle with respect to those in the neighbouring planes. Figure 10.1.

Table 10.1 *Parameters for some commercial non-geostationary orbit satellite constellations*

Parameter	**Constellation**						
	Starlink					**OneWeb**	**Kepler**
Type	Mixed					Walker star	Walker star
Number of satellites N	1584	1584	720	348	172	648	140
Number of orbital planes P	72	72	36	6	4	18	7
Altitude h (km)	550	540	570	560	560	1200	575
Inclination δ (°)	53	53.2	70	97.6	97.6	86.4	98.6
Intended service	Broadband					Broadband	Internet of things

10.3 Communication links

The endpoints for communication in an NGSO constellation can be either on the ground or at the satellite level. Therefore, the paths that the data can take in the constellation can be classified into the following four logical links [19].

- Ground to ground (G2G): With both, source and destination, being ground and/or aerial terminals. This is the typical use of the constellation for terrestrial backhauling.
- G2S: With the source being a ground or aerial terminal and the destination being a satellite. This link is mainly used for operations initiated by dedicated ground stations (GS) such as constellation, route, and link establishment and maintenance, tele-control and tele-command, and content caching.
- Satellite to ground: With the source being a satellite and the destination being a ground or aerial terminal. This link is mainly used when the satellites themselves generate application data that must be transmitted to a GS for storage and/or processing. For example, in Earth and space observation, but also for telemetry, handover, link maintenance and adaptation, and fault reporting.
- Satellite to satellite: With the source and destination being satellites, possibly deployed at different altitudes and/or orbits. This link is used for localised network maintenance, updating routing tables, neighbour discovery or other applications such as distributed processing, sensing and inference.

These links must be realised with a moving infrastructure. NGSO satellites move rapidly with respect to each other in higher and lower orbits and in different orbital planes. They also move with respect to the Earth due to the satellite orbital velocity and to the Earth's rotation [20]. Specifically, the orbital velocity of the satellites v_o is determined by the altitude of deployment h as

$$v_o(h) \approx \sqrt{\frac{GM_E}{R_E + h}} \tag{10.1}$$

where, G is the universal gravitational constant; M_E and R_E are the mass and radius of the Earth, respectively. Then, according to Kepler's third law of planetary motion, the orbital period of a satellite can be closely approximated as

$$T_o(h) \approx \frac{2\pi(R_E + h)}{v_o(h)} = \sqrt{\left(\frac{4\pi^2}{GM_E}\right)(R_E + h)^3} \tag{10.2}$$

From here and assuming traditional LEO altitudes, e.g., with $h = 600$ km, it is easy to observe that the orbital velocity of NGSO satellites may exceed 7.6 km/s and that their orbital period is usually around 90 minutes.

Furthermore, satellites in different locations of the constellation may move rapidly w.r.t. each other. Finally, the whole satellite constellation is moving w.r.t. the Earth due to its rotation [2, 20]. Hence, an important aspect to select the altitude of deployment of a constellation is whether it is desired that the orbit is recursive. That is, whether the satellites should pass over the same point in the Earth at a specific time of the day after a given number of days m. For this, we require to find an altitude h_{rec} for which $nT_o = mT_E$, where $T_E = 86164$ s is the equinoctial day [2]. To find the required altitude for the recursive orbit, let us first rewrite the right-hand side of 10.2 as

$$T_o^2 = \left(\frac{4\pi^2}{GM_E}\right)(R_E + h)^3 \tag{10.3}$$

which allows us to define h as a function of T_o as

$$h = \left(\frac{T_o^2 GM_E}{4(\pi)^2}\right)^{1/3} - RE \tag{10.4}$$

Finally, we substitute the period $T_o = mT_E/n$ in (10.4) to find the altitude for a recursive satellite orbit as

$$h_{\text{rec}}(n, m, T_E) = \left(\frac{(mT_E)^2 GM_E}{(2n\pi)^2}\right)^{1/3} - R_E \tag{10.5}$$

From (10.5), we obtain that NGSO satellites at $h = 554$ km, close to Starlink's altitude of deployment, have recursive orbits for $n = 15$ and $m = 1$. That is, these will orbit the Earth exactly 15 times each day. Moreover, those at $h = 1248$ km, close to OneWeb's altitude of deployment, have recursive orbits for $n = 13$ and $m = 1$.

An essential aspect to observe about NGSO satellite constellations is that, even though the relative positions and velocities of satellites w.r.t. other satellites and to the ground terminals are dynamic, the dynamics of the constellation are fully dictated by the physics of the system and, hence, completely predictable. Therefore, the topology of the network (space and terrestrial) at a point in time t can be perfectly predicted with a high level of certainty. Because of this, approaches from ad-hoc networks [21] as well as from fully structured networks have been applied in the context of NGSO satellite constellations.

Furthermore, the different time scales of the various ongoing processes offer opportunities for simplification via time-scale separation. For instance, the orbital period of a satellite is extremely long when compared to most of the communication tasks within the constellation. Therefore, the satellite constellation can be assumed to be static during short periods to simplify the analysis. In the following, we exemplify this latter aspect by calculating the one-hop latency of the different links.

Depending on whether we consider a ground-to-satellite link (GSL) or an inter-satellite link (ISL), the one-hop latency is determined by different factors. Naturally, it depends on the position of the transmitter u and the receiver v at time t, when the packet is ready to be transmitted and also on the packet length p. In the following, we calculate the three main components of the one-hop latency by considering the relative position of u w.r.t. v to be fixed during a period $[t, \Delta t]$.

First, the waiting time at the transmission queue $q_t(u, v)$ is the time elapsed since the packet is ready to be transmitted until the beginning of its transmission. Note that, depending on the communication protocols, e.g., signalling and frame structure, it may occur that $q_t(u, v) > 0$ for all u, v even when there are no more packets in the queue. Second, the transmission time, which is the time it takes to transmit p bits at the selected rate $R_t(u, v)$ bps. Third, the propagation time, which is the time it takes for the electromagnetic radiation to travel the distance $d_t(u, v)$ from u to v . Hence, the latency to transmit a packet of size p from u to v at time t is given by

$$L_t(u, v) = \underbrace{q_t(u, v)}_{\text{Waiting time}} + \underbrace{\frac{p}{R_t(u, v)}}_{\text{Transmission time}} + \underbrace{\frac{d_t(u, v)}{c}}_{\text{Propagation time}} \tag{10.6}$$

Note that all the factors that contribute to the one-hop packet latency depend on the time the packet is generated. Furthermore, due to the movement of the satellites, the set of established links and communication paths (routes) change depending on t. This creates a greatly dynamic network topology that introduces distinctive challenges in the design and implementation of the distinct physical links. In the following, we elaborate on the main technologies for satellite communications: radio frequency (RF) and free-space optical (FSO) links.

RF links occur in frequencies either in the S-band, the Ka-band or the Ku-band. These links are mainly affected by free-space path loss and thermal noise, so additive white Gaussian noise channels are oftentimes considered. The free-space path loss between two terminals u and v at time t is determined by the distance $d_t(u, v)$ between them and the carrier frequency f as

$$\mathcal{L}_t(u, v) = \left(\frac{4\pi d_t(u, v) f}{c}\right)^2 \tag{10.7}$$

where c is the speed of light.

Next, let $P^{(u)}$ be the transmission power of transmitter u – assumed to be constant for simplicity – and σ_v^2 be the noise power at receiver v. Further, let $G_t^{(u,v)}$ and $G_t^{(v,u)}$ be the antenna gain of transmitter u towards receiver v and vice versa. Based on this, the maximum data rate for reliable communication between two satellites and/or a satellite and a ground terminal at time t can be calculated as a function of the signal-to-noise ratio (SNR)

$$R_t(u, v) = B \log_2 \left(1 + \mathrm{SNR}_t\,(u, v)\right) = B \log_2 \left(1 + \frac{P^{(u)} G_t^{(u,v)} G_t^{(v,u)}}{\mathcal{L}_t(u,v)\sigma_v^2}\right) \tag{10.8}$$

Naturally, the achievable data rate in the presence of interference will be lower than (10.8). Nevertheless, the use of directional antennas and/or orthogonal resource allocation [13] greatly reduces interference within constellations. Building on this, the achievable rate mainly depends on the transmission power, the large-scale fading (path loss) and noise power but also on the gain of the communicating antennas in the direction of the receiver/transmitter. Since the constellation is a moving infrastructure, antenna pointing technology is an essential aspect of constellation design.

Throughout this chapter, we evaluate the performance of the RF physical links by assuming the parameters listed in Table 10.2 unless stated otherwise. These

Table 10.2 Parameter configuration for the physical links: ground-to-satellite link and inter-satellite link (ISL)

Parameter	**Symbol**	**NGSO to ground station**	**ISL**
Carrier frequency (GHz)	f	20	26
Bandwidth (MHz)	B	500	500
Transmission power (W)	P_t	10	10
Noise temperature (K)	T_N	150	290
Noise figure (dB)	N_f	1.2	2
Noise power (dBW)	σ^2	−117.77	−114.99
Parabolic antennas			
Antenna diameter (Tx – Rx) (m)	D	(0.26 - 0.33)	(0.26 - 0.26)
Antenna gain (Tx – Rx) (dBi)	$G_{\max}$	(32.13 - 34.20)	(34.41 -34.41)
Pointing loss (dB)	L_p	0.3	0.3
Antenna efficiency (–)	η	0.55	0.55

parameters were selected to focus on comparing the constellation design and not the implemented (envisioned) communication technologies.

FSO links, on the other hand, face different challenges depending on where they are implemented: GSL or ISL [22]. Hence, these challenges will be briefly described in the following sections.

10.3.1 Ground-to-satellite links

Communication between devices deployed at ground level, and the satellites takes place through GSLs. This can occur either by communicating the user devices (e.g., IoT devices) directly or through gateways. The gateways can not only be deployed at ground level but also in the air, such as unmanned aerial vehicles (UAVs) or high-altitude platforms (HAPs). For simplicity, we use the term ground terminal to refer to any device deployed at ground level. The area where G2S communication is possible is called the coverage area, and the time the satellite and a terrestrial terminal can communicate is called the duration of the satellite pass.

In the following, we provide the expressions to calculate the coverage area and, hence, to determine whether a ground terminal is able to communicate with a specific satellite at a given time t.

The distance between an NGSO satellite and a device located on the Earth's surface within line-of-sight (LoS) at time t is determined by the altitude h and the elevation angle of the satellite w.r.t. the device ε_t. Specifically, the distance of the GSL can be calculated from these parameters using the Pythagorean theorem in a triangle with sides of length: a) $R_E + h$; b) $R_E + d_{\mathrm{GSL}}(h, \varepsilon_t) \sin \varepsilon_t$; and c) $d_{\mathrm{GSL}}(h, \varepsilon_t) \cos \varepsilon_t$ and then applying the quadratic formula to obtain:

$$d_{\mathrm{GSL}}(h,\varepsilon_t) = \sqrt{R_E^2 \sin^2(\varepsilon_t) + 2R_E h + h^2} - R_E \sin(\varepsilon_t) \tag{10.9}$$

A similar procedure can be applied to the case of devices above the Earth's surface (e.g., UAVs, HAPs, etc.) by simply substituting the length of side (b) of the triangle to be $R_E + h_u + d_{\mathrm{GSL}}(h, \varepsilon_t) \sin \varepsilon_t$, where h_u is the altitude of the user above the sea level R_E. For notation simplicity, the rest of the equations presented are for satellites deployed at the Earth's surface only. However, a the substitution described above can be used to adapt the following equations to devices deployed above the Earth's surface.

Once $d_{\mathrm{GSL}}(h, \varepsilon_t)$ has been found, we calculate the Earth central angle [14] $\alpha(h, \varepsilon_t)$ as

$$\alpha(h,\varepsilon_t) = \arccos\left(\frac{(R_E + h)^2 + R_E^2 - d_{\mathrm{GSL}}(h,\varepsilon_t)^2}{2(R_E^2 + hR_E)}\right) \tag{10.10}$$

which determines the shift in the position of the device w.r.t. the satellite's nadir point.

The coverage of an NGSO satellite is usually defined by a minimum elevation angle $\varepsilon_{\mathrm{min}}$. Hence, a device located at an elevation angle $\varepsilon_t \geq \varepsilon_{\mathrm{min}}$ is considered to be within coverage of the satellite at time t. Consequently, the coverage area of a satellite is a function of the altitude of deployment h and of $\varepsilon_{\mathrm{min}}$. By using h and $\varepsilon_{\mathrm{min}}$, we find the angle $\alpha(h, \varepsilon_{\mathrm{min}})$ which allows us to calculate the coverage area as

$$A(h,\varepsilon_{\mathrm{min}}) = 2\pi R_E^2 (1 - \cos(\alpha(h,\varepsilon_{\mathrm{min}}))) \tag{10.11}$$

Furthermore, by assuming a spherical model of the Earth, we can easily determine whether a ground terminal u is within coverage of a satellite v at a given time t; this occurs when the distance $d_t(u, v)$ between them is shorter than $d_{\mathrm{GSL}}(h, \varepsilon_{\mathrm{min}})$.

Next, we calculate the maximum duration of a satellite pass as a function of $\alpha(h, \varepsilon_{\mathrm{min}})$ and $T_o(h)$. For this, let $t = 0$ be the time when the ground terminal enters the coverage area of the satellite. The satellite pass has maximum duration in the case where, at exactly at the middle of the pass, the satellite is exactly located at the zenith point of the ground terminal, and hence, there is an angle $\epsilon_t = 90°$ between the terminal and the satellite w.r.t. the Earth's centre. In such case, the satellite travels $\alpha(h, \varepsilon_{\mathrm{min}})/180°$ of its orbit, and hence, the satellite pass has a duration

$$T_{\mathrm{pass}}(h,\varepsilon_{\mathrm{min}}) \leq \frac{T_o(h)\alpha(h,\varepsilon_{\mathrm{min}})}{\pi} \tag{10.12}$$

For any other cases where the ground terminal and the satellite are not perfectly aligned, we define the angle

$$\alpha_{\mathrm{min}} = \min_t \alpha(h,\varepsilon_t) \quad \text{s.t. } t \in \left[0, T_{\mathrm{pass}}(h,\varepsilon_{\mathrm{min}})\right] \tag{10.13}$$

which determines the misalignment of the GS w.r.t. the orbital plane of the satellite. Naturally, $\alpha_{\mathrm{min}} = 0$ for the perfect alignment case.

The ground coverage of an NGSO satellite is illustrated in Figure 10.2 and the evolution of the achievable data rate along the pass for the altitudes of deployment for Kepler and OneWeb. We considered a typical value for the minimum elevation

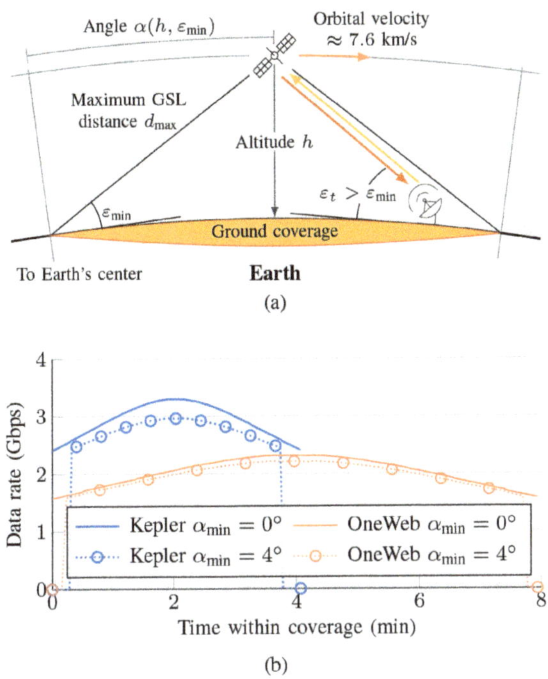

Figure 10.2 *(a) Ground coverage of an NGSO satellite at altitude* h *and (b) the evolution of the achievable data rate along the pass*

angle of $\varepsilon_{min} = 30°$. For devices deployed above the Earth's surface, this angle may be larger as their LoS is less affected by obstacles. From these, it is easy to observe that lower altitudes of deployment result in shorter propagation delays but also in faster orbital velocities, shorter satellite passes and smaller coverage areas.

Note that the coverage area simply defines the area where communication is possible. However, the beams oftentimes present a beamwidth that is much narrower than the coverage area. Therefore, these must be pointed in the desired direction of communication [8]. Because of this, having more than one satellite within the communication range can be beneficial as the access load can be shared among the satellites covering the same areas. Hence, it provides an indicator of the scalability and capacity of the network.

Based on the coverage area and the geometry of a specific constellation, the service availability and the average number of satellites within range can be obtained. Figure 10.3 shows the service availability and the mean number of satellites within coverage for the Kepler and OneWeb constellations, along with the Starlink orbital shell at $h = 550$ km considering the requested modification in the latest FCC filing, where $\varepsilon_{min} = 25°$.

As it can be seen, the density of the Kepler constellation and the considered $\varepsilon_{min} = 30°$ is insufficient to provide full service availability near the Equator, and it increases in near-polar areas. In contrast, the service availability of the Starlink

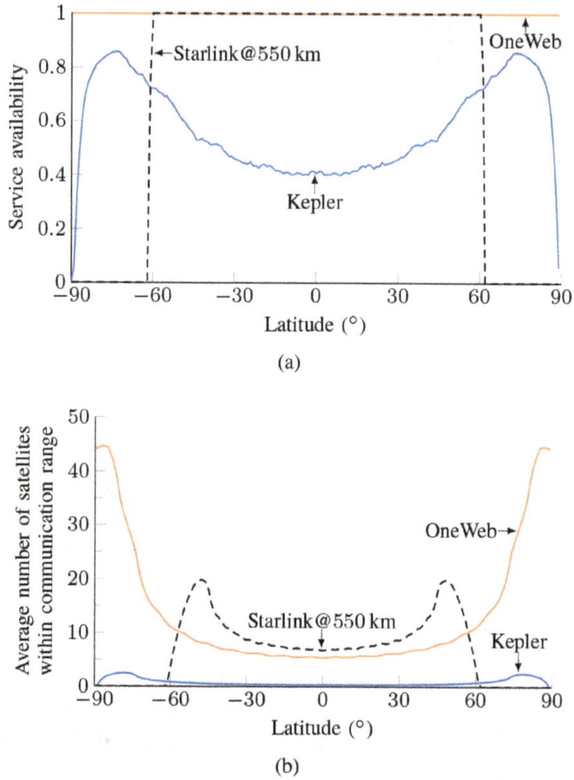

Figure 10.3 *(a) Service availability: probability of being within the coverage area of a satellite as a function of the latitude and (b) average number of satellites within communication range at GSL*

orbital shell is guaranteed between latitudes [−60°, 60°], and the OneWeb constellation provides full-service availability across the globe. Furthermore, it can be seen in Figure 10.3b that there is a significant number of OneWeb satellites within coverage in the polar regions and a considerably lower number in Equatorial regions. This is a distinctive characteristic of Walker star constellations, as the distances between satellites are maximal near the Equator. In contrast, the coverage of the Starlink orbital shell between the latitudes [−60°, 60°] is relatively balanced. To solve the problem of lack of coverage in the polar regions, the Starlink constellation is planned to incorporate satellites in polar orbits, as listed in Table 10.1.

There are many benefits of using RF over FSO for the GSL. For instance, RF links present a wider beamwidth and, hence, a broader coverage. This simplifies the beam switching and allows to provide coverage to several ground terminals simultaneously. Additionally, the use of RF links allows to use the same physical layer technologies as in the terrestrial networks, which simplify the hardware design and enables the integration of satellites and terrestrial networks through mature terrestrial technologies. For instance, the third Generation Partnership Project (3GPP) is aiming to integrate satellites and cellular networks using NB-IoT and 5G New Radio (NR) cellular technologies [1, 3, 8].

In contrast, FSO GSL is mainly affected by the atmosphere. In particular, the atmosphere absorbs and scatters the beam. These effects depend on different factors such as temperature, humidity and the concentration of aerosol particles. Furthermore, the effects vary widely between uplink and downlink communication, with the uplink signals being affected most due to the presence of the atmosphere around the transmitter [22].

Yet another factor that impacts the GSL is the Doppler shift. The latter varies significantly during a satellite pass as a result of the high orbital velocity in combination with the varying relative position and speed with respect to time. That is, the Doppler shift is different between the edge and the centre of the coverage, so this must be taken into account to select an appropriate frequency band and during waveform and antenna design. If information from the Global Navigation Satellite System is available, the Doppler shift can be first pre-compensated at the satellite w.r.t. to a reference point by exploiting the predictable movement of the satellites. Then, the residual frequency offsets are compensated at the ground terminals using traditional Doppler compensation techniques as in terrestrial networks [8].

10.3.2 Inter-satellite links

Inter-satellite communication takes place in 1) the same orbital plane, 2) different orbital planes of the same orbital shell and 3) different orbital altitudes. The dynamics in each of these are significantly different. However, it is essential to establish these links in an efficient manner to maximise the connectivity within the constellation.

Intra-plane ISLs connect satellites in the same orbital plane, usually, at both sides of the roll axis, which is aligned with the velocity vector. In particular, the relative distances between neighbouring satellites within the orbital plane – the intra-plane distance – at an altitude can be considered a constant

$$d_{\text{intra}}(N_{\text{op}}, h) = 2(R_E + h) \sin\left(\frac{\pi}{N_{\text{op}}}\right) \tag{10.14}$$

Hence, intra-plane ISLs are rather stable. Still, the orbital velocity of the satellites must be considered. But this is easily compensated by selecting an appropriate point-ahead-angle; instead of pointing the antennas directly towards the instantaneous position of the receiver at the same time instant t, they are pointed to its position after considering the propagation time $t + d_{\text{intra}}(h)/c$. Because of this, the antennas used for intra-plane communication can be highly directive, and the beams can be fixed to the appropriate direction. Due to the use of narrow beams, FSO links present an interesting option for intra-plane communication, as their power efficiency may be greater than that of RF links [22]. Nevertheless, RF links with either parabolic or patch antenna arrays are also an efficient candidate that combines relatively high gains, cheap components and low-power requirements when compared to FSO.

Inter-plane ISLs, on the other hand, connect satellites within the same orbital shell but in different orbital planes. Usually, satellites will possess either one or two transceivers for inter-plane communication, with antennas pointing towards both sides of the pitch axis. Depending on the constellation geometry, the distances and the velocity vectors between satellites in different orbital planes may be either very similar or vary widely. For instance, in Walker star geometries, the orbital planes are separated by the angle π/P, and the shortest inter-plane distances occur at the crossing points of the orbits near the poles. In contrast, the longest inter-plane distances to the nearest neighbour occur for satellites near the Equator.

Let u and v be a pair of satellites located in neighbouring orbital planes, where v is the closest inter-plane neighbour of u at time t. For simplicity, we assume the same altitude of deployment for both orbital planes to be h. We denote the polar angle of satellites u and v as $\theta_t^{(u)}$ and $\theta_t^{(v)}$, respectively. First, we recall that the distance between two points, u and v, on a sphere of radius $R_E + h$ with azimuth angles ϕ_u and ϕ_v is given as

$$d_{uv}(t) = \sqrt{2\left(\mathrm{R_E}+h\right)^2\left(1-\cos\theta_t^{(u)}\cos\theta_t^{(v)}-\cos\left(\phi_u-\phi_v\right)\sin\theta_t^{(u)}\sin\theta_t^{(v)}\right)} \tag{10.15}$$

The latter can be used to approximate the distance between two satellites adjacent orbital planes in a Walker star constellation assuming perfectly polar orbits. For this, recall that orbital planes in Walker star constellations are separated by π/P; hence, this is also the azimuth angle between satellites in adjacent orbital planes.

If u and v are exactly at the Equator, we have $\theta_t^{(u)} = \theta_t^{(v)} = \pi/2$, and the maximum intra-plane distance for the case where the satellites are perfectly aligned at all times only depends on P and h as

$$d^*_{\text{inter, aligned}}(P) = \sqrt{2(R_E+h)^2\left(1-\cos\left(\tfrac{\pi}{P}\right)\right)} = 2(R_E+h)\sin\left(\tfrac{\pi}{2P}\right) \tag{10.16}$$

However, in a general case where the satellites u and v are not perfectly aligned, we have that, if v is the closest inter-plane neighbour to u, then it follows that $|\theta_t^{(v)} - \theta_t^{(u)}| \in \left[0, \pi/N_{\text{op}}\right]$. Therefore, the maximum inter-plane distance occurs when $\theta_t^{(u)} = \pi/2$ and $\theta_t^{(v)} = \pi/2 \pm \pi/N_{\text{op}}$, which can be approximated as

$$\begin{aligned} d^*_{\text{inter}}(N_{\text{op}}, P) &= \max_t\ d_{uv}(t) \quad \text{s.t. } \theta_t^{(v)} \in \left[-\pi/N_{\text{op}}, \pi/N_{\text{op}}\right] \\ &\approx \left(\mathrm{R_E}+h\right)\sqrt{2-2\cos\left(\tfrac{\pi}{P}\right)\sin\left(\tfrac{\pi}{2}\pm\tfrac{\pi}{N_{\text{op}}}\right)} \end{aligned} \tag{10.17}$$

Hence, to ensure that satellites at the Equator can communicate with at least one of their inter-plane neighbours, it is necessary to ensure that a non-zero data rate can be achieved at this location. To illustrate this aspect in a general case where the satellites are not perfectly aligned, let $\mathcal{R}$ be the set of available rates for communication, which depend on the available modulation and coding schemes (MCSs) and where $0 \notin \mathcal{R}$. Then, to guarantee global ISL connectivity, it is required that any given satellite u can select a rate $R \in \mathcal{R}$ that allows it to achieve reliable communication with the nearest inter-plane neighbour v at all times. Hence, global ISL connectivity is achieved if

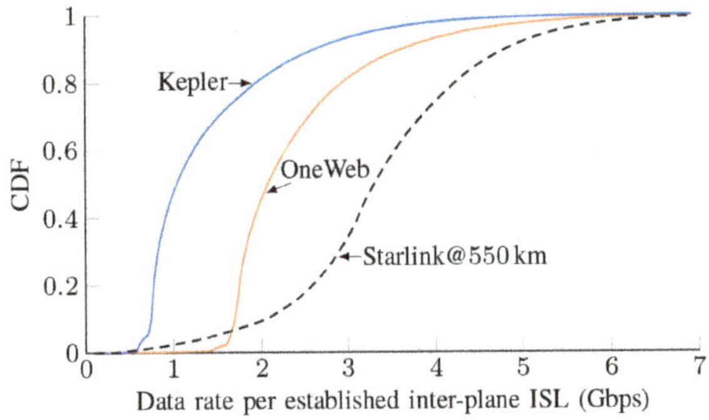

Figure 10.4 *CDF of the achievable data rate per inter-plane ISL with parabolic antennas*

$$\exists R \in \mathcal{R} : 0 < R < B \log_2 \left(1 + \frac{P^{(u)} G_t^{(u,v)} G_t^{(v,u)} c^2}{\left(2\pi\, \sigma_v\, d^*_{\text{inter}}(N,P) f\right)^2} \right) \tag{10.18}$$

As it can be seen, for a fixed set of rates $\mathcal{R}$, global ISL connectivity can be achieved by either increasing the power and/or gains of the antennas or by decreasing the maximum inter-plane distances. The latter is usually achieved by either increasing the number of orbital planes P but also the number of satellites per orbital plane N_{op}. The interested reader is referred to our previous work for a general formulation that considers orbital separation and where the effect of increasing the number of orbital planes P is illustrated [13].

Yet another characteristic of Walker star constellations is that the velocity vectors of the satellites in neighbouring orbital planes usually point in a similar direction. As a result of this, the relative velocities between these satellites are relatively low. However, there are specific pairs of orbital planes where the velocity vectors point in a nearly opposite direction: the so-called cross-seam ISLs. In the latter, the relative velocity of the satellites increases to nearly $2v_o$ and varies along with time.

As a consequence of these great differences, the Doppler shift and the contact times in the inter-plane ISL – the period where two satellites can communicate – also vary widely. Therefore, it is essential to consider the movement of the satellites to select the inter-plane ISL that must be established and to point the beams in the desired directions [13, 23]. Figure 10.4 shows the cumulative distribution function (CDF) of the achievable rate in the inter-plane ISL with a specific link establishment mechanism. While the mechanisms for ISL establishment and beam pointing are described in section 10.4.2, Figure 10.4 shows that the data rate achieved by inter-plane ISL in the Starlink orbital shell is considerably larger than in the OneWeb and Kepler constellations. The main reason for this is the higher density of satellites caused by the low altitude of deployment, the use of Walker delta geometry and, naturally, the large number of satellites.

Finally, inter-orbit ISLs connect satellites between different orbital altitudes [22]. For example, they can connect LEO satellites in different orbital shells or LEO satellites with MEO or even GEO satellites. A clear example are the FSO links in the

European Data Relay System and those envisioned to connect the different orbital shells in the Starlink constellation.

10.4 Functionalities and challenges

10.4.1 Physical layer

Pure LoS connections, high velocities and large transmission distances between satellites and ground terminals introduce some unique characteristics to the physical layer design for NGSO constellations, both in the GSL and the ISL.

In the GSL, it is particularly appealing to maintain the waveforms used in terrestrial systems, e.g., orthogonal frequency division multiplexing (OFDM) in 5G NR and NB-IoT [9, 24]. This would allow full compatibility of terrestrial devices and direct IoT-to-satellite access which, in turn, grants maximum flexibility of deployment following the place-and-play vision. However, the subcarrier spacing in terrestrial OFDM systems is narrow – between 3.75 kHz for NB-IoT and from 15 to 240 kHz for 5G NR [25]. Such narrow subcarrier spacings make OFDM highly sensitive to Doppler shifts, and thus, accurate Doppler compensation is required to achieve reliable communication. To overcome these limitations, several alternatives have been studied intensively in the literature over the past few years, such as Universal Filtered Multi-Carrier, Generalised Frequency Division Multiplexing and Filter Bank Multi-Carrier (FBMC) [26]. These waveforms allow for higher robustness against Doppler shifts and flexible time-frequency resource allocation in exchange for a higher equalisation complexity. However, in case of severe Doppler shifts, factor graph-based equalisation for FBMC transmissions outperforms the OFDM system in terms of complexity and performance [27].

Another challenge for keeping reliable GSL and also ISL is the implementation of adaptive modulation and coding. In 3GPP networks, the users exchange information about the channel quality with the BS [9], which adapts the MCS based on the error rate. Due to the altitude of deployment, the round trip time (RTT) between the ground terminals and a satellite is usually greater than 4 ms. Hence, such a feedback link would introduce a significant delay. Instead, the fully predictable movement of the satellite along the pass, in combination with free-space propagation and the minor impact of atmospheric conditions in RF links, can be exploited to achieve efficient adaptive modulation and coding with minimal signalling.

Furthermore, while multiple-input multiple-output (MIMO) techniques have experienced a dramatic surge of advancements in terrestrial networks, achieving efficient MIMO communication with NGSO satellites is more complicated. In particular, due to the long distances between transmitter and receiver, exploiting the full MIMO gain requires a large array aperture, that is, large distances between transmit and/or receive antennas, that are not feasible in individual satellites [28]. Nevertheless, this separation can be realised by using a group of satellites flying in close formation, usually called a satellite swarm. Specifically, by placing an antenna at each of the satellites in the swarm, these can operate as distributed MIMO arrays. Doing so allows to form extremely narrow beams for GSL, which leads to better

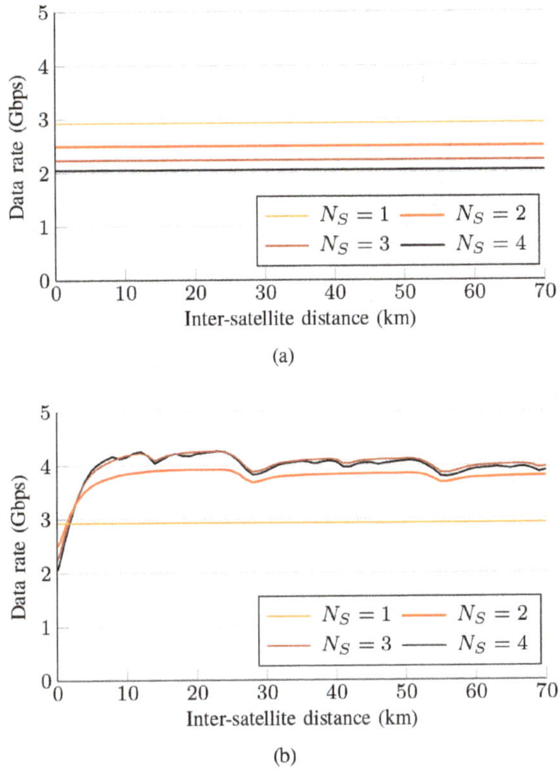

Figure 10.5 *Data rate for satellite swarms as a function of the inter-satellite distance for (a) one and (b) six receiving antennas*

spatial separation via coordinated beamforming and, eventually, to higher spectral efficiency when serving different ground terminals located geographically close to each other [29]. An example of the achievable gain of distributed MIMO in a satellite swarm, with N_S satellites, is shown in Figure 10.5 for $N_r = 1$ and $N_r = 6$ receiving antennas. The overall transmit power as well as the antenna gains are normalised such that they are the same in all scenarios, i.e., the transmit power and antenna gain per satellite are $10/N_S$ W and 32.13 dBi $- 10\log_{10}(N_S)$, respectively, and the receive antenna gain is 34.20 dBi $- 10\log_{10}(N_r)$.Figure 10.5 shows that, despite maintaining the total transmitted power in all cases, the use of distributed MIMO increases the data rate by around 33% with multiple receiving antennas. However, with a single receiving antenna, no MIMO gains can be achieved and resulting in even lower rates because the transmitted signals superimpose constructively or destructively with same probability, reducing the overall received signal energy.

Beam pointing/steering is another essential functionality in NGSO constellations due to the constant and rapid movement of the satellites. The mechanical steering of RF antennas becomes problematic as beams become narrower, which is essential to attain a high SNR. In addition, the ultra-narrow beams present in FSO require high pointing precision and fast repointing to maintain adequate link quality.

A different approach made possible by recent advances in antenna technology is the use of phased antenna arrays, even in small satellites. In these, the antenna elements are separated by a small distance d_e, which is proportional to the wavelength λ, and can be used to produce highly directed beams. This enables efficient interference management due to beamforming, which exploits the spatial domain via Spatial Division Multiple Access or Rate-Splitting Multiple Access, and thus, allows for an efficient use of the bandwidth. Furthermore, these beams can be steered electronically by manipulating the input signals to the antenna elements through variable phase shifters.

Let us consider a satellite u equipped with an $K \times K$ antenna array that attempts to steer the beam towards satellite v at a given time t. To do so, it first needs to calculate the K-dimensional steering vectors for the azimuth angle $\phi_t^{(u,v)}$ as

$$\mathbf{a}_{t,\text{az}}^{(u,v)} = \left[1, e^{\frac{-j2\pi d_e}{\lambda}\sin\left(\phi_t^{(u,v)}\right)}, \ldots, e^{\frac{-j2\pi d_e(K-1)}{\lambda}\sin\left(\phi_t^{(u,v)}\right)}\right]^{\mathsf{T}} \tag{10.19}$$

and for the polar angle $\Theta_t^{(u,v)}$ as

$$\mathbf{a}_{t,\text{pol}}^{(u,v)} = \left[1, e^{\frac{-j2\pi d_e}{\lambda}\cos\left(\Theta_t^{(u,v)}\right)}, \ldots, e^{\frac{-j2\pi d_e(K-1)}{\lambda}\cos\left(\Theta_t^{(u,v)}\right)}\right]^{\mathsf{T}} \tag{10.20}$$

Then, it calculates the overall steering vector $\mathbf{a}_t^{(u,v)} = \mathbf{a}_{t,\text{pol}}^{(u,v)} \otimes \mathbf{a}_{t,\text{az}}^{(u,v)}$. This approach is often called digital beam steering, and it is attractive to combat the fast orbital velocities of the NGSO satellites due to its precision and switching velocity [5]. Nevertheless, it has the main downside that the implementation of the variable phase shifters adds a considerable amount of complexity to the hardware, which might be restrictive for nano-satellites and CubeSats.

Butler matrix beamforming networks offer a simpler mechanism to point the beams and, hence, have gained relevance in terrestrial communications [30, 31]. These are cost-efficient and low-complexity beam switching networks that produce a series of beams in pre-defined directions [32, 33]. In contrast to digital beam steering, the beams in a Butler matrix are switched by simply feeding one or more of the fixed phase shifters (input ports), which offer an interesting trade-off between performance, cost and complexity of operation and implementation that is especially attractive for CubeSats, which oftentimes rely on small and simple dipole antennas with low directivity.

In particular, the steering vector in the polar angle of a Butler matrix is fixed to a specific direction θ

$$\mathbf{b}_{\text{pol}} = \frac{1}{\sqrt{K}}\left[1, e^{\frac{-j2\pi d_e}{\lambda}\cos(\theta)}, \ldots, e^{\frac{-j2\pi d_e(K-1)}{\lambda}\cos(\theta)}\right]^{\mathsf{T}} \tag{10.21}$$

whereas the steering vector of the k-th beam in the azimuth angle is set to

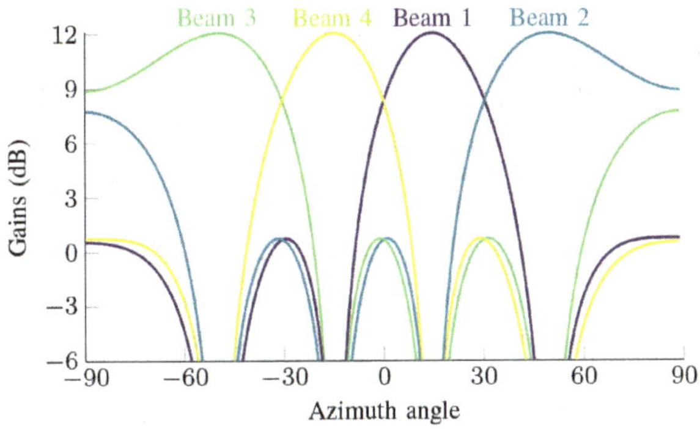

Figure 10.6 Gains for the beams in a 4 × 4 antenna array with a Butler matrix

$$\mathbf{b}_{k,\mathrm{az}} = \frac{1}{\sqrt{K}}\left[1, e^{\frac{-j\pi(2k-1)}{K}}, \ldots, e^{-j\frac{\pi(2k-1)(K-1)}{K}}\right]^{\mathsf{T}} \tag{10.22}$$

The overall steering vector for beam k is $\mathbf{b}_k = \mathbf{b}_{\mathrm{pol}} \otimes \mathbf{b}_{k,\mathrm{az}}$. Figure 10.6 illustrates the gain of $K = 4$ beams in a Butler matrix with 4×4 antenna elements.

Finally, while achieving direct IoT communication with NGSO is feasible with LPWAN technologies, the use of gateways is often beneficial. These gateways may incorporate traditional dish antennas or phased antenna arrays that gather the transmissions from IoT devices with non-directive antennas and then transmit to the satellites with highly directive antennas. However, another option made possible by the predictable movement of the satellites is to deploy intelligent reflecting surfaces. These low-complexity elements modify the characteristics of the incident signals and, hence, can help direct the signals towards the satellites [34].

10.4.2 Frequent link establishment and adaptation

Due to the movement of the satellites, the physical links must be frequently re-established and adapted. This includes selecting the pairs of satellites to establish the ISLs, beam pointing/steering or switching for the Butler matrix case, and rate adaptation. Since the movement of the constellation is fully predictable, these problems can be solved in advance with a specific optimisation objective in mind. These objectives depend on the target service(s) and can be, as listed in Section 10.1, to maximise the transport capacity [35, 36] of the constellation or minimise the E2E latency for a set of specific paths.

Some constellations designs are fully symmetric, with each and every one of the orbital planes containing the same number of satellites and these being deployed at the exact same altitude. In these cases, the orbital period T_o of all the satellites is exactly the same, and hence, these will all be periodically at the exact same position. In these cases, the optimal configuration of the links can be obtained for several instants within the period T_o and applied periodically.

However, asymmetries in the constellation are usually present either 1) to enhance the sustainability of the constellation by considering orbital separation as in

OneWeb [18], 2) to fulfill certain coverage and service availability targets by incorporating several orbital shells as in Starlink or 3) to provide service during the initial phases of deployment of the constellation. In these cases, fixed solutions cannot be used, and the links must be established on the fly.

An essential aspect for link establishment and maintenance is to implement an adequate beam steering technology as discussed in the previous section. Furthermore, the MCS and transmit power may be adapted to maximise throughput and reliability while minimising potential interference. Naturally, the characteristics of the antennas and beams must be considered during link establishment [23].

An option to re-establish the links is to treat the link establishment as a one-to-one matching problem in a dynamic weighted graph $\mathcal{G}_t = (\mathcal{V}, \mathcal{E}_t)$, where the satellite antennas, transceivers or even beams (for the case of beam selection) form a multi-partite vertex set $\mathcal{V}$ and the weighted edge set at time t, denoted as $\mathcal{E}_t$ are the feasible ISLs with non-zero rates. Then, the matching at a time $\mathcal{M}_t$ is the set of pairs of antennas/transceivers/beams and the rates for communication. In this case, the matching $\mathcal{M}_t$ can be calculated periodically, once every Δt s, in a centralised entity with full knowledge of the constellation parameters and dynamics. Then, the solution for the matching for the satellite positions at time t must be propagated through the constellation before this time. With the full predictability of the movement of the constellation, the solution can be calculated sufficiently in advance, and hence, the latency of communicating it to the satellites is irrelevant. Hence, this approach can lead to near-optimal or optimal solutions at the expense of injecting periodic traffic into the network to communicate the solution to all the satellites. An important aspect of the inter-plane ISL link establishment is that the graph $\mathcal{G}$ that represents a single orbital shell is multi-partite, with each subset representing an orbital plane, and hence, traditional algorithms such as the Hungarian algorithm cannot be used, and other solutions are needed.

On the other hand, localised decisions may be implemented, e.g., using distributed algorithms for the matching. An example of these is the deferred acceptance algorithm [37], where the individual agents maintain and inform their preferences to the neighbourhood, and the matching is solved in parallel, after few iterations. While more research is needed to determine the performance of distributed vs. centralised matching solutions, distributed algorithms are required 1) to establish the links during the deployment phase before the constellation is fully operative and 2) in case the connection with the centralised entity is lost.

To solve the inter-plane ISL establishment problem, we have explored the use of greedy matching algorithms with 1) ideal beam pointing (i.e., at each time t) and resource allocation and 2) periodic repointing via digital beamforming and beam switching via Butler matrix beamforming networks with period Δt [13, 23]. Algorithm 1 illustrates the steps of a general greedy matching algorithm for link establishment. The latter can be extended to include orthogonal resource allocation (e.g., frequency sub-bands) to minimise interference [19].

Algorithm 1 Greedy satellite matching with multiple beams.

Input: Set of feasible weighted edges $\mathcal{E}_t$ and $\mathcal{E}_{t+\Delta t}$ and the initial state of the matching $\mathcal{M}_t$
Input: Antenna configuration
Initialise indicator variables
while More edges can be matched **do**
  Select the edge with maximum weight
  if the selected vertices are not in $\mathcal{M}_t$ **then**
    Add the vertices to the matching $\mathcal{M}_t$
    Update the indicator variables
    Remove all adjacent edges to the selected vertices from $\mathcal{E}_t$ and $\mathcal{E}_{t+\Delta t}$
    Update the interference and weights to all edges in $\mathcal{M}_t$, $\mathcal{E}_t$ and $\mathcal{E}_{t+\Delta t}$
  end if
end while

Note that Algorithm 1 attempts to maximise the sum of weights in the matching. Throughout our previous work, we have defined the weights to be the achievable rate for communication at the ISLs in $\mathcal{E}_t$. Following this approach, Figure 10.7a illustrates the increase of the rates per ISL as a function of the number of elements K in a Butler matrix beamforming network for the Kepler constellation. As it can be seen in Figure 10.7a, increasing the number of elements K greatly improves the data rates; however, this also increases the number of beams that must be considered by the matching algorithm and, hence, increases the running time of the algorithm.

Furthermore, Figure 10.7b shows the effect of the re-establishment period Δt on the average data rate per inter-plane ISL with digital beamforming; the data rate achieved with ideal pointing (i.e., with $\Delta t = 0$) and parabolic antennas is included as a reference. It can be seen that increasing the frequency of link re-establishment and adaptation increases the data rates, and phased antenna arrays of $K = 64$ can be used to achieve similar rates as with greatly directional parabolic antennas, even with ideal pointing. However, the re-establishment period cannot be reduced arbitrarily as this can cause problems, e.g., for routing algorithms, due to the frequent changes of the network topology.

Throughout our analyses, we have observed that Butler matrix networks with relatively low dimensions K are an attractive option for the inter-plane link establishment in resource-constrained satellites (i.e., CubeSats and small-sats). However, if large antenna arrays and variable phase shifters can be implemented on the satellites, beamforming offer gains in the data rates that are greater than 200%, and hence, these should be preferred.

It is important to mention that rate maximisation does not directly increases the transport capacity of the network, which is a difficult measure to define. Usually, specific source-destination pairs are defined, and the transport capacity is the maximum amount of data (i.e., flow) that can be transmitted between them [38]. In these cases, calculating the transport capacity usually involves assigning flow to

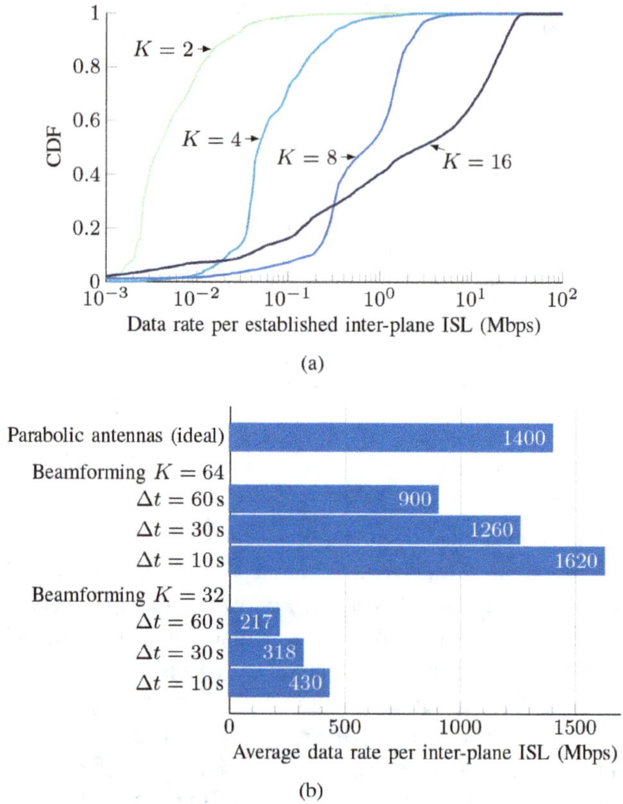

Figure 10.7 *(a) CDF of the rates per inter-plane ISL with Butler matrix arrays and (b) average rates per inter-plane ISL with parabolic antennas with ideal pointing and for digital beam forming for different link re-establishment periods* Δt

all possible paths from the source to the destination, as in the Edmons-Karp algorithm, which has been used to calculate the capacity of constellations with multiple orbital shells [36]. However, this is complicated in dynamic and large networks, so upper bounds based on selecting cuts from the network graph have been used [35]. Yet another hindrance of using the Edmonds-Karp algorithm is that it assumes that ideal mechanisms to redistribute the traffic flows are in place. Instead, in a network with multiple source-destination pairs, the capacity of some links is likely to be shared among them, and the number of alternate paths may be limited depending due to the implemented routing, load balancing and congestion control mechanisms. Furthermore, the distribution of the traffic among the different paths will usually be imbalanced. Therefore, defining the transport capacity of a satellite constellation is complicated.

A simple scenario where it is possible to calculate the maximum (G2G) traffic that can be generated from each GSs is where these have equal traffic characteristics and where unipath source routing is used [39]. In this scenario, we can define $\mathcal{P}_t$ as the set of possible paths at time t. A path $\mathbf{p} \in \mathcal{P}_t$ is an ordered set of edges, denoted as $\mathcal{E}(\mathbf{p}) = (e_1, e_2, e_3, \ldots)$. Here, the load λ of each of the N_{GS}GSs is distributed evenly to the rest of the $N_{\text{GS}} - 1$GSs, using the paths in $\mathcal{P}_t$. Hence, the load assigned to each path $\mathbf{p} \in \mathcal{P}_t$ is

$$\lambda_{\mathbf{p}} = \frac{2\lambda}{N_{\text{GS}}-1} \tag{10.23}$$

The max-flow min-cut theorem states that the maximum flow that can be transmitted through a path is determined by the link (i.e., edge) with minimum capacity (i.e., throughput) [38]. Hence, at time t, we have

$$\sum_{\mathbf{p}\in\mathcal{P}_t} \sum_{uv\in\mathcal{E}_t(\mathbf{p})} \lambda_{\mathbf{p}} = N_{\mathbf{p}}(uv)\lambda_{\mathbf{p}} \leq R_t(u,v), \quad \forall u,v \in \mathcal{V} \tag{10.24}$$

where $N_{\mathbf{p}}(uv)$ is the number of paths in $\mathcal{P}_t$ that contain the edge uv. Naturally, $N_{\mathbf{p}}(uv)$ depends on the routing metric. Building on this, it is possible to calculate the maximum load per GS at time t as

$$\lambda_t^* = \min_{uv\in\mathcal{E}_t} \frac{R_t(u,v)(N_{\text{GS}}-1)}{N_{\mathbf{p}}(uv)} \tag{10.25}$$

10.4.3 Routing, load balancing and congestion control

A general goal to achieve in the design of higher layer algorithms is to account for both the traffic characteristics (load, queues and Quality of Service QoS/Quality of Experience QoE requirements) and the instantaneous state of the links/paths. However, the time variations of the traffic and the channel are different in terrestrial and satellite networks, and for example, the conventional Transmission Control Protocol TCP/Internet Protocol IP stack is ineffective against the long delays, packet losses and intermittent connectivity that characterises NGSO communications. Therefore, specific networking solutions are required.

A routing algorithm is a collaborative process for deciding, in every intermediate node, the directions to reach the destination as soon as possible.[†] This routing problem presents the following unique characteristics in NGSO constellations:

- The topology is highly dynamic, with frequent handovers in the links between ground and NGSO and between NGSOs in different orbital planes (inter-plane ISL).
- The load from the ground terminals (GSs and users) is imbalanced, with 1) some satellites serving, e.g., deserted/ocean areas while other nodes pass above densely populated regions and 2) some source-destination pairs experiencing more intense data flows than others.

[†]In an NGSO constellations and other satellite systems, a second option for delay-tolerant applications is the store-carry-forward strategy where nodes can temporarily store and carry in-transit data until a suitable link becomes available, e.g., until the next pass with a ground stationGS.

- The need to have a reliable and resilient routing solution, which implies that the satellite segment must possess a sufficient degree of autonomy to cope with, e.g., queueing delays or local link or satellite failures and find alternative routes at each time instant. However, this must be achieved while exchanging minimal feedback and routing information to limit the signalling overhead.

A good overview of routing protocols for satellite can be found in Reference 40. Most previous works have oversimplified the ground/space segments geometry and the ISL connectivity to focus on other challenges like the QoS. One prominent exception is Reference 7; although the study is for a specific commercial constellations [39], it takes a more general approach and focuses on two distinctive elements to the routing problem in an NGSO constellation. First, the propagation time has a great impact on the overall latency, contrary to the terrestrial mesh networks. Second, the location of the GSs greatly impacts the traffic load injected to the constellation and the geographic locations where the traffic is injected. As in section 10.4.2, the space and ground infrastructure at a given time t are modelled as a dynamic weighted undirected graph $\mathcal{G}_t = (\mathcal{V}, \mathcal{E}_t)$. However, by adding the GSs, the vertex set is now defined as $\mathcal{V} = \mathcal{U} \bigcup_{a \in \mathcal{P}} \mathcal{V}_a$, where $\mathcal{U}$ is the set of GSs and $\mathcal{V}_a$ is the set of satellites deployed in orbital plane a, and $\mathcal{P} = \{1, 2, \ldots, P\}$ is the set of orbital planes. The edge set $\mathcal{E}_t$ represents the wireless links available for communication. For instance, the satellites might deploy four ISLs at all times: two intra-plane ISLs and two inter-plane ISLs. In this case, the intra-plane ISLs within an orbital plane a constitute the fixed set of edges $\mathcal{E}^{(a)} = \{uv : u, v \in \mathcal{V}_a\} \subset \mathcal{E}_t$. On the other hand, the inter-plane ISLs between orbital plane a and orbital plane b constitute the set of edges $\mathcal{E}_t^{\text{inter}} = \{uv : u \in \mathcal{V}_a, v \in \mathcal{V}_b, a \neq b\} \subset \mathcal{E}_t$; as mentioned in section 10.4.2, these must be frequently re-established due to the movement of the satellites. Furthermore, the GSs maintain one GSL with their closest satellite at all times. These GSLs constitute the set of edges

$$\mathcal{E}_t^G = \{uv : u \in \mathcal{U}, v \in \mathcal{V}_a, a \in \mathcal{P}\}$$

Finally, we define the edge set as

$$\mathcal{E}_t = \mathcal{E}_t^G \cup \mathcal{E}_t^{\text{inter}} \bigcup_{a \in \mathcal{P}} \mathcal{E}_t^a$$

The route of a single packet transmitted at time t is then a weighted path $\mathbf{p}$ in $\mathcal{G}_t = (\mathcal{V}, \mathcal{E}_t)$ with edge set $\mathcal{E}(\mathbf{p})$. The weights $w(e)$ for all $e \in \mathcal{E}_t$ are defined by the routing metric to account to, e.g., the path loss and/or the communication latency. Specifically, capturing the non-linearity of the path loss in the ISL will favour paths with high-data rates and consequently reduce the waiting times in the buffers. Rather than complex feedback mechanisms to collect up-to-date network status information, this simpler approach has proven to provide a good trade-off between complexity and performance.

The degree of integration of the NGSO constellation with the terrestrial infrastructure has also an effect on the traffic load. Not in vain, a prominent application of 5G satellite communications is to offload the terrestrial networks in congested urban areas, either with direct satellite access or through a gateway [10]. In both cases, it will further exacerbate the load imbalance. A subsidiary case is the use of the constellation as a backhaul that transparently carries the payload between the two communication extremes. This is typically used to connect isolated BSs to the core network.

Regarding resilience, the classical approach to space routing is to centrally compute all the paths in a location register and then broadcast the information to all the satellites. Satellites forward the packets according to the on-board routing tables, which are configured based on the central computations. In the case of an NGSO, the central location register can be a terrestrial station or a GEO satellite. In any case, this approach scales poorly due to the highly dynamic topology, with frequent handover events between nodes and terminals causing significant signalling overhead. Moreover, the current status of the satellites (load, buffers and batteries) should be included in the decision, but this requires an enormous amount of feedback from each node in the graph to the central location register. The alternative is to move towards more distributed solutions. From semi-distributed to fully autonomous algorithms, the idea is that each satellite decides the next hop for each received packet, taking into consideration all the available information, including the prior knowledge (past) and the predicted paths (future).

As in terrestrial networks, the space network might be shared by several services with heterogeneous requirements. For example, some broadband users require high rates, as provided by the GEO segment, whereas IoT devices are sensitive to delays or freshness of the information [10], better provided by the NGSO segment, or some services demand extra satellite computation. In general, there are multiple paths for most source-destination pairs, and this diversity should be exploited to meet the heterogeneity of requirements.

The example in Figure 10.8 illustrates the performance of different routing metrics, taking the latency as the KPI of interest. Three different configurations are considered: (a) the Kepler constellation with the communication parameters listed in Table 10.2; (b) the Kepler constellation with transmission power $P_t = 1$ W; and (c) a Walker-star constellation with $P = 5$ orbital planes at height $h = 600$ km and $N_{\text{op}} = 40$ satellites per orbital plane and $P_t = 1$ W. The time-varying data rate follows the channel variations. The ground segment consists of $N_{\text{GS}} = 23$GSs placed accordingly to the KSAT GS service[‡]

The compared metrics are: (1) a classical hop-count approach that merely minimises the number of hops to reach the destination; (2) a path loss metric that considers the non-linearity of the ISL; (3) a latency metric that takes into account the propagation and transmission times but skips the need for a feedback channel by having a statistical model of the queueing times. As expected, the latency metric effectively selects the routes with the shortest propagation and transmission times in all three cases. However, the waiting times are shorter with the pathloss metric. This is because the pathloss metric emphasises the selection of high data rate links

[‡]https://www.ksat.no/services/ground-station-services/. The details of the simulations can be found in Reference [39].

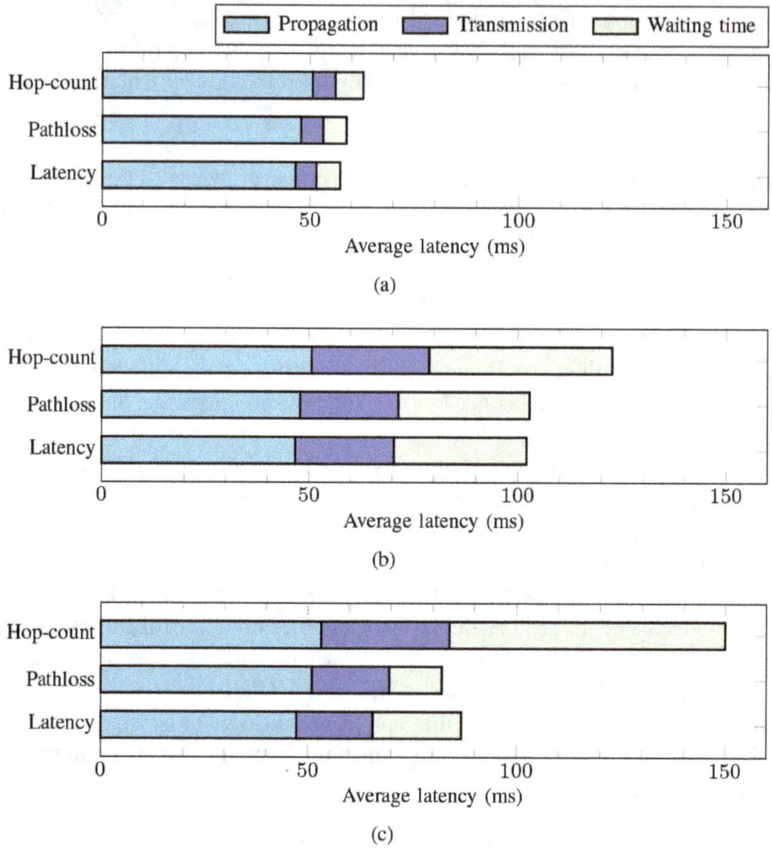

Figure 10.8 *Average routing latency per packet due to propagation, transmission and waiting times for three topology-aware metrics with: (a) Kepler constellation; (b) Kepler constellation with Pt = 1 W; and (c) constellation with P = 5, Nop = 40, and Pt = 1 W*

over short routes, which support greatest traffic load. As a consequence, the pathloss metric leads to the lowest overall latency with configuration (c) and to a closely similar latency to the latency metric in the other two cases. The reason for this is that the configuration with $P = 5$ and $N_{op} = 40$ has a greater density of satellites along the orbital planes, which lead to a much greater data rate at the intra-plane when compared to the inter-plane ISLs. These links are prioritised by the pathloss metric. Furthermore, it can be observed that, while the propagation delay changes slightly, the choice of communication and constellation parameters greatly affects the transmission and waiting times. Finally, Figure 10.8 illustrates the need for an advanced routing metric: even when the number of satellites with configuration (c) is greater than that with the other two configurations, the latency achieved by the hop-count metric is greater for this case.

A complementary function to routing is congestion control, which aims at ensuring high bandwidth utilisation while avoiding network congestion. This is done at the transport layer by regulating the rate at which traffic sources inject packets into the network. However, the standard TCP assumes that the bottleneck link will stay the same over time and that changes in its capacity are erratic. This is not true in the satellite network case, in which the capacity of a link is predictable, and therefore a location-aware congestion control mechanism can improve the throughput and latency. In this direction, several works have proposed variations of TCP for space networks. Although the topic is definitely not new [41], the initial works were targeting space networks very different from NGSO constellations, where delay- and disruption-tolerant satellite applications and large distances Earth-GSO were the norms. For example, the Space Communications Protocol Specification-Transport Protocol, mainly developed by NASA and the US Department of Defence, has a selective negative acknowledgement to accommodate asymmetric channels and explicit congestion notification [41]. Another option that does not modify the underlying protocol is the Delay and Disruption Tolerant Networking architecture, which provides long-term information storage on intermediate nodes to cope with disrupted or intermittent links [42]. A more recent alternative is the use of QUIC (Quick User Datagram Protocol UDP Internet Connections), the general purpose transport protocol defined by Google [43] to combine the advantages of connected-oriented TCP and low-latency UDP. The NGSO networks can benefit from QUIC [44] when there is a high RTT and a poor bandwidth. Moreover, QUIC introduces a connection ID instead of IP addresses as identification which inherently avoids re-connections with frequently changeable topological space networks.

10.5 Conclusions

In this chapter, we described relevant aspects of NGSO constellation design to achieve global connectivity. That is, to provide global service availability to ground terminals but also to ensure inter-satellite connectivity can be achieved along the constellation. We emphasised that the constellation geometry, the altitude of deployment and the density of satellites have a major impact on these and other relevant KPIs and compared the performance of three commercial designs: Kepler, OneWeb and the Starlink orbital shell at 550 km. We observed that, while the Starlink orbital shell has a greater number of satellites than the other two constellations, it still requires an additional orbital shell with nearly-polar orbital planes to provide connectivity near-polar regions. On the other hand, around 45 satellites from the OneWeb constellation are simultaneously within communication range in the near-polar regions, which may lead to waste of communication resources. Finally, the Kepler constellation may suffer from coverage holes near the Equator where, on average, less than one satellite is within communication range from the Earth's surface. To provide ubiquitous global coverage, a constellation similar to Kepler but with slightly larger number of orbital planes and satellites would be sufficient. Still, the NGSO constellations that aim to provide broadband services would benefit from further increasing the density of deployment, which would lead to greater data rates both in the inter- and intra-plane RF ISLs.

Besides the impact of the main parameters for constellation design, we elaborated on the major challenges and technologies to achieve global connectivity at the physical layer for link establishment and routing. These arise from the distinctive characteristics of NGSO constellations, which are greatly dynamic yet fully predictable large-scale infrastructures.

References

[1] 3GPP *5G; Study on scenarios and requirements for next generation access technologies*. TR 38.913 V16.0.0; 2020.

[2] Qu Z., Zhang G., Cao H., Xie J. LEO satellite constellation for Internet of Things. *IEEE Access*. 2017, vol. 5, pp. 18391–401.

[3] Liberg O., Lowenmark S.E., Euler S, *et al.* Narrowband Internet of Things for non-terrestrial networks. *IEEE Communications Standards Magazine*. 2020, vol. 4(4), pp. 49–55.

[4] Di B., Song L., Li Y., Poor H.V. 'Ultra-dense LEO: integration of satellite access networks into 5G and beyond'. *IEEE Wireless Communications*. 2019, vol. 26(2), pp. 62–69.

[5] Su Y., Liu Y., Zhou Y., Yuan J., Cao H., Shi J. Broadband Leo satellite communications: architectures and key technologies. *IEEE Wireless Communications*. 2019, vol. 26(2), pp. 55–61.

[6] del Portillo I., Cameron B.G., Crawley E.F. A technical comparison of three low Earth orbit satellite constellation systems to provide global broadband. *Acta Astronautica*. 2021, vol. 159, pp. 123–35.

[7] Handley M. 'Delay is not an option: low latency routing in space'. *Proceedings of the 17th ACM Workshop on Hot Topics in Networks*; 2018. pp. 85–91.

[8] 3GPP *Solutions for NR to support non-terrestrial networks (NTN)*. TR 38.821 V16.0.0; 2019.

[9] Guidotti A., Vanelli-Coralli A., Conti M, *et al.* Architectures and key technical challenges for 5G systems incorporating satellites. *IEEE Transactions on Vehicular Technology*. 2021, vol. 68(3), pp. 2624–39.

[10] Soret B., Leyva-Mayorga I., Cioni S., Popovski P. 5G satellite networks for Internet of Things: offloading and backhauling. *International Journal of Satellite Communications and Networking*. 2021, vol. 39(4), pp. 431–44. Available from https://onlinelibrary.wiley.com/toc/15420981/39/4

[11] Xie R., Tang Q., Wang Q., Liu X., Yu F.R., Huang T. 'Satellite-terrestrial integrated edge computing networks: architecture, challenges, and open issues'. *IEEE Network*. 2020, vol. 34(3), pp. 224–31.

[12] Razmi N., Matthiesen B., Dekorsy A., Popovski P. Ground-assisted federated learning in LEO satellite constellations. *IEEE Wireless Communications Letters*. 2022, vol. 11(4), pp. 717–21.

[13] Leyva-Mayorga I., Soret B., Popovski P. 'Inter-plane inter-satellite connectivity in dense LEO constellations'. *IEEE Transactions on Wireless Communications*. 2021, vol. 20(6), pp. 3430–43.

[14] Kak A., Akyildiz I.F. 'Large-scale constellation design for the internet of space things/cubesats'. *IEEE Globecom Workshops (GC Wkshps)*; Waikoloa, HI, USA, 2019.

[15] Walker J.G. 'Circular orbit patterns providing continuous whole earth coverage'. *Royal Aircraft Establishment, Technical Report 702011*. 1970, pp. 369–84.

[16] Walker J.G. 'Satellite constellations'. *Journal of the British Interplanetary Society*. 1984, vol. 37, pp. 559–71.

[17] Federal Communications Commission (FCC) SEH. 'LLC request for modification of the authorization for the spacex NGSO satellite system'. *IBFS File No. SAT-MOD-20200417- 00037*. 2021, p. 00037.

[18] Lewis J.T., Maclay H.G., Sheehan J.P., Lindsay M. 'Long-term environmental effects of deploying the oneweb satellite constellation'. *70th International Astronautical Congress (IAC)*; 2019. pp. 21–25.

[19] Leyva-Mayorga I., Soret B., Roper M, *et al.* 'LEO small-satellite constellations for 5G and beyond-5G communications'. *IEEE Access*. 2020, vol. 8, pp. 184955–64.

[20] Ye J., Pan G., Alouini M.S. 'Earth rotation-aware non-stationary satellite communication systems: modeling and analysis'. *IEEE Transactions on Wireless Communications*. 2021, vol. 20(9), pp. 5942–56.

[21] Marcano N.J.H., Norby J.G.F., Jacobsen R.H. 'On AD hoc on-demand distance vector routing in low earth orbit nanosatellite constellations'. *IEEE 91st Vehicular Technology Conference (VTC2020-Spring)*; Antwerp, Belgium, 2020.

[22] Kaushal H., Kaddoum G. 'Optical communication in space: challenges and mitigation techniques'. *IEEE Communications Surveys & Tutorials*. 2017, vol. 19(1), pp. 57–96.

[23] Leyva-Mayorga I., Roper M., Matthiesen B., Dekorsy A., Popovski P., Soret B. 'Inter-plane inter-satellite connectivity in LEO constellations: beam switching vs. beam steering'. *IEEE Global Communications Conference*; Madrid, Spain, GLOBECOM, 2021.

[24] Kodheli O., Guidotti A., Vanelli-Coralli A. 'Integration of satellites in 5G through LEO constellations'. *IEEE Global Communications Conference (GLOBECOM)*; Singapore, GLOBECOM, 2017.

[25] 3GPP *Physical channels and modulation*. TS 38.211 V16.1.0; 2020.

[26] Wunder G., Jung P., Kasparick M, *et al.* 5GNOW: Non-orthogonal, asynchronous waveforms for future mobile applications. *IEEE Communications Magazine*. 2014, vol. 52(2), pp. 97–105.

[27] Woltering M., Wubben D., Dekorsy A. 'Factor graph-based equalization for two-way relaying with general multi-carrier transmissions'. *IEEE Transactions on Wireless Communications*. 2018, vol. 17(2), pp. 1212–25.

[28] Schwarz R.T., Delamotte T., Storek K.-U., Knopp A. MIMO applications for multibeam satellites. *IEEE Transactions on Broadcasting*. 2019, vol. 65(4), pp. 664–81.

[29] Röper M., Dekorsy A. 'Robust distributed MMSE precoding in satellite constellations for downlink transmission'. *IEEE 2nd 5G World Forum*; Dresden, Germany, 2019. pp. 642–47.

[30] Chang C.-C., Lee R.-H., Shih T.-Y. Design of a beam switching/steering Butler matrix for phased array system. *IEEE Trans Antennas Propag*. 2010, vol. 58(2), pp. 367–74.

[31] Yu X., Zhang J., Letaief K.B. 'A hardware-efficient analog network structure for hybrid precoding in millimeter wave systems'. *IEEE Journal of Selected Topics in Signal Processing*. 2017, vol. 12(2), pp. 282–97.

[32] El Zooghby A. Smart antenna engineering. Norwood, MA: Artech House, Inc; 2005.

[33] Wang Y., Ma K., Jian Z. 'A low-loss butler matrix using patch element and honeycomb concept on SISL platform'. *IEEE Transactions on Microwave Theory and Techniques*. 2017, vol. 66(8), pp. 3622–31.

[34] Matthiesen B., Bjornson E., De Carvalho E., Popovski P. Intelligent reflecting surface operation under predictable receiver mobility: a continuous time propagation model. *IEEE Wireless Communications Letters*. 2017, vol. 10(2), pp. 216–20.

[35] Liu R., Sheng M., Lui K.S., Wang X., Zhou D., Wang Y. Capacity of two-layered satellite networks. *Wireless Networks*. 2017, vol. 23(8), pp. 2651–69.

[36] Jiang C., Zhu X. Reinforcement learning based capacity management in multi-layer satellite networks. *IEEE Transactions on Wireless Communications*. 2000, vol. 19(7), pp. 4685–99.

[37] Gu Y., Saad W., Bennis M., Debbah M., Han Z. Matching theory for future wireless networks: fundamentals and applications. *IEEE Communications Magazine*. 2000, vol. 53(5), pp. 52–59.

[38] Ahlswede R., Cai N., Li S.-Y.R., Yeung R.W. 'Network information flow'. *IEEE Transactions on Information Theory*. 2000, vol. 46(4), pp. 1204–16.

[39] Rabjerg J.W., Leyva-Mayorga I., Soret B., Popovski P. Exploiting topology awareness for routing in Leo satellite constellations. *IEEE Global Communications Conference (GLOBECOM)*; Madrid, Spain, 2021.

[40] Ruiz de Azua J.A., Calveras A., Camps A. Internet of satellites (IoSat): analysis of network models and routing protocol requirements. *IEEE Access*. 2000, vol. 6, pp. 20390–411.

[41] Durst R.C., Miller G.J., Travis E.J. TCP extensions for space communications. *Wireless Networks*. 1997, vol. 3(5), pp. 389–403.

[42] Caini C., Cruickshank H., Farrell S., Marchese M., Disruption-Tolerant Networking An alternative solution for future satellite networking applications. *Proceedings of the IEEE*. 2011, vol. 99(11), pp. 1980–97.

[43] Hamilton R., Iyengar J., Swett I., Wilk A. QUIC: a UDP-Based secure and reliable transport for HTTP/2. 2016. Available from https://tools.ietf.org/html/draft-tsvwg-quic-protocol-02

[44] Yang S., Li H., Wu Q. Performance analysis of QUIC protocol in integrated satellites and terrestrial networks. *International Wireless Communications & Mobile Computing Conference (IWCMC)*; Limassol, Cyprus, 1997. pp. 1425–30.

Chapter 11

Massive MIMO transmission for non-geostationary orbit

Ke-Xin Li[1], Li You[1], and Xiqi Gao[1]

11.1 Introduction

In recent years, non-geostationary orbit (NGSO) satellites, e.g., low-earth-orbit (LEO) and medium-earth-orbit (MEO) satellites, have been an interesting research topic due to their superiority in shorter round-trip delay, reduced path loss, and lower launch costs [1–3]. Since the LEO satellites are the most representative examples of NGSO ones, this chapter mainly focuses on the LEO satellites, although the proposed transmission approaches are applicable to other NGSO satellites as well.

As an indispensable part of satellite communications (SATCOM), multibeam satellites can serve a number of user terminals (UTs) within the coverage area by using spot beams [4]. For the LEO satellites, phased-array antennas are more often used to generate spot beams due to their wide-angle coverage capabilities [5], e.g., Globalstar [6] and Starlink [7]. In current satellite systems, the inter-beam interference can be suppressed by using a multiple color reuse scheme, in which different frequency bands and orthogonal polarizations are assigned to adjacent beams [8]. As a result, the frequency bands can be reused among sufficiently isolated beams, and the system capacity is improved substantially.

To make full use of the scarce spectrum, full frequency reuse (FFR) scheme has been proposed, which allows all the beams to share the frequency band so that the spectral efficiency can be further enhanced [9, 10]. In this case, it is imperative to use advanced signal processing techniques to alleviate the serious inter-beam interference. Nowadays, the precoding techniques originated from multiuser multiple-input multiple-output (MIMO) communications have been adopted in multibeam satellite systems to handle the inter-beam and inter-user interferences [11–15].

In the previous works, it is usually assumed that the beamforming network (BFN) at the satellite side is fixed [11–15]. The conventional BFN can only be modified at a very slow pace [10], and cannot adapt to the dynamic link conditions of UTs.

[1]National Mobile Communications Research Laboratory, Southeast University, Nanjing, China

In the last decade, massive MIMO has been one of the pivotal techniques in terrestrial 5G communications [16]. With a large number of antennas at the base station, massive MIMO can provide high-resolution in the beam domain, and can significantly improve the spectrum and energy efficiency [17]. With the rapid development of 5G communications, it becomes possible to use a more flexible and versatile fully digital BFN at the satellite [18], which can cater to the dynamic link conditions of UTs. This chapter considers that the LEO satellite is equipped with a massive antenna array, namely a massive MIMO LEO satellite, and the BFN at the LEO satellite is assumed to be digitally reconfigurable in real-time, which is expected to enhance the throughput in wideband LEO SATCOM systems.

Notice that the performance of multiuser MIMO/massive MIMO precoding hinges on the quality of the channel state information (CSI) available at the transmitter. For most previous works on multibeam satellites, it is usually assumed that the transmitter can acquire the instantaneous CSI (iCSI) [11–13, 15]. However, in practical SATCOM systems, the inherent channel impairments, e.g., large propagation delays and Doppler effects, will make it difficult to acquire the iCSI at the transmitter. In particular, for time-division duplexing systems, the estimated uplink (UL) iCSI is used for the downlink (DL) transmission, which could be outdated due to the long propagation delays. At the same time, for frequency-division duplexing systems, each UT first estimates the DL iCSI and then feeds it back to the satellite, which could consume a lot of channel estimation and feedback overhead. Moreover, the feedback would also be outdated due to the long propagation delays. In comparison with the iCSI, statistical CSI (sCSI) is stable for longer time intervals [19] and thus can be more easily acquired at the transmitter side. Hence, a more practical scenario is considered in this chapter where only sCSI is available at the satellite to perform the DL transmit design in massive MIMO SATCOM.

The transmit design using sCSI at the transmitter (sCSIT) has become an attractive topic in massive MIMO terrestrial wireless communications. Up to now, many transmit strategies have been presented, e.g., the two-stage precoder design [20], the beam domain transmission [21], and the robust precoder design [22]. However, the aforementioned works do not consider the special massive MIMO LEO satellite channel characteristics. Besides, the limited satellite payloads bring considerable restrictions on the transmit design. Thus, it is of great importance to seek out more efficient DL transmit designs with sCSIT tailored for massive MIMO LEO SATCOM.

In this chapter, a massive MIMO LEO SATCOM system is considered where the satellite and the UTs are both equipped with uniform planar arrays (UPAs), and the primary target is to achieve high data rates of the whole system using only the slow-varying sCSIT via the proper design of DL transmit strategies. For this purpose, the DL massive MIMO LEO satellite channel model with the UPA configurations at the satellite and each UT is first derived. The adverse Doppler and delay effects are compensated by performing frequency and time synchronization at each UT to facilitate the DL wideband transmission. Then, based on the massive MIMO LEO satellite channel characteristics, the DL transmit design is investigated to maximize the ergodic sum rate by exploiting sCSIT.

It is proved that the single-stream transmit strategy for each UT is optimal for the linear transmitters in the sense of maximizing the system's ergodic sum rate, even though each UT has multiple antennas. This result is important and favorable because the complicated design of transmit covariance matrices can be simplified into that of precoding vectors without any loss of optimality. To reduce the computational complexity, another transmit design is formulated by approximating the ergodic sum rate with its upper bound. In this case, it is shown that the optimality of the single-stream transmit strategy still holds. More importantly, the design of precoding vectors is further simplified to that of scalar variables. Simulation results demonstrate the effectiveness of the proposed approaches and show remarkable performance gains over the existing schemes.

The remainder of this chapter is organized as follows. Section 11.2 introduces the system model, where the channel model is presented for the satellite and the UTs equipped with UPAs. In section 11.3, the DL transmit design and the low-complexity implementations are presented. The user grouping strategies are discussed in section 11.4. Section 11.5 provides the simulation results, and section 11.6 concludes this chapter.

Notations: Throughout this chapter, lower case letters denote scalars and boldface lower (upper) letters denote vectors (matrices). The set of all n-by-m complex (real) matrices is denoted as $\mathbb{C}^{n\times m}$ ($\mathbb{R}^{n\times m}$). tr(·), det(·), rank(·), $(\cdot)^*$, $(\cdot)^T$, and $(\cdot)^H$ denotes the trace, determinant, rank, conjugate, transpose, and conjugate transpose operations for the matrix argument, respectively. The $|.|$ denotes the absolute value. The Euclidean norm of a vector $\mathbf{x}$ is denoted as $\|\mathbf{x}\| = \sqrt{\mathbf{x}^H\mathbf{x}}$ $\otimes$ denotes the Kronecker product. $[\mathbf{A}]_{n,m}$ denotes the (n, m)th element of matrix $\mathbf{A}$. diag $(\mathbf{a})$ denotes the diagonal matrix with $\mathbf{a}$ along its main diagonal. $\mathbb{E}\{\cdot\}$ denotes the expectation operator. $\mathcal{CN}(\mathbf{0}, \mathbf{C})$ denotes the circular symmetric complex Gaussian random vector with zero mean and covariance matrix $\mathbf{C}$.

11.2 System model

An FFR massive MIMO LEO SATCOM system operating over lower frequency bands, e.g., L/S/C bands, is considered. The mobile UTs are served by a single LEO satellite at an altitude of H as shown in Figure 11.1. The satellite and the mobile UTs are equipped with the UPAs of digital active antennas [18], which means that the amplitude and phase of each antenna element of the UPAs can be digitally controlled. The satellite has a large-scale UPA with M_{x} and M_{y} elements in the x-axis and y-axis, respectively. The total number of antennas at the satellite is $M_{\text{x}}M_{\text{y}} \triangleq M$. We assume that each antenna element of the UPA at the satellite is directional. On the other hand, each UT's UPA consists of $N_{\text{x}'}$ and $N_{\text{y}'}$ omnidirectional elements in the x$'$-axis and y$'$-axis, respectively, and the total number of antennas at each UT is $N_{\text{x}'}N_{\text{y}'} \triangleq N$. The approach in this chapter can be directly extended to the cases where the UTs have different numbers of antenna elements.

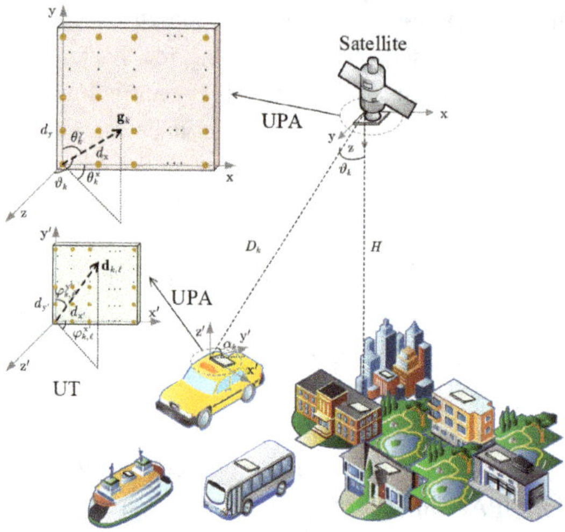

Figure 11.1 The FFR massive MIMO LEO SATCOM system

11.2.1 DL signal and channel models in analog baseband

The DL received signal of UT k at the time instant t can be written as

$$\mathbf{y}_k(t) = \int_{-\infty}^{\infty} \check{\mathbf{H}}_k(t, \tau)\mathbf{x}(t-\tau)\,\mathrm{d}\tau + \mathbf{z}_k(t) \tag{11.1}$$

where $\check{\mathbf{H}}_k(t, \tau) \in \mathbb{C}^{N\times M}$, $\mathbf{x}(t) \in \mathbb{C}^{M\times 1}$ and $\mathbf{z}_k(t) \in \mathbb{C}^{N\times 1}$ are the channel impulse response, transmit signal, and additive noise signal of UT k at time instant t, respectively. More specifically, the LEO satellite channel impulse response $\check{\mathbf{H}}_k(t, \tau)$ can be expressed as

$$\check{\mathbf{H}}_k(t, \tau) = \sum_{\ell=0}^{L_k-1} a_{k,\ell} e^{j2\pi \nu_{k,\ell} t} \delta\left(\tau - \tau_{k,\ell}\right) \mathbf{d}_{k,\ell} \mathbf{g}_{k,\ell}^H \tag{11.2}$$

where $j \triangleq \sqrt{-1}$, $\delta(x)$ is the Dirac delta function, L_k is the multipath number of UT k's channel, $a_{k,\ell}$, $\nu_{k,\ell}$, $\tau_{k,\ell}$, $\mathbf{d}_{k,\ell} \in \mathbb{C}^{N\times 1}$, and $\mathbf{g}_{k,\ell} \in \mathbb{C}^{M\times 1}$ are the DL channel gain, Doppler shift, propagation delay, array response vector at the UT side, and array response vector at the satellite side, respectively, associated with the ℓth path of UT k 's channel.

For simplicity, it is assumed that the channel matrices are fixed within each coherence time interval, and change from block to block according to some ergodic process. In the following, the LEO satellite channel characteristics are described one by one, which mainly include the Doppler shifts, propagation delays, and array response vectors.

11.2.1.1 Doppler shifts

For LEO satellite channels, the Doppler shifts are much larger compared with those in terrestrial wireless channels, due to the large relative velocity between the satellite and the UTs. At the 4 GHz carrier frequency, the Doppler shift can be 80 kHz for an LEO satellite at an altitude of 1000 km [23]. The Doppler shift $\nu_{k,\ell}$ for the ℓth path of UT k's channel mainly consists of two parts [24], i.e., $\nu_{k,\ell} = \nu^{\text{sat}}_{k,\ell} + \nu^{\text{ut}}_{k,\ell}$, where $\nu^{\text{sat}}_{k,\ell}$ and $\nu^{\text{ut}}_{k,\ell}$ are the Doppler shifts relevant to the movement of the satellite and UT k, respectively. The first part $\nu^{\text{sat}}_{k,\ell}$ is nearly identical for different paths of UT k's channel, because of the high altitude of the satellite [24]. Hence, $\nu^{\text{sat}}_{k,\ell}$ can be rewritten as $\nu^{\text{sat}}_{k,\ell} = \nu^{\text{sat}}_{k}$ for $0 \leq \ell \leq L_k - 1$. The variation of ν^{sat}_{k} with time behaves rather deterministically, and it can be estimated and compensated at each UT. On the other hand, $\nu^{\text{ut}}_{k,\ell}$'s are usually distinct for different paths.

11.2.1.2 Propagation delays

For LEO satellites, the propagation delay is a more serious problem than that in terrestrial wireless channels, due to the long distance between the satellite and the UTs. For an LEO satellite at an altitude of 1000 km, the round-trip delay can be about 17.7 ms [25]. Besides, let $\tau^{\min}_k = \min_\ell \tau_{k,\ell}$ and $\tau^{\max}_k = \max_\ell \tau_{k,\ell}$ denote the minimal and maximal propagation delays of UT k's channel, respectively.

11.2.1.3 Array response vectors

Define $\boldsymbol{\theta}_{k,\ell} = (\theta^{\text{x}}_{k,\ell}, \theta^{\text{y}}_{k,\ell})$ and $\boldsymbol{\varphi}_{k,\ell} = (\varphi^{\text{x}'}_{k,\ell}, \varphi^{\text{y}'}_{k,\ell})$ as the paired angles-of-departure (AoDs) and angles-of-arrival (AoAs) for the ℓth path of UT k's channel, respectively. The array response vectors $\mathbf{g}_{k,\ell}$ and $\mathbf{d}_{k,\ell}$ in (11.2) can be written as $\mathbf{g}_{k,\ell} = \mathbf{g}(\boldsymbol{\theta}_{k,\ell})$ and $\mathbf{d}_{k,\ell} = \mathbf{d}(\boldsymbol{\varphi}_{k,\ell})$, respectively. For arbitrary $\boldsymbol{\theta} = (\theta_{\text{x}}, \theta_{\text{y}})$ and $\boldsymbol{\varphi} = (\varphi_{\text{x}'}, \varphi_{\text{y}'})$, $\mathbf{g}(\boldsymbol{\theta})$ and $\mathbf{d}(\boldsymbol{\varphi})$ can be expressed as $\mathbf{g}(\boldsymbol{\theta}) = \mathbf{a}_{M_{\text{x}}}\left(\sin\theta_{\text{y}}\cos\theta_{\text{x}}\right) \otimes \mathbf{a}_{M_{\text{y}}}\left(\cos\theta_{\text{y}}\right)$ and $\mathbf{d}(\boldsymbol{\varphi}) = \mathbf{a}_{N_{\text{x}'}}\left(\sin\varphi_{\text{y}'}\cos\varphi_{\text{x}'}\right) \otimes \mathbf{a}_{N_{\text{y}'}}\left(\cos\varphi_{\text{y}'}\right)$. Here, $\mathbf{a}_{n_{\text{v}}}(x) \in \mathbb{C}^{n_{\text{v}} \times 1}$ is given by

$$\mathbf{a}_{n_{\text{v}}}(x) = \frac{1}{\sqrt{n_{\text{v}}}}\left(1, e^{-j\frac{2\pi d_{\text{v}}}{\lambda}x}, \ldots, e^{-j\frac{2\pi d_{\text{v}}}{\lambda}(n_{\text{v}}-1)x}\right)^T \tag{11.3}$$

where $\lambda = c/f$ represents the carrier wavelength, c is the speed of light, f is the carrier frequency, d_{v} is the antenna spacing along v-axis with $\text{v} \in \{\text{x}, \text{y}, \text{x}', \text{y}'\}$. In satellite channels, the scattering on the ground takes place only within a few kilometers around each UT. Thus, the paired AoDs for different paths of UT k's channel are nearly identical due to the long distance between the satellite and UT k [26], i.e., $\boldsymbol{\theta}_{k,\ell} = \boldsymbol{\theta}_k$, $0 \leq \ell \leq L_k - 1$. Therefore, $\mathbf{g}_{k,\ell} = \mathbf{g}_k = \mathbf{g}(\boldsymbol{\theta}_k)$, where $\boldsymbol{\theta}_k = (\theta^{\text{x}}_k, \theta^{\text{y}}_k)$ is referred to as the physical angle pair of UT k. Due to the long distance between the satellite and UT k, $\mathbf{g}_k$ changes quite slowly, and it is assumed that $\mathbf{g}_k$ can be perfectly known at the satellite. The space angle pair $\tilde{\boldsymbol{\theta}}_k = (\tilde{\theta}^{\text{x}}_k, \tilde{\theta}^{\text{y}}_k)$ of UT k is defined as $\tilde{\theta}^{\text{x}}_k = \sin\theta^{\text{y}}_k \cos\theta^{\text{x}}_k$ and $\tilde{\theta}^{\text{y}}_k = \cos\theta^{\text{y}}_k$, which reflects the space domain property of UT k's channel [26]. The physical angle pair $\boldsymbol{\theta}_k$ and nadir angle ϑ_k of UT k are related by $\cos\vartheta_k = \sin\theta^{\text{y}}_k \sin\theta^{\text{x}}_k$. The space angle pair $\tilde{\boldsymbol{\theta}}_k = (\tilde{\theta}^{\text{x}}_k, \tilde{\theta}^{\text{y}}_k)$ should satisfy $(\tilde{\theta}^{\text{x}}_k)^2 + (\tilde{\theta}^{\text{y}}_k)^2 \leq \sin^2\vartheta_{\max}$ due to the relation $\cos\vartheta_k = \sin\theta^{\text{y}}_k \sin\theta^{\text{x}}_k = \sqrt{1 - (\tilde{\theta}^{\text{y}}_k)^2 - (\tilde{\theta}^{\text{x}}_k)^2} \geq \cos\vartheta_{\max}$, where $\vartheta_{\max}$ is the maximum nadir angle of UTs.

11.2.2 DL signal and channel models for OFDM-based transmission

The orthogonal frequency division multiplex (OFDM) is used to facilitate the wide-band transmission in the LEO SATCOM systems, due to its benefits on robustness to frequency selective fading and efficient implementations. The number of subcarriers is $N_{\rm sc}$, and the cyclic prefix (CP) length is $N_{\rm cp}$. Let $T_{\rm s}$ be the system sampling period. The time duration of CP is $T_{\rm cp} = N_{\rm cp}T_{\rm s}$. The OFDM symbol time duration without and with CP is given by $T_{\rm sc} = N_{\rm sc}T_{\rm s}$ and $T = T_{\rm sc} + T_{\rm cp}$, respectively.

Let $\{\mathbf{x}_{s,r}\}_{r=0}^{N_{\rm sc}-1}$ be the $M \times 1$ frequency-domain DL transmit signal within the s th OFDM symbol. Then, the time-domain DL transmit signal in OFDM symbol s can be expressed as [27]

$$\mathbf{x}_s(t) = \sum_{r=0}^{N_{\rm sc}-1} \mathbf{x}_{s,r} e^{j2\pi r \Delta f t}, \quad -T_{\rm cp} \le t - sT < T_{\rm sc} \tag{11.4}$$

where $\Delta f = 1/T_{\rm sc}$. The time-domain received signal of UT k in the OFDM symbol s can be written as

$$\mathbf{y}_{k,s}(t) = \int_{-\infty}^{\infty} \check{\mathbf{H}}_k(t,\tau)\mathbf{x}_s(t-\tau)\,\mathrm{d}\tau + \mathbf{z}_{k,s}(t) \tag{11.5}$$

where $\mathbf{z}_{k,s}(t)$ is the additive noise signal of UT k at the OFDM symbol s. Next, we perform the Doppler and delay compensation at each UT. Let $\nu_k^{\rm cps} = \nu_k^{\rm sat}$ and $\tau_k^{\rm cps} = \tau_k^{\rm min}$. Based on the results in Reference 26, the compensated time-domain received signal of UT k in the OFDM symbol s is given by

$$\mathbf{y}_{k,s}^{\rm cps}(t) = \mathbf{y}_{k,s}(t + \tau_k^{\rm cps})e^{-j2\pi \nu_k^{\rm cps}(t+\tau_k^{\rm cps})} \tag{11.6}$$

After the Doppler and delay compensation, the well-designed OFDM parameters can be chosen to combat the multipath fading effect. Hence, the frequency-domain received signal of UT k over the subcarrier r in the OFDM symbol s can be written as [27]

$$\mathbf{y}_{k,s,r} = \frac{1}{T_{\rm sc}} \int_{sT}^{sT+T_{\rm sc}} \mathbf{y}_{k,s}^{\rm cps}(t)e^{-j2\pi r\Delta f t}\,\mathrm{d}t \tag{11.7}$$

Let us denote $\tau_{k,\ell}^{\rm ut} = \tau_{k,\ell} - \tau_k^{\rm min}$, and define the effective channel frequency response of UT k after the Doppler and delay compensation as

$$\mathbf{H}_k(t,f) = \mathbf{d}_k(t,f)\mathbf{g}_k^H \tag{11.8}$$

where $\mathbf{d}_k(t,f) = \sum_{\ell=0}^{L_k-1} a_{k,\ell} e^{j2\pi\left(\nu_{k,\ell}^{\rm ut}t - f\tau_{k,\ell}^{\rm ut}\right)} \mathbf{d}_{k,\ell} \in \mathbb{C}^{N\times 1}$. Then, the received signal $\mathbf{y}_{k,s,r}$ in (11.7) can be further expressed as

$$\mathbf{y}_{k,s,r} = \mathbf{H}_{k,s,r}\mathbf{x}_{s,r} + \mathbf{z}_{k,s,r}, \tag{11.9}$$

where $\mathbf{H}_{k,s,r}$ and $\mathbf{z}_{k,s,r}$ are the channel matrix and additive Gaussian noise of UT k over the subcarrier r in the OFDM symbol s. Note that $\mathbf{H}_{k,s,r}$ in (11.9) can be written as

$$\mathbf{H}_{k,s,r} = \mathbf{H}_k(sT, r\Delta f) = \mathbf{d}_k(sT, r\Delta f)\,\mathbf{g}_k^H = \mathbf{d}_{k,s,r}\mathbf{g}_k^H \tag{11.10}$$

Since the Doppler and the delay effects are compensated at each UT, the time and frequency at the satellite and the UTs are assumed to be perfectly synchronized in the following.

11.2.3 DL satellite channel's statistical properties

For convenience to describe the statistical properties of the satellite channel, the subscripts of the OFDM symbol s and subcarrier r in $\mathbf{H}_{k,s,r} = \mathbf{d}_{k,s,r}\mathbf{g}_k^H$ are omitted. Let $\mathbf{H}_k = \mathbf{d}_k\mathbf{g}_k^H$ denote the DL channel matrices of UT k over a specific subcarrier. The channel matrix $\mathbf{H}_k$ is supposed to be Rician distributed as follows:

$$\mathbf{H}_k = \mathbf{d}_k\mathbf{g}_k^H = \sqrt{\frac{\kappa_k\beta_k}{\kappa_k+1}}\mathbf{H}_k^{\mathrm{LoS}} + \sqrt{\frac{\beta_k}{\kappa_k+1}}\mathbf{H}_k^{\mathrm{NLoS}} \tag{11.11}$$

where $\beta_k = \mathbb{E}\left\{\mathrm{tr}(\mathbf{H}_k\mathbf{H}_k^H)\right\} = \mathbb{E}\left\{\|\mathbf{d}_k\|^2\right\}$ is the average channel power, κ_k is the Rician factor, $\mathbf{H}_k^{\mathrm{LoS}} = \mathbf{d}_{k,0}\mathbf{g}_k^H$ is the deterministic line-of-sight (LoS) part, $\mathbf{H}_k^{\mathrm{NLoS}} = \tilde{\mathbf{d}}_k\mathbf{g}_k^H$ is the random scattering part. Besides, $\tilde{\mathbf{d}}_k$ is distributed as $\tilde{\mathbf{d}}_k \sim \mathcal{CN}(\mathbf{0}, \mathbf{\Sigma}_k)$ with $\mathrm{tr}(\mathbf{\Sigma}_k) = 1$. The channel parameters $\mathcal{H} \triangleq \{\beta_k, \kappa_k, \mathbf{g}_k, \mathbf{d}_{k,0}, \mathbf{\Sigma}_k\}_{\forall k}$ are related to the operating frequency bands, the practical link conditions, and so on [5]. It is assumed that the satellite and the UTs move within a certain range, such that the channel parameters $\mathcal{H}$ can be considered as nearly unchanged. Whenever the satellite or some UT steps out of this range, the channel parameters $\mathcal{H}$ should be updated at the satellite accordingly.

The channel correlation matrices of UT k at the satellite and the UT sides are given by

$$\mathbf{R}_k^{\mathrm{sat}} = \mathbb{E}\{\mathbf{H}_k^H\mathbf{H}_k\} = \beta_k\mathbf{g}_k\mathbf{g}_k^H \tag{11.12a}$$

$$\mathbf{R}_k^{\mathrm{ut}} = \mathbb{E}\{\mathbf{H}_k\mathbf{H}_k^H\} = \frac{\kappa_k\beta_k}{\kappa_k+1}\mathbf{d}_{k,0}\mathbf{d}_{k,0}^H + \frac{\beta_k}{\kappa_k+1}\mathbf{\Sigma}_k \tag{11.12b}$$

respectively. The matrix $\mathbf{R}_k^{\mathrm{sat}}$ is rank-one, which implies that the signals on different antennas at the satellite are highly correlated. Meanwhile, the rank of matrix $\mathbf{R}_k^{\mathrm{ut}}$ depends on the specific propagation environment around UT k.

11.3 DL transmit design

In this section, the DL transmit design is investigated for the examined massive MIMO LEO SATCOM system based on the established signal and channel models in section 11.2. First, by exploiting the LEO satellite channel characteristics, it is proved that the rank of transmit covariance matrix of each UT must be no greater than one to maximize the ergodic sum rate. This indicates that the optimal DL transmission strategy is to transmit a single data stream to each UT, even if each UT has multiple antennas. This result is particularly important since the original design of transmit covariance matrices can be simplified into that of the precoding vectors without any loss of optimality. To reduce the computational complexity, the ergodic sum rate is approximated with its closed-form upper bound. Interestingly, it is shown that the optimality of transmitting single data stream to each UT also holds. In this case, it is manifested that the design of precoding vectors can be further simplified into that of scalar variables.

11.3.1 Rank-One property of transmit covariance matrices

By dropping the subscripts of OFDM symbol s and subcarrier r in $\mathbf{x}_{s,r}$ for simplicity, let $\mathbf{x} \in \mathbb{C}^{M\times 1}$ denote the transmit signal at the satellite over a specific subcarrier. It is assumed that K UTs are simultaneously served in the DL transmission. The set of UT indices is denoted as $\mathcal{K} = \{1, \ldots, K\}$. The transmit signal $\mathbf{x}$ can be expressed as

$$\mathbf{x} = \sum_{k=1}^{K} \mathbf{s}_k \tag{11.13}$$

where $\mathbf{s}_k \in \mathbb{C}^{M\times 1}$ is the transmit signal related to UT k. In this chapter, the most general design of the transmit signals $\{\mathbf{s}_k\}_{k=1}^{K}$ is considered, where $\mathbf{s}_k$ is a Gaussian random vector with zero mean and covariance matrix $\mathbf{Q}_k = \mathbb{E}\{\mathbf{s}_k\mathbf{s}_k^H\}$. For simplicity, it is assumed that the DL transmit signal satisfies the total power constraint as in Reference 11, i.e., $\sum_{k=1}^{K} \mathrm{tr}(\mathbf{Q}_k) \leq P$. The DL received signal at UT k is given by

$$\mathbf{y}_k = \mathbf{H}_k \sum_{i=1}^{K} \mathbf{s}_i + \mathbf{z}_k \tag{11.14}$$

where $\mathbf{z}_k \in \mathbb{C}^{N\times 1}$ is the additive complex Gaussian noise at UT k distributed as $\mathbf{z}_k \sim \mathcal{CN}\left(0, \sigma_k^2\mathbf{I}_N\right)$. The DL ergodic rate of UT k is defined as

$$\begin{aligned} \mathcal{I}_k &= \mathbb{E}\left\{\log\det\left(\sigma_k^2\mathbf{I}_N + \mathbf{H}_k\sum_{i=1}^{K}\mathbf{Q}_i\mathbf{H}_k^H\right)\right\} - \mathbb{E}\left\{\log\det\left(\sigma_k^2\mathbf{I}_N + \mathbf{H}_k\sum_{i\neq k}\mathbf{Q}_i\mathbf{H}_k^H\right)\right\} \\ &\overset{(a)}{=} \mathbb{E}\left\{\log\left(1 + \frac{\mathbf{g}_k^H\mathbf{Q}_k\mathbf{g}_k\mathbf{d}_k{}^2}{\sum_{i\neq k}\mathbf{g}_k^H\mathbf{Q}_i\mathbf{g}_k\mathbf{d}_k{}^2 + \sigma_k^2}\right)\right\}, \end{aligned} \tag{11.15}$$

where (a) follows from $\mathbf{H}_k = \mathbf{d}_k\mathbf{g}_k^H$ and $\det(\mathbf{I} + \mathbf{AB}) = \det(\mathbf{I} + \mathbf{BA})$ [28]. The DL sum rate maximization problem can be formulated as

$$\mathcal{P} : \max_{\{\mathbf{Q}_k\}_{k=1}^{K}} \sum_{k=1}^{K} \mathcal{I}_k, \text{ s.t. } \sum_{k=1}^{K} \mathrm{tr}(\mathbf{Q}_k) \leq P,\ \mathbf{Q}_k \succeq \mathbf{0},\ \forall k \in \mathcal{K} \tag{11.16}$$

Theorem 11.3.1: The optimal $\{\mathbf{Q}_k\}_{k=1}^{K}$ to problem $\mathcal{P}$ must satisfy $\mathrm{rank}(\mathbf{Q}_k) \leq 1$, $\forall k \in \mathcal{K}$.

Theorem 11.3.1 shows that owing to the particularities of LEO satellite channels, the rank of the optimal transmit covariance matrix of each UT should be no larger than one. Since $\mathrm{rank}(\mathbf{Q}_k)$ represents the number of independent data streams transmitted to UT k, Theorem 11.3.1 reveals that the single-stream precoding strategy for each UT is optimal for linear transmitters even though each UT has multiple antennas. From the rank-one property of the transmit covariance matrices, $\mathbf{Q}_k$ can be written as $\mathbf{Q}_k = \mathbf{w}_k\mathbf{w}_k^H$, where $\mathbf{w}_k \in \mathbb{C}^{M\times 1}$ is the precoding vector of UT k. Since $\{\mathbf{w}_k\}_{k=1}^{K}$ denote the linear precoding vectors, the transmit signal $\mathbf{s}_k$ in (11.13) is expressed as $\mathbf{s}_k = \mathbf{w}_k s_k$, where s_k is the desired data symbol for UT k with zero mean and unit variance. Henceforth, the design of the transmit covariance matrices $\{\mathbf{Q}_k\}_{k=1}^{K}$ is now simplified into that of the precoding vectors $\{\mathbf{w}_k\}_{k=1}^{K}$. Substituting $\mathbf{Q}_k = \mathbf{w}_k\mathbf{w}_k^H$ into (11.15) yields:

$$\mathscr{I}_k = \mathbb{E}\left\{\log\left(1 + \frac{\mathbf{w}_k^H \mathbf{g}_k^{\ 2} \mathbf{d}_k^{\ 2}}{\sum_{i\neq k} \mathbf{w}_i^H \mathbf{g}_k^{\ 2} \mathbf{d}_k^2 + \sigma_k^2}\right)\right\} \triangleq \mathscr{R}_k. \tag{11.17}$$

Here, $\mathscr{I}_k$ is replaced with $\mathscr{R}_k$ to represent the DL ergodic rate of UT k, since $\mathscr{R}_k$ is now a function of the linear precoding vectors $\{\mathbf{w}_k\}_{k=1}^K$. Thus, the complicated transmit covariance matrix optimization problem $\mathscr{P}$ in (11.16) can be reformulated as follows:

$$\mathscr{S}: \max_{\mathbf{W}} \sum_{k=1}^{K} \mathscr{R}_k, \quad \text{s.t.} \sum_{k=1}^{K} \|\mathbf{w}_k\|^2 \leq P, \tag{11.18}$$

where $\mathbf{W} = [\mathbf{w}_1 \ \cdots \ \mathbf{w}_K] \in \mathbb{C}^{M\times K}$ denotes the collection of the precoding vectors. The power inequality in (11.19) must be met with equality at the optimum, i.e., $\sum_{k=1}^K \|\mathbf{w}_k\|^2 = P$. Otherwise, $\{\mathbf{w}_k\}_{k=1}^K$ can be scaled up, which increases the DL sum rate and contradicts the optimality.

In the following subsection, the optimal linear receivers that maximize their corresponding DL ergodic rates are derived.

11.3.2 Optimal linear receivers

According to Theorem 11.3.1, the satellite can send at most one data stream to each UT. Hence, each UT just needs to decode at most one data stream, and only diversity gain is obtained with multiple antennas at the UT sides. Let $\mathbf{c}_k \in \mathbb{C}^{N\times 1}$ be the linear receiver of UT k. Then, the recovered data symbol at UT k can be written as

$$\hat{s}_k = \mathbf{c}_k^H \mathbf{y}_k = \mathbf{c}_k^H \mathbf{d}_k \mathbf{g}_k^H \mathbf{w}_k s_k + \sum_{i\neq k}^{K} \mathbf{c}_k^H \mathbf{d}_k \mathbf{g}_k^H \mathbf{w}_i s_i + \mathbf{c}_k^H \mathbf{z}_k \tag{11.19}$$

Thus, the signal-to-interference-plus-noise ratio (SINR) of UT k can be expressed as

$$\text{SINR}_k = \frac{\left|\mathbf{w}_k^H \mathbf{g}_k\right|^2 \left|\mathbf{c}_k^H \mathbf{d}_k\right|^2}{\sum_{i\neq k} \left|\mathbf{w}_i^H \mathbf{g}_k\right|^2 \left|\mathbf{c}_k^H \mathbf{d}_k\right|^2 + \sigma_k^2 \left\|\mathbf{c}_k\right\|^2} \tag{11.20}$$

Because $\frac{ax}{bx+c}$ is a monotonically increasing function of x for $a, b, c > 0$, it can be derived that

$$\text{SINR}_k \overset{(a)}{\leq} \frac{\left|\mathbf{w}_k^H \mathbf{g}_k\right|^2 \mathbf{d}_k^{\ 2}}{\sum_{i\neq k} \left|\mathbf{w}_i^H \mathbf{g}_k\right|^2 \mathbf{d}_k^{\ 2} + \sigma_k^2} \triangleq \underline{\text{SINR}}_k \tag{11.21}$$

where (a) follows from the Cauchy-Schwarz inequality $|\mathbf{c}_k^H \mathbf{d}_k|^2 \leq \mathbf{c}_k^{\ 2} \mathbf{d}_k^{\ 2}$, and the equality holds if and only if $\mathbf{c}_k = \alpha \mathbf{d}_k$ for any nonzero $\alpha \in \mathbb{C}$. The receivers satisfying $\mathbf{c}_k = \alpha \mathbf{d}_k$ for different α will have the same value of SINR_k. Thus, the receivers with the form $\mathbf{c}_k = \alpha \mathbf{d}_k$ are optimal for UT k. Now, let us return to the precoding vector design in the following subsection.

11.3.3 Precoding vector design

Considering that the precoding vector optimization problem $\mathcal{S}$ in (11.19) is a non-convex program, it is generally intractable to obtain its globally optimal solutions. However, the existing optimization techniques allow us to derive the locally optimal precoding vectors for the problem $\mathcal{S}$, e.g., minorization-maximization (MM) algorithm [29] and successive convex approximation (SCA). For brevity, the detailed derivations for solving the problem $\mathcal{S}$ are omitted. The readers interested in it can refer to the derivations in Reference 22.

Due to the expectation in the ergodic rate $\mathcal{R}_k$, the Monte-Carlo method with exhaustive sample average is required to compute the precoding vectors, which is a computational demanding task when a large number of samples are considered on the averaging procedure. Next, the low-complexity transmit designs that avoid the sample average are presented.

11.3.4 Low-complexity implementations

To avoid the exhaustive sample average, the ergodic sum rate is approximated by its closed-form upper bound. First, it is proved that in this case, the optimal transmit covariance matrices are still rank-one. Therefore, the design of the transmit covariance matrices can also be boiled down to that of the precoding vectors. Then, it is shown that the design of the precoding vectors can be further converted into that of the scalar variables.

Notice that $f(x) = \log\left(1 + \frac{ax}{bx+c}\right)$ is a concave function of $x \geq 0$ for $a, b, c \geq 0$ [30]. By invoking Jensen's inequality [31], the DL ergodic rate $\mathscr{I}_k$ of UT k can be upper bounded by

$$\begin{aligned} \mathscr{I}_k &= \mathbb{E}\left\{\log\left(1 + \frac{\mathbf{g}_k^H \mathbf{Q}_k \mathbf{g}_k \mathbf{d}_k{}^2}{\sum_{i\neq k} \mathbf{g}_k^H \mathbf{Q}_i \mathbf{g}_k \mathbf{d}_k{}^2 + \sigma_k^2}\right)\right\} \\ &\leq \log\left(1 + \frac{\mathbf{g}_k^H \mathbf{Q}_k \mathbf{g}_k \beta_k}{\sum_{i\neq k} \mathbf{g}_k^H \mathbf{Q}_i \mathbf{g}_k \beta_k + \sigma_k^2}\right) \triangleq \mathscr{I}_k^{\text{ub}}. \end{aligned} \tag{11.22}$$

The problem of maximizing the upper bound of the DL ergodic sum rate can be formulated as

$$\mathscr{P}^{\text{ub}}: \max_{\{\mathbf{Q}_k\}_{k=1}^K} \sum_{k=1}^{K} \mathcal{I}_k^{\text{ub}}, \text{ s.t. } \sum_{k=1}^{K} \text{tr}(\mathbf{Q}_k) \leq P,\ \mathbf{Q}_k \succeq \mathbf{0},\ \forall k \in \mathscr{K} \tag{11.23}$$

Theorem 11.3.2: The optimal $\{\mathbf{Q}_k\}_{k=1}^K$ to the problem $\mathscr{P}^{\text{ub}}$ must satisfy $\text{rank}(\mathbf{Q}_k) \leq 1$, $\forall k \in \mathscr{K}$.

According to Theorem 11.3.2, the rank of the optimal transmit covariance matrices to the problem $\mathscr{P}^{\text{ub}}$ should be no greater than one, which manifests that the single-stream precoding strategy for each UT suffices to maximize the upper bound on the ergodic sum rate. Thus, $\mathbf{Q}_k$ can be written as $\mathbf{Q}_k = \mathbf{w}_k \mathbf{w}_k^H$, and once more the design of the transmit covariance matrices $\{\mathbf{Q}_k\}_{k=1}^K$ can be reduced to that of the precoding vectors $\{\mathbf{w}_k\}_{k=1}^K$. Hence, the $\mathscr{I}_k^{\text{ub}}$ expression in (11.23) can be further written as

$$\mathscr{I}_k^{\text{ub}} = \log\left(1 + \frac{\left|\mathbf{w}_k^H \mathbf{g}_k\right|^2 \beta_k}{\sum_{i\neq k} \left|\mathbf{w}_i^H \mathbf{g}_k\right|^2 \beta_k + \sigma_k^2}\right) \triangleq \mathscr{R}_k^{\text{ub}}. \tag{11.24}$$

Here, $\mathscr{I}_k^{\text{ub}}$ is replaced with $\mathscr{R}_k^{\text{ub}}$, because $\mathscr{R}_k^{\text{ub}}$ has become a closed-form expression of the precoding vectors $\{\mathbf{w}_k\}_{k=1}^K$. Then, the transmit covariance matrix optimization problem $\mathscr{P}^{\text{ub}}$ in (11.23) can be reformulated as

$$\mathscr{S}^{\text{ub}}: \max_{\mathbf{W}} \sum_{k=1}^{K} \mathscr{R}_k^{\text{ub}}, \quad \text{s.t.} \sum_{k=1}^{K} \|\mathbf{w}_k\|^2 \le P \tag{11.25}$$

Note that the problem $\mathscr{S}^{\text{ub}}$ is in the similar form to the sum rate maximization problem in DL multi-user MISO channels [32]. The optimal precoding vectors to the problem $\mathscr{S}^{\text{ub}}$ must satisfy $\sum_{k=1}^{K} \mathbf{w}_k{}^2 = P$, because any precoding vectors with $\sum_{k=1}^{K} \mathbf{w}_k{}^2$ can be scaled up to increase the objective value.

For the problem $\mathscr{S}^{\text{ub}}$, only the channel parameters $\{\beta_k/\sigma_k^2, \tilde{\boldsymbol{\theta}}_k\}_{k=1}^K$ are required at the satellite to compute the precoding vectors, which can be determined by the location information and average channel power of UTs. When the UPA placement is fixed, the space angle pairs $\{\tilde{\boldsymbol{\theta}}_k\}_{k=1}^K$ can be obtained by exploiting the location information of the satellite and UTs, which can be known via the global positioning system. The satellite can estimate the average channel power $\{\beta_k\}_{k=1}^K$ by exploiting the UL sounding signals and the reciprocity of sCSI [30].

Next, it is shown that the design of high-dimensional precoding vectors in the problem $\mathscr{S}^{\text{ub}}$ can be transformed into that of K scalar variables. For ease of statement, an optimization problem is formulated as follows:

$$\mathscr{M}^{\text{ub}}: \max_{\boldsymbol{\lambda}} \sum_{k=1}^{K} r_k, \quad \text{s.t.} \sum_{k=1}^{K} \lambda_k = P, \lambda_k \ge 0, \forall k \in \mathscr{K} \tag{11.26}$$

where $\boldsymbol{\lambda} = [\lambda_1 \ldots \lambda_K]^T \in \mathbb{R}^{K\times 1}$ and r_k is a function of $\{\lambda_k\}_{k=1}^K$ given by

$$\begin{aligned} r_k(\lambda_1, \ldots, \lambda_K) &= \log\det\left(\sum_{i=1}^{K} \frac{\lambda_i\beta_i}{\sigma_i^2}\mathbf{g}_i\mathbf{g}_i^H + \mathrm{I_M}\right) \\ &- \log\det\left(\sum_{\mathrm{i}\neq \mathrm{k}} \frac{\lambda_\mathrm{i}\beta_\mathrm{i}}{\sigma_\mathrm{i}^2}\mathbf{g}_\mathrm{i}\mathbf{g}_\mathrm{i}^\mathrm{H} + \mathrm{I_M}\right) \end{aligned} \tag{11.27}$$

The relationship between the problems $\mathscr{S}^{\text{ub}}$ and $\mathscr{M}^{\text{ub}}$ will be established in the following.

Denote $\{\mathbf{w}_k^{\text{opt}}\}_{k=1}^K$ and $\{\lambda_k^{\text{opt}}\}_{k=1}^K$ as the optimal solutions to problems $\mathscr{S}^{\text{ub}}$ and $\mathscr{M}^{\text{ub}}$, respectively. As described in the following theorem, as long as the scalar variables $\{\lambda_k^{\text{opt}}\}_{k=1}^K$ are known, the precoding vectors $\{\mathbf{w}_k^{\text{opt}}\}_{k=1}^K$ can be derived in the closed form immediately.

Theorem 11.3.3: The precoding vectors $\{\mathbf{w}_k^{\text{opt}}\}_{k=1}^K$ can be written as

$$\mathbf{w}_k^{\text{opt}} = \sqrt{q_k^{\text{opt}}} \cdot \frac{(\mathbf{V}^{\text{opt}})^{-1}\mathbf{g}_k}{\|(\mathbf{V}^{\text{opt}})^{-1}\mathbf{g}_k\|}, \forall k \in \mathscr{K} \tag{11.28}$$

In (11.28), the matrix $\mathbf{V}^{\text{opt}} \in \mathbb{C}^{M\times M}$ and q_k^{opt} are given by

$$\mathbf{V}^{\text{opt}} = \sum_{k=1}^{K} \frac{\lambda_k^{\text{opt}}\beta_k}{\sigma_k^2}\mathbf{g}_k\mathbf{g}_k^H + \mathbf{I}_M \tag{11.29a}$$

$$q_k^{\text{opt}} = \frac{\lambda_k^{\text{opt}}\beta_k(\gamma_k^{\text{opt}}+1)}{\mu^{\text{opt}}\sigma_k^2}\|(\mathbf{V}^{\text{opt}})^{-1}\mathbf{g}_k\|^2, \tag{11.29b}$$

where the parameters γ_k^{opt} and μ^{opt} in (11.29b) are also determined by $\{\lambda_k^{\text{opt}}\}_{k=1}^K$ as follows

$$\gamma_k^{\mathrm{opt}} = \frac{1}{1-(\lambda_k^{\mathrm{opt}}\beta_k/\sigma_k^2)\mathbf{g}_k^H\left(\mathbf{V}^{\mathrm{opt}}\right)^{-1}\mathbf{g}_k} - 1, \tag{11.30a}$$

$$\mu^{\mathrm{opt}} = \frac{1}{P}\sum_{k=1}^{K}\frac{\lambda_k^{\mathrm{opt}}\beta_k(\gamma_k^{\mathrm{opt}}+1)}{\sigma_k^2}\|\left(\mathbf{V}^{\mathrm{opt}}\right)^{-1}\mathbf{g}_k\|^2. \tag{11.30b}$$

In massive MIMO LEO SATCOM systems, the dimension of the precoding vectors $\{\mathbf{w}_k\}_{k=1}^K$ might be extremely large. Theorem 11.3.3 indicates that thanks to the massive MIMO LEO satellite channel properties, the design of the high-dimensional precoding vectors $\{\mathbf{w}_k\}_{k=1}^K$ in the problem $\mathcal{S}^{\mathrm{ub}}$ can be simplified into that of K scalar variables $\{\lambda_k\}_{k=1}^K$ in the problem $\mathcal{M}^{\mathrm{ub}}$, with which the precoding vectors $\{\mathbf{w}_k\}_{k=1}^K$ can be calculated in closed form.

Even though the non-convexity of the problem $\mathscr{M}^{\mathrm{ub}}$ makes it difficult to derive its globally optimal solutions, many optimization algorithms can guarantee convergence to a locally optimal solution to $\mathscr{M}^{\mathrm{ub}}$, such as MM algorithm, SCA, concave-convex procedure, block coordinate descent (BCD) [33], and so on. For brevity, the detailed procedures for solving the problem $\mathscr{M}^{\mathrm{ub}}$ are omitted.

11.4 User grouping

In SATCOM systems, the number of UTs is usually much larger than the number of antennas equipped with the satellite. Therefore, it is of practical importance to investigate the user grouping strategy, such that the proposed DL transmit design for massive MIMO LEO SATCOM systems can fully realize its potential. In this section, a novel user grouping strategy is developed by only exploiting the space angle information of UTs.

Notice that $\tilde{\theta}_k^{\mathrm{x}}$ and $\tilde{\theta}_k^{\mathrm{y}}$ are both located in the range $[-\sin\vartheta_{\max}, \sin\vartheta_{\max})$. Hence, it is assumed that the space angle range $[-\sin\vartheta_{\max}, \sin\vartheta_{\max})$ is divided into $M_{\mathrm{x}}G_{\mathrm{x}}$ and $M_{\mathrm{y}}G_{\mathrm{y}}$ equal parts in the x-axis and y-axis, respectively. The 2D space angle interval is represented by

$$\begin{aligned}\mathscr{A}_{(u,v)}^{(m,n)} &= \Big\{(\phi_{\mathrm{x}},\phi_{\mathrm{y}}) : \phi_{\mathrm{x}} \in \left[\phi_{u,m}^{\mathrm{x}} - \tfrac{\delta_{\mathrm{x}}}{2}, \phi_{u,m}^{\mathrm{x}} + \tfrac{\delta_{\mathrm{x}}}{2}\right), \\ &\qquad \phi_{\mathrm{y}} \in \left[\phi_{v,n}^{\mathrm{y}} - \tfrac{\delta_{\mathrm{y}}}{2}, \phi_{v,n}^{\mathrm{y}} + \tfrac{\delta_{\mathrm{y}}}{2}\right)\Big\}\end{aligned} \tag{11.31}$$

where $\phi_{a,b}^{\mathrm{v}}$ is defined as

$$\phi_{a,b}^{\mathrm{v}} = -\sin\vartheta_{\max} + \frac{\delta_{\mathrm{v}}}{2} + (a+bG_{\mathrm{v}})\delta_{\mathrm{v}}, \tag{11.32}$$

with $0 \le a \le G_{\mathrm{v}}-1$, $0 \le b \le M_{\mathrm{v}}-1$ and $\delta_{\mathrm{v}} = \frac{2\sin\vartheta_{\max}}{M_{\mathrm{v}}G_{\mathrm{v}}}$ with $\mathrm{v} \in \{\mathrm{x},\mathrm{y}\}$. Thus, UT k is scheduled into the (u,v)th group, if and only if there exists some $0 \le m \le M_{\mathrm{x}}-1$ and $0 \le n \le M_{\mathrm{y}}-1$ such that

$$\tilde{\boldsymbol{\theta}}_k = (\tilde{\theta}_k^{\mathrm{x}}, \tilde{\theta}_k^{\mathrm{y}}) \in \mathscr{A}_{(u,v)}^{(m,n)} \tag{11.33}$$

Table 11.1 Simulation parameters

Parameters	**Values**
Earth radius R_e	6378 km
Orbit altitude H	1000 km
Central frequency f_c	4 GHz
Bandwidth B	50 MHz
Noise temperature T_n	290 K
Number of antennas $M_x, M_y, N_{x'}, N_{y'}$	12, 12, 6 and 6
Antenna spacing $d_x, d_y, d_{x'}, d_{y'}$	$\lambda, \lambda, \frac{\lambda}{2}$ and $\frac{\lambda}{2}$
Per-antenna gain G_{sat} and G_{ut}	6 dBi and 0 dBi
Maximum nadir angle ϑ_{max}	30°
Number of UTs K	60
Transmit power P	10 dBW–30 dBW

Let $\mathscr{K}_{(u,v)}^{(m,n)} = \{k : \tilde{\boldsymbol{\theta}}_k \in \mathscr{A}_{(u,v)}^{(m,n)}\}$ denote the set of UTs whose space angle pair is located in $\mathscr{A}_{(u,v)}^{(m,n)}$. For simplicity, it is assumed that $|\mathscr{K}_{(u,v)}^{(m,n)}| \leq 1$. Hence, the set of UTs in the (u, v)th group is given by

$$\mathscr{K}_{(u,v)} = \bigcup_{\substack{0 \leq m \leq M_x - 1 \\ 0 \leq n \leq M_y - 1}} \mathscr{K}_{(u,v)}^{(m,n)}. \tag{11.34}$$

In addition, the UTs allocated in the same group use the same time-frequency resources, while the UTs in different groups are assigned with different time-frequency resources.

11.5 Simulation results

In this section, the simulation results are presented to verify the performance of the proposed DL transmits designs in a massive MIMO LEO SATCOM system. The simulation parameters are summarized in Table 11.1. In the simulations, the space angle pairs of UTs are generated within the circle region $\{(x, y) : x^2 + y^2 \leq \sin^2 \vartheta_{max}\}$. The per-antenna gains at the satellite and UTs are denoted as G_{sat} and G_{ut}, respectively. The details of the antenna gain computation can be found in Reference 4. For simplicity, it is assumed that each antenna element at the satellite has the ideal directional power pattern $R(\theta_x, \theta_y) = G_{sat}$, if $(\sin\theta_y \cos\theta_x)^2 + (\cos\theta_y)^2 \leq \sin^2 \vartheta_{max}$, and otherwise, $R(\theta_x, \theta_y) = 0$, which is in accord with the coverage area seen at the satellite. The elevation angle of UT k can be computed by $\alpha_k = \cos^{-1}\left(\frac{R_s}{R_e} \sin\vartheta_k\right)$ [5], where R_e is the earth radius, $R_s = R_e + H$ is the orbit radius. The distance between the satellite and UT k is given by $D_k = \sqrt{R_e^2 \sin^2 \alpha_k + H^2 + 2HR_e} - R_e \sin\alpha_k$ [34]. The random vector

$\mathbf{d}_k = \sqrt{\frac{\kappa_k \beta_k}{\kappa_k+1}}\mathbf{d}_{k,0} + \sqrt{\frac{\beta_k}{\kappa_k+1}}\tilde{\mathbf{d}}_k$ in (11.11) is simulated in terms of $\mathbf{d}_k(t,f)$ in (11.8), where the first path is used to produce the LoS direction $\mathbf{d}_{k,0} = \mathbf{d}(\boldsymbol{\varphi}_{k,0})$ and the remaining $L_k - 1$ paths are used for $\tilde{\mathbf{d}}_k$. For simplicity, each UT's UPA is assumed to be placed horizontally, which implies that $\boldsymbol{\varphi}_{k,0}$ satisfies $\sin\varphi_{k,0}^{y'}\sin\varphi_{k,0}^{x'} = \sin\alpha_k$ (e.g., $\varphi_{k,0}^{x'} = 90°$ and $\varphi_{k,0}^{y'} = \alpha_k$). To simulate $\tilde{\mathbf{d}}_k$, the path gains $\{a_{k,\ell}\}_{\ell=1}^{L_k-1}$ are generated by using the exponential power delay profile, while the paired AoAs $\{\boldsymbol{\varphi}_{k,\ell}\}_{\ell=1}^{L_k-1}$ are produced according to the wrapped Gaussian power angle spectrum, as described in the 3rd Generation Partnership Project (3GPP) technical report on non-terrestrial networks [34, section 6]. Moreover, the path loss, shadow fading and Rician factors are computed in accordance with the suburban scenarios, and the ionospheric loss is set as 1 dB approximately [34, section 6]. The average channel power β_k is simulated by $\frac{1}{N_S}\sum_{n=1}^{N_S}\|\mathbf{d}_{k,n}\|^2$, where $\mathbf{d}_{k,n}$ is the nth sample of $\mathbf{d}_k$ and the number of channel samples is set as $N_S = 1000$. The noise variance is given by $\sigma_k^2 = k_B T_n B$ where $k_B = 1.38 \times 10^{-23}$ J $\cdot$ K^{-1} is the Boltzmann constant, T_n is the noise temperature and B is the system bandwidth.

In Figure 11.2, the sum rate performance of proposed DL transmit designs is depicted for different numbers of user groups. It is shown that the difference in the sum rate performance between the proposed DL precoding and its low-complexity implementations is negligible. In addition, both the proposed DL precoding and its low-complexity implementations show substantial performance superiority compared with the conventional scheme using fixed beams. Since the proposed precoding vector design strategies only exploit the slow-varying sCSI, which is identical

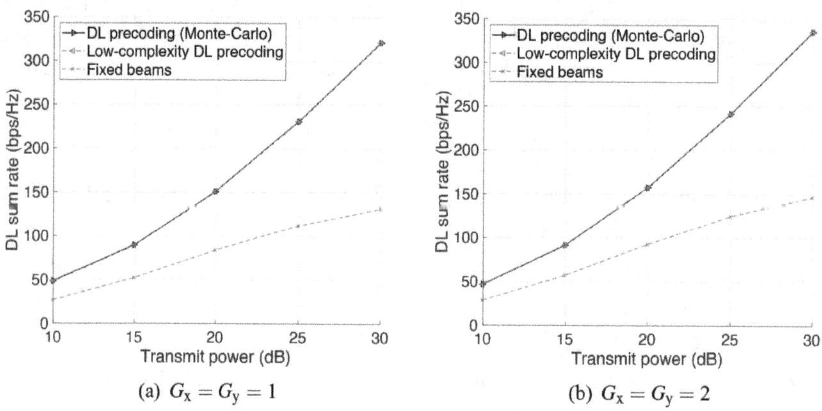

Figure 11.2 Sum rate performance of proposed DL transmit designs

for different subcarriers and OFDM symbols within a stable sCSI period, the implementation complexity at the satellite could be extremely low. Therefore, the proposed approaches constitute a promising candidate for high-throughput massive MIMO LEO SATCOM systems.

11.6 Conclusions

In this chapter, the DL transmit design with sCSIT was investigated for massive MIMO LEO SATCOM systems. First, the DL massive MIMO LEO satellite channel model was derived, where the satellite and the UTs are both equipped with UPAs. Then, it was shown that the single-stream precoding for each UT is able to maximize the ergodic sum rate for the linear transmitters. To reduce the computational complexity, another transmit design was formulated by using an upper bound on the ergodic sum rate, for which the optimality of single-stream precoding also holds. Moreover, it was revealed that the design of precoding vectors can be simplified into that of scalar variables. The effectiveness and the performance gains of the proposed DL transmit designs were verified via the simulation results. There are still many potential challenges for future high-throughput LEO SATCOM systems, which are briefly summarized as follows, e.g., low-complexity hybrid precoding, real-time resource management, multiple satellite cooperation, the coexistence of NGSO and GSO satellites.

References

[1] Kodheli O., Lagunas E., Maturo N, *et al.* 'Satellite communications in the new space era: A survey and future challenges'. *IEEE Communications Surveys & Tutorials*. 2009, vol. 23(1), pp. 70–109.

[2] Su Y., Liu Y., Zhou Y., Yuan J., Cao H., Shi J. 'Broadband leo satellite communications: architectures and key technologies'. *IEEE Wireless Communications*. 2019, vol. 26(2), pp. 55–61.

[3] Kourogiorgas C.I., Lyras N., Panagopoulos A.D. 'Capacity statistics evaluation for next generation broadband MEO satellite systems'. *IEEE Transactions on Aerospace and Electronic Systems*. 2009, vol. 53(5), pp. 2344–58.

[4] Maral G., Bousquet M. *Satellite Communications Systems*. 5th ed. Chichester: Wiley; 2009 Dec 18. Available from http://doi.wiley.com/10.1002/9780470834985

[5] LutzE., Werner M., JahnA. *Satellite systems for personal and broadband communications*. Berlin, Heidelberg: Springer; 2009 Dec 18. Available from http://link.springer.com/10.1007/978-3-642-59727-5

[6] Metzen P.L. 'Globalstar satellite phased array antennas'. Presented at 2000 IEEE International Conference on Phased Array Systems and Technology; Dana Point, CA. USA,207-210,

[7] del Portillo I., Cameron B.G., Crawley E.F. 'A technical comparison of three low earth orbit satellite constellation systems to provide global broadband'. *Acta Astronautica*. 2019, vol. 159, pp. 123–35.

[8] Fenech H., Tomatis A., Amos S., Soumpholphakdy V., Serrano Merino J.L. 'Eutelsat HTS systems'. *International Journal of Satellite Communications and Networking*. 2019, vol. 34(4), pp. 503–21.

[9] Vazquez M.A., Perez-Neira A., Christopoulos D., *et al.* 'Precoding in multibeam satellite communications: present and future challenges'. *IEEE Wireless Communications*. 2016, vol. 23(6), pp. 88–95.

[10] Perez-Neira A.I., Vazquez M.A., Shankar M.R.B., Maleki S., Chatzinotas S. 'Signal processing for high-throughput satellites: challenges in new interference-limited scenarios'. *IEEE Signal Processing Magazine*. 2019, vol. 36(4), pp. 112–31.

[11] Gan Z., Chatzinotas S., Ottersten B. 'Generic optimization of linear precoding in multibeam satellite systems'. *IEEE Transactions on Wireless Communications*. 2012, vol. 11(6), pp. 2308–20.

[12] Christopoulos D., Chatzinotas S., Ottersten B. 'Multicast multigroup precoding and user scheduling for frame-based satellite communications'. *IEEE Transactions on Wireless Communications*. 2015, vol. 14(9), pp. 4695–707.

[13] Joroughi V., Vazquez M., Perez-Neira A. 'Precoding in multigateway multibeam satellite systems'. *IEEE Transactions on Wireless Communications*. 2016, pp. 1–1.

[14] Wang W., Liu A., Zhang Q., You L., Gao X., Zheng G. 'Robust multigroup multicast transmission for frame-based multi-beam satellite systems'. *IEEE Access*. 2018, vol. 6, pp. 46074–83.

[15] Schwarz R.T., Delamotte T., Storek K.-U., Knopp A. 'MIMO applications for multibeam satellites'. *IEEE Transactions on Broadcasting*. 2019, vol. 65(4), pp. 664–81.

[16] Marzetta T.L. 'Noncooperative cellular wireless with unlimited numbers of base station antennas'. *IEEE Transactions on Wireless Communications*. 2010, vol. 9(11), pp. 3590–600.

[17] Ngo H.Q., Larsson E.G., Marzetta T.L. 'Energy and spectral efficiency of very large multiuser MIMO systems'. *IEEE Transactions on Communications*. 2013, vol. 61(4), pp. 1436–49.

[18] Hong W., Jiang Z.H., Yu C., *et al.* 'Multibeam antenna technologies for 5G'. *IEEE Transactions on Antennas and Propagation Wireless Communications*. 2017, vol. 65(12), pp. 6231–49.

[19] Gao X., Jiang B., Li X., Gershman A.B., McKay M.R. 'Statistical eigenmode transmission over jointly correlated MIMO channels'. *IEEE Transactions on Information Theory*. 2009, vol. 55(8), pp. 3735–50.

[20] Adhikary A., Junyoung N., Junyoung A., Caire G. 'Joint spatial division and multiplexing—the large-scale array regime'. *IEEE Transactions on Information Theory*. 2013, vol. 59(10), pp. 6441–63.

[21] Sun C., Gao X., Jin S., Matthaiou M., Ding Z., Xiao C. 'Beam division multiple access transmission for massive MIMO communications'. *IEEE Transactions on Communications*. 2015, vol. 63(6), pp. 2170–84.
[22] Lu A.-A., Gao X., Zhong W., Xiao C., Meng X. 'Robust transmission for massive MIMO downlink with imperfect CSI'. *IEEE Transactions on Communications*. 1998, vol. 67(8), pp. 5362–76.
[23] Ali I., Al-Dhahir N., Hershey J.E. 'Doppler characterization for LEO satellites'. *IEEE Transactions on Communications*. 1998, vol. 46(3), pp. 309–13.
[24] Papathanassiou A., Salkintzis A.K., Mathiopoulos P.T. 'A comparison study of the uplink performance of W-CDMA and OFDM for mobile multimedia communications via LEO satellites'. *IEEE Personal Communications*. 1998, vol. 8(3), pp. 35–43.
[25] Guidotti A., Vanelli-Coralli A., Foggi T., *et al.* 'LTE-based satellite communications in LEO mega-constellations'. *International Journal of Satellite Communications and Networking*. 2019, vol. 37(4), pp. 316–30. Available from https://onlinelibrary.wiley.com/toc/15420981/37/4
[26] You L., Li K.-X., Wang J., Gao X., Xia X.-G., Ottersten B. 'Massive MIMO transmission for LEO satellite communications'. *IEEE Journal on Selected Areas in Communications*. 2020, vol. 38(8), pp. 1851–65.
[27] Hwang T., Yang C., Wu G., Li S., Ye Li G. 'OFDM and its wireless applications: A survey'. *IEEE Transactions on Vehicular Technology*. 2004, vol. 58(4), pp. 1673–94.
[28] Horn R.A., Johnson C.R. *Matrix Analysis*. 2nd ed. New York, NY: Cambridge University Press; 2013.
[29] Hunter D.R., Lange K. 'A tutorial on MM algorithms'. *The American Statistician*. 2004, vol. 58(1), pp. 30–37.
[30] Sun C., Gao X., Jin S., Matthaiou M., Ding Z., Xiao C. 'Beam division multiple access transmission for massive MIMO communications'. *IEEE Transactions on Communications*. 2004, vol. 63(6), pp. 2170–84.
[31] Boyd S., Vandenberghe L. *Convex Optimization*. New York, NY: Cambridge University Press; 2004 Mar 8. Available from https://www.cambridge.org/core/product/identifier/9780511804441/type/book
[32] Bjornson E., Bengtsson M., Ottersten B. 'Optimal multiuser transmit beamforming: A difficult problem with A simple solution structure [lecture notes]'. *IEEE Signal Processing Magazine*. 2014, vol. 31(4), pp. 142–48.
[33] Shi Q., Razaviyayn M., Luo Z.-Q., He C. 'An iteratively weighted MMSE approach to distributed sum-utility maximization for a MIMO interfering broadcast channel'. *IEEE Transactions on Signal Processing*. 2011, vol. 59(9), pp. 4331–40.
[34] 3GPP TR 38 811. 'Study on new radio (NR) to support nonterrestrial networks (release 15)'.2020.in press. Sophia Antipolis Valbonne, France.

Chapter 12

Internet of Things over non-geostationary orbit system and random access aspects

Riccardo De Gaudenzi[1], Nader Alagha[1], and Stefano Cioni[1]

This chapter deals with the Internet of things (IoT) over non-geostationary orbit (NGSO) system design as depicted in Figure 12.1 and random access (RA) aspects. A large number of selected references are provided for the reader interested to have a more in-depth understanding of the aspects touched in the text. The chapter is organized as follows:

- Section 12.1 reviews IoT over NGSO system aspects. In particular, it deals with IoT frequency bands, satellite orbit effects (Doppler, propagation aspects), land mobile satellite (LMS) channel, Doppler, and path loss (PL) compensation techniques.
- Section 12.2 discusses the rationale and the challenges for RA in NGSO networks. In particular, it deals with IoT traffic models, RA versus demand assignment multiple access (DAMA), slotted versus unslotted RA solutions, RA schemes trade-off, and signal processing aspects.
- Section 12.3 deals with NGSO RA schemes design aspects. In particular, it covers the design of the forward and return link, the RA key performance indicators, and how to perform detailed and simplified RA analysis.
- Section 12.4 is covering NGSO RA standard and proprietary solutions and provides examples of system implementations. It covers RA solutions like S-band mobile interactive multimedia (S-MIM), very high frequency (VHF) data exchange (VDE), narrowband IoT (NB-IoT), and universal network for IoT (UNIT). This section also highlights in-orbit demonstrations of emerging satellite IoT systems and associated opportunities and challenges ahead.

[1]European Space Agency, European Space and Technology Centre, Noordwijk, The Netherlands

Figure 12.1 NGSO systems (Source: *https://www.esa.int*)

12.1 IoT over NGSO system aspects

IoT services are becoming an essential part of digital transformation that modernize many aspects of industry allowing to track, monitor, and manage assets while improving their remote operations. The socioeconomic prospects of such transformation have rightfully created an expectation that the IoT wireless networks would readily surpass billions of connections in 2020s. Today's terrestrial communication infrastructure can only provide coverage to less than 20% of the planet surface. Hence, it is inevitable to consider nonterrestrial access solutions to maintain the service coverage continuity. The IoT service offering via satellite is a natural solution to extend the terrestrial service coverage.

Traditionally, satellite communication systems have been supporting supervisory control and data acquisition (SCADA) services. However, there are several

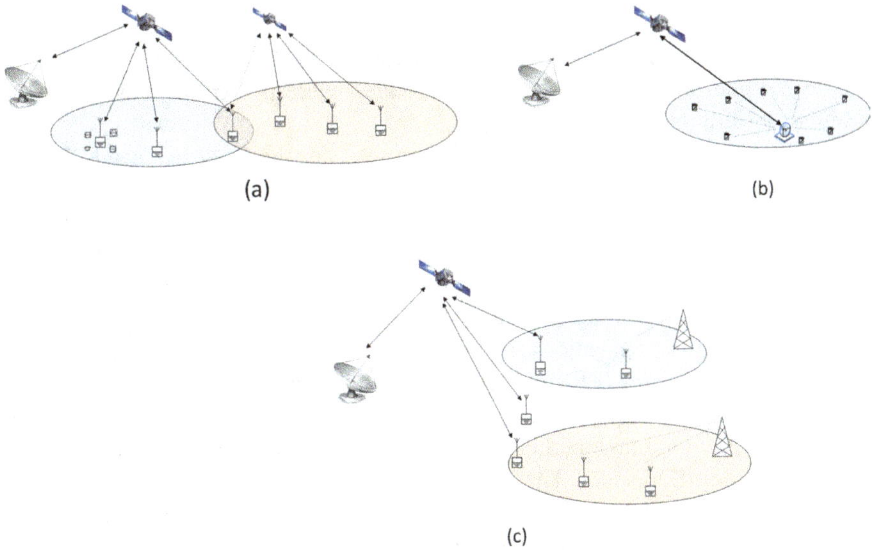

Figure 12.2 *Satellite IoT network connectivity scenarios: (a) direct satellite access by user equipments, (b) indirect access via a satellite access node, and (c) hybrid satellite-terrestrial access*

other emerging system scenarios for satellite access to and from the IoT end nodes (IoT devices). Figure 12.2 illustrates three network scenarios for the satellite IoT connectivity. An IoT service can be provided directly between a satellite and end devices as a stand-alone service. One can envisage another category of IoT services where IoT devices communicate via terrestrial links to a common satellite terminal that collects information and provides access to multiple IoT devices via satellite. There are also emerging scenarios, particularly for mobile services, for which the use of hybrid terrestrial-satellite IoT terminals is envisaged. In the latter case, depending on the availability and required quality of service, the end user terminal may use terrestrial access wherever such service is available and seamless roaming to satellite access in remote areas lacking terrestrial infrastructure.

The largest growth in the satellite IoT is related to mobile services to devices with relatively low throughput and low power and relaxed latency requirements. Solutions based on NGSO satellite constellation with low or moderately low number of satellites are naturally suitable for such services due to the potential link budget advantage* and a possible reduction in the total cost of ownership due to a lower cost of the space segment, the launch segment, and more frequent launch opportunities.

*This is assuming similar satellite antenna aperture size for the NGSO and GSO satellites. Otherwise, a GSO satellite can compensate for the extra path losses adopting larger antenna reflectors. This is the case for the L/S-band GSO fleet of mobile satellite operators such as Inmarsat, Thuraya and Dish Network.

12.1.1 Operating frequency bands

Despite a large number of technologies reported by various industry players who plan to provide IoT services via NGSO satellites, there is a critical barrier to get access to spectrum to run such services.

There are various factors that contribute to selecting the frequency bands for the IoT access via NGSO satellites. While, as mentioned in section 12.1, there are many different IoT applications, two major categories of services can be envisaged. In the first category, a large number of devices for fixed or mobile services are typically deployed to collect intermittent and small data volume in an extended geographical area. The total cost of the service is sensitive to the cost of the IoT devices. For this category of IoT applications, low-cost and low-complexity technologies, particularly radio frequency (RF) front-end and antenna subsystems, are key elements in the overall system design trade-offs. Hence, the use of lower frequency bands (below 6 GHz) with omnidirectional antenna is considered the most suitable solution for satellite IoT devices. The popularity of frequency bands below 6 GHz for mass-market applications facilitates the adoption of affordable commercial-off-the-shelf (COTS) components with potential dual use for both terrestrial and satellite IoT access.

For the second category of IoT applications, the high reliability and availability of services are essential features. For such systems, providing an uninterrupted service with a high probability of timely data delivery could be the key performance factor. For example, monitoring and supervisory systems for critical infrastructure belong to this category. Compared with the first category, the number of end devices are lower, and the total cost of service is less sensitive to the complexity and associated cost of the end devices; hence, more sophisticated antenna and RF subsystems may be deployed. This could include the use of higher frequency bands (above 6 GHz) with a higher user equipment (UE) antenna directivity and gain. Such device may be equipped with active antenna at the end node able to track the NGSO satellites and ensure continuity of service during the satellite handover. Under the second category of IoT services, one may also include IoT traffic backhauling, where the satellite node at the user segment aggregates traffic collected from terrestrial IoT networks and connects them to the core network via a satellite link. Under this category, the satellite link may be established on a permanent basis and according to a dedicated access protocol. This category of service is already well-established exploiting conventional satellite exclusive bands (e.g., C, Ku, and Ka bands). However, for this kind of applications, particularly with fixed user terminals, geosatellite orbit (GSO) satellites represent a serious competitor to NGSO constellations due to the user terminals' reduced antenna and RF front-end complexity and cost. For this reason in the following, we will focus on the NGSO satellite systems that operate at frequencies below 6 GHz.

The International Telecommunication Union (ITU) radio regulations (RR) [1] specify spectrum allocations to the mobile satellite services (MSSs) in three ITU regions. Some spectrum allocations for MSSs are applicable either as primary or secondary allocations in all regions. Table 12.1 provides examples of frequency band allocations to MSSs.

Table 12.1 Examples of regulated frequency spectrum that could potentially be used for NGSO satellite IoT

Frequency (MHz)	Service allocation	Description
137–137.025	Space-to-Earth	Global primary MSS allocation
137.025–137.175	Space-to-Earth	Global secondary MSS allocation
137.175–137.825	Space-to-Earth	Global primary MSS allocation
149.9–150.05	Earth-to-space	Global primary MSS allocation
157.1875–157.3375	**Bidirectional**	Maritime terrestrial and satellite VDES
161.7875–161.9375	**Bidirectional**	Maritime terrestrial and satellite VDES
399.9–400.05	Earth-to-space	Global primary MSS allocation
400.15–401	Space-to-Earth	Global primary MSS allocation
1 518–1 525	Space-to-Earth	Global primary MSS allocation
1 670–1 675	Earth-to-space	Global primary MSS allocation
1 980–2 010	Earth-to-space	Global primary MSS allocation
2 170–2 200	Space-to-Earth	Global primary MSS allocation

In Europe, the Electronic Communication Committee has recently reported on the IoT operation via satellite [2] and the lack of suitable frequency bands below 5 GHz for emerging satellite IoT systems. In [2], the use of overlapping frequency bands below 3 GHz for terrestrial mobile and satellite mobile services is recognized. However, frequency sharing, not only between terrestrial and satellite services but also among satellite networks, is considered as a challenging task. This would require cooperation among existing operators, taking into account the exploitation of frequency bands for existing systems featuring global or regional coverage.

Due to the scarcity of frequency allocations to the MSS and ever increasing deployment of IoT devices in so-called "unlicensed bands," an approach pursued by some commercial entities is to reuse frequency allocations of short-range devices. This approach is successfully adopted for low-power wide area networks (LPWAN) that are under a general authorization or exempt from individual authorization for providing NGSO satellite IoT services. Examples of these frequency bands in Europe are 433.05–434.79 MHz, 862–870 MHz as well as 2.4 GHz ([2] Table 12.5). It should, however, be noted that the use of such frequencies for satellite IoT could potentially introduce a business risk since there are no specific provisions or recognition under the international RR that could prevent excessive interference caused by existing or new terrestrial services.

Concerning the frequency sharing among different NGSO systems for the satellite uplink access, there are two different strategies. A conventional approach is to impose a transmit radiated power limit for each device and allow for coexistence of multiple systems without the need for further coordination. As stated above, this approach could lead to a risk on service availability and the quality of service since the level of interference may not be known in advance. The second approach is based on a more proactive collaboration among different NSGO satellite systems. This approach is more challenging to implement. However, given the scarcity of the spectrum, particularly in the frequency bands of interest, it would be in-line with the common interest of growing number of NGSO satellite service providers to deliver a higher quality of service that would be worth the challenge.

The need for a harmonized usage of frequency for the emerging satellite IoT services was recognized by several administrations and their interested industry in preparation and during the World Radio Communication Conference (WRC) held in 2019. Resolution 811 of the WRC 2019 approved a new Agenda Item 1.18 [3] to consider studies to assess new spectrum allocations to the MSS for future development of narrowband mobile satellite systems. This agenda item considers the band (2 010–2 025 MHz) in ITU Region 1 and frequency bands (1 690–1 710 MHz, 3 300–3 315 MHz, and 3 385–3 400 MHz) in Region 2. The agenda item is mainly intended for NGSO satellite systems and particularly for providing low-data rate services via satellite. There are, however, existing terrestrial, and in some regions also satellite, services already assigned to these frequency bands. The use of these bands requires sharing studies with the existing primary services in the considered bands. Resolution 248 invites administrations to participate in studies concerning the spectrum needs and potential new allocations. Furthermore, linked to the outcomes of WRC in 2023, there is already provision for a follow-up agenda item for the WRC in 2027 for a possible global frequency allocation for low-data rate services in frequency band 1.5–5 GHz (see Resolution 812, Item 2.3 in [3]).

In the remaining chapter, the reference satellite spectrum allocation for IoT services is S-band (i.e., 2 GHz), unless specifically mentioned otherwise.

12.1.2 NGSO orbit mechanics effects

12.1.2.1 Doppler shift and Doppler rate

The Doppler shift and the Doppler rate in an NGSO system depend on the relative speed between the satellite and the mobile terminal on-ground, and the transmission carrier frequency, f_c. In particular, the Doppler shift refers to the change of received carrier frequency with respect to the transmitted carrier frequency due to the relative motion between the source and the destination. The Doppler rate is the Doppler shift variations over time, i.e., its first derivative. Figure 12.3 recalls some basic geometry of a generic NGSO orbit with respect to a mobile terminal on-ground, in order to derive the analytical formulas derived in [4]. Specifically, three angles are relevant for the following analytical derivation: θ is the *elevation* angle measured

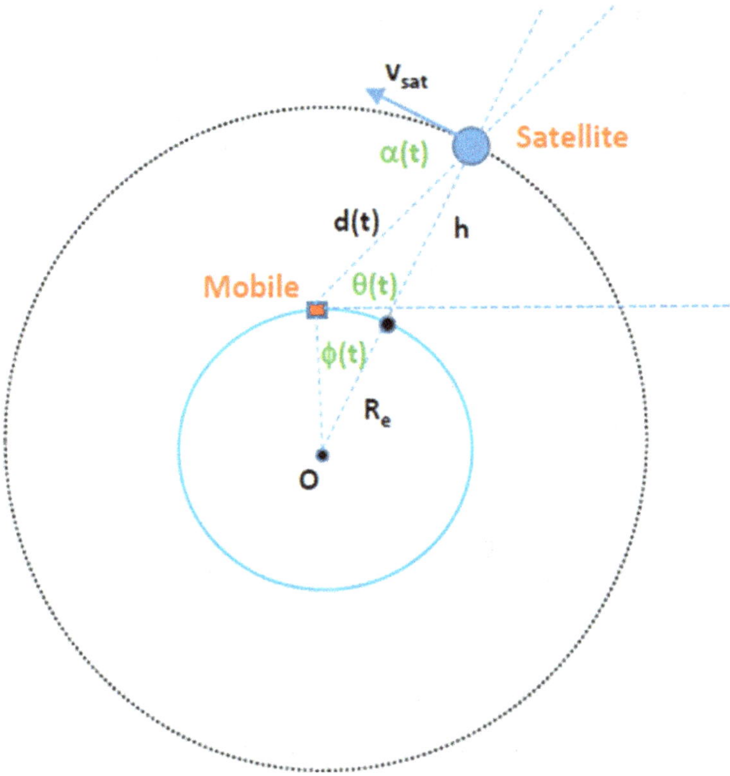

Figure 12.3 *Example for the Doppler geometry computation*

at the mobile terminal on-ground between the local horizon and the satellite; ϕ is the *Earth central* angle measured at the center of the Earth between the subsatellite point (i.e., nadir point on-ground) and the mobile terminal; and finally α is the *relative motion* angle measured between the mobile satellite direction and the tangential satellite direction. It is remarked that the angle $\eta = 90° - \alpha$ is generally identified as the *scanning* angle, measured at the satellite between the subsatellite point and the mobile terminal. Finally, it is important to recall the following time-dependent relationship among these angles during the satellite pass [4].

$$\phi(t) + \theta(t) = \alpha(t). \tag{12.1}$$

The satellite linear speed, v_{sat}, is constant over time, and it is solely a function of the orbital height. It can be expressed as a function of the satellite angular speed, ω_{sat}, as follows

$$v_{sat} = (R_e + h)\omega_{sat}, \quad \omega_{sat} = \sqrt{\frac{G \cdot M_e}{(R_e + h)^3}}, \tag{12.2}$$

where R_e is the Earth's radius (i.e., $R_e = 6\,371$ km), h is the satellite altitude, G is the gravitational constant (i.e., $G = 6.671 \cdot 10^{-11}$ m^3/kg/s^2), and finally M_e represents the Earth's mass (i.e., $M_e = 5.98 \cdot 10^{24}$ kg). As long as the mobile terminal is quasistatic or assuming a speed negligible with respect to the Satellite 1, the generated Doppler shift frequency $f_d(t)$ can be computed by the classical equation

$$f_d(t) = \frac{f_c}{c} \cdot v_{\text{sat}} \cdot \cos\alpha(t), \tag{12.3}$$

where c denotes the speed of light. In-line with the notation adopted in Figure 12.3, the Doppler shift is positive when the satellite is approaching the mobile user from the right side (i.e., $0° \leq \theta \leq 90°$), and it becomes negative in the left-side direction (i.e., $90° \leq \theta \leq 180°$). From the system geometry and the sine/cosine theorems [4, 5], we can deduce the following formula

$$\cos\alpha(t) = \frac{R_e \cdot \sin\phi(t)}{\sqrt{R_e^2 + R_s^2 - 2R_eR_s\cos\phi(t)}}, \tag{12.4}$$

where $R_s = (R_e + h)$ is introduced to simplify the notation. Recalling equations (12.3) and (12.4), the Doppler frequency shift generated by the solely NGSO satellite can be computed as follows

$$f_d(t) = \frac{f_c}{c} \cdot \frac{v_{\text{sat}} R_e \sin\phi(t)}{\sqrt{R_e^2 + R_s^2 - 2R_eR_s\cos\phi(t)}}. \tag{12.5}$$

while the Doppler rate, namely the derivative of f_d with respect to time t, can be computed as

$$\frac{df_d}{dt}(t) = -\frac{f_c}{c} \cdot \frac{v_{\text{sat}} R_e(R_eR_s\cos^2\phi(t) - (R_e^2 + R_s^2)\cos\phi(t) + R_eR_s)}{(R_e^2 + R_s^2 - 2R_eR_s\cos\phi(t))^{3/2}} \cdot \omega_{\text{sat}}, \tag{12.6}$$

where ω_{sat} has been introduced in (12.2).

Figure 12.4 shows the Doppler shift as a function of the elevation angle for two different satellite altitudes and $f_c = 2$ GHz assuming that the UE is located on the satellite orbit trajectory projection. Similarly, the Doppler rate as a function of the elevation for the same three cases has been shown in Figure 12.5. As expected, the lower the elevation angle the higher the Doppler shift, while the Doppler rate has the opposite trend. Moreover, the Doppler shift and Doppler rate decrease as a function of the satellite altitude.

Unlike the carrier frequency shift introduced by up/down conversions in the transmitter and receiver chain, the impact of the Doppler shift is not only affecting the center frequency but also the signal bandwidth and symbol rate. As illustrated in Figure 12.6, the received signal bandwidth is proportionally expanded or contracted. The impact of symbol rate is particularly important for the symbol timing estimation of rather long-duration bursts where the accumulation of symbol timing offset between the transmitter and the receiver could cause significant performance degradation. As a countermeasure, the demodulator often requires not only to recover the initial timing offset estimate but also to track the Doppler induced symbol clock

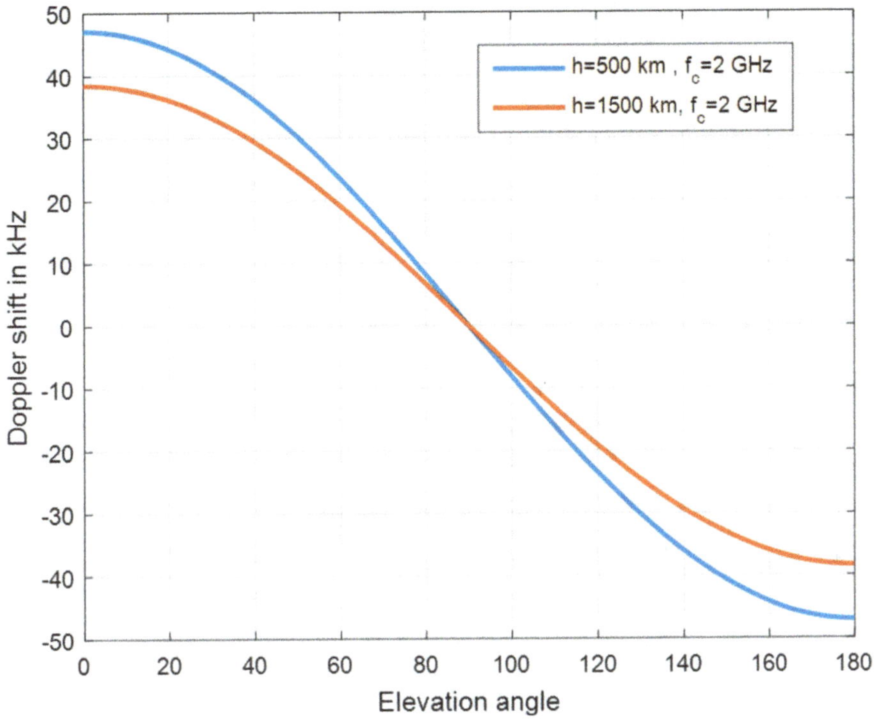

Figure 12.4 Example for the Doppler shift values as a function of the elevation angle for $f_c = 2$ GHz and two different orbital heights

drift, across the incoming burst. This is to ensure the accumulation of the timing error does not significantly impact the demodulator performance.

12.1.2.2 Satellite distance and beam size

The distance between the terminal on-ground and the satellite (also known as *slant range*) is a function of the altitude and of the elevation angle. Again, referring to Figure 12.3, the slant range is indicated with $d(t)$ and it can be computed as

$$d(t) = \sqrt{R_e^2 \sin^2\theta(t) + h^2 + 2R_e h} - R_e \sin\theta(t). \tag{12.7}$$

Given this distance, it is then straightforward to computate the propagation delay between the satellite and the user by dividing with the speed of light. The other additive component of the delay is the distance between the satellite and the serving gateway on-ground, and the computation follows the same equations.

Another useful parameter is the distance of the mobile terminal from the subsatellite point and it can be derived from

$$\rho_{\mathrm{SSP}}(t) = R_e \sin[\phi(t)]. \tag{12.8}$$

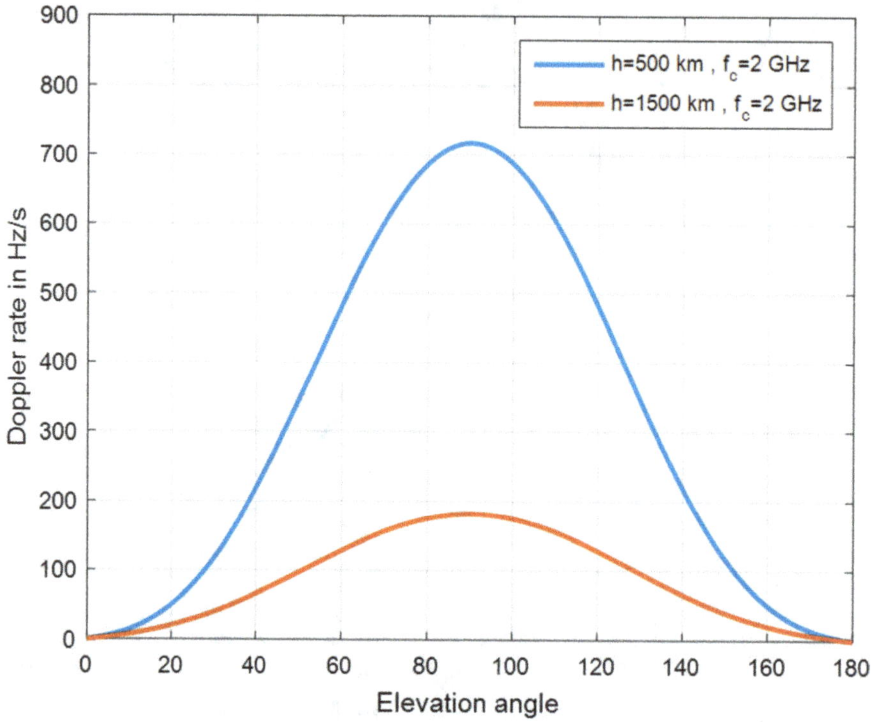

Figure 12.5 *Example for the Doppler rate values as a function of the elevation angle for $f_c = 2$ GHz and two different orbital heights*

Interestingly, from the previous equation it is also possible to compute the radius of the entire field of view from a given satellite altitude by recalling that the maximum value of $\phi(t)$ is $\phi_0 = \cos^{-1}\left(\frac{R_e}{R_e+h}\right)$ (i.e., elevation angle θ equal to 0° or 180°), therefore, it results in $\rho_{\mathrm{FoV}} = R_e \sin\phi_0$. In addition, recalling that the satellite angular

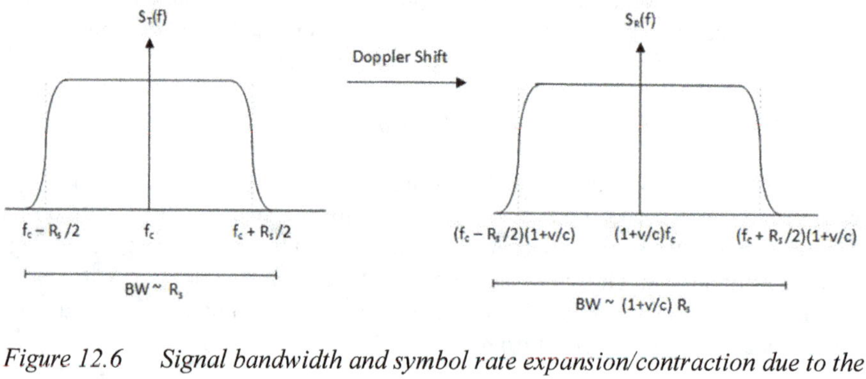

Figure 12.6 *Signal bandwidth and symbol rate expansion/contraction due to the Doppler*

speed is constant and given this maximum angle of the field of view, it is easy to derive the maximum visibility time of the satellite as a function of the altitude:

$$T_{\text{max-vis}} = \frac{2\phi_0}{\omega_{\text{sat}}}. \tag{12.9}$$

where adopting the values $h = 500$ km and $h = 1\,500$ km, it yields to $T_{\text{max-vis}} \approx 692$ seconds and $T_{max-vis} \approx 1\,338$ seconds, respectively.

In a multibeam satellite system, rather than deriving the distance from the subsatellite point in (12.8), it is more relevant to compute the distance from the beam center (i.e., the *beam radius*) at a given elevation of the terminal on-ground. Figure 12.7 shows the information required for the computation of the beam radius under the simplified assumption of a flat Earth. The same angle notation introduced in section 12.1.2.1 is used, and the subscripts *c* and *e* refer to the beam center and the

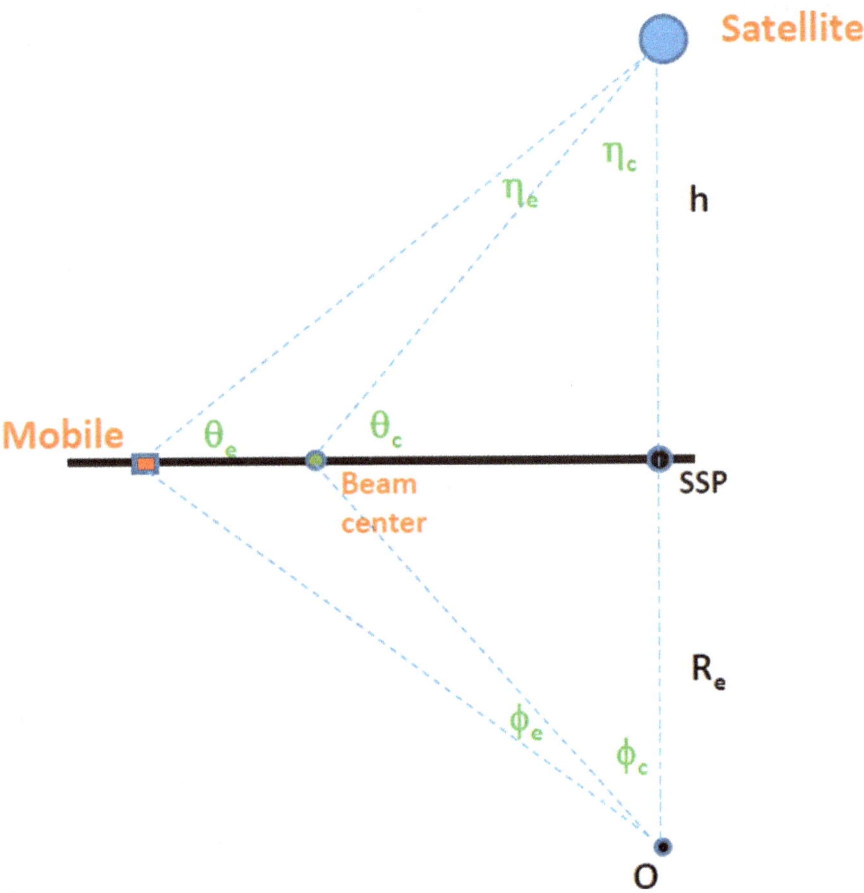

Figure 12.7 Simplified geometry for deriving the satellite beam radius on-ground

beam edge, respectively. Given the elevation angle at the beam center, θ_c, the other two angles associated to the beam center can be computed as follows

$$\eta_c = \sin^{-1}\left[\left(\frac{R_e}{R_{e+h}}\right)\cdot\cos\theta_c\right], \phi_c = \frac{\pi}{2} - \eta_c - \theta_c. \tag{12.10}$$

Typically, the beams on-ground are associated to the half power beamwidth (HPBW) of the satellite antenna; therefore, the difference between η_e and η_c equals the half value of that. Consequently, by knowing η_e, the other two angles associated to the beam edge location can be derived by applying the previous formulas. Finally, the beam radius can be approximated by $R_e(\sin\phi_e - \sin\phi_c)$. For example, Figure 12.8 shows the satellite beam diameter size, expressed in kilometer, depending on the satellite elevation angle seen at the center of the beam, θ_c, and assuming HPBW = 5°, f_c = 2 GHz, and two altitude values. As expected, the minimum diameter value is obtained at θ_c = 90° with approximately 45 km and 130 km for the two distinct satellite altitudes, while it increases up to 1 000 km rather rapidly for very low elevation angles.

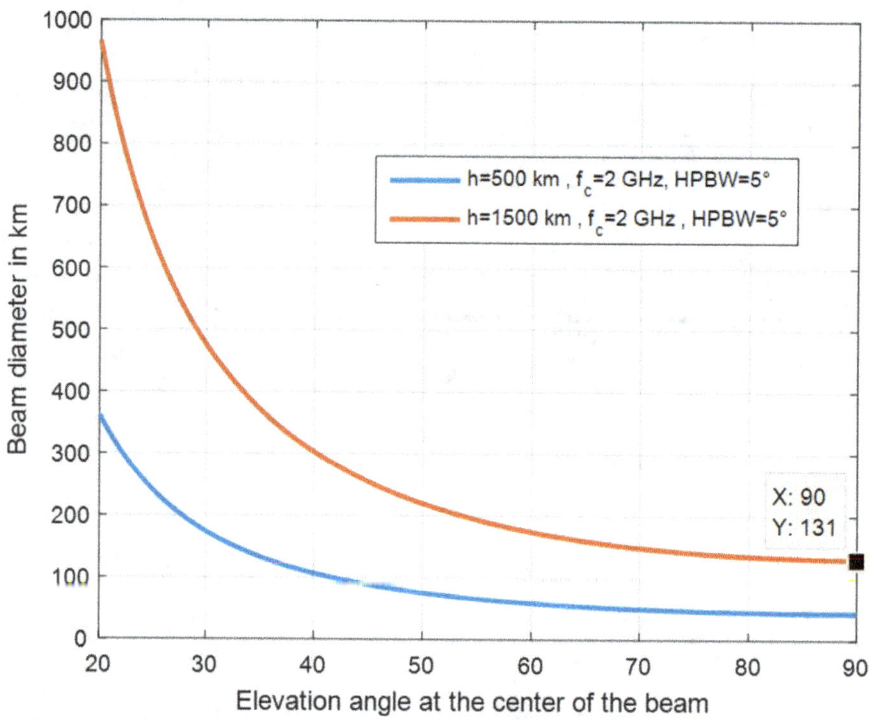

Figure 12.8 *Example for the satellite beam diameter on-ground as a function of the elevation angle, θ_c, assuming HPBW of 5°, f_c = 2 GHz, and two different satellite altitudes*

12.1.2.3 Line-of-sight signal attenuation aspects

The signal path between the satellite and an on-ground terminal undergoes several stages of attenuation due to link geometry and propagation effects [6]. Commonly, the total PL in decibel can be computed as follows

$$PL(\text{dB}) = FSPL(\text{dB}) + PL_g(\text{dB}) + PL_r(\text{dB}) + PL_s(\text{dB}) + PL_e(\text{dB}). \quad (12.11)$$

where $FSPL$ denotes the free space path loss, PL_g is the attenuation due to atmospheric gasses, PL_r is the attenuation due to rain and clouds, PL_s is the attenuation due to either ionospheric or tropospheric scintillation, and finally the building entry loss is accounted in PL_e. The FSPL value in dB is given in

$$FSPL(\text{dB}) = 32.45 + 20\log_{10}(f_c) + 20\log_{10}(d). \quad (12.12)$$

and it is a function of the carrier frequency (in Hertz) and the slant range (in meter), as shown in (12.7). The attenuation by atmospheric gases depends mainly on the frequency, elevation angle, altitude above sea level, and water vapor density (absolute humidity). However, at frequencies below 10 GHz, it is normally neglected. Its relevance increases at frequency above 10 GHz, especially for low elevation angles. Annex 1 of Recommendation ITU-R P.676 gives a complete method for calculating gaseous attenuation.

Rain and cloud attenuation is considered negligible for frequencies below 6 GHz. The general method to predict attenuation due to precipitation and clouds along a slant propagation path is presented in section 2.2 of [6]. The interested reader can find a comprehensive review of rain fade models for Earth–space telecommunication links in [7].

Scintillation corresponds to rapid fluctuations of the received signal amplitude and phase. Ionosphere propagation shall only be considered for frequencies below 6 GHz, while tropospheric propagation shall only be considered for frequencies above 6 GHz. The ionospheric scintillation effects differ at low and high latitudes, and they are not observed at midlatitudes except during strong geomagnetic storms. Reference [8] introduces a full characterization of ionospheric scintillation, and the model introduced in section 4.8 of [8] is valid for the regions located approximately 20° North and South of the magnetic equator. At high latitudes (e.g., above 60°), this model is not applicable, whereas for other latitude locations the ionospheric scintillation can be neglected. The amplitude of tropospheric scintillation depends on the magnitude and structure of the refractive index variations along the propagation path. Amplitude scintillation increases with frequency and path length and decreases as the antenna beamwidth decreases due to aperture averaging. The reader can find a prediction model in section 2.4 of [6].

Finally, if the terminal on-ground is indoor, the additional loss varies greatly with the location and construction details of buildings, and a statistical evaluation is required. Reference [9] gives a suitable building entry–exit loss model for this purpose, while experimental results are collected in [10].

12.1.3 Mobile channel aspects

The design of a satellite communication system is highly dependent on the frequency band used, the UE type, and of course on the environment in which the UE is going to be used. In the following, we focus on the mobile channel aspects related to the terrestrial environment [11].

12.1.3.1 NGSO LMS channel

The modeling of the LMS channel has been extensively covered in the last decades due to the interest for satellite digital broadcasting, personal communications, and navigation applications. Most of the focus has been for the L and S band, although some results are also available for the higher frequency bands. It is important to appreciate the importance of stochastic models, which allows to model and simulate the signal amplitude and phase evolution in the time domain. This modeling allows an accurate characterization of the physical layer performance under fading/shadowing conditions, thus supporting the system design optimization and a proper sizing of the required link margins.

Concerning the LMS channel models, an overview of the ones proposed up to 1999 can be found in [12]. One of the first widely adopted stochastic models was proposed by Loo in [13] covering the rural environment. The model assumes that the line-of-sight (LOS) component under foliage attenuation (shadowing) is log-normally distributed, and that the multipath effect is Rayleigh distributed resulting in an overall Rice/lognormal distribution. Typically, shadowing is a slower process compared with the fading due to the above-mentioned two different phenomena originating them. The two processes are correlated, and the Loo's model was shown to match fairly well with field measurements. Despite this good match with experimental data, the Loo's model applicability is limited to the rural (tree shadowed) environment.

A more complete LMS model covering different environments, different satellite elevations, both narrowband and wideband signals, and different frequency bands (L, S, and Ka bands) was proposed by Perez-Fontan in [14]. The key innovation was the introduction of a three-state Markov chain to model LOS conditions (State 1), moderate shadowing conditions (State 2), and deep shadowing conditions (State 3). For narrowband signals, in each state, a Loo's distribution was assumed with state-dependent parameters. For wideband model, less relevant to our IoT application, a large number of rays with exponential multipath power profile were assumed. Based on very extensive satellite measurement campaigns, the Loo's channel parameters, the Markov process transition matrix probabilities, and the minimum state length were provided in tabular form for a large number of cases.

An evolution of the three-state LMS channel model described above is represented by the two-state one (good/bad) proposed in [15]. The new model proposes a more versatile selection of Loo's distribution parameters within a given state. Both states are allowed to take up a wide range of possible values for the different parameters. Instead, the three-state model assumed a unique Loo's distribution with fixed parameters for a given state and environment, and elevation angle of interest. This model is improving the non-LOS modeling, which may be important when diversity can be exploited or link margins may be large like in the forward link of a broadcasting system. The channel model parameters provided are based on a set of experimental campaigns covering different mobile environments but limited to L and S bands.

These two models represent the most complete and generic models available and recommended for analyzing the mobile IoT performance. It is remarked that the more recently derived third-generation partnership project (3GPP) nonterrestrial network channel model [16] for flat fading is also reusing the two-state channel model [15] mentioned before.

12.1.3.2 Diversity aspects

One of the key advantages of NGSO for IoT is not only related to the PL reduction but also to the possible exploitation of the time variant link geometry (elevation and azimuth angle) for a single satellite in view or the time diversity† or spatial diversity in case of a constellation providing multiple satellites in a simultaneous view.

Reference [17] is one of the few references available on LMS channel modeling for multisatellite reception. This goal is achieved by introducing correlation between channel states and statistical parameters for fast and slow variations. In line with 15, two states per satellite are assumed. A new master–slave method for state sequence generation is described. In practice, it is assumed that each slave satellite parameters depend only on a master satellite.

The LMS channel model and parameters dependency on the satellite geometry versus UE movement direction have been studied in [18]. In this chapter, it is found that the satellite azimuth position with respect to the mobile trajectory needs to be modeled explicitly. A statistical LMS channel model whose parameters are obtained via an image-based state estimation method is proposed, and its validity is verified by a comparison with measured S-band satellite measurement campaign. The proposed method allows obtaining a complete statistical description of the channel for arbitrary satellite elevation and azimuth angles, thus extending the previous versatile two-state model to the multisatellite case.

12.1.3.3 Doppler and Doppler rate precompensation

The techniques presented in this section are applicable to both moving beams and Earth-fixed beams NGSO systems.

12.1.3.3.1 GNSS-based solution

A first solution to precompensate the NGSO satellite orbit Doppler effects is based on the knowledge of the satellites' ephemeris and the UE location. One or multiple satellites may broadcast the ephemeris of the constellation periodically, while the current UE location can be accurately derived using a global navigation satellite service (GNSS) receiver integrated in the UE. By doing so, the UE can predict

†In this case, for non delay sensitive services like is typically the case for IoT, possible link obstructions have a temporary impact as the satellite to UE geometry will evolve during the pass thus hopefully overcoming the shadowing/blockage condition.

the downlink carrier frequency Doppler and Doppler rate and precorrect the uplink carrier frequency to eliminate the Doppler effects seen by the satellite when transmitting. The main drawback of this solution is related to the need to have a GNSS receiver on-board the UE, which drains a non-negligible amount of power for an IoT mobile device‡ on top of increasing the bill of materials. Furthermore, to achieve a position fix, the GNSS receiver needs a good angle of view to receive the satellites (ideally a hemispherical coverage or at least a good portion of it unobstructed).

Another implementation challenge in adopting this solution is the downlink carrier synchronization in the presence of the unknown carrier frequency offset. The main sources of carrier frequency uncertainty are the local oscillator drifts at the IoT device and the downlink Doppler shift. To cope with a large carrier frequency offset, relative to the downlink symbol rate, more complex techniques such as multiple hypothesis testing or frequency sweeping techniques may be required. The complexity of such techniques will be even higher at a low operating signal-to-noise plus interference ratio (SNIR) due to a low-gain antenna gain of IoT devices and power constraints of the satellite transmitter.

12.1.3.3.2 GNSS-independent solutions

To limit the above-mentioned GNSS-based precompensation drawbacks, an alternative solution is the one adopted by existing mobile satellite constellations, based on center of beam Doppler precompensation. This precompensation can be implemented on-board of the satellite as in the case of Iridium [19] or at the ground gateway as it is the case for Globalstar. The main drawback of this solution is related to the fact that the Doppler compensation is valid at the center of beam; hence, some residual Doppler remains when the UE is located at different locations.

The precompensation worst-case error clearly occurs when the terminal is at the beam edge, which corresponds to the maximum distance from the beam center. Denoting with α_c and α_e the motion angle at the beam center and the beam edge, respectively, it is easy to compute the residual Doppler shift as follows:

$$\Delta f_d(t) = \frac{f_c}{c} \cdot v_{\text{sat}} \cdot [\cos\alpha_c(t) - \cos\alpha_e(t)]. \tag{12.13}$$

Typically, the difference between these two angles is also known, and it equals the half value of the HPBW. For example, assuming the HPBW of 5°, $f_c = 2$ GHz, and two distinct satellite altitudes, Figure 12.9 shows the absolute value of the residual Doppler shift as a function of the elevation angle at the center of the beam, θ_c. The residual Doppler shift not only introduces a carrier frequency offset at the UE receiver, it also has an indirect impact on the carrier frequency alignment of the UE uplink transmission. For the GNSS-independent solution, the UE receiver must rely on the downlink signal from the satellite to extract a clock reference in

‡For a fixed UE it can only be activated at installation time.

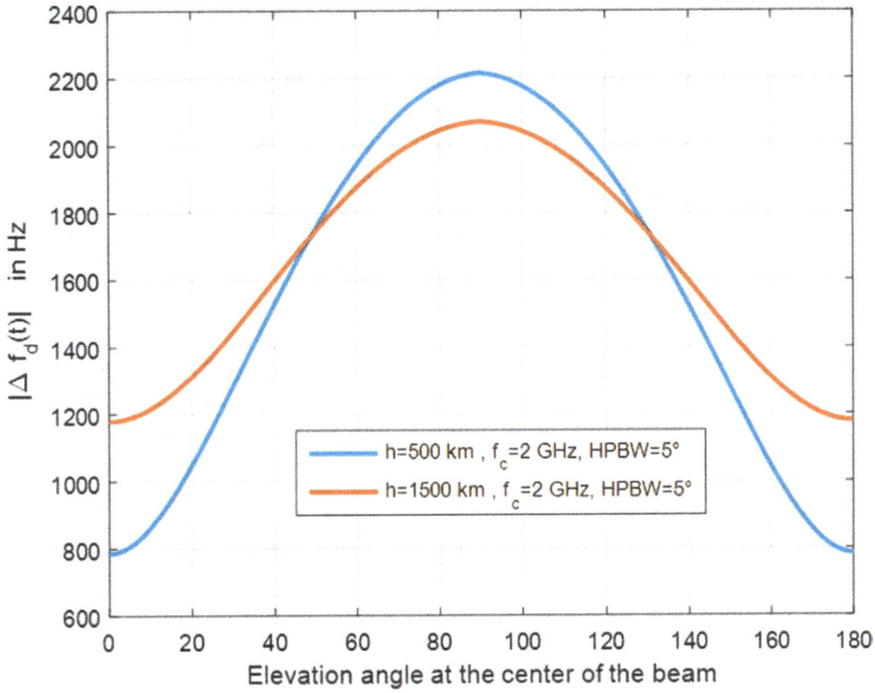

Figure 12.9 *Example for the residual Doppler shift values as a function of the elevation angle assuming a beam with HPBW of 5°, f_c = 2 GHz, and two different satellite altitudes*

order to adjust the timing and carrier frequency of the uplink transmission. The residual Doppler shift of the received signal can introduce an error in the IoT device reference timing extraction. This is particularly important for the orthogonal multiple access (MA) schemes such as single-carrier frequency-division multiple access (SC-FDMA) where the frequency misalignment of signals from different users at the receiver (on-board satellite or on-ground gateway station) can degrade the performance. This may lead to a system trade-off in terms of the beam size and the maximum frequency misalignment that can be tolerated at the gateway demodulator.

12.1.3.4 Path and link losses variability compensation (isoflux antenna design, power control)

Because of the spherical Earth surface shape, the NGSO orbit generates a sizeable beam shape distortion when projecting on the Earth surface a regular satellite antenna beams grid. For this reason, the design of multibeam layouts for lowe earth orbiting (LEO) orbits is much more difficult due to a considerable slant range variation from the nadir to the edge of the coverage. The slant range d can be computed using (12.7). For example, at an altitude 1 200 km, the slant range varies by 7.2 dB from the nadir to 15° elevation (i.e., 75° nadir angle). This extra PL increases to 8.7 dB for a LEO at 600 km.

Using active antennas on-board with proper beam-forming design, it is possible to compensate this elevation angle PL dependency by means of the so-called isoflux design. An example of multibeam LEO isoflux antenna design can be found in [20]. In addition to antenna isoflux-type compensation, other countermeasures are possible. For IoT systems having a downlink carrier present, as it is normally the case, the open loop transmission and power control described in [21, 22], represent an interesting solution. A pilot-aided downlink SNIR estimation based on the signal-to-noise ratio estimation algorithm [23] is performed first. By storing the maximum medium-term estimated SNIR the demodulator computes the LOS SNIR reference. In this way, the UE may estimate in real time the current link margin by simply subtracting the current SNIR from the minimum system operating SNIR (both expressed in decibel). When a packet has to be transmitted, the UE can then determine if there is sufficient margin to allow successful packet reception and which range of maximum power randomization can be used to further boost the system throughput. Simulation results reported in [21] demonstrate the good performance of the proposed packet transmission control algorithm in the three states of the LMS channel model and for speeds up to 170 km/h. This open loop transmission power control has been experimentally validated during field trials campaign reported in [24].

12.1.4 Frequency reference determination for the UE and satellite

In this section, we review the needs and the solutions for network synchronization and the impact of clock stability on-board for the satellite as well as the UE.

Clock errors on-board of the satellites and the UEs have direct impact on the long- and short-term stability of carrier frequency, symbol clock timing as well as the frame transmission, and reception timing. The uncertainty of the clock at the transmitters and receivers contributes to the access network performance. More importantly, some network access protocols demand tight timing and frequency synchronization of incoming signals from multiple sources to maintain their orthogonality (such as time-division multiple access (TDMA), multifrequency TDMA (MF-TDMA), or orthogonal FDMA (OFDMA)). The stability and synchronicity of the clock at the receiver and transmitters will play a key role in the overall system performance. In fact, the sensitivity of the MA protocol to clock timing and frequency uncertainty and jitters could be a contribution factor in selecting an access scheme.

12.1.4.1 GNSS-based solution

The use of GNSS receivers has been a game changer in providing a stable timing reference both for the satellites and the UEs. A stable reference, such as a pulse per second (PPS) signal from a GNSS receiver, is often used to discipline the local oscillator and correct the clock drifts.

There are already GNSS receivers with flight heritage for NGSO satellites that can provide reference clock to on-board computers as well as communication payloads. While GNSS clock reference is highly effective in removing long-term clock errors, the short-term stability of the local oscillators is not affected by the GNSS reference. Such short-term variation could manifest itself as phase noise that could hamper the performance of phase coherent detection. This is particularly relevant for the IoT uplink low bit rate. A careful selection of key components such as oscillators and design of the timing feedback loops are important to control the phase noise characteristics.

As discussed in section 12.1.3.3, a GNSS-equipped UE could extract its location and a stable time reference (PPS). This would reduce the uncertainty in the carrier frequency of the transmitter and the symbol timing. However, as noted in section 12.1.3.3, for certain class of IoT devices, the power consumption of the GNSS receiver could be exceeding the available power and/or energy budgets.

12.1.4.2 GNSS-independent solutions

For certain IoT service applications, the utilization of GNSS receiver on-board of the satellite may not be desired due to the nature of the service and a requirement to avoid dependency on external systems. While the duplication of full-fledged atomic clocks (similar to GNSS) for NGSO IoT satellites is an expensive proposition, alternative solutions based on newly developed chip scale atomic clock (CSAC) could be an option. This may lead to a lower accuracy in clock stability, which could still be acceptable for the IoT network synchronization. Although the use of CSAC does not completely rectify the need to correct the on-board clock, the correction may be applied once per orbit or even less frequent, depending on the required accuracy.

Assuming that the NGSO satellite has access to a stable timing reference, the satellite may distribute the clock reference to UEs via a beacon signal. Alternatively, the downlink in-band signal from the satellite to UEs may contain a network clock reference encapsulated in a downlink frame format. This is similar to the network synchronization approach adopted in digital video broadcasting (DVB) standard for interactive services (i.e., digital video broadcasting second generation return channel via satellite (DVB-RCS2)). However, this solution has the drawback of requiring a wideband downlink signal, thus representing a power hungry solution for the UE.

For this reason, unslotted (asynchronous) RA solutions not requiring precise frequency and time reference at the UE are preferred for satellite IoT applications.

12.2 Rationale and challenges for RA in NGSO networks

12.2.1 Why and when to use RA in satellite networks?

12.2.1.1 IoT traffic features

Without any doubt, the characterization of IoT traffic profiles along with the amount of data generated in IoT or machine-type communications (MTC) networks is influenced by several factors, which makes difficult to define a universal model. Surely, it is possible to identify the peculiar characteristics of the IoT data pattern in order to design protocols with properties to efficiently support IoT services.

More specifically, the traffic generated by IoT nodes is influenced by

- **The density of the nodes** usually quantified in terms of number of nodes per unit area. This factor mainly influences the total amount of traffic exchanged via satellite while connecting the sensors and the network center. Typically, the expected number of sensors per unit area might vary from few hundreds to hundreds of thousands per square kilometer, and it depends on the environments (e.g., dense urban, rural areas, or even oceans).
- **The frequency of generated packets,** usually measured in terms of number of packets generated per time unit, or alternatively as the time interval between the generation of two consecutive packets. Again, the expected frequency of traffic generated by the single node can vary from one packet per day to several packets per hour.
- **The size of generated packets,** expressed in bits or bytes, is the third factor that influences the generated traffic. This quantity influences mainly the time needed for the transmission of a single packet. Typically, the expected size of traffic generated by the single node varies from tens to thousands of bytes per packet. For example, low packet sizes range from 50 to 100 bytes, as happens in applications of tracking coordinates, or metering data; on the other hand, high packet sizes are in the range of 1–2 kB in applications like management and control data, alarms, or electronic transaction data.
- **The periodicity of generated packets** identifies the variability of the traffic generation. More specifically, if the traffic generated by the node is not periodic (as happens, e.g., in the case of critical events or alarms triggering the generation and transmission of messages), the frequency of packet generation for each node is variable in time, and therefore also the global amount of traffic can also be highly variable in time.

Already today, billions of IoT devices are connected, and this number will increment rapidly in the near future. This massive amount of data appears in the network periodically or irregularly, and the transmission rate is also variable. A model capable to characterize this vast heterogeneous traffic would be important for managing, optimizing, and regulating IoT networks. Nevertheless, in the literature, there is a unique model that can effectively define all these features of MTC networks. For instance, traffic generation based on the Poisson process is suited to a very large set of data sources that may input packets randomly to the network. However, this solution always inaccurately estimates heavy-tailed data traffic and changes in long-term correlation of packet arrival time.

To address these issues and other related aspects, the reader can find interesting classifications and traffic characterization in [25, 26], and [27]. For instance, [25] has identified some parameters based on a classification of the type of IoT device (e.g., *telemetry and tracking*, or *alarm device*, or *traffic aggregators*). In [26], the traffic characteristics have been collected and summarized on the basis of IoT applications (e.g., smart grid, intelligent living and buildings, smart environment). Finally [27], deals with analytical models to quantify the error introduced by the assumption of a Poisson process in case of aggregated period traffic.

12.2.1.2 Satellite versus terrestrial network peculiarities

The bursty and low duty cycle IoT traffic nature previously outlined has pushed terrestrial networks to opt for shared type of channel access solution. This type of approach faces the challenge to control the access to the common channel resource while providing a good level of performance combined with affordable implementation complexity. As reported in [28], to cope with this issue, many of terrestrial RA techniques have been devised exploiting channel sensing and distributed reservation solutions. Unfortunately, these techniques are not suitable for satellite networks. This is because the propagation delay is much larger than the time taken to transmit a packet, and a sender may have sent several packets before the receiver starts receiving (or not in case of collisions) the first packet. This intrinsic, yet unavoidable, system latency, makes less efficient the adoption of possible retransmission techniques as the latency and the channel load may be further increased. Furthermore, the size of satellite cell (beam) is typically much larger than the terrestrial one. This may lead to a larger number of potential UEs to be served by a satellite beam. This effect is typically compensated by the lower density of traffic typically targeted by satellite networks.

12.2.1.3 Random access versus demand assignment multiple access

To cope with the above satellite specific issues, a number of MA solutions have been developed in support of satellite very small aperture terminals (VSATs) supporting retail point-of-sale transactions and SCADA. Special protocols based on fast reservation were developed on top of conventional DAMA (see [28]). However, these solutions were shown to be inefficient in the presence of bursty type of traffic typical of IoT [29]. Furthermore, the overhead generated by the forward signaling, makes the solution based on DAMA less attractive when a massive number of UEs are deployed in the network.

12.2.1.4 Synchronous versus asynchronous access

The other aspect of enhanced DAMA solutions is that they require a slotted MA typical of MF-TDMA commonly used in the return link of VSAT networks. All DAMA schemes introduce control subframes that are needed by UEs to make capacity requests and to remain tightly time synchronized with the satellite network reference (e.g., for MF-TDMA). The corresponding signaling represents a sizeable overhead in the inbound channel of large size networks composed of low-duty traffic UEs. In summary, the combination of a DAMA protocol requiring some resource allocation upon UE request and the need for UEs to remain synchronized with the network to allow exploitation of MF-TDMA were the key drivers for looking at more efficiency MA techniques for IoT type of traffic.

12.2.2 From ALOHA to modern non orthogonal multiple access schemes

12.2.2.1 IoT RA solutions

From previous discussion, it is apparent that an efficient satellite IoT solution cannot be based on the conventional MA techniques. Before describing more in detail recently developed high performance and scalable RA solutions for satellite IoT, we shortly review the most common resource sharing techniques to highlight their advantages and drawbacks. Basically, we focus our discussion on the three classical MA techniques namely:

- FDMA;
- TDMA;
- Code-division multiple access (CDMA).

Clearly, combination of these access techniques is possible and often adopted in practical systems. In FDMA, the available system resources are shared in the frequency domain subdividing the uplink spectrum in sub-bands. For the IoT applications, frequency sub-bands are sized taking into account the UE's baud rate, which is normally modest, although typically increased by the presence of a forward error correcting (FEC) scheme. Clearly, for a fixed information bit rate, using a lower coding rate FEC increases the occupied signal transmission bandwidth, hence, reducing the overall MA spectral efficiency. In this case, the MA is obtained assigning (or selecting) one of the FDMA sub-bands to transmit the wanted information. Clearly, if multiple UEs are accessing the same sub-band at the same time, this approach would lead to packet's collisions. In order to maintain a constant power flux density at the satellite receiver for the different data rates supported, the UE transmitted power needs to be proportional to the transmitted bandwidth. This may lead to oversizing of the UE power amplifier or inefficient use of the power amplifier (i.e., large output back-off) that may not be suitable for IoT applications. In terms of UE's synchronization it is required only for the carrier frequency (not the timing synchronization since there is no time slot assignment). Techniques to extract the frequency reference at the UE are discussed in section 12.1.4.

As mentioned before, TDMA typically combined with FDMA, is often adopted for VSAT networks. Each carrier is divided into frames and subsequently each frame is subdivided into time slots. The slots can be allocated for some time to a specific UE in a DAMA fashion or alternatively used in an uncoordinated RA fashion by the UEs. An important aspect is that the UE has a slot, i.e., a fraction of the frame duration, to transmit the required data. Hence, although the average UE data rate over the frame may be modest, its slot data rate is artificially increased by its limited time duration. This means that in TDMA, the UE's transmission power has to be sized to the aggregate TDMA bit rate rather than the individual UE one. For this reason, satellite VSAT networks have adopted MF-TDMA, which allows to split the uplink spectrum into N_F frequency sub-bands. Each FDMA sub-band is then organized in TDMA fashion thus reducing the UE peak bit rate required by a factor N_F compared with pure TDMA. Another negative feature of (MF-)TDMA is the previously mentioned need to keep UE synchronized both in time and frequency even if the network access is not frequent[§]. This has negative impact in terms of signaling overhead and UE's power consumption, particularly for an NGSO-based system where the delay is highly time variant.

The last MA technique considered is CDMA. In CDMA, typically the available frequency spectrum is shared among all UEs, which are asynchronously accessing to it by DAMA or RA. The UEs are typically using a different spreading sequence belonging to a family of spreading codes, which provide good cross-correlation and autocorrelation properties. Low-rate FEC, in this case, has a positive impact on the spectral efficiency as it allows to better cope with thermal noise and cochannel interference [29]. The only drawback is the reduction of the spreading factor (number of chips/symbol) given a constant occupied bandwidth. As for FDMA, the transmit power is in line with the single UE bit rate requirement. No time and frequency synchronization is required as the access is asynchronous and some carrier frequency error has no appreciable impact except for the extension of the demodulator acquisition range.

12.2.2.2 Overall trade-off

As we have seen, each access technique has peculiarities which may be best matching different use cases. In the following, we summarize the key MA characteristics that are relevant to IoT applications. For comparing the MA techniques we made the following assumptions:

- The total bandwidth, B_w, and the resource allocation window, T_{frame}, are fixed.
- The same quantity of single active UE information $N_b = \overline{R}_b \cdot T_{\text{frame}}$ (bits) is assumed to be transferred over the frame duration T_{frame}.
- The same physical layer FEC is adopted (in general, this may not be the case).
- The available overall number of multidimensional resources N_R (number of TDMA slots N_T, FDMA carriers N_T, CDMA spreading sequences N_C) are kept constant, i.e., $N_T \cdot N_F \cdot N_C = N_R$.
- For TDMA, $N_F = N_C = 1$ thus $N_R = N_T$, for MF-TDMA $N_C = 1$ thus $N_R = N_T \cdot N_F$ (with FDMA as special case when $N_T = 1$) and for CDMA $N_T = N_F = 1$ thus $N_R = N_C$.

[§]A possible alternative consists of repeating the network synchronization process each time some data has to be transmitted. However, this approach is creating a large overhead and increase the UE power consumption.

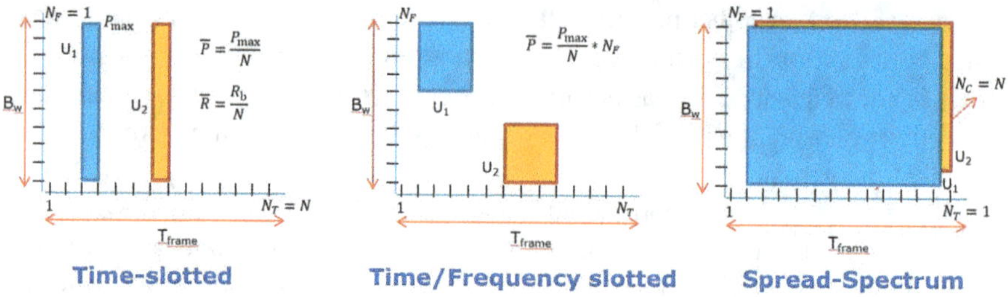

Figure 12.10 *FDMA, MF-TDMA, and CDMA MA comparison*

Figure 12.10 provides a graphical representation of the above assumptions. It is easy to see that the following relations hold

$$\overline{P}_{\text{TDMA}} = \frac{P^{\max}_{\text{TDMA}}}{N_T} = \frac{P^{\max}_{\text{TDMA}}}{N_R}, \quad \overline{R_b} = \frac{[R_b]_{\text{TDMA}}}{N_T}, \tag{12.14}$$

$$\overline{P}_{\text{MF-TDMA}} = \frac{P^{\max}_{\text{MF-TDMA}}}{N_R} N_F = \frac{P^{\max}_{\text{MF-TDMA}}}{N_T}, \quad \overline{R_b} = \frac{[R_b]_{\text{MF-TDMA}}}{N_T}, \tag{12.15}$$

$$\overline{P}_{\text{CDMA}} = P^{\max}_{\text{CDMA}}, \quad \overline{R_b} = [R_b]_{\text{CDMA}}, \tag{12.16}$$

Thus comparing the peak power required to transmit the N_b bits in the frame, we get

$$\frac{P^{\max}_{\text{TDMA}}}{P^{\max}_{\text{CDMA}}} = N_R, \quad \frac{P^{\max}_{\text{MF-TDMA}}}{P^{\max}_{\text{CDMA}}} = \frac{N_R}{N_F}, \tag{12.17}$$

It is clear that if we want to save UE transmit peak power, we should prefer CDMA or FDMA solutions. An additional advantage of CDMA is the lack of time/frequency synchronization needs, which is a key feature for IoT. Instead, MF-TDMA represents an intermediate case between TDMA and CDMA in terms of peak power requested by the terminal to send the N_b bits. Finally, FDMA requires the same transmit power than CDMA.

12.2.2.3 The beauty and limits of ALOHA and SS-ALOHA

The well-known ALOHA RA scheme proposed by Abramsom [30], initially devised to link the computers belonging to the sparse Hawaiian's university sites, has the nice advantage to do not require network synchronization. The main drawback being its limited throughput at acceptable packet loss rate (PLR). Despite the often cited peak throughput of 0.368, at a more practical PLR¶ of 10^{-3} the normalized throughput amounts to just 10^{-3} bits/symbol. This represents a very low efficiency considering the target to economically serve a large number of UEs exploiting the scarce available bandwidth.

¶This value of PLR is typically required in satellite networks to avoid too frequent packet re-transmissions.

An exception may be represented by terrestrial systems, which may increase the frequency reuse increasing the base stations density according to the traffic needs as it is the case for LoRa and SigFox.

The LoRa is based on a proprietary M-ary chirp spread-spectrum (CSS) ALOHA type of RA described in [31]. The CSS modulation allows to multiplex different users on the same bandwidth with different data rates exploiting different values of the so-called spreading factor parameter SF. According to [32], colliding packets having different SF are seen as additive noise. Instead collisions among packets with same SF may lead to packet loss unless power unbalance is larger than 6 dB. In terrestrial networks, the link losses are heavily dependent on the distance from the base station; hence, different SFs are required to close the link for different cell locations. This is typically not the case for satellite networks. LoRa CSS ALOHA RA scheme has interesting features, however, seems to have limited physical layer code protection and not supporting collision effect mitigation based on interference cancellation. Some LoRa low-power wide area terrestrial network capacity analysis is reported in [33]. No similar analysis for satellite networks is available to the authors' knowledge; however, it is expected that the performance will be similar to spread-spectrum ALOHA (SSA) discussed in the following.

Another interesting option may be represented by SSA [34] which when combined with a powerful low-rate FEC achieves a throughput of 0.5 bits/chip PLR = 10^{-3} [29]. This result is valid when received packets' power is perfectly balanced. The SSA Achilles' heel resides in its high sensitivity to packets' power imbalance because of the well-known CDMA near-far problem. As shown in [28], the SSA throughput is diminished by several orders of magnitude when the received packets power is lognormally distributed with standard deviation of 2–3 dB. This SSA behavior is opposed to ALOHA (or slotted ALOHA (S-ALOHA) [35]) where power imbalance results in improved performance because of the packets' capture effect.

We can conclude that spread-spectrum ALOHA (SS-ALOHA) represents an interesting RA scheme for uncoordinated satellite IoT. However, its fragility to power unbalance needs to be overcome to make it a truly appealing solution. ALOHA is instead very poor in terms of throughput for relatively low PLR target applications such as satellite IoT. Compared with ALOHA, S-ALOHA provided some very modest throughput improvement at practical PLR values but requires UEs time synchronization.

12.2.3 RA signal processing aspects

12.2.3.1 Pragmatic solutions to approaching the MAC capacity

The previous discussion showed that none of the ALOHA solutions is fully satisfactory for our application. One major recent years' innovation has been to devise relative simple solutions to mitigate the packet collision in ALOHA type of RA. Chronologically, the first solution was found for S-ALOHA and more specifically for its diversity S-ALOHA (DSA) variant repeating twice the transmission of the same packet in two randomly selected slots to increase the probability of having one replica received [36]. The contention resolution diversity S-ALOHA (CRDSA)

[37] idea has been to exploit the correctly decoded DSA's packet replica to cancel its "twin". The twin packet location in the frame is contained in a signaling field embedded with the packet payload. Another novelty is that the frame received signal samples are kept in the packet demodulator memory to allow repeating the slot detection process more times to maximize the packet collision resolution process. This simple add-on to DSA allows to boost the S-ALOHA performance by 450 times, thus achieving 0.45 bits/symbol throughput at PLR = 10^{-3} (see Figure 12.8 in [28]). An asynchronous version of CRDSA-dubbed asynchronous contention resolution diversity ALOHA (ACRDA) was also devised in [38] achieving the remarkable throughput of 1 bit/symbol for equipowered packets. Other schemes have been proposed along the same line and summarized in [28]. However, several of these schemes are increasing the demodulator complexity without a real performance improvement when adopting a realistic system model [39]. The demodulator complexity is another important element to consider, in particular when, as it is often the case in NGSO system, it has to be implemented on-board the satellite.

The next line of improvement is related to SSA. With the enhanced SSA (E-SSA) [40], the main weakness of SSA has been overcome by adopting ideas borrowed from CRDSA/ACRDA, customized to the SSA case which is not featuring the use of packet replicas. The adoption on a three packets window-based iterative successive interference cancellation (iSIC) in E-SSA allows to achieve a throughput of 1.2 bits/chip at PLR = 10^{-3} and even larger with unbalanced packets' power. This corresponds to three orders of magnitude improvement compared with classical ALOHA. In fact, thanks to the iSIC processing, E-SSA is taking advantage of power unbalance, as this condition is easing the packet cancellation once decoded. In this case, the E-SSA demodulator is starting to decode the packets with highest SNIR to take advantage of the higher-quality channel decoding prior cancellation. An approximate, yet accurate analysis of the E-SSA iSIC performance in the presence of imperfect cancellation has been provided in [41].

The E-SSA scheme is particularly attractive since it requires minimal changes in processing the UE, and the iSIC processing is carried out at the central (on-board or on-ground) demodulator side. In other words, from the waveform design, there is iSIC, which can be applied to SSA transmitted waveform to improve the aggregate throughput. It is also inherent in the design that the transmitters do not need to use a network synchronization. Another powerful feature of the E-SSA is that all users can share the same spreading sequence. Hence, no user identification is required prior to user packet detection and decoding. In fact, a true asynchronous nature of transmission along with the use of a long spreading code (not repeating within the packet) allows for individual packet detection despite the fact that all users may share the same preamble known symbols and the same spreading sequence. A single spreading code for all users is typically used in E-SSA systems in order to reduce the receiver complexity. In this case, a single preamble search is performed to detect the presence of packets on air. It should be noted that the packet acquisition is the most computationally intensive part of the receiver implementation. Hence, the use of a single signature would allow to maintain the complexity of the receiver within an acceptable level.

12.2.3.2 Is MMSE on top of SIC worthwhile?

Despite a significant performance improvement of iSIC scheme for SSA-type random packet transmission, theoretically there is a residual gap between the MA channel capacity and the performance reached by the iSIC scheme. The E-SSA scheme uses a conventional single-user matched filter (SUMF) receiver for despreading the received packet. A linear minimum mean square error (MMSE) detector is known to enhance the achievable performances of spread-spectrum schemes. Furthermore, the combination of MMSE and iSIC processing could reach the MA channel capacity [42]. However, the implementation of an MMSE detector would require the inversion of a covariance matrix that is computationally heavy hardware implementation. An alternative approach based on multistage detector design that approximates the MMSE is definitely more suitable for hardware implementation. As shown in [43], the complexity of multistage detector is marginally higher than that of an SUMF detector. However, there are other aspects such as dynamic update of the weighting coefficients that could increase the complexity of MMSE-iSIC implementation. Considering the overall complexity of the receiver, the packet acquisition (i.e., preamble search and detection) typically represents the largest contributor to the demodulator complexity. Hence, the overall contribution of MMSE-iSIC implementation at the receiver can be considered acceptable compared with that of the SUMF-based iSIC. In terms of the number of iterations, the required number of iterations for iSIC algorithms are often larger than what is needed for MMSE-SIC detector and decoder.

In recent work on implementation of MMSE-iSIC algorithms [44], improvement in the achievable aggregate throughput has been reported. However, the accuracy and performance of the packet acquisition and identification of existing packets within a processed window could impact the resulting performance of the MMSE-iSIC demodulator. For this reason, the use of MMSE on top of iSIC is generally not recommended when the demodulation takes place on-board the satellite.

12.2.3.3 FEC for RA channel and SIC at the receiver

For RA schemes that exploit iSIC at the receiver, the FEC code optimization is more involved than in conventional communication systems with additive white Gaussian noise (AWGN) channel. This is because, as observed in [45], the SIC mechanism can already be triggered by operating even at high values of the packet error rate (PER) (e.g., > 0.5). Unlike AWGN channels for which more powerful FEC having steeper PER curve characteristics, for interference-limited channel with iSIC operation at the receiver, the performance of the FEC at high PER regions can impact the overall throughput. In fact, it was found that the smooth PER curve of 3GPP wideband CDMA (W-CDMA) turbo codes with relatively small packet size was giving superior or equal RA performance than more powerful FEC (e.g., bigger block size or low-density parity check codes (LDPC)). In [42], the possible advantage of using convolutional codes instead of turbo or LDPC codes for small packet in RA schemes with SIC was investigated. The investigations have shown that for RA with iSIC (based on hard decisions) the turbo code is outperforming convolutional codes for short packets (few hundred bits). The convolutional codes showed some slight advantage over turbo codes only when using soft-IC schemes.

12.2.3.4 When not to iSIC?

The use of iSIC process relies on buffering the incoming samples at the receiver, carrying out packet detection, decoding, reconstruction and removal from the memory window samples, and reiterating further the process to detect and decode further packets after each round of packet removal. The iterative nature of this process allows to detect the strongest packets, i.e., highest SNIR ones, first. Once those packets are removed from the memory, other packets will become detectable. If the arriving packet experiences power unbalance, iSIC maximum throughput performance is further boosted even performing less iterations. In fact as shown in [41], the iSIC performance improves considerably if the arriving packet power has inherently a uniform distribution in logarithmic scale.

If the total spectral efficiency is not a system design driver, and for incoming packets the power balance is modest and/or the complexity of the (on-board) demodulator has to be minimized, one can progressively reduce the iSIC number of iterations down to zero.

Another aspect to be considered is the demodulation latency, as the iSIC scheme increases the processing delay in the packet detection since the process is carried out sequentially. For certain application, such delay may not be acceptable. However, for most IoT applications, the trade-off between the performance enhancement and processing delay may lead toward sum rate performance enhancement, which may indirectly reduce the delay in receiving packets by avoiding retransmission of the packets.

12.2.4 Congestion control aspects

As mentioned in section 12.2.1.1, the burst nature of IoT traffic requires attention to avoid short-term overload situation which can lead to unstable RA operations. As mentioned in [28], the congestion control algorithm should ensure that the short-term average traffic load allows to keep the RA PER below a target value (typically $<10^{-2}$ to avoid an excessive number of retransmissions). Typical techniques employed for congestion control are based on a *p*-persistent algorithm, exponential back-off, or a combination of the two. These techniques are widely used in Ethernet networks. In satellite IoT networks, the congestion control algorithm parameters are typically broadcasted on the forward channel and are adapted dynamically as a function of the average channel load. RA schemes, such as SSA and E-SSA, are aggregating more users on the same band, enhancing the load averaging effect, thus easing the congestion control algorithm operations [46].

12.3 NGSO RA scheme design

12.3.1 Forward link design for IoT

12.3.1.1 Key requirements

Before starting the search for an existing forward link solution or the design of a new one, it is extremely important to have clear requirements to be fulfilled. Based on the specific characteristics of IoT services and recalling that we are interested in NGSO systems below 6 GHz, the following general requirements are presented:

- Support of NGSO operations (e.g., to cope with high Doppler shift values);
- Support of limited channelization bandwidth (e.g., few hundred of kiloHertz);
- Enable operations with inexpensive and simple terminals;
- Support both fixed and mobile sensors;
- Support different types of messaging (e.g., unicast, broadcast, or multicast);
- Implement different activity modes to minimize the UE power consumption;
- Implement loops and procedures (e.g., power control, and automatic repeat request, congestion control) to increase the system capacity.

12.3.1.2 Possible solutions

Several satellite communication standards are available and commercially widely adopted. However, none of them has been specifically designed for narrowband IoT applications in NGSO constellations.

The DVB-S2 waveform [47], although frequency band agnostic, is focused on satellite broadcast/broadband systems on GSO satellites typically operating above 6 GHz bands. Some of the DVB-S2 strong aspects are the provision of a very powerful FEC schemes, adaptive coding, and modulation combined with a large set of digital modulation formats. However, being narrowband mobile services not the primary target scenario, some weak points can be easily highlighted such as: no countermeasures for the LMS channel (e.g., absence of medium/long time interleavers), variable frame duration (i.e., making difficult the initial synchronization in case of sporadic access to the network), and large spacing among pilot symbols (not suitable for low symbol rates in the presence of mobile channel fast carrier phase variations and phase noise).

The DVB-SH standard [48] has tackled a number of these issues by introducing a programmable convolutional interleaver in the time domain and a regular frame structure and a pilot structure suitable to mobile channels. Nevertheless, the long data packet format (i.e., 12 282 bits) and the large channelization bandwidth (e.g., 5 MHz) still remain incompatible with the IoT narrowband requirements.

Finally, for the sake of completeness, the GEO-mobile radio-1 [49] and broadband global area network [50] are designed for supporting voice calls and medium/high terminal data rates over GSO satellites operating in L-band frequencies.

12.3.1.3 A possible robust design solution

Since a unique standard solution fulfilling the narrowband IoT requirements was not present, a project was initiated by European Space Agency in 2017 to design a suitable forward link waveform to be paired with RA schemes like E-SSA in the return link [51]. The new air interface has been designed capitalizing on the most suitable technology solutions adopted in satellite standards, as discussed in the section 12.3.1.2. In particular, the relevant key design aspects are as follows:

- Channel coding based on the 3GPP long term evolution (LTE) turbo codes [52], since they provide a good trade-off between performance and complexity for short packet transmissions (e.g., up to few thousands of bits).
- Channel programmable length time interleaver employed to counteract outages due to shadowing or short blockages in mobile scenarios. It is based on convolutional interleaving (like in DVB-SH), since it provides a reduction in the memory occupation by a factor 2 compared with block interleavers.
- Data and control information organized in equal length frames with a constant pilot symbol spacing to ease the acquisition process at the terminal.
- The concept of physical layer pipes (PLP) [53] introduced for supporting different quality of services. Each of them can be programmable and identified by a combination of physical layer parameters (modulation order, coding rate, and convolutional interleaver parameters) and mapped into the transmitted frame.
- A programmable spreading up to factor 4, common for all PLPs in the frames, applied in order to improve the minimum signal-to-noise ratio (SNR) demodulation threshold, when necessary.
- Minimal link layer functions introduced to transport traffic of different types (e.g., unicast or broadcast). The essential layer-2 protocol loops implemented are: congestion control, packet acknowledgment, and power control.
- The generic stream encapsulation protocol [54] adopted.

Figure 12.11 shows a pictorial view of the forward link functionalities proposed in [51].

12.3.2 Return link design for IoT

12.3.2.1 Key requirements

As mentioned in section 12.2.1.1, IoT traffic has peculiarities that imply a profound impact on the RA design for return link communications. In particular, typical satellite IoT requirements can be summarized as follows:

1. Efficiently and reliably support a very large number of users, sporadically transmitting small- to medium-sized packets, typical of satellite-based IoT applications.
2. Capable of operating in systems with a limited channelization bandwidth per service area (e.g., from few tens of kilohertz to a few megahertz);

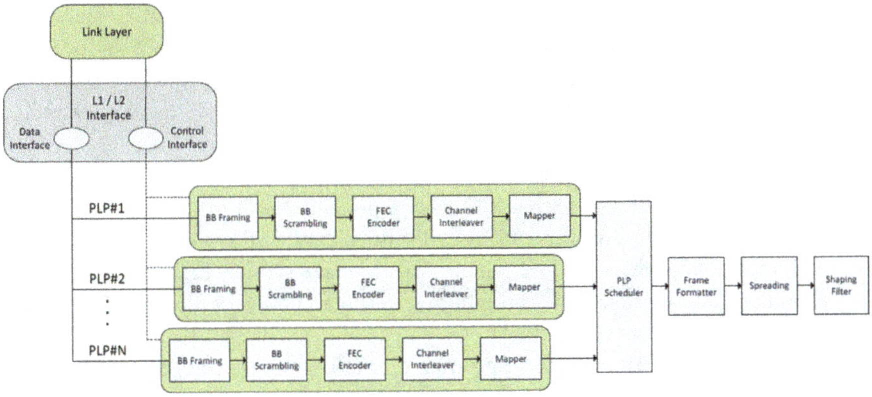

Figure 12.11 Functional block diagram of the forward link transmitter

3. Optimized for small transactions, typical of objects communication, minimizing overheads such as IP headers, bandwidth assignment demands, and/or synchronization signaling.
4. Energy-efficient solution allowing unattended terminal operation for long time.
5. High spectral efficiency, as the available spectrum is limited.
6. Massive scalability, i.e., capability to handle a very large number of UEs.
7. Reliable performance when operating in typical LMS channels.
8. Low-cost easy-to-install technology for the UEs.
9. Low-cost service.

12.3.2.2 Possible solutions

As mentioned before, different RA solutions suitable for IoT over NGSO have been proposed in the recent years. In the following, we dwell on our preferred option based on E-SSA RA in the return link and on a time-division multiplex downlink. The E-SSA choice is based on its best match to the requirements expressed in section 12.3.2.1. Other solutions (being) implemented will be briefly illustrated in section 12.4. More specifically, referring to the requirements listed in section 12.3.2.1, the E-SSA solution:

- allows asynchronous uncoordinated RA very well matched to IoT traffic nature;
- allows to easily adapt the available service bandwidth choosing the best suitable spreading factor;
- provides an energy-efficient solution, thanks to the very high packet correct reception probability with minimum signaling and transmission time;
- achieves unprecedented spectral efficiency up to 2 bps/chip in pure uncoordinated RA mode;
- simple congestion control ensuring stable network operations with overhead minimization;

- supports easy network scalability by simply adding extra FDMA channels when the current channel reaches saturation.
- provides quasi error-free RA solution that combined with simple uplink transmission control, such as the one described in section 12.1.3.4, to guarantee reliable performance when operating in typical LMS channels;
- reuses a physical layer closely derived from the terrestrial widely adopted 3GPP W-CDMA and allows to reuse COTS components for both RF and digital UE elements. Furthermore, the E-SSA is limiting the UE RF power requirements and does not require tight UE synchronization (see section 12.2.2.2);
- provides a low-cost service being high spectral and power efficient while minimizing signaling overhead.
- is combined with IP protocols optimized for IoT small transactions.

12.3.3 Critical demodulation aspects

12.3.3.1 Waveform attributes

Considering the NGSO satellite channel, the attributes of the waveform used in the RA channel should be carefully selected to facilitate the signal detection, demodulation, and decoding at the receiver while maintaining the constraints related to the transmitted power, equivalent, isotropically radiated power, and spectral masks at the transmitter station. Similar to any communication transmission, the transmitted packet is structured in a way to assist the packet detection, carrier synchronization, and coherent demodulation of the receiver packet.

Each packet is composed of a preamble with a carefully selected sequence of symbols. The length of the known symbols and the actual sequence is selected such that in the presence of carrier frequency uncertainty, the preamble selection can result in a reliable detection of packets present. A trade-off between the probability of false detection and probability of miss detection and the actual operating point would determine the required preamble length and the detection strategy.

There are two distinct approaches to combine information part of the message (data channel) and the signaling (control channel). Section 12.4.1 introduces an open standard known as S-MIM in which the data channel and control channel are combined as in-phase and quadrature components of a composite signal prior to long spreading. In this approach, the assigned power to data and control channels is adjusted according to the severity of the channel impairments and the need to strengthen the known segments in each symbol. The resulting constellation corresponds to an asymmetric eight phase shift keying constellation.

Alternatively, in a recent implementation of E-SSA like scheme, the use of quadrature phase shift keying (QPSK) modulation for data symbols with inserted time multiplexed pilot symbols has been investigated for better supporting the adoption of MMSE in front of the iSIC [44]. Similar approach was also utilized in the VDE RA channel as described in section 12.4.2.

The selection of the FEC encoding for the transmitted waveform could also impact the performance of the iSIC scheme as discussed in section 12.2.3.3.

Another important attribute of the transmit waveform is the reduction of the signal envelope peak-to-average ratio and creation of the quasi-constant envelope. This is an important feature to improve the power amplifier efficiency at the UE while maintaining the spectral mask and out-of-band emission at the transmitter under control. As described in [55], a combination or QPSK/binary phase shift keying modulation with a careful selection of spreading sequence phases at the edge of each symbol would allow to minimize the phase jumps at the symbol edge transitions.

As reported in [44], additional degrees of flexibility for the transmit waveform include the support of multiple spreading factors, packet size, FEC coding rate, and the chip rates. Such flexibility would allow to adapt the transmission to channel interference and noise conditions as well as reduce the overhead in case is sporadic traffic volume at the UE by adjusting the information packet size.

12.3.3.2 Demodulation aspects

The packet demodulator in an NGSO system can be implemented on-board or on-ground. Considering that often the LEO constellations do not have a continuous connection to terrestrial hub stations for cost and suitable site availability (e.g., over the oceans), on-board demodulation is a kind of must unless an on-board store and forward approach are followed. Instead, for MEO constellations, it may be possible to use a bent-pipe transponder as the terrestrial gateway coverage may be sufficient to support the service. Clearly, in terms of processing power, the on-board implementation represents the most challenging option.

The demodulator should be able to reliably detect and demodulate the incoming time asynchronous packets in the presence of carrier frequency offset (residual Doppler and Doppler rate, see section 12.1.2.1), random arrival time, and large amount of cochannel MAI. The key burst demodulator main challenges are:

- Packet preamble detection;
- Channel estimation for packet demodulation;
- Channel refined estimation for decoded packet cancellation.

As we discussed in section 12.3.2.1, for satellite IoT, it is pivotal to achieve high packet detection probability with minimum miss-detection events. Therefore, the packet preamble design and its detector represents a key driver in the system performance. A good discussion about preamble design for E-SSA can be found in section 4.1 of [22]. A robust noncoherent nondecision-directed maximum likelihood preamble detector for CDMA has been described in [56]. This scheme provides constant false alarm rate detection property and can be implemented with single-bit quantization with a 2 dB implementation loss. A computationally efficient technique to cope with the carrier frequency error has been described in [57] and section 4.5.1 of [22]. The different frequency hypotheses on which the incoming signal is tested are implemented by means of a fast Fourier transform, which can be applied to the preamble detection scheme proposed in [56]. It should be remarked that the E-SSA iSIC process works on a sliding window memory, thus in the first pass it is important to be able to detect a subset of the packets present in the digital memory. As soon as one packet is detected it will be removed from memory, thus easing successive preamble detection steps.

The channel estimation for detection is based on the preamble and the pilot channel in S-MIM (see section 4.2 of [22]). As mentioned before, the pilot, similarly to 3GPP W-CDMA standard, is multiplexed in quadrature to the packet payload. The payload and pilot components are kept orthogonal through the use of Walsh Hadamard channelization sequences. The continuous in-quadrature pilot component provides a packet long reference for data-aided channel estimation (carrier amplitude and phase). However, to limit the overhead, typically, the pilot channel power is set at a reasonably lower level compared with the payload component; hence, the estimates are noisy in particular when the averaging time has to be limited due to the time variant nature of the channel. As explained in section 4.5.2 of [22], the pilot power setting represents a system trade-off heavily dependent on the channel assumptions.

In case of very fast time variant channel due to fading or UE phase noise, one may consider to adopt a noncoherent detection scheme. However, the demodulation loss (typically of about 2 dB) may heavily impact the RA scheme performance, hence, shall be avoided as much as possible.

As mentioned before, once the packet has been detected, it has to be locally reconstructed for successively being removed out of the digital memory. For this purpose, the preamble and pilot channel estimation exploited for the packet detection is typically not good enough. Once the packet has been decoded and successfully verified through the cyclic redundancy check (CRC), the data modulation affecting the payload signal samples can be wiped out using the locally re-encoded and remodulated packet bits. In this way, a more accurate packet amplitude, carrier frequency, and phase can be reconstructed for canceling the packet from memory. More analytical details about the refined channel estimation process for packet cancellation can be found in [58, 59]. The impact of imperfect packet cancellation has been investigated for the E-SSA demodulator [41]. In this chapter, a semianalytical model is developed allowing to estimate the impact of inaccurate interference cancellation on the iSIC process. It is evident that the refined channel estimation described above is key to achieve good performance. The impact of channel estimation errors is growing with the average packets SNR.

12.3.3.3 Example of an E-SSA practical design

In addition to the demodulator algorithmic aspects discussed in the previous section, in the following, we provide some information about E-SSA demodulator implementation for both on-ground and in-space case. In particular, section 4.3 of [60] provides an overview of a terrestrial gateway implementation. Section 5.4.1 of the same reference is presenting an on-board E-SSA demodulator implementation specifically conceived for LEO small (cubesat) satellites using COTS-based solutions. As previously mentioned, to limit the on-board demodulator complexity, it may be wise to reduce the number of iSIC iterations. This is a system-specific complexity performance trade-off that should take into account the specific satellite implementation constraints. The implementation of E-SSA and E-SSA-like algorithms is also reported in [44] and deployed also for VDE signal detection and decoding as reported in [61] and also discussed in section 12.4.2.

12.3.4 Performance assessment

A fair and meaningful performance assessment of the RA schemes is pivotal to achieve a proper optimization and system performance analysis. In the following, we will give some short suggestions on how to accomplish this challenging and important task.

12.3.4.1 Key Performance indicators

There are a number of key performance indicators for assessing RA schemes performance [39]. Often in literature, the emphasis is given to the throughput versus the (average) MAC load and in particular about its peak value. Although very relevant, this information is incomplete as it does not provide immediate evidence about the amount of packets lost. For this purpose, it is preferable to also provide the PER or PLR results as a function of the (average) MAC load. In any system, and in particular in a satellite system, it is important to keep the PLR low (e.g., $<10^{-2}$) to limit the amount of re-transmissions which may cause unwanted congestion situations and unacceptable latency. Packet delivery latency and energy efficiency also represent key performance indicator for certain applications. The average MAC load is often used because, being the traffic bursty, its aggregation, in particular for Internet type of traffic, shows non negligible amount of randomness. For this reason, the MAC aggregate load is time variant and its average value is typically used. It may also be handy to normalize the MAC load and the RA throughput to ease the comparison of different system configurations (e.g., different bandwidth allocated or bit rate transmitted). For this reason, it is suggested to use bits/symbol or bits/chip for both the MAC load and the RA throughput. The reader is referred to [39] for detailed definition of the suggested key performance indicators.

12.3.4.2 Detailed MAC layer analysis

To get reliable and accurate RA performance results, the preferred approach is to develop a RA simulator able to properly model the following aspects:

- The users' traffic;
- The physical layer including the selected RA scheme;
- The channel model (in particular in case of mobile systems);
- The demodulator key functions (e.g., possible preamble detection, demodulation errors, FEC scheme, possible signal processing functions to reduce the impact of packet collisions).

The last point is of particular relevance as an oversimplified model may lead to erroneous results and wrong trade-off conclusions. As an example [39], provides evidence that a simplified physical layer modeling the effects of packet collisions detection failure based on collision event or SNIR threshold may lead to incorrect numerical findings and design decision. The importance of a faithful physical layer model can be explained by the fact that under loaded conditions there will be a large number of packets colliding causing a relative low SNIR. However, even operating at a SNIR corresponding to a PLR = 0.8 means that 20% of the packets can be correctly decoded. This is sufficient to ignite the iSIC process resulting in a potential successful detection** for all the other packets contained in the demodulator sliding window at the end of the process. This SNIR threshold is physical layer configuration dependent (e.g. FEC, modulation, spreading factor) and can be derived by simulation. To this SNIR threshold experimentally derived (see [41]), corresponds a PER typically much higher than the final target PER after iSIC for the reasons explained above. Therefore, an oversimplified physical layer model based on SNIR threshold or even worst on a collision based packet detection failure will not be able to properly model the iSIC operation. This accurate physical layer modeling is also necessary for semi-analytical RA models such as the one applicable to E-SSA described in [41] or the CRDSA one reported in [62].

12.3.4.3 Simplified system performance analysis

RA systems are by nature affected by co-channel interference due to the randomness of the time UEs are accessing to the shared channel resource to transmit a packet. In particular, channel sensing techniques often used in terrestrial networks, are of no use in satellite due to the inherent system latency. Busy tone congestion control techniques [63], which avoids MAC overload conditions monitoring the aggregated traffic interference, are also suitable for satellite IoT networks. However, the system sizing shall account for co-channel interference on the MAC. The presence of packet collision mitigation techniques (e.g., iSIC) requires some adaptation of the standard link budget calculation by proper high-level modeling of the RA demodulator behaviour. As we will see in the link budget example described in the section 12.3.4.4, in practice based on the detailed simulation findings, one can derive the minimum SNIR at which a packets can be detected. In case of RA schemes adopting iSIC, this detection condition corresponds to the $[\mathrm{SNIR}]_{\min}^{\mathrm{boot}}$ threshold at which the iSIC process boots (see previous discussion in section 12.3.4.2). Naming r the FEC coding rate, M the modulation order and SF the spreading factor, the corresponding $[E_b/(N_0+I_0)]_{\min}^{\mathrm{boot}}$ value is derived from $[\mathrm{SNIR}]_{\min}^{\mathrm{boot}}$ as

$$\left[\frac{E_b}{(N_0+I_0)}\right]_{\min}^{\mathrm{boot}} = \frac{SF}{r\log_2 M}[\mathrm{SNIR}]_{\min}^{\mathrm{boot}}. \tag{12.18}$$

We note that because of the different PLR target value, $[E_b/(N_0+I_0)]_{\min}^{\mathrm{boot}}$ is lower than the $[E_b/N_0]_{\min}^{\mathrm{phy}}$ at which packets can be detected satisfying the required PLR, when assuming the co-channel interference has been removed by the iSIC. Therefore, two link budgets should be computed in two distinct conditions:

**Corresponding to achieving the target PER typically of 10^{-2} or lower.

- A co-channel free condition to check the resulting packet E_b/N_0 received at the gateway station assuming the iSIC has been completely[††] removing the cochannel multiple access interference (MAI). The computed E_b/N_0 in the absence of MAI shall be greater than $[E_b/N_0]_{\min}^{\text{phy}}$ required to achieve the target PER. If this is not the case some system parameter like UE RF power, UE bit rate, satellite antenna gain over noise temperature shall be modified to satisfy this link closure condition. A sizeable link margin in the absence of MAI will allow to accommodate UEs link budget differences and create a spread in the received packets E_b/N_0, which is beneficial for increasing the achievable MAC throughput [41].
- A distinct link budget computing the SNIR before despreading at the gateway demodulator in the presence of the selected MAC average load (proportional to the number of UEs). The MAC load shall be tuned to ensure that the resulting SNIR is greater than $[\text{SNIR}]_{\min}^{\text{boot}}$ for all UEs to be served. Then, for estimating the maximum system load allowed, or equivalently the maximum number of active UEs, $N_{\text{UE}}^{\max}$, one can modify the number of active UEs in order to satisfy this SNIR link margin condition. The latter can also be performed in terms of $E_b/(N_0 + I_0)$ using (12.18). In the case of a multi-beam system a multi-dimensional link budget accounting for antenna beam pattern will allow to compute the maximum number of active UEs/beam $N_{\text{UE}}^{\max}(b)$. For simplicity, we can assume that the UE received power at the satellite antenna beam port is the same for all UEs. The impact of possible power randomization on the throughput performance will be approximated in a following step through a corrective coefficient derived by simulation (see [41] for details).

The above link budgets are computed for the bandwidth occupied by a single UE and for all satellite beams, if applicable. The overall amount of UEs which can be supported by the network can then be estimated extending the formula provided in [25]

$$\left[N_{\text{UE}}^{\max}\right]_{\text{tot}} = \frac{N_{\text{FDM}}\psi_{ps}}{\eta_a}\sum_{b=1}^{N_b} N_{\text{UE}}^{\max}(b), \tag{12.19}$$

where N_{FDM} represents the number of FDMs available in the beam, N_b represents the number of beams, ψ_{ps} is the estimated throughput increase factor due to the power spreading [41], and $0 < \eta_a < 1$ is the IoT traffic activity factor.

[††]In practice, one can more realistically assume that a certain percentage of the MAI remains due the iSIC demodulator imperfect channel estimation.

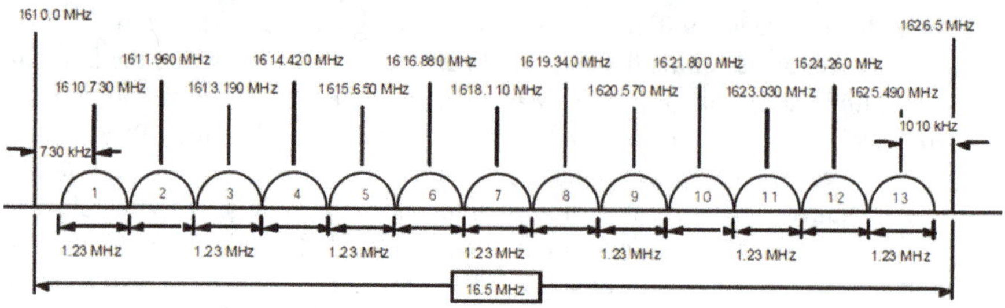

Figure 12.12 Globalstar L-band frequency plan

12.3.4.4 Link budget assumptions and examples

Hereafter, we present a link budget evaluation in L-band with state of the art NGSO satellite payload assumptions and considering the E-SSA waveform implemented in the mobile IoT terminal. As far as the satellite payload is concerned, the relevant parameters are in line with the current Globalstar constellation at 1500 km altitude. Each Globalstar satellite uses 16.5 MHz in S-band for downlink transmissions and in L-Band for uplink communications. This entire bandwidth is divided into 13 channels of 1.23 MHz, and Figure 12.12 shows the channel allocation in the uplink direction. The link budget impact of the feeder downlink (satellite to the gateway) is considered negligible as it is typically the case. Assuming the following IoT terminal requirements, such as a peak data rate of 5 kbps and maximum transmission power of 200 mW along with omni-directional antenna, the link budget analysis has been summarized in Figure 12.13 with the relevant computations. In order to fit the available channel bandwidth, the optimal E-SSA spreading factor

Waveform Characteristics		
Terminal bit-rate	bps	5000
Spreading factor		64
Terminal chip-rate	kcps	960
Occupied bandwidth	kHz	1170
Terminal Characteristics		
Transmission frequency	GHz	1.6
Transmission power	dBW	-7.0
Total EIRP	dBW	-7.0
Satellite Characteristics and Propagation		
Satellite altitude	km	1500
Elevation angle	deg	30
Free Space Path-Loss	dB	164.5
Fading Margin	dB	3.0
Satellite G/T	dB/K	-10.0

Link Budget Results		
Received packet SNR (no MAI) (before despreading)	dB	-15.6
Received packet E_s/N_0 (no MAI)	dB	2.0
Received packet E_b/N_0 (no MAI)	dB	6.8
Required packet $[E_b/N_0]_{\min}^{\text{phy}}$ for PER=1E-3	dB	0.5
Link margin for packet detection after ISIC (ideal)	dB	6.3
Required packet $[\text{SNIR}]_{\min}^{\text{boot}}$ for SIC booting (including implementation loss)	dB	-23.8
Link margin for MAI unloaded system configuration	dB	8.2
Max number of simultaneous active terminals (assuming equal power transmission)		78
Received packet $E_b/(N_0+I_0)$ before ISIC with MAI	dB	-0.96
Required packet $\left[\frac{E_b}{N_0+I_0}\right]_{\min}^{\text{boot}}$ for SIC booting	dB	-1.0
ISIC booting link margin loaded configuration	dB	0.04
System spectral efficiency	b/s/Hz	0.33
Max number of simultaneous active terminals (assuming optimal power distribution)		182
System spectral efficiency	b/s/Hz	0.77

Figure 12.13 Example of Globalstar uplink budget and simplified system performance with E-SSA waveform

is equal to 64. Firstly, focusing on the single user transmission (i.e., no MAI), it can be observed that the received E_b/N_0 is 6.8 dB, and it is confirmed that it is well above the detection threshold for PER=10−3 corresponding to $[E_b/N_0]_{\min}^{\text{phy}} \approx 0.5$ dB. Secondly, recalling the methodology described in section 12.3.4.3 in case of MAI, the required $[\text{SNIR}]_{\min}^{\text{boot}}$ with the chosen spreading factor is approximately −23.8 dB and that allows about 8 dB margin with respect to the case of a single terminal transmission (i.e., $C/N = -15.6$ dB). Aiming at reducing to zero this MAI margin by increasing the number of active IoT terminals, it has been reported that up $N_{\text{UE}}^{\max} = 78$ simultaneous users can transmit with equal power distribution and still keeping the received $E_b/(N_0 + I_0)$ slightly greater than $[E_b/(N_0 + I_0)]_{\min}^{\text{boot}}$. Further, approximately $\psi_{ps} \cdot N_{\text{UE}}^{\max} = 182$ users might be successfully decoded in case of optimal power distribution [41]. Assuming now $\eta_a = 2.7 \times 10^{-4}$ (i.e., one transmission every hour), $N_b = 16$, $N_{\text{UE}}^{\max}(b) = N_{\text{UE}}^{\max}$, and optimal power distribution, we get $\left[N_{\text{UE}}^{\max}\right]_{\text{tot}} = 1.362 \times 10^8$ users supported by a single satellite and exploiting all available channels (i.e., for $N_{\text{FDM}} = 13$).

12.4 NGSO RA (standard and proprietary) solutions and system implementation examples

12.4.1 S-band mobile interactive multimedia

The S-MIM is an European Telecommunication Standard Institute (ETSI) standard published in 2004 [64–69]. The S-MIM aims to two main services: a robust downlink for broadcasting audio/video information and signaling information based on the ETSI Satellite-to-Handheld (DVB-SH) [48] standard and an uplink to provide ubiquitous messaging services over GSO or NGSO satellites using low-power terminals operating in S-band. The return link is largely based on the 3GPP Wideband CDMA physical layer with some specific adaptations to support packet mode transmission for IoT applications. The return link exploits the E-SSA RA with open loop uplink power/transmit control described in previous sections. A good overview of the S-MIM standard is provided in [70]. The S-MIM offers a low-cost bandwidth and power efficient solution for short messages with modest power requirements on the terminal side. S-MIM standard prototype UEs and gateways were implemented by Eutelsat and went through field trials in the frame of the S-band Solaris initiative [24]. The experimental campaign was using the S-band payload on-board the Eutelsat 10A GSO satellite, was located at 10° East. The field trials carried out during this campaign were performed under real mobile channel environments (highway, tree shadowing, suburban, etc.). The return link performances were assessed with and without background traffic in such a way that the comparison of the two considered scenarios could be as fair as possible. The test campaign confirmed the effectiveness of the S-MIM messaging protocol and more particularly, of the transmit power control algorithm. The latter permits a smart use of the satellite bandwidth, allowing the simultaneous transmission of thousands of packets by low-power mobile UEs, hence, demonstrating the S-MIM suitability for low-cost consumer IoT products.

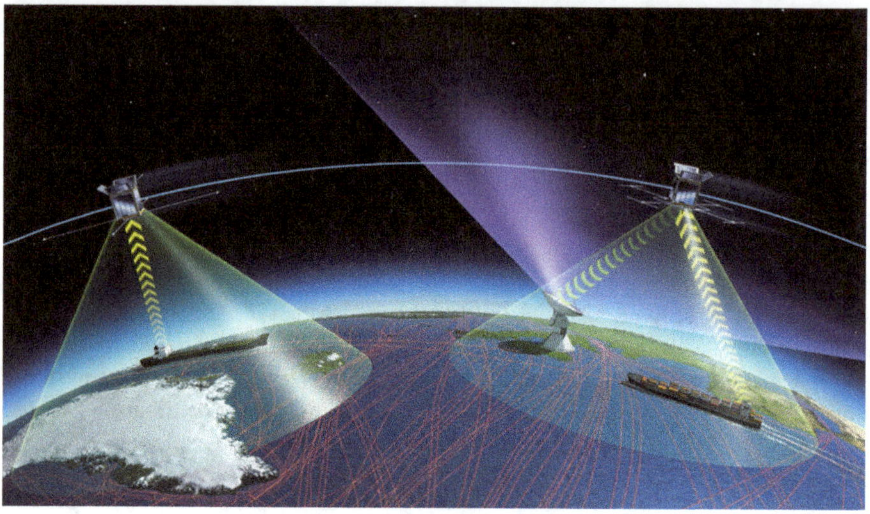

Figure 12.14 VDES Scenario. Source: [73]

Unfortunately, the incomplete Eutelsat 10A S-band antenna deployment did not allow for a commercial S-MIM service roll-out. However, the S-MIM technology has been evolved by Eutelsat as a proprietary standard dubbed F-SIM for Ku/Ka-band applications and commercially deployed in several areas of the world [71].

12.4.2 VHF data exchange

The WRC in 2019 modified Appendix 18 of the RR [1], allocated maritime VHF frequencies for two-way VDE via satellite, making space communication an integrated component of the VDE system (VDES) [72]. Figures 12.14 and 12.15 illustrate the frequency usage for terrestrial and satellite components of VDE. According to the ITU Recommendation [74], the VDE signal transmission occupies one or multiple VHF channels, carrying a waveform with a bandwidth of 25, 50, 100, or 150 kHz. In time domain, each transmission follows a frame structure that is synchronized

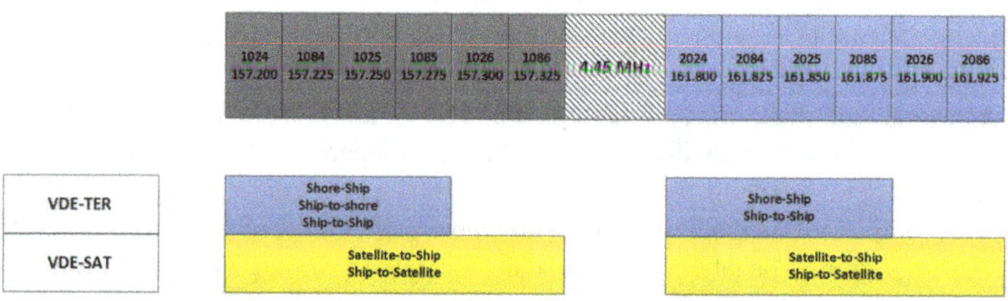

Figure 12.15 VDES frequency plan

to the coordinated universal time with a time duration of one or multiple slots of 26.667 ms.

Among several modes of operation, the ITU Recommendation [74] defines an RA channel for the VDE satellite uplink, known as link ID 20. The actual allocation of RA channels with the VDE frame structure follows specific patterns. As default setting of slots mapped in [74], within each frame of 2250 slots (60 seconds), three sets of RA channels, each with a duration of 179 slots are allocated. The physical layer for the RA channel was carefully designed to support resolving overlapping reception of multiple packets. Some initial consideration for the RA physical layer design was reported reported in [55]. The final specification of the link ID 20 was further evolved to include the following main attributes:

- Information block contains 80 bits, followed by 16 bits of CRC, forming 96 bits payload at the input of a turbo encoder.
- A turbo FEC encoder (similar to that of LTE and DVB-SH Standard) is applied with an effective coding rate 1/4 (after puncturing) and 18 FEC tail bits.
- Bit-wise scrambling of resulting 402 codewords for energy dispersal followed by bit to QPSK symbol mapping.
- Preamble and pilot insertion. A pilot symbol is inserted every 16 data symbols, for a total of 12 pilot symbols. Preamble length contains 48 known symbols. A burst contains a total of 261 symbols.

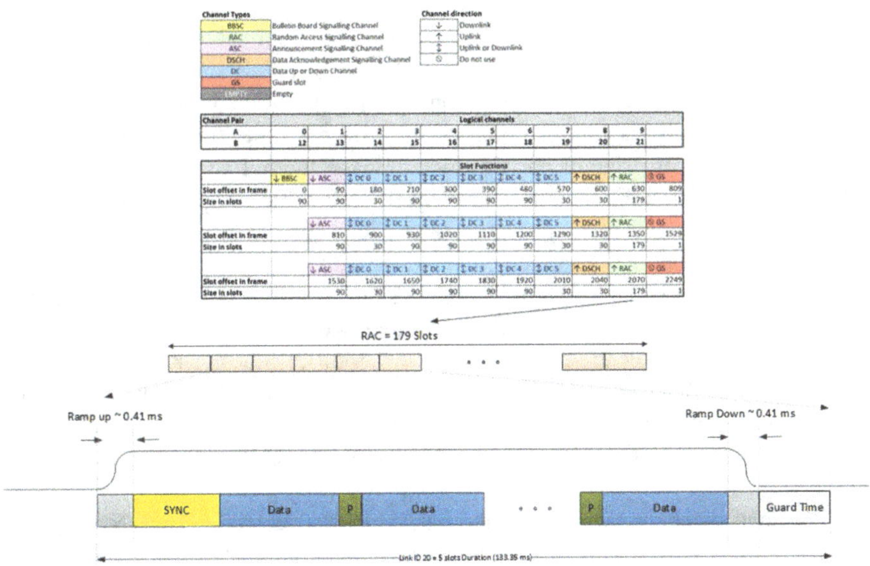

Figure 12.16 VDE-SAT uplink ID 20 – frame structure

- A spreading procedure, corresponding to a spreading factor 16 is applied in such a way to create a minimum phase transition at the edge of each QPSK symbol[‡‡].
- The corresponding signal is transmitted at a chip rate of 33.6 kchips/s, occupying two VDE channels with a total bandwidth of 50 kHz (including the guard band).

Figure 12.16 illustrates the link ID 20 frame structure and the slot constituents. The results of over-the-air campaign of RA channel in VDE channels according the link ID 20 captured by the NorSat-2 LEO satellite have been reported in [61]. The test trial was set up based on a single uplink station that emulates a population of mobile stations (multiple vessels) occupying simultaneously the RA channel, taking into account arrival time delay as well as Doppler frequency shift due to geometrical distributions of the real scenario. Receiver performance of receiver with iSIC is analyzed and compared with conventional SSA. Also the impact of external sources of interference in VDE satellite uplink channel is reported and discussed.

12.4.3 Narrowband-IoT

The 3GPP has standardized in the last 20 years the terrestrial mobile broadband technologies for 3G, 4G, and now 5G cellular systems. As far as MTC services are concerned, the latest 3GPP solutions are called NB-IoT, and it has been made available around mid-2016 [75]. NB-IoT uses SC-FDMA in uplink and OFDMA in downlink.

The RA procedure is based on the exchange of four messages between the terminal and the base station, as shown in Figure 12.17, and it is necessary for the logon procedure (i.e., from idle to connected mode). Of course, the most critical step is the transmission/reception of the first message, where time and frequency resources are shared among all terminals attempting the connection to the network. The first message is simply an RA preamble composed of four symbol groups, and each symbol group has a cyclic prefix followed by five symbols [76]. Frequency hopping is applied on symbol group granularity, i.e., each symbol group is transmitted on a different subcarrier. The first subcarrier of the first symbol group is chosen randomly, while the following ones are determined according to a deterministic sequence that depends on the initial subcarrier. Since the narrowband physical RA channel is composed by a maximum of 48 subcarriers, only 48 orthogonal preambles can be uniquely identified. In other words, two terminals selecting the same initial subcarrier will consequently collide for the entire preamble sequence.

The analysis about adapting the terrestrial NB-IoT technologies to NGSO systems has been reported in [77]. Some modifications are inevitably necessary (e.g., timing relationships during the RA procedure), nevertheless the physical layer aspects associated with the preamble generation remain unchanged in Release 17.

[‡‡]Further detail regarding the quasi constant envelope spreading can be found in [55].

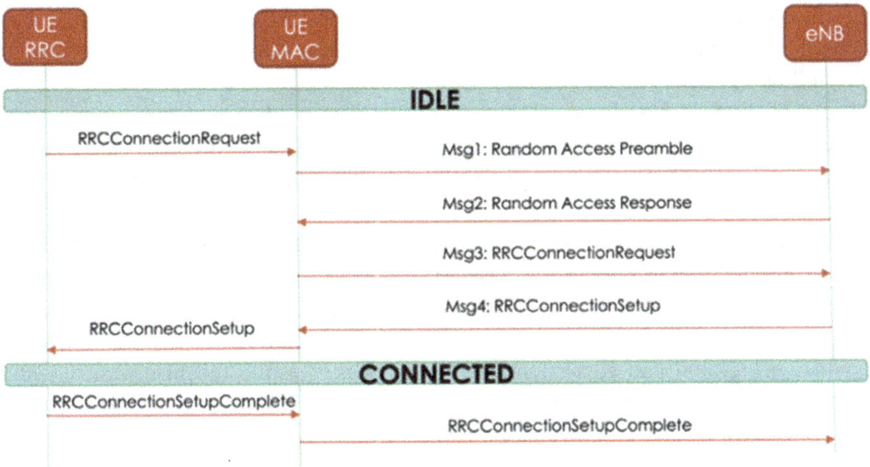

Figure 12.17 Example of message exchange in NB-IoT between the user terminal (UE) and the base station

Some NB-IoT protocol performance assessments and parameter optimization for NGSO constellations can be found in [78]. In this reference it is shown that even after NB-IoT optimization for the NGSO case the number of supported users is rather limited mainly because of the ALOHA RA procedure adopted. This should stimulate the development of more suitable RA solutions for 5G IoT applications.

12.4.4 Universal network for IoT

A proprietary solution addressing the MTC services in terrestrial LPWAN is owned by SigFox [79], and it is based on an ultra narrowband (UNB) signal transmission. Anteur *et al.* 80] have investigated the performance of this UNB technology in NGSO systems, and lately a proprietary adaptation has been announced by Airbus in [81] with the name universal network for IoT (UNIT). This air interface is adopted in the AstroCast system, but unfortunately there is no public information available on its waveform characteristics.

12.4.5 In-orbit demonstrations and beyond

While the idea of NGSO satellite constellations became a reality more than 20 years ago with pioneering examples such as Globalstar and Iridium, a new wave of in-orbit IoT technology demonstration gained popularity in the recent years.

Since 2018, there have been several demonstration missions planned and executed worldwide to provide the satellite IoT concept. In most cases, the demonstration started with single LEO satellite, performing store and forward of short messages. In a recently published survey ([82], section V), authors have provided examples of publicly known companies who are actively pursuing IoT solutions via LEO small satellites. Their target IoT technologies range from proprietary solutions, industry standards such as Lora as well as open NB-IoT standards, adapted to satellite.

As of the first quarter of 2021, several satellite IoT LEO constellations (such as Kepler[§§], Swarm[¶¶] and AstroCast[***]) have already reached multiple satellites in orbit, going beyond the research phase and entering into (pre)operational phase for providing IoT services.

Compared with conventional GSO satellites, the NGSO satellite IoT solutions aim to provide global services for massive number of devices in an affordable price. The "new space" paradigm in flight segment design, qualification, and lifetime settings has created new momentum and expectations for agility and flexibility in establishing global and cost-effective IoT services. Despite recent advancements and exciting opportunities, there are challenges ahead that require further technology development, coordination, and planning.

- **Scarce spectrum and regulatory aspects:** Access to the spectrum is a severe barrier for many new players. As discussed in section 12.1.1, the current spectrum allocations and service line-ups in different regions make it extremely competitive for new players to establish satellite IoT services. Some initiatives for allocation of new spectrum and possible sharing the spectrum among multiple players have already envisaged as new agenda items for the upcoming WRCs to address this issue.
- **Intersystem interference:** As the number of satellite communications grow, the likelihood of intentional or unintentional RF interference among such system also increases significantly. The traditional approach to split the frequency resources among different players would restrict the deployment of the individual service and prevent the service growth in a global scale. A new paradigm for resource sharing among different satellite systems as well as satellite and terrestrial systems is essential for sustainability of services in a global scale.
- **Scalability:** The technology demonstration phase starts typically with a very limited number of UEs deployed in the network. This arrangement may fail to reveal hidden overhead in system resource consumption when the number of end users grows significantly. For example, signaling overhead for network synchronization or limited aggregate throughput as well as constraints on the traffic load could prevent the system scalability and commercial viability of the offered service. This is an important aspect that requires careful attention at the design phase. It is also important to establish a clear bridging between the in-orbit demonstration phase, that is typically based on a limited number of space assets, and the operational phase to avoid oversimplifying the design approach.
- **UE size, weight, and power (and) cost and service continuity:** For many IoT service applications, the size, weight, and power and cost of the UEs are

[§§]http://kepler.space
[¶¶]http://swarm.space
[***]http://astrocast.com

extremely critical for sustainability of the service offering. It is often critical to see a seamless integration of the satellite and terrestrial service to maintain continuity of service in a large geographical scale. Although, this does not necessarily mean identical terrestrial and satellite access solutions, from the end-user perspective, the same piece of equipment should be able to maintain service continuity and service quality.

References

[1] *Radio regulations*. International Telecommunication Union; 2020.

[2] *M2M/IoT operation via satellite*. The European Conference of Postal and Telecommunications Administrations (CEPT), Electronic Communications Committee; 2020.

[3] *World radiocommunication conference 2019 final acts*. ITU; 2019.

[4] Wertz J.R., Larson W.J. *Space mission analysis and design*. 3rd ed. Microcosm Press and Kluwer Academy Publishers; 1999.

[5] Zheng C., Chen X., Huang Z. 'A comprehensive analysis on doppler frequency and doppler frequency rate characterization for GNSS receivers'. *2016 2nd IEEE International Conference on Computer and Communications (ICCC)*; Chengdu, China, 2016. pp. 2606–10.

[6] *Propagation data and prediction methods required for the design of earth-space telecommunication systems* [ITU-R P618-13 Recommendation]. International Telecommunication Union; 2017.

[7] Abdus Samad M., Debo Diba F., Choi D.Y. 'A survey of rain fade models for earth–space telecommunication links—taxonomy, methods, and comparative study'. *MDPI Remote Sensing*. 2021, vol. 13(1965), pp. 1–44.

[8] *Ionospheric propagation data and prediction methods required for the design of satellite services and systems* [ITU-R P531-13 Recommendation]. International Telecommunication Union; 2016.

[9] *Prediction of building entry loss* [ITU-R P2109-1 recommendation]. International Telecommunication Union; 2019.

[10] *Compilation of measurement data relating to building entry loss* [ITU-R P2346-14 Recommendation]. International Telecommunication Union; 2021.

[11] *Propagation data required for the design systems in the land mobile-satellite service* [ITU-R P681-11 Recommendation]. International Telecommunication Union; 2019.

[12] Karaliopoulos M.S., Pavlidou F.N. 'Modelling the land mobile satellite channel: a review'. *Electronics & Communication Engineering Journal*. 1999, vol. 11(5), pp. 235–48.

[13] Loo C. 'A statistical model for a land mobile satellite link'. *IEEE Transactions on Vehicular Technology*. 1999, vol. 34(3), pp. 122–27.

[14] Fontan F.P., Vazquez-Castro M., Cabado C.E., Garcia J.P., Kubista E. 'Statistical modeling of the LMS channel'. *IEEE Transactions on Vehicular Technology*. 1999, vol. 50(6), pp. 1549–67.
[15] Prieto-Cerdeira R., Perez-Fontan F., Burzigotti P., Bolea-Alamañac A., Sanchez-Lago I. 'Versatile two-state land mobile satellite channel model with first application to DVB-SH analysis'. *International Journal of Satellite Communications and Networking*. 2010, vol. 28(5–6), pp. 291–315. Available from http://doi.wiley.com/10.1002/sat.v28%3A5/6
[16] '3rd generation partnership project; technical specification group radio access network; study on new radio (NR) to support non-terrestrial networks'. 2019, vol. Release 15(5G 3GPP;. 3GPP TR 38.811).
[17] Arndt D., Ihlow A., Heuberger A., Eberlein E. 'State modelling of the land mobile propagation channel with multiple satellites'. *International Journal of Antennas and Propagation*. 2012, vol. 2012, pp. 1–15.
[18] Rieche M., Ihlow A., Arndt D., *et al.* 'Modeling of the land mobile satellite channel considering the terminal's driving direction'. *Hindawi International Journal of Antennas and Propagation*. 2015, vol. 2015, pp. 1–21.
[19] Siwiak K. *Method for compensating for frequency shifts for satellite communication system*. Motorola; IL: Inventor; Inc; 1997.
[20] Sherman K.N. 'Phased array shaped multi-beam optimization for LEO satellite communications using a genetic algorithm'. *IEEE International Conference on Phased Array Systems and Technology*; Dana Point, CA, IEEE, 2012. pp. 501–04.
[21] Del Rio Herrero O., De Gaudenzi R. 'High efficiency satellite multiple access scheme for machine-to-machine communications'. *IEEE Transactions on Aerospace and Electronic Systems*. 2014, vol. 48(4), pp. 2961–89.
[22] De Gaudenzi R., del Rio Herrero O., Gallinaro G. 'Enhanced spread ALOHA physical layer design and performance'. *International Journal of Satellite Communications and Networking*. 2014, vol. 32(6), pp. 457–73. Available from http://doi.wiley.com/10.1002/sat.v32.6
[23] 'Signal-to-noise monitoring. JPL space programs summary'. 1976, pp. 37–169.
[24] Hermenier R., Del Bianco A., Marchitti M.-A, *et al.* 'S-MIM field trials results'. *International Journal of Satellite Communications and Networking*. 2014, vol. 32(6), pp. 535–48. Available from http://doi.wiley.com/10.1002/sat.v32.6
[25] Cioni S., De Gaudenzi R., Del Rio Herrero O., Girault N. 'On the satellite role in the era of 5G massive machine type communications'. *IEEE Network*. 2014, vol. 32(5), pp. 54–61.
[26] Li Y., Tu W. 'Traffic modelling for IoT networks'. *Proceedings of the 2020 10th International Conference on Information Communication and Management (ICICM)*; Paris, France, New York, NY, 2020. pp. 4–9. Available from https://dl.acm.org/doi/proceedings/10.1145/3418981

[27] Nguyen-An H., Silverston T., Yamazaki T., Miyoshi T. 'IoT traffic: modeling and measurement experiments'. *IoT*. 2020, vol. 2(1), pp. 140–62. Available from https://www.mdpi.com/2624-831X/2/1/8

[28] De Gaudenzi R., Del Rio Herrero O., Gallinaro G., Cioni S., Arapoglou P.-D. 'Random access schemes for satellite networks, from VSAT to M2M: a survey'. *International Journal of Satellite Communications and Networking*. 2020, vol. 36(1), pp. 66–107.

[29] del Rio Herrero O., Foti G., Gallinaro G. 'Spread-spectrum techniques for the provision of packet access on the reverse link of next-generation broadband multimedia satellite systems'. *IEEE Journal on Selected Areas in Communications*. 2020, vol. 22(3), pp. 574–83.

[30] Abramson N. 'The ALOHA system – another alternative for computer communications'. *Proceedings of 1970 Fall Joint Conference AFIPS*; AFIPS, 1970. pp. 281–85.

[31] Chiani M., Elzanaty A. 'On the lora modulation for IoT: waveform properties and spectral analysis'. *IEEE Internet of Things Journal*. 1994, vol. 6(5), pp. 8463–70.

[32] What are LoRa® and LoRaWAN®. [Tecnical Documentation]. SEMTECH 2019. Available from https://lora-developers.semtech.com/documentation/tech-papers-and-guides/lora-and-lorawan

[33] Mikhaylov K., Petaejaejaervi J., Haenninen T. '*22nd European Wireless Conference*'. 2016. pp. 1–6.

[34] Abramson N. 'Multiple access in wireless digital networks'. *Proceedings of the IEEE*. 1994, vol. 82(9), pp. 1360–70.

[35] Abramson N. 'The throughput of packet broadcasting channels'. *IEEE Transactions on Communications*. 1994, vol. 25(1), pp. 117–28.

[36] Choudhury G.L., Rappaport S. 'Diversity ALOHA--a random access scheme for satellite communications'. *IEEE Transactions on Communications*. 1983, vol. 31(3), pp. 450–57.

[37] Casini E., De Gaudenzi R., Herrero Od. R. 'Contention resolution diversity slotted ALOHA (CRDSA): an enhanced random access scheme for satellite access packet networks'. *IEEE Transactions on Wireless Communications*. 1983, vol. 4, pp. 19–1408.

[38] De Gaudenzi R., Del Rio-Herrero O., Acar G, *et al*. 'Asynchronous contention resolution diversity ALOHA: making CRDSA truly asynchronous'. *IEEE Trans on Wireless Communications*. 2007, vol. 6(4), pp. 1408–19.

[39] Mengali A., De Gaudenzi R., Stefanovic C. 'On the modeling and performance assessment of random access with SIC'. *IEEE Journal on Selected Areas in Communications*. 1983, vol. 36(2), pp. 292–303.

[40] Del Rio Herrero O., De Gaudenzi R. 'High efficiency satellite multiple access scheme for machine-to-machine communications'. *IEEE Transactions on Aerospace and Electronic Systems*. 1983, vol. 48(4), pp. 2961–89.

[41] Collard F., De Gaudenzi R. 'On the optimum packet power distribution for spread ALOHA packet detectors with iterative successive interference

cancelation'. *IEEE Transactions on Wireless Communications*. 2004, vol. 13(12), pp. 6783–94.

[42] Caire G., Muller R.R., Tanaka T. 'Iterative multiuser joint decoding: optimal power allocation and low-complexity implementation'. *IEEE Transactions on Information Theory*. 2004, vol. 50(9), pp. 1950–73.

[43] Gallinaro G., Alagha N., De Gaudenzi R., Kansanen K., Muller R., Salvo Rossi P. 'ME-SSA: an advanced random access for the satellite return channel'. *2015 IEEE International Conference on Signal Processing for Communications (ICC)*; London, IEEE, 2004. pp. 856–61.

[44] Isca A., Alagha N., Andreotti R., Andrenacci M. 'Recent advances in design and implementation of satellite gateways for massive uncoordinated access networks'. *Sensors (Basel, Switzerland)*. 2022, vol. 22(2), 565.

[45] del Rio Herrero O., Pijoan Vidal J. 'Design guidelines for advanced satellite random access protocols'. *30th AIAA International Communications Satellite System Conference (ICSSC)*; Ottawa, Canada, Reston, VA: AIAA, 2004. pp. 24–27. Available from https://arc.aiaa.org/doi/book/10.2514/MICSSC12

[46] del Rio Herrero O., Pijoan Vidal J. 'Design guidelines for advanced satellite random access protocols'. *30th AIAA International Communications Satellite System Conference (ICSSC)*; Ottawa, Canada, Reston, VA, 2012. Available from https://arc.aiaa.org/doi/book/10.2514/MICSSC12

[47] 'European telecommunication standards institute'. *Part*. 2014, vol. 1(ETSI Technical Specification EN 302 307-1 V1.4.1), p. DVB–S2.

[48] 'Digital video broadcasting (DVB); framing structure, channel coding and modulation for satellite services to handheld devices (SH) below 3 ghz'. [ETSI Technical Specification EN] *European Telecommunication Standards Institute*. 2011.

[49] Route des Lucioles L. 'GEO-mobile radio interface specifications (release 3); third generation satellite packet radio service; part 1: general specifications; sub-part 3: general system description'. *European Telecommunication Standards Institute*. 2015, p. 101.

[50] 'Satellite earth stations and systems (SES); family SL satellite radio interface (release 1); part 2: physical layer specifications; sub-part 1: physical layer interface'. *European Telecommunication Standards Institute*. 2015, p. 102.

[51] 'Final report on gateway demonstrator for E-SSA-based machine-to-machine applications'. *European Space Agency - Telecommunications and Integrated Applications (TIA)*. 2020. Available from https://artes.esa.int/projects/gemma

[52] 'Evolved universal terrestrial radio access (E-UTRA); multiplexing and channel coding. 3rd generation partnership project (3GPP)'. *Technical Specification TS*. 2016, vol. 36.

[53] 'Satellite earth stations and systems (SES); satellite digital radio (SDR) systems; outer physical layer of the radio interface'. [ETSI Technical Specification TS] *European Telecommunication Standards Institute*. 2008, p. 102.

[54] 'Digital video broadcasting (DVB); generic stream encapsulation (GSE) protocol'. [ETSI Technical Specification TS] *European Telecommunication Standards Institute*. 2007, p. 102.

[55] Gallinaro G., Alagha N., Müller R, *et al.* 'Quasi constant envelope CDMA for VHF maritime communications via satellite'. *Ka and Broadband Communications Conference (Ka)*; 2017.

[56] De Gaudenzi R., Giannetti F., Luise M. 'Signal recognition and signature code acquisition in CDMA mobile packet communications'. *IEEE Transactions on Vehicular Technology*. 1998, vol. 47(1), pp. 196–208.

[57] Sust M.K., Kaufmann R.F., Molitor F., Bjornstrom G.A. 'Rapid acquisition concept for voice activated CDMA communication'. *GLOBECOM '90*; San Diego, CA, USA, IEEE, 1998. pp. 1820–26.

[58] Casini E., Del Rio Herrero O., De Gaudenzi R. *Method of packet model digital communication over a transmission channel shared by a plurality of user* [United States patent 8,094,672 B2]. 10 January 2012.

[59] Liva G. 'Graph-based analysis and optimization of contention resolution diversity slotted ALOHA'. *IEEE Transactions on Communications*. 1998, vol. 59(2), pp. 477–87.

[60] Arcidiacono A., Finocchiaro D., De Gaudenzi R, *et al.* 'Is satellite ahead of terrestrial in deploying NOMA for massive machine-type communications?'. *Sensors (Basel, Switzerland)*. 2021, vol. 21(13), pp. 4290–321.

[61] Andreotti R., Nanna L., Andrenacci M., Isca A., Haugli H., Alagha N. 'On-field test campaign performance of VDE-SAT link ID 20 over norsat-2 LEO satelli4te'. *International Journal of Satellite Communications and Networking*. 2021.

[62] del Rio Herrero O., De Gaudenzi R. 'Generalized analytical framework for the performance assessment of slotted random access protocols'. *IEEE Transactions on Wireless Communications*. 2014, vol. 13(2), pp. 809–21.

[63] Corazza G.E., Cioni S., Padovani R. 'Application of closed loop resource allocation for high data rate packet transmission'. *IEEE Transactions on Wireless Communications*. 2007, vol. 6(11), pp. 4049–59.

[64] *Satellite earth stations and systems (SES); air interface for S-band mobile interactive multimedia (S-MIM) MIM); part 1: general system architecture and configurations. 650 Route des Lucioles, F-06921 Sophia Antipolis CEDEX* [ETSI Technical Specification TS]. FRA: European Telecommunication Standards Institute; 2013. p. 102.

[65] *Satellite earth stations and systems (SES); air interface for S-band mobile interactive multimedia (S-MIM); part 2: forward link subsystem requirements. 650 Route des Lucioles, F-06921 Sophia Antipolis CEDEX* [ETSI Technical Specification TS]. FRA: European Telecommunication Standards Institute; 2013. p. 102.

[66] *Satellite earth stations and systems (SES); air interface for S-band mobile interactive multimedia (S-MIM); part 3: physical layer specification, return link asynchronous access. 650 Route des Lucioles, F-06921 Sophia Antipolis CEDEX* [ETSI Technical Specification TS]. FRA: European Telecommunication Standards Institute; 2013. p. 102.

[67] *Satellite earth stations and systems (SES); air interface for S-band mobile interactive multimedia (S-MIM): part 4: physical layer specification, return link*

synchronous access. 650 Route des Lucioles, F-06921 Sophia Antipolis CEDEX [ETSI Technical Specification TS]. FRA: European Telecommunication Standards Institute; 2013. p. 102.

[68] *Satellite earth stations and systems (SES); air interface for S-band mobile interactive multimedia (S-MIM): part 5: protocol specifications, link layer. 650 Route des Lucioles, F-06921 Sophia Antipolis CEDEX* [ETSI Technical Specification TS]. FRA: European Telecommunication Standards Institute; 2013. p. 102.

[69] *Satellite earth stations and systems (SES); air interface for S-band mobile interactive multimedia (S-MIM): part 6: protocol specifications, system signalling. 650 Route des Lucioles, F-06921 Sophia Antipolis CEDEX* [ETSI Technical Specification TS]. FRA: European Telecommunication Standards Institute; 2013. p. 102.

[70] Scalise S., Niebla C.P., De Gaudenzi R., Del Rio Herrero O., Finocchiaro D., Arcidiacono A. 'S-MIM: A novel radio interface for efficient messaging services over satellite'. *IEEE Communications Magazine*. 2013, vol. 51(3), pp. 119–25.

[71] Arcidiacono A., Finocchiaro D., De Gaudenzi R, *et al.* 'Is satellite ahead of terrestrial in deploying NOMA for massive machine-type communications?'. *Sensors (Basel, Switzerland)*. 2021, vol. 21(13), p. 13.

[72] 'ITU-R M.2435 technical characteristics for an automatic identification system using time division multiple access in the VHF maritime mobile frequency band'. *ITU*. 2018.

[73] 'From satellites to the sea: VDES offers global link for ships'. *ESA*. 2018. Available from https://www.esa.int/Enabling_Support/Space_Engineering_Technology/From_satellites_to_the_sea_VDES_offers_global_link_for_ships

[74] 'ITU-R M.2092 technical characteristics for a VHF data exchange system in the VHF maritime mobile band'. *ITU*. 2021.

[75] Wang Y.-P.E., Lin X., Adhikary A, *et al.* 'A primer on 3GPP narrowband Internet of Things'. *IEEE Communications Magazine*. 2017, vol. 55(3), pp. 117–23.

[76] 'Evolved universal terrestrial radio access (E-UTRA); physical channels and modulation. 3rd generation partnership project (3GPP);'. *Technical Specification TS*. vol. 36(211). 2016

[77] 'Study on narrow-band internet of things (NB-iot) / enhanced machine type communication (EMTC) support for non-terrestrial networks (NTN)'. [Technical Report TR 36.763] *3rd Generation Partnership Project (3GPP)*. 2021, vol. V17.

[78] Amatetti C., Conti M., Guidotti A., Vanelli-Coralli A. '*Submitted to IEEE International Communication Conference (ICC)*'. 2022. pp. 1–6.

[79] Fourtet C., Bailleul T., assignee S.A. *Terminals and telecommunication system* [inventors; Sigfox]. 2013. Available from https://www.google.com/patents/US20130142191

[80] Anteur M., Thomas N., Deslandes V., Beylot A. 'On the performance of UNB for machine-to-machine low earth orbit (LEO) satellite communications'. *International Journal of Satellite Communications and Networking*. 2019, vol. 37(1), pp. 56–71.

[81] 'Delivering internet of things (iot) services world-wide'. [Press Release] *Airbus*. 2021. Available from https://www.airbus.com/newsroom/news/en/2021/05/Delivering-Internet-of-Things-IoT-services-worldwide.html

[82] Centenaro M., Costa C.E., Granelli F., Sacchi C., Vangelista L. 'A survey on technologies, standards and open challenges in satellite IoT'. *IEEE Communications Surveys & Tutorials*. 2021, vol. 23(3), pp. 1693–720.

Chapter 13

Virtual network embedding for non-geostationary orbit-terrestrial systems: parallel computation and software defined networking testbed implementation

Mario Minardi[1], Fabian Mendoza[1], Lei Lei[1], Thang X Vu[1], and Symeon Chatzinotas[1]

Beyond 5G and 6G networks are expected to meet ambitious performance parameters of coverage, data rates, latency, etc., with the objective of exploiting as many as possible physical network resources, such as capacity, to get the maximum achievable performance out of them. In this context, the smart and efficient use of available resources, as well as ubiquitous and continuous coverage provided by satellite networks, have become a must. Network virtualization (NV) has been proved to be a key enabling technology to fulfill the challenging requirements of the upcoming telecommunication networks. NV is based on algorithms that can instantiate virtualized services on the substrate infrastructure, optimizing the embedding, according to a specific objective. This kind of algorithms is known as virtual network embedding (VNE). The aim of this chapter is to focus on two main aspects of the VNE. First, an efficient parallel approach for the VNE problem is considered. More precisely, the aim is to show how a parallel computation for the resource mapping allows to further improve the performance of the algorithm. Second, the chapter introduces a practical implementation of the VNE algorithm in a software defined networking (SDN)-based testbed. An experimental testbed to support the VNE algorithm for non-geostationary orbit (NGSO) constellations is presented. The laboratory testbed has been developed and validated, consisting of a Mininet-based simulator, a Ryu SDN controller with an end-to-end (E2E) traffic engineering (TE) application for the virtual networks (VNs) establishment, and the VNE algorithm implemented in MATLAB®.

[1]Interdisciplinary Centre for Security, Reliability and Trust (SnT), University of Luxembourg, Luxembourg

13.1 Introduction

Nowadays, the complex and heterogeneous upcoming telecommunication networks, together with the challenging network performance requirements (e.g., data rates in the order of Tbps, latency lower than 1 ms, very high reliability $\sim$ 99.9999%, etc.), have brought the need for new optimization techniques/strategies to effectively maximize the utilization of the physical network, also defined as substrate network. 5G and the upcoming 6G networks are expected to support the scenarios with a large variety of applications and services in terms of quality of service (QoS), by adopting the technology of "network slicing" [1, 2].

Network slicing has become very attractive due to an ever-increasing demand for very differentiated services. The main aim of network slicing is to allocate these demanded services (slices) onto the same substrate network. Each service has its own pool of resources to be allocated, and accordingly, a given QoS has to be satisfied during the embedding. Therefore, the role of network providers has become to investigate the new virtualization techniques and find the one which offers the best performance. The ambitious requirements from the scenario described above have made the presence of satellite networks always more and more crucial for fulfilling the demanded QoS. Satellites have always proved to offer an ubiquitous and anytime coverage, with the minimization of terrestrial infrastructures' usage. Furthermore, nowadays, NGSO communication systems have attracted network providers for being one of the key enabling technologies to support an increasing service requests, offering reduced latency and link loss [3, 4]. Given these premises, the main objective of this chapter is to provide some implementation details of network slicing, pointing out the challenges and advantages of each considered scenario. In particular, two main aspects of network slicing are considered.

On one side, network slicing is based on optimization algorithms with the aim of fulfilling the challenging allocation of the demanded resources by each service onto the physical network, known as the VNE problem. Each service is typically seen as a VN request (VNR), composed by nodes, links, and nodes/links requirements (e.g., computational node capacity, link bandwidth, maximum latency, etc.). VNE is the resource allocation process, node and link mapping, of an incoming VNR, or a multitude of them, in the physical network.

On the other side, the efficiency of network slicing optimization algorithms for NGSO satellite mega constellations rely on the support of novel technologies, such as SDN, to facilitate the management and routing of an always higher amount of traffic, thanks to its network awareness capabilities and well-investigated level of programmability. Furthermore, the SDN concept has been proved to be an efficient technology to cope with a highly dynamic environment, such as the NGSO satellite networks. These two different but related aspects of network slicing constitute the main motivations of this chapter. In the following, a more detailed introduction to the two topics is presented.

In this chapter, section 13.2 gives an insight to the VNE problem and introduces a parallel formulation for the link mapping, one of the two subprocesses the VNE is subdivided into. The parallel computation over a set of demanded services is considered in order to prove the increase in performance. Afterwards, section 13.3 introduces the NV aspect for dynamic environment such as the integrated satellite/terrestrial networks. In this section, the enabling technologies are also presented. Section 13.4 discusses the VNE implementation in the built-in SDN testbed. More specifically, the testbed is emulating a simplified NGSO satellite network integrated with a terrestrial network. The VNE algorithm computes the real-time embedding for the dynamic physical network over the time. Those services which have been embedded over satellites are expected to undergo some recomputations due to the time-limited visibility between the ground station and NGSO satellites. An SDN controller has been implemented to support the interaction between the algorithm and the substrate network. Finally, section 13.5 concludes this chapter with an analysis of the performance for both the theoretical simulations and the practical results from the integration between the algorithm and the testbed.

13.2 Virtual network embedding

VNE is proved to be an NP-hard (Non-deterministic Polynomial-time Hard) problem [5]; hence, this brings to an unavoidable trade-off in the solution between complexity/computing time and optimality. As previously mentioned, given that substrate network and VNRs are generally described by a graph composed of nodes and links, VNE problem is the composition of the two following resource mappings. The mapping of virtual nodes onto the substrate ones is defined as node mapping, and the embedding of virtual links onto the physical ones is called link mapping. From a mathematical point of view, these mappings correspond to functions. Therefore, we can describe them in the following way. Given N_S the set of substrate nodes, N_V the set of virtual nodes, E_S the set of substrate links, and E_V the set of virtual links, the node and link mapping functions, M_N and M_L, respectively, are described as

- $M_N : N_V \longrightarrow N_S$
- $M_L : E_V \longrightarrow E_S$

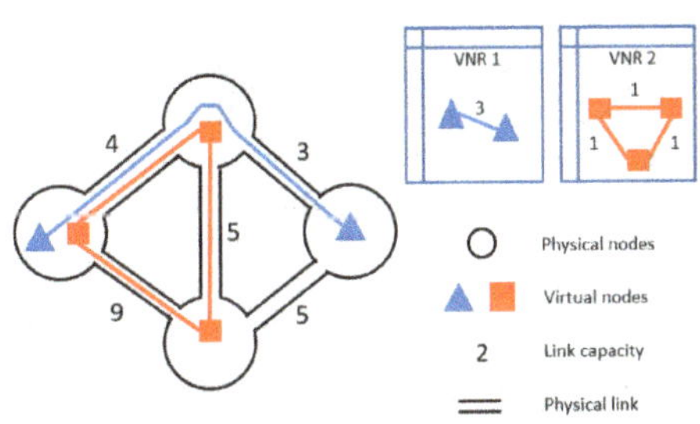

Figure 13.1 Illustration of VNE

Figure 13.1 shows a simplified example of VNE where two VNRs, with their respective resource requirements, are mapped to the same substrate network.

Due to the limited physical resources, the mapping problem is a challenging field in the telecom sector, and with the increase in network services' demand, it became more and more complex and attractive. This is the reason why, in the literature, many related works can be found, surveys such as Reference 6 and several proposed and different solutions, which will be taken into consideration later on in this section. The complexity of the problem has inspired different perspectives and solving methods. Indeed, depending on the importance given to time and quality of the solution, a different approach can be proposed. Before describing in more details the state-of-the-art for VNE, it is important to underline some main features of this problem [6]. A VNE algorithm is usually defined as:

- Static if the mapping is fixed and does not accept any modification or dynamic when variations of previously computed mappings (process of resources' reallocation) are considered, for example, to improve the average network load and physical resources exploitation.
- Centralized or distributed depending on whether the decisions are taken from a centralized entity or in a more distributed way.
- Concise or redundant whether the algorithm assigns the resources in a strict or a more failure tolerant way, respectively. A redundant VNE assigns some additional resources to each service in order to have some back-ups in case of real-time failures.

Since these features are considered independent, every mapping algorithm can be developed following any combination of them.

As previously mentioned, VNE involves the assignment of several node- and link-related resources. It is worth underlining that the physical characteristics of each of the substrate and virtual resources should be considered. For instance, algorithms have to tackle with consumable and static resources. Central Processing Unit (CPU), bandwidth, and delay are some of the main mapped consumable resources, i.e., they decrease with the increasing number of mappings. Consumable resources are released when the service is not available anymore. On the opposite, there are also static resources such as the link delay and the link loss probability, which are independent of the number of mappings. The efficiency of VNE solutions has become more and more important due to the scarcity of the available physical resources; the evaluation of the algorithm's performance, the satisfaction of the demanded requests, and the maximization of their QoS have become crucial. Consequently, in the following, some common objective functions/metrics are presented:

- The acceptance ratio which is the ratio between the amount of accepted and embedded VNRs over the total amount of received VNRs.
- The revenue/cost ratio computed as the ratio between the allocated requests over the substrate resources spent from the infrastructure provider (InP). It is

clear that the aim of the InP is to exploit its physical resources in the best possible way.

- Load-balancing which evaluates how much the traffic is spread over the substrate network. In some scenarios, this metric is very important because the more the traffic is spread, the less the physical links are overloaded, and the higher the probability of successfully coping with unexpected peaks of traffic.
- Energy saving which evaluates the ratio between the active substrate nodes (effectively involved in any mapping) over the total amount of the substrate nodes. This metric is the opposite of load-balancing.
- The computing time which plays a relevant role because it determines the interval of time between the arrival of a VNR and the instant at which it is served. This amount of time is very much dependent on the formulation of the problem, and the trade-off with the quality of the solution will be analyzed in this chapter.
- QoS metrics, such as delay, jitter, throughput, and network element (NE) utilization, should respect the required QoS of the VNR after the embedding is completed and over the entire duration of the service.
- Metrics to describe how reliable the VNE is, such as the presence of back-up resources, path redundancy and traffic migrations due to links' instability.

In general, VNE solutions present a subset of the above mentioned features. In the following subsections, an example of a proposed static, concise, and centralized solution for the VNE problem is presented, together with the state-of-the-art, motivations, and results. In the second part of the chapter, instead, a dynamic, concise, and centralized embedding algorithm is shown, where the time and substrate network variations come into play.

13.2.1 Related works on VNE

Many trends can be highlighted in the literature for VNE, and in the following, those that have motivated this work are presented. As previously mentioned, the VNE takes in consideration both node and link mapping. However, it is very common in the literature to find the optimization of the node mapping rather than link mapping [7–9]. After the computation of the node mapping, the link mapping is commonly computed with k-shortest paths or multicommodity flow algorithm. These mentioned algorithms are preferred due to their low-computing time, but they reduce the quality of the link mapping solution since they do not always provide an optimal solution in terms of metrics different from the path length. This chapter focuses to the link mapping computation, while the node mapping is already assigned (source and destination of the VNR).

In addition, a service mapping problem is by nature a sequential computation over time. In fact, as soon as a service provider is demanding for new substrate resources, the InP is asked to find available resources satisfying the QoS requirements. Therefore, VNRs usually enter and leave the network at different instants of time. Given these premises, this work intents to show that a sequential approach affects the performance of the mapping due to the fact that the solution, in a sequential case, is strongly dependent on the order in which the VNRs are elaborated with. In the network slicing era, the heterogeneity of services' requirements has brought the need to look for parallel solutions. Despite the increase in complexity, the parallel VNE problem formulation has become extremely relevant for the development of an efficient services mapping. It is widely known [10–12] that the sequential approach can achieve only a local near-optimal solution since the optimization is applied independently for each VNR. On the contrary, considering the simultaneous embedding of a set of VNRs would bring to a global near-optimal solution. A heterogeneous scenario, in terms of demanded resources, is considered to further demonstrate the efficiency of a parallel embedding.

Parallel computations have already been used in many different fields. Even for the VNE problem, some parallel implementations have been developed. In References 10, 11, authors focus on the relevance of the link mapping optimization rather than the node mapping and propose a parallel approach based on the genetic algorithm (GA). The use of GA seems to be the most common approach to solve a parallel VNE formulation. The parallel method proposed in References 10, 11, however, only considers the link mapping parallel computation (supposing of course that a VNR is composed by several links) and embeds one VNR at a time. The proposal presented in Reference 12 goes one step further, where authors proposed the use of GA for a parallel link mapping applied to a set of VNRs instead of just one. The idea is that the parallel computation can increase the coordination among all VNRs' embeddings and, consequently, achieve better performance. In the literature, among the proposed parallel approaches, GA seems to be the most promising one because it reduces the computing time, thanks to parallel computation, and, at the same time, improves the quality. The computing time is reduced due to the fact that parallel machines, exploiting a priori knowledge, are initially fed with a set of link mappings (typically k-shortest paths) from which they produce new feasible solutions. This approach is very efficient when the network is either static or dynamic with a priori known changes over time. However, this might not always be the case. For example, for very dynamic networks (e.g., low-earth orbit satellite networks), online embedding algorithms might be required in real time, especially if the connectivity variations are not predictable due to the unavailability of some embeddings.

Dynamic satellite networks constitute the second main part of this chapter. Indeed, while at the beginning the focus is on the resource mapping efficiency, considering a static scenario (stable physical connections without variations over time); sections 13.3 and 13.4 consider a satellite networks oriented scenario where the physical network varies over time, following a defined time development scheme, and we show how the VNE is interfaced with the physical network's variation.

13.2.2 Proposed VNE solution

The proposed solution is an integer linear programming (ILP) formulation for the link mapping, while considering the well-known load-balancing and energy-saving objective functions [13]. This scenario will be proved through some software simulations without involving the SDN-based testbed. In order to avoid any confusion in the terminology, it is worth underlining that, in this chapter, the terms sequential and parallel refer to how a set of VNRs is embedded, rather than the order of node and link mapping computation, as it is usually presented in the literature. Therefore, sequential approach, here, means that one VNR is embedded per time. On the contrary, in parallel approach, a set of VNRs is embedded per each iteration. This solution aims to illustrate that:

- The increase in the level of parallelism for the VNE computation corresponds to a higher quality embedding. In this case, the average substrate link utilization is the metric to evaluate the efficiency of the algorithm with load-balancing as the objective function;
- The impact of parallelism level and network size on the algorithm's computing time;
- The more the heterogeneity among the demanded services the more benefits the parallel approach will bring with respect to the sequential case.
- It might be worth to consider intermediate solutions for parallel computations because very good results can be provided without increasing too much complexity of the problem and, consequently, the computing time.

13.2.3 Problem initialization and formulation

The substrate network is modeled as a directed weighted graph: $G_S = (N_S, E_S)$, with N_S the set of substrate nodes and E_S the set of substrate edges. Every substrate node n_s is assigned with the power consumption $p(n_S)$, and every substrate link (u, v) is assigned with the capacity $c(u, v)$ representing the sum of all the substrate nodes' power consumption. The general n-th VNR is modeled as a directed graph: $G_v^n = (N_v^n, E_v^n)$, with N_v^n the set of virtual nodes and E_v^n the set of virtual edges. In addition, $bw(n)$ is the demanded bandwidth from the n-th VNR. For the n-th VNR, the node mapping is considered known a priori; hence, s^n and d^n are the source and destination substrate nodes, respectively. The node mapping is initially computed via the D-ViNE approach proposed in Reference 7. Finally, the set VN represents all the VNRs to be embedded. It is important underlining that an initial set of VNR(s) is supposed to be known. This is not necessarily always true, but some scenarios can be very realistic, for example, when some services have been embedded over an NGSO satellite link and suddenly that link is not available anymore. In this case, a set of already embedded services, which means their details are known, has to be embedded again.

In the following, the objective and constraints of the problem are presented. The objective function (13.1) represents the combination of load-balancing and energy-saving objectives. The load-balancing objective aims to spread the traffic in the substrate network as more efficiently as possible. To do so, it needs to minimize the overall load in each substrate link. On the contrary, energy saving minimizes the amount of power consumed in the network by all the active substrate nodes. The combined objective can be written as:

Table 13.1 Variables in (13.1)–(13.7)

Variable	Description
z^n_{uv}	Binary flow variable to indicate if the n-th VNR is embedded inuv
y_{n_S}	Binary variable to indicate an active node

$$\min_{z^n_{uv}, y_{n_S}} (\alpha \cdot \sum_{n \in VN} \sum_{uv \in E_S} \frac{z^n_{uv} \cdot bw(n)}{c(u,v)+\epsilon} + (1-\alpha) \cdot \frac{1}{p_{sum}} \cdot (\sum_{n_S \in N_S} y_{n_S} \cdot p(n_s))) \quad (13.1)$$

where the first term represents the load-balancing objective, and the second term defines the energy-saving objective. The parameter α and its complementary version are the weights for the load-balancing and energy-saving functions, respectively. A substrate node, hosting one or multiple virtual nodes, is considered active when traffic is passing through it. Therefore, with the energy-saving objective, more priority is given to the less power-consuming nodes in order to reduce the sum of the total consumption.

While load-balancing objective spreads the traffic, energy saving tends to concentrate it in order to reduce the amount of active nodes. Table 13.1 and Table 13.2 present the problem variables and parameters, respectively. In the following, the constraints of the problem are formulated and explained.

$$\sum_{n \in VN} z^n_{uv} \cdot bw(n) \leq c(u,v), \quad \forall (u,v) \in E_s \quad (13.2)$$

$$\sum_{w \in N_S} z^n_{s^n w} - \sum_{w \in N_S} z^n_{w s^n} = 1, \quad \forall n \quad (13.3)$$

$$\sum_{w \in N_S} z^n_{d^n w} - \sum_{w \in N_S} z^n_{w d^n} = -1, \quad \forall n \quad (13.4)$$

$$\sum_{v \in N_S} z^n_{vu} - \sum_{v \in N_S} z^n_{uv} = 0, \quad \forall n, \forall u \in N_s \setminus \{s^n, d^n\} \quad (13.5)$$

$$z^n_{uv} \in \{0,1\}, \quad \forall (u,v) \in E_S, \forall n, \quad (13.6)$$

Table 13.2 Parameters in (13.1)–(13.7)

Parameter	Description
s^n	Source node for the$n$-th VNR
d^n	Destination node for the$n$-th VNR
$bw(n)$	Demanded bandwidth by the$n$-th VNR
$c(u,v)$	Residual capacity of the substrate linkuv
$p(n_s)$	Power consumption of the substrate node n_s
p_{sum}	Total power consumption when all substrate nodes are active

$$y_{n_s} = \begin{cases} 1, & \text{if} \sum_n z^n_{un_s} > 0 \text{ or } \sum_n z^n_{n_s u} > 0 \\ 0, & \textit{otherwise} \end{cases} \tag{13.7}$$

In (13.2), for each substrate link (u,v), the sum of the bandwidth occupied by all virtual links mapped in (u,v) is upper-bounded by the residual capacity of the substrate link. Constraints (13.3), (13.4), and (13.5) ensure the flow's conservation law, which means that for each VNR, the overall flow is zero for all substrate intermediate nodes except for its source and destination nodes. Equations (13.6) and (13.7) define the binary constraints for the two variables of the problem. The algorithm is initiated with the actual status of the substrate network and the VNR(s) definition (graph description and demanded resources). If a feasible solution exists, the mapping is computed. Then, the current substrate network is updated, and the algorithm will run again for the following VNR.

13.2.4 Simulation setup

The objective of the simulation is to embed a predefined set, defined as VN, of generated VNRs. The set contains N VNRs to be embedded. Since this proposed work wants to show the advantage of parallel computations with respect to the sequential ones, from the main set VN, smaller subsets of cardinality K VNRs are created and given as input to the algorithm (K represents the parallelism level). This means that the algorithm embeds K VNRs at the same time. The problem is formulated and compiled using GNU Linear Programming Kit (GLPK) solver.

The value K is an integer parameter between 1 and N. When $K = 1$, the algorithm is sequential because it embeds one VNR per time. On the opposite, when $K = 30$, the algorithm is fully parallel. Intermediate values are also considered.

MATLAB is the main tool where all processes are managed, including the creation of substrate graph, the random generation of VNRs and their demanded resources, the report of the current status of the network over time, and the computation and update of the embedding results, coming from GLPK.

In the solver, the problem is formulated following the objective and constraints (13.1)–(13.7). If there is a solution, GLPK will return the results back to MATLAB to update the substrate network resources and compute the performance required. In total, the algorithm runs for N/K times. The pseudocode is presented in Table 13.3. The nodes of the substrate network are randomly distributed in a 100 × 100 grid.

Given u and v pair of substrate nodes, the probability to be connected is given by

$$P(u, v) = \gamma \cdot \exp\left(-\frac{d}{\delta \cdot d_{max}}\right) \tag{13.8}$$

where d is the geometric distance in the grid between the two nodes, and d_{max} is the maximum distance between any pair of substrate nodes. In addition, γ and δ are two design parameters which allow to control the complexity of the network. We use $\gamma = \delta = 0.5$ as the "simple case" considered in Reference 8. The capacity of each substrate link is randomly generated in the interval [60, 80]. The virtual bandwidth

Table 13.3 Parallel link mapping algorithm

Begin
Generation of substrate graph according to (13.8)
Generation of the set of VNRs randomly (Table 13.4)
Precomputed node mapping with D-ViNE algorithm
Subdivision in subsets with K VNRs
Input
Objective function
Substrate link capacity $c(u, v), \forall (u, v) \in E_s$
Power consumption $p(n_s), \forall n_s \in N_s$
$s^n, d^n, bw(n), \forall n \in VN$
Output
Link mapping for each VNR
If branch-and-cut method finds a solution
MATLAB updates the substrate resources
Else link mapping failed
End if

required from the n-th VNR is a random number in a defined interval. Different cases of this interval are considered (Table 13.4). Finally, a total amount of $N = 30$ (cardinality of the set VN) VNRs are considered for the embedding.

13.2.5 Performance evaluation

MATLAB with GLPK is used for the simulations. The substrate network for the results in Figure 13.2 and Figure 13.3 (a) is composed by 30 nodes while in Figure 13.3 (b) the amount of substrate nodes varies to demonstrate the scalability of the algorithm. In the first two case studies, we analyze the performance of the algorithm with the variation of the parallelism level K. The two extremes are the sequential case, with $K = 1$, and the fully parallel one with $K = 30$.

Figure 13.2 shows the advantage of the parallel computations over the sequential ones. In particular, Figure 13.2 (a) studies the efficiency of the parallel approach with different levels of heterogeneity among the demanded services. In this case, the heterogeneity is given by the demanded bandwidth. Three different scenarios are considered (Table 13.4), where the range for the generation of demanded bandwidth changes. The aim is to compare the performance in less heterogeneous

Table 13.4 VNR demanded bandwidth

Scenarios	(min, max)
Use case 1	(5, 20)
Use case 2	(5, 30)
Use case 3	(5, 40)

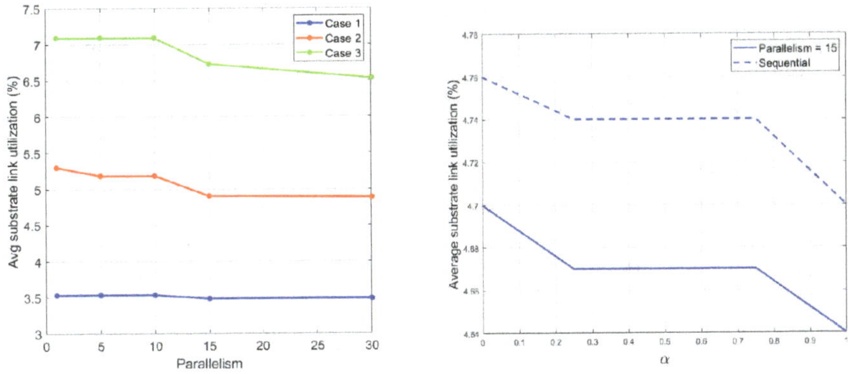

(a) Average link utilization vs Parallelism

(b) Average link utilization vs α

Figure 13.2 Simulation results for average substrate link utilization

scenarios (use case 1) to more heterogeneous scenarios (use case 3). The scenarios differ on each other in the random generation interval. In fact, the range of random demanded bandwidth changes for the different scenarios. It is proved that the efficiency of parallel embeddings improves with the increase in the range of virtual demanded bandwidth. This is due to the fact that if there is more heterogeneity in the demanded bandwidth, the algorithm is able to better manage the link mapping in order to achieve the objective with respect to the sequential approach. Therefore, the VNR which demands for more bandwidth will be embedded in the path with more available bandwidth. For example, for use case 1, since there is almost no difference among all VNRs, the efficiency of fully parallel approach compared to the sequential case is almost none. While in use case 3, the fully parallel case performs much better. The load-balancing objective is considered for this scenario, with average substrate link utilization as the metric. The efficiency of the parallel approach can also be seen in Figure 13.2 (b). In this case, different values of $\alpha \in [0, 1]$ are considered, with intervals of 0.25. With reference to the objective function (13.1),

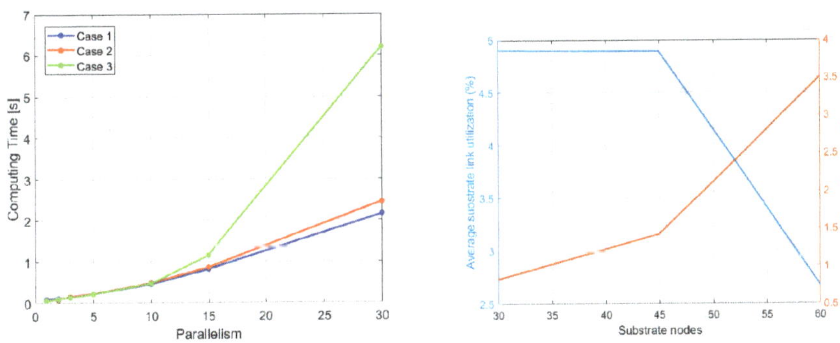

(a) Computing time vs Parallelism

(b) Computing time vs Substrate nodes

Figure 13.3 Simulation results for computing time

when $\alpha = 1$, the objective is load balancing. On the other hand, $\alpha = 0$ depicts the energy-saving objective. Intermediate values represent a weighted combination of those two objectives. For the purpose of the comparison, a level of parallelism is fixed to 15 and compared to the sequential case.

For all values of α, the parallel approach (solid lines) always performs better (lower values) than the sequential case (dashed lines). In general, we can also appreciate that, for lower values of α, the average link utilization is higher because the energy-saving objective drives the performance. This means that more priority is given to the minimization of the number of active nodes, and the more the traffic is concentrated. This brings to higher average substrate link utilization.

Figure 13.3, instead, shows an analysis of the computing time. Given the NP-hardness of the VNE problem, its computing time is expected to grow exponentially with the increasing size of the problem. In this solution, the size of the problem is increased by three different factors.

- The demanded load in the network;
- The increasing level of parallelism (more VNRs embedded per time);
- Higher number of substrate nodes and links in the network.

Figure 13.3(a) shows the exponential relation between the computing time (y-axis) and the parallelism level (x-axis). For all presented cases, a main trend can be highlighted. The computing time and, consequently, the complexity increase exponentially with the increase of the level of the parallelism. Figure 13.3(a) shows also that, given a certain level of parallelism (e.g., 15), the higher is the load (i.e. use case 3), the more computing time will be needed. It can also be noticed that when the parallelism level is high, the complexity of the problem is very sensitive to the load of the network. Indeed, the gain from the higher level of parallelism in use case 3 (highest load), with respect to use case 1 and use case 2, is quite significant when the parallelism level is the maximum, i.e., 30 with respect to lower values such as 15 or 10. The third factor which increases the complexity of the problem, as depicted in Figure 13.3(b), is the scale of the substrate network. Indeed, the algorithm is tested with a higher number of substrate nodes, and as expected, the computing time (right y-axis) increases due to the increased size of the network, which raises the complexity of the problem. For these results, the level of parallelism has been set to 15, and load balancing has been considered. The average link utilization (left y-axis) decreases with the increase in the number of nodes because the traffic to be embedded is still the same, but there are more available substrate resources due to the larger size of the physical network.

13.3 NV approach for dynamic SDN-based satellite-terrestrial networks

As the integration of satellite-terrestrial networks is becoming more and more attractive, the introduction of new technological enablers to facilitate the process has become unavoidable. Satellite and terrestrial networks traditionally have been considered independent systems, hampering their interoperability, scalability, and programmability, making it impossible to dynamically execute virtualization schemes. To solve this, in recent years, SDN has been employed as part of seamless terrestrial-satellite integration for the applicability of virtualization schemes, considering a federated control for managing heterogeneous network segments for multiple network infrastructures (e.g., 5G and next generations) [14–16]. Compared to the traditional multiprotocol label switching/Traffic Engineering (MPLS/TE) mechanisms used in today's transport networks, the centralized SDN framework for the realization of TE solutions allows a holistic view of the network, accompanied by mechanisms to enforce network policies in a centralized way. In this line, relevant advances have been carried out for the analysis of the potential use cases, requirements, and definitions of functional frameworks for the exploitation of SDN/network function virtualization (NFV) technologies in satellite networks [17–19]. Some developments include network architecture designs for the exploitation of SDN/NFV technologies for the seamless integration of satellite-terrestrial 5G networks [19, 20]. Some works have developed SDN/NFV practical applications like testbeds/proof-of-concepts [21–24]. As we show later in section 13.4, the SDN programmability also facilitates the automated execution of virtualization schemes for this type of networks, even under highly dynamic scenarios. As this chapter discusses the implementation of the SDN for E2E network slicing, we leave to further implement the introduction of the implemented concepts into the NFV management and orchestration architecture [25].

13.3.1 SDN: a network slicing enabler

Traditionally, virtualization in legacy networks consisted of establishing overlay networks, where a small set of nodes uses tunnels to form their own topology on top of the network. The overlay networks were made through labeling packets when entering the network, encapsulating and sending them through the network and decapsulating them when leaving the network (e.g., MPLS networks). This process involves several disadvantages, such as the addition of headers to packets, the manual configurations by administrators, and the need for additional network equipment, running specific routing protocols, which reduce the efficiency and increase the overall costs for the integration of new technologies. On the opposite, SDN allows to add, update, and delete flow entries in flow tables (routing tables), through the OpenFlow (OF) protocol, with a faster and more scalable process; for each network node (OF-enabled switch), the SDN controller is connected. Each flow entry consists of matching fields, counters, and a set of instructions and actions (e.g., packet forwarding action) to apply to matching packets. In this regard, the matching rules can be configured based on multiple packet header fields such as ingress port, source/destination medium access control-Internet protocol (IP) addresses, etc. This information allows to identify each packet in the network (eventually which VNR belongs to), differentiating the routing schemes, according to the VNR's Service Level Agreement. To ease the notations, in this section, the VNR is simply denoted by VN.

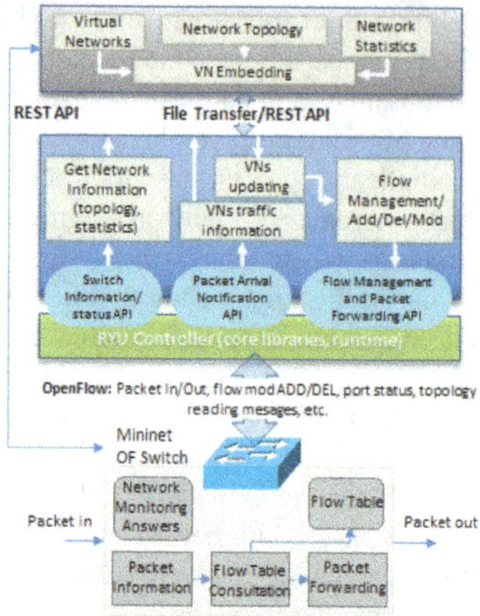

Figure 13.4 Illustration of an SDN-based TE application for virtualization

13.3.2 SDN-based TE application approach for VNs

An illustration of an SDN-based TE application is shown in Figure 13.4. The flows are configured by the SDN controller. The controller exposes application programming interfaces (APIs) to deploy network management and control applications. APIs are a collection of programming libraries that give access to the previously mentioned mechanisms supported by OF protocol. The OF switches information such as network topology, network state (e.g., switch status, port status, traffic load per port/flow, etc.), and flow table information (e.g., flows information) are visible at the application level. The exposed API capabilities allow to program and demonstrate the operation of an SDN-based TE application, which is able to set dynamically the embedded VNs in the substrate network and deliver the input information to the VNE algorithm module. More specifically, the implemented SDN-based TE application is able to: (1) learn the network topology, (2) real-time monitor the network/port status, (3) create monitoring network statistics, (4) identify the VN's new traffic based on a set of user information (e.g., origin IP address) at input/output ports, (5) set the forwarding path based on a VN embedding information by populating the flow tables of the OF switches, and (6) enforce the maximum rate per VN with rate limiters.

13.4 Implementation of an SDN-based testbed for dynamic VNE

In this section, an illustrative example of practical implementation for a VNE algorithm is presented to validate the feasibility of an SDN-based implementation in real networks for hybrid satellite-terrestrial networks under highly dynamic scenarios.

The goal of this section is to show preliminary results for this kind of dynamic networks, with the focus on the SDN support to manage real-time network slicing allocation. Consequently, the considered use case is a simplified version of a real satellite constellation. The proposed SDN-based testbed executes a dynamic installation of VNs. To achieve this, some essential capabilities have been added to the system such as maximum rate limiters per VN, dynamic network topology learning, and network status changes reports.

13.4.1 Experimental testbed

A high-level view of the experimental testbed is depicted in Figure 13.5. The testbed comprises a Personal Computer (PC) that hosts the network emulator for the SDN-enabled hybrid satellite-terrestrial network, composed of OF switches. Satellite and terrestrial links are differentiated by the emulated latency. Any link in the substrate network can be programmed to be periodically modified in order to simulate the typical non-GEO satellites orbital movements. A second PC hosts both the external Ryu SDN controller and the VNE algorithm script, which runs in MATLAB. During the emulation, the VNE algorithm simulates the random arrival of VNRs (Poisson distribution). Each arrival triggers the computation of the VNE algorithm. The embedding information for each VN consists of two end nodes (with host connection capability), the lifetime defined as the time that the VN will be active, the maximum rate per VN, and the list of hops of the computed path between source and destination. Based on the embedding information, the TE application, programmed in the SDN controller, creates the path for each VN, sets the rate limiters, obtains the network information statistics, reads the topology, and monitors any network topology variation in real time. Each VN is considered as a simple E2E service, composed of two nodes and only one path. The node mapping of the two endpoints is defined a priori, thus, only the link mapping has to be computed. The ILP formulation for the link mapping, presented in section 13.2, with the load-balancing objective function,

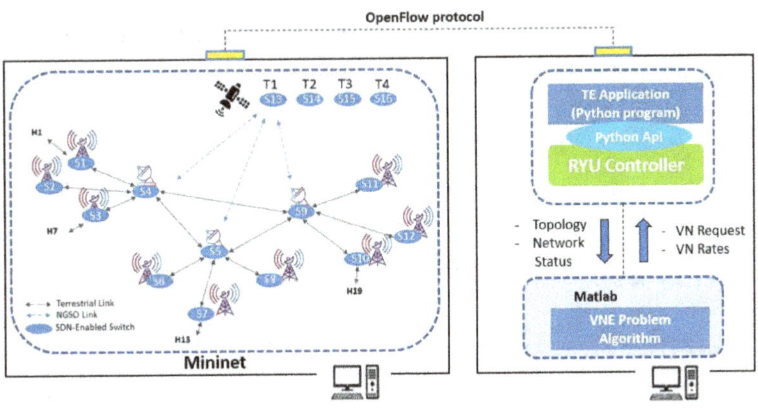

Figure 13.5 High-level view of the SDN-based testbed components

is used. As mentioned earlier, load-balancing function aims at reducing, as much as possible, the average bandwidth utilization in the substrate network.

13.4.2 Operational validation

The testbed operation is validated through the execution of three illustrative examples. The first one, for the VNs installation, validates the implemented SDN-based TE application to enforce a desired routing scheme, over the satellite-terrestrial network, according to the VNE algorithm's output. The second one presents the dynamic VN reconfiguration given to the NGSO network's variation over time. In this scenario, when the topology changes, the VNE is recomputed for the VN(s) affected by the variation. Finally, the third example presents the dynamic configuration when several established VNs experience a link failure. For demonstration purposes, we consider the substrate network illustrated in Figure 13.5, emulated on Mininet. The switches S13–S16 represent Medium Earth Orbit (MEO) satellites, while the switches S4, S5, and S9 are the backhaul NEs with satellite link availability. The remaining nodes simulate the access NEs with terrestrial hosts connectivity. The available capacity on terrestrial and satellite links is set to 800 and 400 kbps, respectively. For each satellite link, a latency of 27 ms is also introduced to simulate MEO links.

13.4.2.1 VNs implementation

The first example considers the embedded VNs depicted in Figure 13.6 (a). It can be observed that in node S13, four forwarding rules are created for VN1 and VN2, to properly forward their traffic over the assigned paths. For each VN, one flow rule is needed to forward the traffic from source to destination and the second one from destination back to source. The same process is repeated for the second VN with different output ports since the traffic of VN2 is forwarded to switch S5, instead of S9. The TE application manages the installation of the right forwarding rules by recognizing the incoming traffic and assigning it to the VN it belongs to. Once the packet is assigned to the VN, the correspondent forwarding rules (output port) are installed.

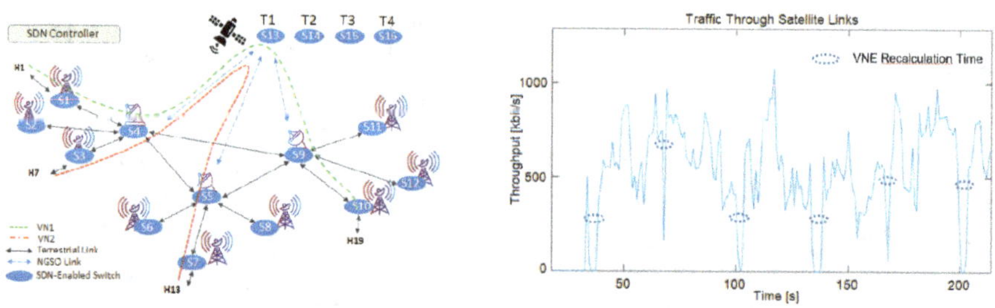

(a) VNs Configuration (b) Traffic load over satellite links

Figure 13.6 Dynamic emulation scenario

13.4.2.2 Dynamic VNE

a) Dynamic non-GEO satellite-terrestrial topology: For the dynamic VN computation/establishment procedure, the testbed is started with the configuration depicted in Figure 13.6 (a), at time T1, where only satellite S13, among the NGSO nodes, is available. Consequently, the two VNs are embedded through the satellite link (S13). Every 30 s, the current satellite connection is deactivated, and the following satellite link is activated in order to emulate a MEO satellite constellation. This process is repeated among nodes S13–S16 consecutively. It is worth clarifying that the selected interval of time does not represent a real line of sight (LoS) duration between a MEO satellite and a ground station, but a reasonable low value is selected for demonstration purposes. To validate the dynamic configurations, a video streaming is started, and User Datagram Protocol traffic begins to flow between source and destination hosts. Every time a satellite link status changes, the two involved OF switches (the one not visible anymore and the new one) send a notification to the controller ("OFPT_PORT_STATUS" message) with the "port down" flag, triggering the process for the VNE recalculation. Figure 13.6 (b) presents the generated traffic by VN1. Every 30 s, the throughput goes down to 0 kbps or similar values, but after a small interval of time, it is restored to initial values. For this scenario, despite the link drops are predictable and the traffic drops can be avoided, for demonstration purposes, the testbed recalculates the VNE after a topology change notification.

b) Dynamic VNE recomputations: The scenario consists of six VNs, with a required rate of 200 kbps each. Three of them are depicted in Figure 13.7(a). Under this configuration, we simulate a terrestrial backhaul link failure between the nodes S4 and S9. Then, we observe the system's reaction in order to reconfigure the VNs. After the terrestrial link fails, we observe, as illustrated in Figure 13.7(b), three different embedding reconfigurations. The VN1 (S1-S10) changes from the failed terrestrial path to a different terrestrial one, the VN4 (S3-S6) keeps the same terrestrial path since it is not affected by the failed link, and the VN5 (S3-S10) is migrated from the failed path to a satellite backhaul path. This configuration can also be validated by the measured latency presented in Figure 13.8. The figure illustrates the round trip time (RTT) for each VN before and after the terrestrial link failure (after 70 s). It is worth underlining that each time there is a change in the topology, the flows in the network are deleted to establish the new VN paths. This has an unavoidable impact

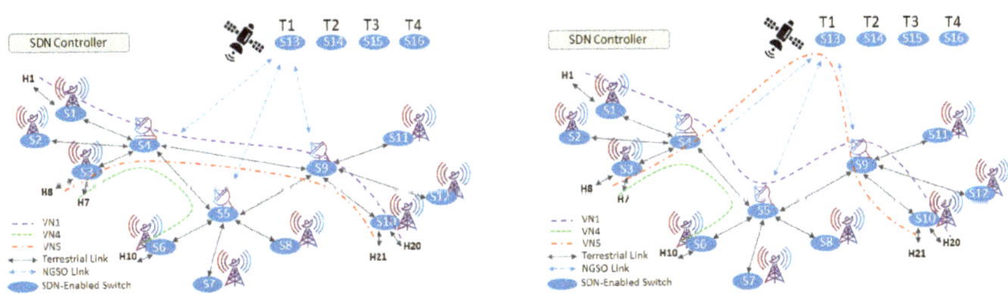

(a) before a link failure

(b) after a link failure

Figure 13.7 VNE configuration

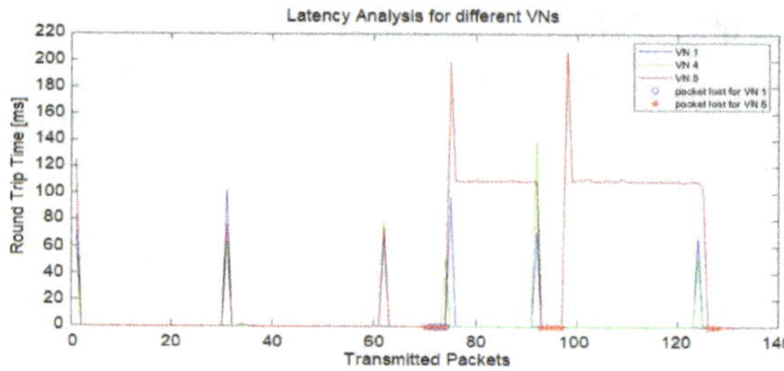

Figure 13.8 Latency by VNs

on the RTT, as Figure 13.8 shows. In fact, the peaks of RTT, visible each time a satellite link changes or a terrestrial link failure occurs, represent the time taken by the system to create the flows in all switches for the new VN's path.

13.5 Conclusions

This chapter has considered a well-known NP-hard problem, known under the name of VNE. Two different but related aspects of VNE have been analyzed. We first proposed and described a theoretical formulation of the link mapping for VNE, highlighting the advantages of using a parallel computation rather than a sequential one. The theoretical results have shown that, for a static scenario, the parallel computations are more efficient than the sequential ones. Furthermore, the gain obtained from parallel computations is directly proportional to the heterogeneity level within the demanded resources. This means that the more heterogeneity within the demanded requests, the more the final performance will benefit from parallel computations. In addition, we have validated the VNE algorithm over a real testbed. The main objective is to show the real-time interaction between a VNE algorithm and a time-varying physical network. For this purpose, a dynamic physical network has been emulated over the testbed using Mininet. Thanks to the SDN controller, the testbed has proved to correctly cope with the dynamic scenario of the NGSO network where connectivity's variations may happen due to either their nature of temporal LoS visibility between terrestrial segment and satellites or for link degradation/failure. Indeed, during the simulation, the combined work of the VNE algorithm together with the SDN controller has allowed the connection between the endpoints of the service, almost without interruptions. It is worth underlining that there are not many theoretical works on VNE which provide also a practical implementation of the proposed algorithm. These experimentation results demonstrated the efficiency of the SDN implementation for this kind of algorithms, especially when many recomputations are expected to happen, such as for very dynamic networks.

References

[1] Afolabi I., Taleb T., Samdanis K., Ksentini A., Flinck H. 'Network slicing and softwarization: A survey on principles, enabling technologies, and solutions'. *IEEE Communications Surveys & Tutorials*. 2018, vol. 20(3), pp. 2429–53.

[2] Lei L., Yuan Y., Vu T.X., Chatzinotas S., Minardi M., Montoya J.F.M. 'Dynamic-adaptive ai solutions for network slicing management in satellite-integrated B5G systems'. *IEEE Network*. 2021, vol. 35(6), pp. 91–97.

[3] Kodheli O., Lagunas E., Maturo N, *et al.* 'Satellite communications in the new space era: A survey and future challenges'. *IEEE Communications Surveys & Tutorials*. 2021, vol. 23(1), pp. 70–109.

[4] Al-Hraishawi H., Chatzinotas S., Ottersten B. 'Broadband non-geostationary satellite communication systems: research challenges and key opportunities'. *2021 IEEE International Conference on Communications Workshops (ICC Workshops)*; Montreal, QC, Canada, 2021. pp. 1–6.

[5] Rost M., Schmid S. 'On the hardness and inapproximability of virtual network embeddings'. *IEEE/ACM Transactions on Networking*. 2020, vol. 28(2), pp. 791–803.

[6] Fischer A., Botero J.F., Beck M.T., de Meer H., Hesselbach X. 'Virtual network embedding: A survey'. *IEEE Communications Surveys & Tutorials*. 2013, vol. 15(4), pp. 1888–906.

[7] Chowdhury M., Rahman M.R., Boutaba R. 'ViNEYard: virtual network embedding algorithms with coordinated node and link mapping'. *IEEE/ACM Transactions on Networking*. 2012, vol. 20(1), pp. 206–19.

[8] Pham M., Hoang D.B., Chaczko Z. 'Congestion-aware and energy-aware virtual network embedding'. *IEEE/ACM Transactions on Networking*. 2020, vol. 28(1), pp. 210–23.

[9] Cao H., Zhu Y., Zheng G., Yang L. 'A novel optimal mapping algorithm with less computational complexity for virtual network embedding'. *IEEE Transactions on Network and Service Management*. 2018, vol. 15(1), pp. 356–71.

[10] Nguyen K.T.D., Huang C. 'An intelligent parallel algorithm for online virtual network embedding'. *2019 International Conference on Computer, Information and Telecommunication Systems (CITS)*; Beijing, China, IEEE, 2019. pp. 1–5. Available from https://ieeexplore.ieee.org/xpl/mostRecentIssue.jsp?punumber=8850888

[11] Lu Q., Huang C. 'Distributed parallel vn embedding based on genetic algorithm'. *2019 IEEE Symposium on Computers and Communications (ISCC)*; Barcelona, Spain, 2019. pp. 1–6. Available from https://ieeexplore.ieee.org/xpl/mostRecentIssue.jsp?punumber=8961334

[12] Zhou Z., Chang X., Yang Y., Li L. 'Resource-aware virtual network parallel embedding based on genetic algorithm'. *2016 17th International Conference on Parallel and Distributed Computing, Applications and Technologies (PDCAT)*; Guangzhou, China, 2016. pp. 81–86.

[13] Minardi M., Sharma S.K., Chatzinotas S., Vu T.X. 'A parallel link mapping for virtual network embedding with joint load-balancing and energy-saving'. *2021 IEEE International Mediterranean Conference on Communications and Networking (MEDITCOM)*; Athens, Greece, 2021. pp. 1–6.

[14] Giambene G., Kota S., Pillai P. 'Satellite-5g integration: A network perspective'. *IEEE Network*. 2018, vol. 32(5), pp. 25–31.

[15] Xu S., Wang X.-W., Huang M. 'Software-defined next-generation satellite networks: architecture, challenges, and solutions'. *IEEE Access: Practical Innovations, Open Solutions*. 2018, vol. 6, pp. 4027–41.

[16] Li T., Zhou H., Luo H., Yu S. 'SERvICE: A software defined framework for integrated space-terrestrial satellite communication'. *IEEE Transactions on Mobile Computing*. 2016, vol. 17(3), pp. 703–16.

[17] Bertaux L., Medjiah S., Berthou P. 'Software defined networking and virtualization for broadband satellite networks'. *IEEE Communications Magazine*. 2016, vol. 53(3), pp. 54–60.

[18] Ahmed T., Dubois E., Dupé J.-B., Ferrús R., Gélard P., Kuhn N. 'Software-defined satellite cloud ran'. *International Journal of Satellite Communications and Networking*. 2018, vol. 36(1), pp. 108–33. Available from http://doi.wiley.com/10.1002/sat.v36.1

[19] Ferrús R., Koumaras H., Sallent O, *et al.* 'SDN/nfv-enabled satellite communications networks: opportunities, scenarios and challenges'. *Physical Communication*. 2016, vol. 18(2), pp. 95–112.

[20] *VIrtualized hybrid satellite-terrestrial systems for resilient and flexible future networks. VITAL, project*. 2015. Available from https://cordis.europa.eu/project/id/644843

[21] Mendoza F., Minardi M., Chatzinotas S., Lei L., Vu T.X. 'An SDN based testbed for dynamic network slicing in satellite-terrestrial networks'. *2021 IEEE International Mediterranean Conference on Communications and Networking (MEDITCOM)*; Athens, Greece, 2021. pp. 1–6.

[22] Vergütz A., G Prates N., Henrique Schwengber B., Santos A., Nogueira M. 'An architecture for the performance management of smart healthcare applications'. *Sensors (Basel, Switzerland)*. 2020, vol. 20(19), p. 19.

[23] Esmaeily A., Kralevska K., Gligoroski D. 'A cloud-based SDN/NFV testbed for end-to-end network slicing in 4G/5G'. *2020 6th IEEE International Conference on Network Softwarization (NetSoft)*; Ghent, Belgium, 2020. pp. 29–35. Available from https://ieeexplore.ieee.org/xpl/mostRecentIssue.jsp?punumber=9158198

[24] Meklkis P.-V., Ramantas K., Antonopoulos A, *et al.* 'NFV-enabled experimental platform for 5G tactile internet support in industrial environments'. *IEEE Transactions on Industrial Informatics*. 2020, vol. 16(3), pp. 1895–903.

[25] ETSI 'Network functions virtualisation (NFV); architectural framework'. *ETSI Gs NFV*. 2013, vol. 2, p. 2.

Chapter 14

3rd Generation Partnership Project integration of non-geostationary orbit satellites

Thomas Heyn[1], Arman Ahmadzadeh[1], and Alexander Hofmann[1]

In recent years, an ever-growing connectivity demand is experienced in wireless communications. Practically, everyone and everything need to be connected. This trend is supported by the rich variety of applications available on the market. This is a challenging situation for the terrestrial telecommunications infrastructure requiring extensions in the system architecture. Therefore, the 3rd Generation Partnership Project (3GPP) started in 2017 to study the integration of satellites as a part of the 5G ecosystem involving both cellular and satellite stakeholders. The substantial value added by the satellite segment as part of the access technologies mix for 5G is now becoming clear, especially for mission critical and other applications where ubiquitous coverage is crucial [1]. For example, 5G nonterrestrial networks (NTNs) can broaden service delivery to unserved or underserved areas by complementing and extending terrestrial networks.

Satellite integration has been accepted as part of the 5G New Radio (NR) roadmap in the cellular standardization organization 3GPP, and for the first time since its establishment, 3GPP supports satellites. After two study items (SIs), specification work in Release 17 is now ongoing to define the necessary adaptations of the 5G NR Standard to support satellite use cases. The use of satellite-based networks to provide connections to different user equipment (UE) is referred to in the 3GPP community as 5G NTN. In such a network, the satellite employs either a transparent (bent pipe) payload or a regenerative payload and can be placed into geostationary Earth orbit (GEO), medium Earth orbit (MEO), or low Earth orbit (LEO). Transparent satellite works as a relay between the UEs and the base station, also known as next-generation NodeB (gNB), implemented on the gateway (GW) side. In contrast, the regenerative satellite acts as either an entire flying gNBs or a gNB-DU (distributed unit). In case of a gNB-DU, many different options are available to split the functionality of an entire gNB into a gNB-CU on ground and gNB-DU inside the satellite depending on the layer to be split.

[1]Fraunhofer IIS, Erlangen, Germany

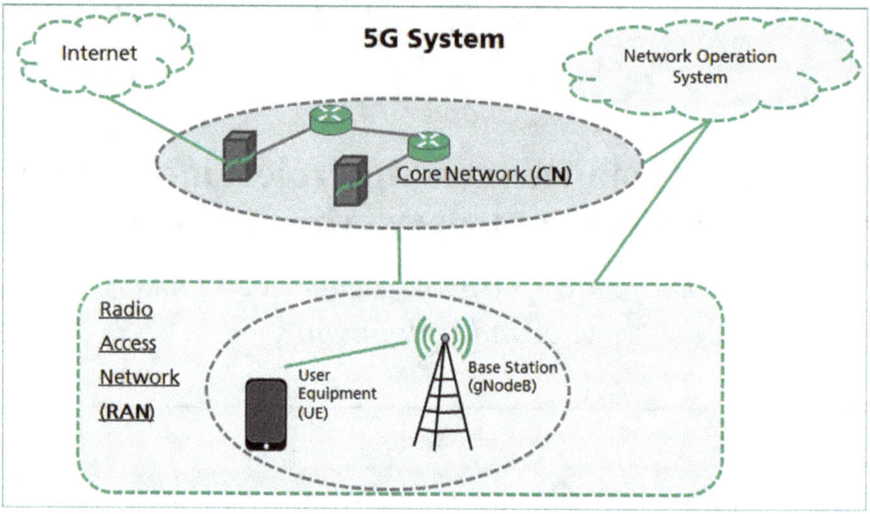

Figure 14.1 5G architecture

3GPP Release 17 is a crucial working point in the 5G standardization groups in order to develop and to approve the technical specifications (TS) to enable direct access technologies via satellite links. The vision is to deploy NTN as part of 5G by 2025 in order to meet the challenges of mobile network operators in terms of ability to reach, availability, and resilience. The satellite communication industry is gaining increasing interest in 5G NTN, and several companies contribute to the 3GPP standardization process. In this chapter, we will elaborate on the current standardization of NTN in 5G especially for non-geostationary orbit (NGSO) satellites.

14.1 5G system and 3GPP procedures

In order to understand the satellite integration into 5G systems, an overview of the 5G system architecture is given. A 5G system is consisting in general of two different parts (Figure 14.1): core network (CN) and radio access network (RAN). The CN in 5G is a cloud-oriented and service-based architecture that handles network-related functions of the system, e.g., authentication, subscriber management, security. The CN also interfaces to the data network, e.g., Internet, the operator network, and the RAN. The RAN is responsible for the radio access of the UE using radio frequencies via a base station, which in the context of 5G is called gNB.

In the 3GPP organization, the standardization process is split into three different working groups, so-called technical specification groups, which handle and standardize the different parts of the 5G system:

- Service and system aspects (SA)
- CN and terminals (CT)
- RAN

Overall 5G system-related aspects are handled in the "SA" group, CN-related aspects are handled in the "CT" group, and the radio access-related aspects are handled in the "RAN" group.

In general, within the 3GPP new topics will be studied first during a so-called SI phase with a technical report capturing the outcome of the SI. Within the SI, all expected challenges will be identified as well as possible technical solutions, which might overcome the identified issues. After an SI has successfully finished, a work item (WI) phase will be started to work on the identified solutions and assess the necessary changes to the standard. As a result of that, the outcome of the WI will be incorporated into the final specifications of 5G called TS.

14.2 Architecture options for 5G NTN

In 5G, generally three different architecture options are possible: backhauling, indirect access, and direct access to include satellites, which are described in the following sections.

14.2.1 Backhauling

The more classical approach, which could be also realized in 4G networks, is the satellite backhauling. Here, the satellite connects a terrestrial RAN (gNB) with the 5G CN via a satellite GW station as seen in Figure 14.2.

14.2.2 Indirect access

A second architecture option is the connection of a so-called "relay" on ground via a 5G satellite RAN, which then connects classical UEs (cf. Figure 14.3). A relay is a special type of terminal in 5G, which was not yet completely specified as the NTN

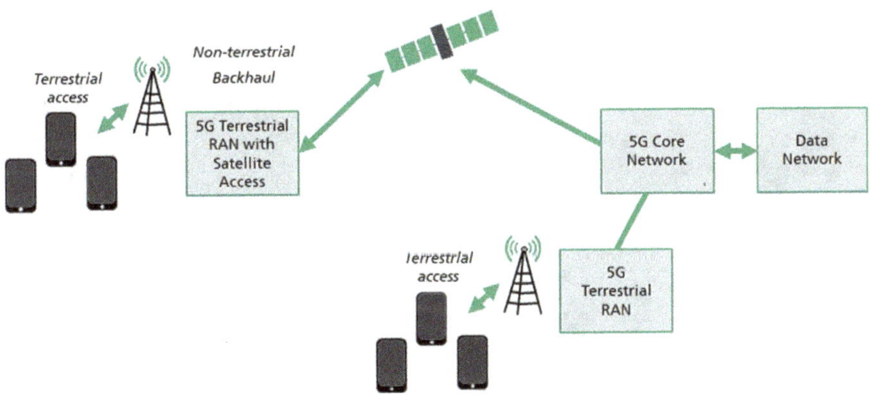

Figure 14.2 Backhauling architecture in 3GPP

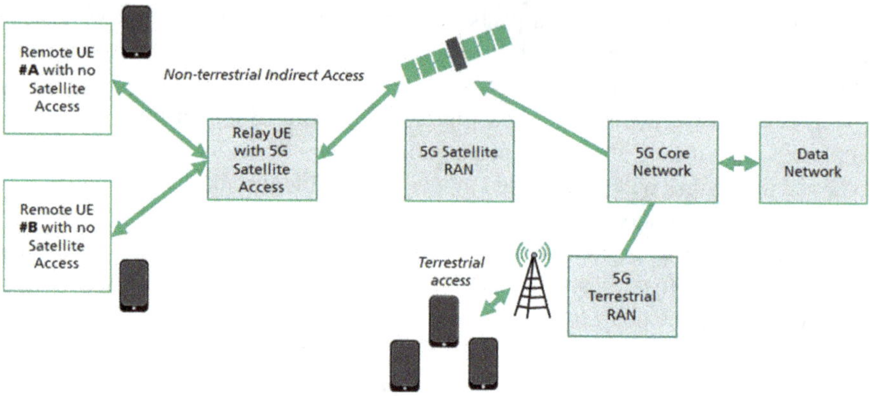

Figure 14.3   Indirect access architecture in 3GPP

SI in the RAN groups, has been started. This is why in the current Release 17, this type of architecture could not yet been taken into account. There is a possibility that we will see this type of architecture to be included in future releases, e.g., in Release 18.

14.2.3   Direct access

The direct access, as the name is pointing out, connects UEs directly via 5G satellite RAN (cf. Figure 14.4). Within the RAN group this architecture option is the focus, as here the most technical challenges are arising, which are currently being solved in Release 17. As this architecture was followed with the highest priority in the RAN working groups, we will focus on the direct access architecture in the next sections.

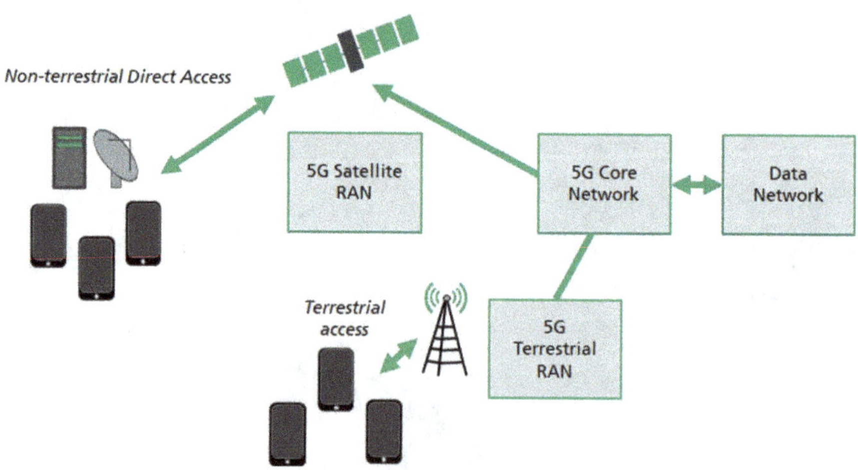

Figure 14.4   Direct access architecture in 3GPP

14.3 Standardization of NTNs in 5G

The use of satellite-based networks to provide connections to different UE is also referred to as 5G NTN in the 3GPP community. These so-called NTNs cover satellites including high-altitude platform stations (HAPS) like balloons or unmanned aerial vehicles. The satellite employs either a transparent (bent pipe) payload or, in future, 3GPP releases a regenerative payload and can be placed into GEO, MEO, or LEO. The HAPS operate at a height between 8 and 50 km. A transparent satellite works as a relay between the UEs and the base station, also known as gNB, implemented on the GW side on ground. In contrast, a regenerative satellite with onboard processor acts as a flying gNB, with a backhaul link to the 5G CN on ground.

Figure 14.5 gives an overview of the SIs and WIs within the 3GPP in the SA and RAN working groups over the different releases. Start of the first SIs in SA and RAN has been in Release 15.

In the 3GPP SA working group, a study on using satellite access in 5G summarizes the use cases including satellite in Reference 1. Another study report [2] "Study on architecture aspects for using satellite access in 5G" details the role of satellite links in 5G networks. From an architectural point of view, satellites either act as backhaul for gNBs on ground or provide direct access with 5G NR to UEs. The Release 17 WI in working group SA2 [3] specifies the integration of satellites in 5G both for backhaul and for direct access.

The satellite direct access with 5G NR has been accepted to the roadmap in the cellular standardization organization 3GPP in the RAN working groups. After finalizing two SIs in Release 15 [4] and Release 16 [5], the 3GPP RAN working group currently specifies the extension of 5G NR to support NTNs as part of the Release 17. For the first time, satellite communication with direct access will be supported by the 3GPP standards, which were formerly limited to terrestrial cellular networks. The RAN WI [6] covers a frequency range from 2 to 30 GHz and GEO, MEO, and LEO satellite constellations and states that these extensions are implicitly compatible with HAPS. Different terminal types are considered, either smartphone type with

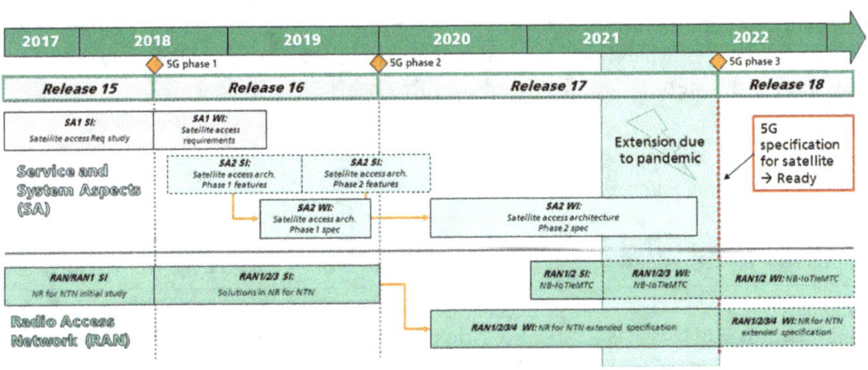

Figure 14.5 3GPP roadmap for 5G NTN

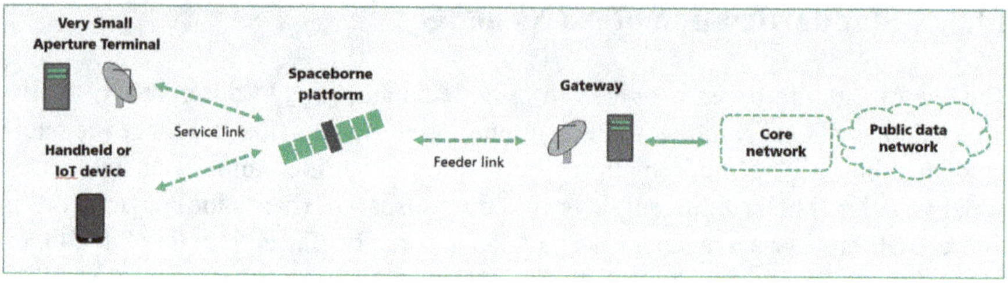

Figure 14.6 5G NTN architecture in 3GPP for very small aperture terminals as well as for handheld and Internet of Things (IoT) devices

the regular transmit power of 200 mW (UE Power Class 3) and omni-directional antennas or very-small-aperture terminals (VSAT) with directional antennas as seen in Figure 14.6.

3GPP has identified in Reference 5 four NTN reference scenarios, which are depicted in Table 14.1.

In Release 17, the focus of the standardization works has been on transparent (nonregenerative) payload satellite systems for both LEO and GEO scenarios. For LEO, satellite fixed (C2, D2) or steerable (C1, D1) beams result, respectively, in Earth moving or Earth fixed beam footprint (and thus NR cells) on the ground. Scenarios, where a LEO constellation generates Earth moving beams (C2, D2), are challenging to deploy due to frequently required satellite handovers due to the fast speed of the satellites at approximately 7.5 km/s. However, Earth fixed beam scenarios are especially suited for narrow beams and broadband handheld applications, thanks to well-proven phased array technology on board satellites.

A detailed link budget analysis for various system constellations as a combination of GEO and LEO satellites, VSAT and handheld terminals, and frequency bands is included in Reference 5. Reference 7 includes a comprehensive high-level description of the adaptations in 5G RAN to support NTN. To complement the upcoming 5G NR broadband standard for satellites, another SI is carried out in 3GPP Release 17 on the adaptation of the LTE-based technologies for massive machine-type communication technologies, Narrow Band Internet of Things (NB-IoT) and enhanced Machine Type Communication (eMTC), to support low data rate use cases with satellites [8], followed by a Release 17 WI [9].

Table 14.1 NTN reference scenarios

Nonterrestrial access network	Transparent Satellite	Regenerative Satellite
GEO based	Scenario A	Scenario B
LEO based: steerable beams	Scenario C1	Scenario D1
LEO based: beams move with satellite	Scenario C2	Scenario D2

14.4 5G NR physical layer enhancements for NGSO

In this section, we review the main enhancements introduced in the 3GPP RAN physical layer working group (RAN1) to support NTN as part of Release 17 5G NR. In particular, 3GPP classifies the necessary enhancements for NTN into four categories. The first category is related to the enhancements necessary for existing timing relationships in 5G NR, mainly to cope with the large and variable propagation delays in NTN. The second category deals with the challenges related to uplink (UL) time and frequency synchronization for NGSO satellites*. The third category discusses the enhancements related to HARQ†, and the last category deals with any other enhancement necessary for reliable functioning of NTN, which is not included in the previous categories such as polarization signaling issues, random access channel (RACH), etc. In the following, we only review the main enhancements and achievements for the first and second categories, i.e., timing relationship, and UL time and frequency synchronization. Moreover, we note that a major part of the discussion provided in this section is based on the authors' contributions in References 10–15.

14.4.1 Timing relationships enhancements

This section is organized as follows. In the first part, we provide an overview of the existing timing relationships in 5G NR, specifically for a reader, who is not familiar with the existing timing relationships, with the focus on summarizing the key driving factors for the choice of the range of the values of the existing timing offsets, i.e., K_1, K_2, k. Then, in the second part, we explain 3GPP RAN1 enhancements of the existing timing offsets via an offset value, i.e., K_{offset}.

14.4.1.1 5G NR timing relationships overview

Below, we review the main timing relationships that require an enhancement for supporting NTN.

1. **Transmission timing for Hybrid Automatic Repeat Request - Acknowledgement (HARQ-ACK) on Physical Uplink Control Channel (PUCCH):** For this case, upon reception of a Physical Downlink Shared Channel (PDSCH), UE sends a valid HARQ-ACK message on PUCCH, carrying the HARQ-ACK information. The first UL symbol of PUCCH that carries

*We note that the main difference between the timing issues introduced in categories one and two is related to their use-cases. While the timing issues that are studied in the first category mainly deal with scheduling procedures (e.g., the scheduling of data shared channels (PDSCH or PUSCH) via PDCCH, or the scheduling of HARQ acknowledgment transmission for PDSCH via PUCCH), the timing issue that is studied in the second category deal only with UL synchronization. Due to the importance of UL time synchronization in NTN, 3GPP decided to allocate a separate agenda item to it. In this contribution, we also adopt the same categorization that introduced in 3GPP, with respect to timing issues.

†We note that scheduling aspects of HARQ is considered and studied in timing relationship category. However, 3GPP allocated a dedicated third category for other issues and enhancements related to HARQ, which is out of the scope of this chapter.

Table 14.2 Table 5.3-1 in [16]: PDSCH processing time for UE processing capability 1; PDSCH decoding time N_1 [symbols]

μ	dmrs-AdditionalPosition = pos0 in DMRS-DownlinkConfig in both of dmrs-DownlinkForPDSCH-MappingTypeA, dmrs-DownlinkForPDSCH-MappingTypeB	dmrs-AdditionalPosition≠ pos0 in DMRS-DownlinkConfig in either of dmrs-DownlinkForPDSCH-MappingTypeA, dmrs-DownlinkForPDSCH-MappingTypeB or if the higher layer parameter is not configured
0	8	$N_{1,0} = 13, 14$
1	10	13
2	17	20
3	20	24

HARQ-ACK information is determined by HARQ-ACK timing parameter K_1 and the assigned time-domain resource including the effect of timing advance (TA). However, UE can only send a valid HARQ-ACK if the corresponding first UL symbol of PUCCH starts on or after symbol L_1. Here, L_1 is defined as the next UL symbol, including its cyclic prefix (CP), starting after UE PDSCH processing time, followed by the end of the last symbol of the PDSCH carrying a transport block being acknowledged. In particular, UE PDSCH processing time can be evaluated as follows:

$$T_{\text{proc},1} = (N_1 + d_{1,1})(2048 + 144) \times \kappa 2^{-\mu} \times T_c \tag{14.1}$$

where N_1 is based on μ of Table 14.2 and Table 14.3 for UE processing capabilities 1 and 2, respectively, where μ corresponds to the one of (μ_{PDCCH}, μ_{PDSCH}, μ_{UL}) resulting with the largest $T_{\text{proc},1}$. Parameter $d_{1,1}$ is obtained based on the PDSCH mapping type, i.e., mapping type A or mapping type B, see Reference 16 for further details.

In terrestrial networks, common rationale is to adopt the minimum value of K_1 such that $K_1 \geq N_1$. In other words, the minimum value of K_1 is chosen such that UE PDSCH processing time is preserved. Given the current range of the values of parameters N_1 and $d_{1,1}$, the UE PDSCH processing time, in a very

Table 14.3 Table 5.3-2 in [16]: PDSCH processing time for UE processing capability 2; PDSCH decoding time N_1 [symbols]

μ	dmrs-AdditionalPosition = pos0 in DMRS-DownlinkConfig in both of dmrs-DownlinkForPDSCH-MappingTypeA, dmrs-DownlinkForPDSCH-MappingTypeB
0	3
1	4.5
2	9 for FR1

coarse approximation, can range between 0.5 and 2 slots. However, the range of the values of K_1 is between 0 and 15. This provides the freedom to choose from a dynamic range of approximately 13 slots for the gNB to schedule HARQ-ACK messages.

2. **Timing relationships for PUSCH scheduled by Downlink Control Information (DCI), PUSCH scheduled by random access response (RAR) grant, and transmission timing for channel state information (CSI) on PUSCH:** In the following, we study the timing relationships for PUSCH scheduled by DCI, PUSCH scheduled by RAR grant, and transmission timing for CSI on PUSCH altogether, since the above-mentioned timing relationships are highly correlated. Potentially, UE can be scheduled to transmit a transport block and no CSI report, or to transmit a transport block and CSI report, or to transmit a PUSCH only for the purpose of a CSI report(s). For the first two cases, i.e., transmission of a transport block with/without CSI report, the time domain resource assignment field value m of the DCI gives the row index $m + 1$ to an allocation table, where the indexed row defines the slot offset K_2, along with other parameters such as the start and length indicator value, the PUSCH mapping type, and the number of repetitions. The current range of the values of K_2 is between $\{0, 1, \cdots, 32\}$. On the other hand, when PUSCH is scheduled for transmitting only a CSI report(s) via CSI request field of the DCI, the time domain resource assignment field value m provides a row index $m + 1$ to an allocation table, and the value of K_2 is obtained as $K_2 = \max_j Y_j(m + 1)$, where $Y_j, j = 0, \cdots, N_{\text{rep}} - 1$ are the corresponding list entries of the higher layer parameter *reportSlotOffsetList* in *CSI-ReportConfig* for N_{rep} triggered CSI reporting settings, and $Y_j(m + 1)$ is the $(m + 1)$th entry of Y_j. The range of the values of K_2 is $\{0, 1, \cdots, 32\}$. Furthermore, when UE receives the DCI, the corresponding scheduled PUSCH is transmitted only if the first UL symbol in PUSCH (including the DM-RS), which is determined via parameters K_2, start and length indicator value, and including the effect of the TA, is not earlier than at symbol L_2. Here, symbol L_2 is defined as the next UL symbol with its CP that comes $T_{\text{proc},2}$ after the end of the last symbol of the PDCCH carrying the DCI scheduling the PUSCH. In particular, $T_{\text{proc},2}$ is the PUSCH preparation procedure time and can be evaluated as follows:

Table 14.4 Table 6.4-1 in [16]: PUSCH preparation time for PUSCH timing capability 1

μ	PUSCH preparation time N_2 [symbols]
0	10
1	20
2	23
3	36

Table 14.5 Table 6.4-2 in [16]: PUSCH preparation time for PUSCH timing capability 2

μ	PUSCH preparation time N_2 [symbols]
0	5
1	5.5
2	11 for FR1

$$T_{\text{proc},2} = \max((N_2 + d_{2,1})(2048 + 144) \times \kappa 2^{-\mu} \times T_c, d_{2,2}) \tag{14.2}$$

N_2 is based on μ of Table 14.4 and Table 14.5 for UE processing capabilities 1 and 2, respectively, where μ corresponds to the one of $(\mu_{\text{DL}}, \mu_{\text{UL}})$ resulting with the largest $T_{\text{proc},2}$.

- if the first symbol of the PUSCH allocation consists of DMRS only, then $d_{2,1} = 0$, otherwise $d_{2,1} = 1$.
- if the scheduling DCI triggered a switch of bandwidth part (BWP), $d_{2,2}$ equals to the switching time as defined in [17], otherwise $d_{2,2} = 0$.

Specifically, the value of K_2 is chosen such that the PUSCH preparation time is preserved, i.e., $K_2 \geq N_2$. Given the range of the values of the parameter N_2, the maximum number of slots required for PUSCH preparation is approximately 2.5 slots. This gives the gNB the huge margin of around 30 slots for scheduling PUSCH.

3. **CSI reference resource timing:** When UE is scheduled to transmit a CSI report(s) on PUSCH in slot n' via the CSI request field on a DCI, its corresponding "CSI reference resource" appears in a single downlink slot $n - n_{\text{CSI\_ref}}$, where $n = \lfloor n' \frac{2^{\mu_{\text{DL}}}}{2^{\mu_{\text{UL}}}} \rfloor$, and μ_{DL} and μ_{UL} are the Subcarrier Spacing (SCS) for DL and UL, respectively. The value of $n_{\text{CSI\_ref}}$ is determined based on the properties of the CSI reporting setting. In particular, for periodic and semi-persistent CSI reporting, $n_{\text{CSI\_ref}}$ is the smallest value greater than or equal to $4 \times \mu_{\text{DL}}$ and $5 \times \mu_{\text{DL}}$ for a single CSI-RS resource and multiple CSI-RS resources, respectively. For the case where aperiodic CSI reporting is configured, $n_{\text{CSI\_ref}}$ is the smallest value greater than or equal to $\lfloor \frac{Z'}{N_{\text{symbol}}^{\text{slot}}} \rfloor$, where Z' corresponds to UE CSI computation time.

4. **Aperiodic sounding reference symbols (SRS) transmission timing:** For the case of SRS transmission, and in particular aperiodic SRS, after receiving the DCI in slot n, the UE transmit aperiodic SRS in each of the triggered SRS resource set(s) in slot

$$\lfloor 2 \times \frac{2^{\mu_{\text{SRS}}}}{2^{\mu_{\text{PDCCH}}}} \rfloor + k \tag{14.3}$$

where, k is configured via higher layer parameter *slotOffset* for each of the triggered SRS resource set and is based on the subcarrier spacing (SCS) of the triggered SRS transmission. The current range of the values of k is between $\{0, 1, \cdots, 32\}$. Furthermore, it is worth mentioning that each SRS resource set can be configured with a certain usage condition. Specifically, when SRS resource set usage is set to either "*codebook*" or "*antennaSwitching*," for each SRS in the SRS resource set, the minimum time interval, in terms of the number of symbols, between the last symbol of PDCCH triggering the aperiodic SRS transmission and the first symbol of SRS resource is N_2. For other usage scenarios, the minimum time interval is $(N_2 + 14)$. From the above discussion, we can observe that gNB has the flexibility of around 18 slots for scheduling UE aperiodic SRS transmission, taking the maximum preparation time $(N_2 + 14)$ into account.

14.4.1.2 Timing relationships enhancements in NTN

In this section, we provide 3GPP analysis regarding the enhancement of timing relationships for NTN. The propagation delays in terrestrial mobile systems are usually less than 1 ms. In contrast, the propagation delays in NTN are much longer, ranging from several milliseconds to hundreds of milliseconds depending on the altitudes of the spaceborne or airborne platforms and payload type in NTN. In an NTN, a UE may need to apply a large TA value that leads to a large offset in its DL and UL frame timing, see Figure 14.7. Here, we assumed that UE1 is closer to the beam center of the satellite, while UE2 is closer to the beam edge. As a consequence of this, it can be observed that the offset between the UL and DL frames is larger for UE2 compared to that for UE1.

Based on the discussion provided in the previous section, we can see that one of the purposes of the existing timing offsets is to provide flexibility for the gNB scheduler. In NTN, due to the large propagation delays, and as a result of that, large TA, the gNB's scheduling flexibility is substantially reduced (see UE2 in Figure 14.7). Thus, in order to preserve the current gNB's scheduling margin, considered in 5G NR, RAN1 introduced an offset value, denoted by K_{offset}, to be added to the existing timing offsets. In other words, the value range of all the existing timing offsets

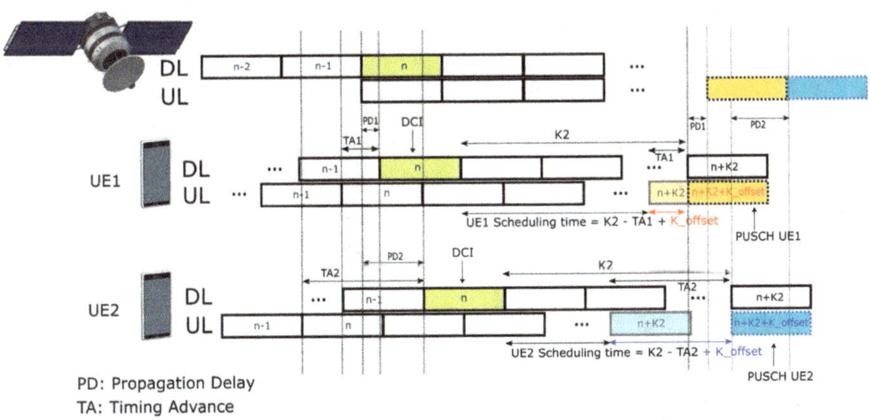

Figure 14.7 Impact of large propagation delay in NTN

is extended by an offset value K_{offset}. This leads to the following enhancements for timing relationships in 5G NR NTN:

- For the transmission timing of HARQ-ACK on PUCCH, the UE provides corresponding HARQ-ACK information in a PUCCH transmission within slot $n + K_1 + K_{\text{offset}}$.
- For the transmission timing of DCI scheduled PUSCH (including CSI on PUSCH), the slot allocated for the PUSCH can be modified to be $\lfloor n \times \frac{2^{\mu_{\text{PUSCH}}}}{2^{\mu_{\text{PDCCH}}}} + K_2 + K_{\text{offset}} \rfloor$.
- For the transmission timing of RAR grant scheduled PUSCH, the UE transmits the PUSCH in slot $n + \Delta + K_2 + K_{\text{offset}}$.
- For the CSI reference resource timing, the CSI reference resource is given in the downlink slot $n - n_{\text{CSI\_ref}} - K_{\text{offset}}$.
- For the transmission timing of aperiodic SRS, the UE transmits aperiodic SRS in each of the triggered SRS resource set(s) in slot $\lfloor 2 \times \frac{2^{\mu_{\text{SRS}}}}{2^{\mu_{\text{PDCCH}}}} \rfloor + k + K_{\text{offset}}$.

As stated above, the main reason for the introduction of K_{offset} is the large propagation delay and consequently large TA applied by NTN UEs. As a result of this, the particular value of K_{offset} is closely tight to the specific value of TA applied by each UE. One example scenario where DCI scheduled PUSCH timing relationship is enhanced via K_{offset}, where the value of K_{offset} for each UE is adopted based on its corresponding TA value, is presented in Figure 14.7. Here, we assumed that the SCS of PUSCH and PDCCH are the same. As can be seen in Figure 14.7, the UL packets of UE1 and UE2 arrive at the gNB without any overlap.

Although the choice of K_{offset} based on the UE-specific TA is a straightforward solution, but this approach cannot be adopted for the transmission timing of RAR grant scheduled PUSCH. This is because the UE-specific TA is not known before RACH procedure (or initial access). Consequently, the application of K_{offset} is divided into two regimes. The first regime is before initial access, and the second regime is after RACH, where UE obtains its UE-specific TA. In particular, before initial access, the value of K_{offset} is chosen to be common for all UEs in the cell and broadcast via the system information block (SIB). Here, the value of K_{offset} is identified based on the maximum round-trip-time (RTT) experienced by UEs in the cell. After initial access, when UE is in CONNECTED mode, the value of K_{offset} can be updated, if UE receives updated value of K_{offset} in SIB. Furthermore, for UE specific update of K_{offset}, after initial access, UE TA report to gNB is also required. Currently, 3GPP RAN1 WG is working on the design of the content of the UE TA report. The reason behind the need for UE TA report is became clear in section "UL Time Synchronization."

14.4.2 UL time and frequency synchronization

In this section, we review the main NTN enhancements with respect to UL time and frequency synchronization. In particular, this section is divided into two parts. In the first part, we explain the issues related to UL time synchronization. Then, the second part discusses the enhancements introduced for UL frequency synchronization.

14.4.2.1 UL time synchronization

In traditional terrestrial systems, e.g., 5G NR, the UL synchronization is achieved via the RACH procedure. In the RACH procedure, UE first sends a randomly chosen random access preamble, the so-called message 1 (MSG1), from a pool of preambles shared with other UEs. This indicates that multiple UEs may select the same preamble and experience contention. In the next step, gNB calculates the propagation delay of the corresponding UE and sends the TA command as part of message two (MSG2); this message is also referred to as RAR. In RAR, the frequency and time resources for transmission of message three (MSG3) are provided to the UE. Generally, the main purpose of exchanging MSG3 and MSG4 is to resolve the contention event if multiple UEs send the same preamble for MSG1. The schematic presentation of RACH procedure is depicted in Figure 14.8.

As stated above, the TA command is transmitted via the gNB to the UE as part of RAR. In the following, we first review the TA procedure in 5G NR, especially for a reader not familiar with TA calculation. Subsequently, we discuss the enhancement introduced in TA calculation for NTN UE.

1. **TA in 5G NR** TA is used to adjust the UL transmission timing of individual UEs, so that UL transmissions from all UEs are synchronized when received by the gNB. In particular, the ultimate effect of UL synchronization is to overcome the symbol interference between consecutive UL transmissions of individual UEs,

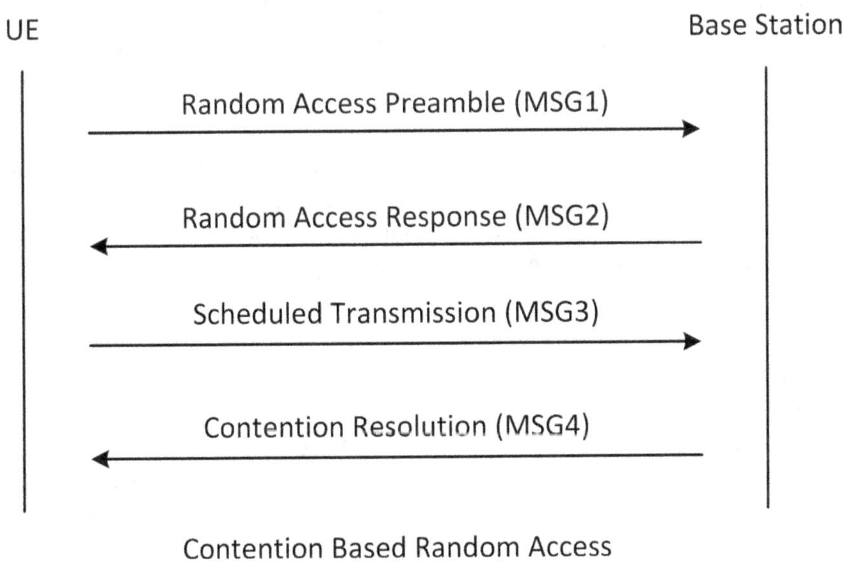

Figure 14.8 RACH procedure

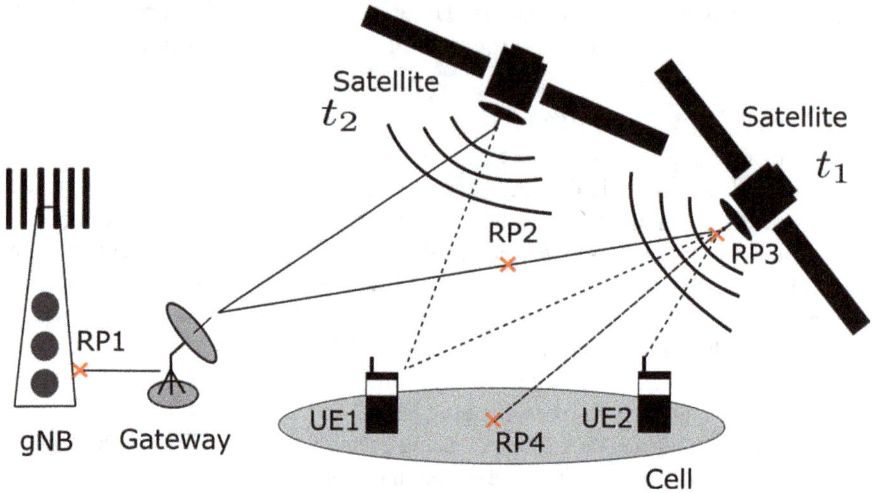

Figure 14.9: UE-to-gNB delay components for NTN UE

as well as among UL transmissions of multiple UEs. The TA is closely related to the propagation delay experienced by the UE. The larger the value of the propagation delay, the larger the value of TA. As a consequence, UEs located at the cell edge must apply a larger TA, see, e.g., Figure 14.9 for the impact of TA on the UL DL frame timing of two sample UEs. In 5G NR, the TA is calculated as $(2 \times \text{propagation delay}) + N_{\text{TA\_offset}} \times T_C$, where $T_C = 1/(480000 \times 4096)$ [s] is a reference time unit. Here, the term $N_{\text{TA\_offset}} \times T_C$ accounts for proper margin, so that the UL radio frame finishes before the start of subsequent DL radio frames. This margin is necessary for a Time Division Duplex (TDD) base station to account for the delay associated with activating its transmitter. The corresponding activation delay for TDD base station, which is considered in Reference 18, is 10 μs in FR1 and 3 μs in FR2. Another use case of $N_{\text{TA\_offset}} \times T_C$ margin is to overcome BTS-to-BTS interference, which can occur due to the nonideal BTS synchronization [17]. It specifies that the maximum allowed timing error for base stations with overlapping coverage is 3 μs. Keeping the discussion above in mind, the TA applied by the UE in 5G NR can be calculated as follows:

$$T_{\text{TA}} = (N_{\text{TA}} + N_{\text{TA\_offset}}) \times T_C \tag{14.4}$$

where $N_{\text{TA}} \times T_C$ takes the round trip delay between the UE and the base station into account. Here, UE calculates the value of N_{TA} in two steps. First, UE receives its first TA command as part of RAR or MSG2. There, a set of 12 bits is used to provide a value range of 0–3846 for the so-called variable T_A. Subsequently, UE uses the obtained value of T_A to derive N_{TA} as follows:

$$N_{\text{TA}} = T_A \times 16 \times \frac{64}{2^{\mu}} \tag{14.5}$$

where $\mu = \{0, 1, 2, 3\}$ is correspond to SCS = $\{15, 30, 60, 120\}$ kHz, respectively. This indicates that TA uses a time resolution, which is proportional to the SCS.

2. **TA in 5G NR for NTN UE:** As discussed in the previous section, TA is closely related to the RTD experienced by the UE. Thus, in the following, we first discuss how the end-to-end (UE-to-gNB) delay is accounted for in the NTN scenario. In other words, we discuss the NTN UE-to-gNB delay components. Then, we provide the TA calculation for NTN UE. Moreover, it is assumed that, for Release 17, the NTN UE is equipped with a global navigation satellite system (GNSS) unit. In particular, the GNSS unit together with the satellite ephemeris data assist the UE to estimate its distance to the satellite. The importance of UE-to-satellite distance estimation becomes clear later in this section, when we discuss the TA calculation for NTN UE.

   UE-gNB RTT/Delay Components: Generally, the end-to-end delay experienced by NTN UE can be split into two major parts: UE-specific delay and UE-common delay. Calculation of both UE-specific and UE-common delay depends on the choice of the so-called reference point (RP). In particular, 3GPP RAN1 WG considers the RP as the point with respect to which the DL and UL frames are aligned, after UE applies the TA command in RACH procedure and/or autonomously obtains TA. As a result of this, the value of TA is calculated with respect to RP. Typically, RP can be chosen to be at gNB, at feeder link, at the satellite, or at a point located at service link. It is decided in RAN1 that the choice of RP is arbitrary, and it must be under control of the network and should at least include the RP at gNB, see Figure 14.9. For instance, when the RP is chosen to be at satellite (RP3 in Figure 14.9), the UL and DL frames are aligned at satellite and gNB has to deal with not aligned UL and DL frame timing and applies a post timing compensation based on RTT of the feeder link. On the other hand, the choice of RP at gNB (RP1 in Figure 14.9) leads to frame timing in UL and DL that are aligned at gNB. Given the definition of the RP above, we can define the UE-specific delay and UE-common delay as follows:

- UE-specific delay: It is defined as the delay of the UE to the satellite. In Release 17, NTN UE is assumed to be equipped with a GNSS unit. As a result of this, the GNSS enabled UE estimates the distance to satellite together with the assistance of satellite ephemeris and calculates the UE-Sat delay.
- UE-common delay: It is defined as the delay of a satellite to the RP (Sat-RP). Depending on the location of RP, UE-common delay can be evaluated as follows:
  - It can be set to zero. This is the case when RP is chosen to be at the satellite, e.g., RP 3 in Figure 14.9.
  - It can capture the partial delay of the service link, when RP is chosen on the service link, e.g., RP 4 in Figure 14.9.
  - It can capture the entire feeder link delay, i.e., gNB-GW-satellite delay, when the RP is chosen to be at gNB, e.g. RP 1 in Figure 14.9.

In addition to the common delay, we also define the feeder link delay as the delay of gNB to the satellite. It is important to emphasize that many procedures in RAN1 and RAN2 require the knowledge of end-to-end UE-gNB delay. Given the definition of the UE-specific and UE-common delay as above, unless for the case of RP at the gNB, for calculation of UE-gNB delay, signaling of both common delay and feeder link delay from network to UE is required.

Note: In the remainder of this chapter, we refer to the feeder link delay and common delay together, for conciseness of presentation, as common delay. In other words, we assume that RP is located at gNB. However, the procedures introduced in the following sections are also valid for other choices of RP as well. **NTN TA:** Taking the components of UE-to-RP into account, the TA for NTN UE is evaluated as follows:

$$T_{TA} = (N_{TA} + N_{TA\_offset} + N_{TA,UE\_Specific} + N_{TA,UE\_Common}) \times T_C \tag{14.6}$$

where $N_{TA,UE\_Specific}$ is referred to as UE-specific TA and captures the delay/RTT of the service link (UE-Sat), while N_{TA,UE_Common} is also referred to as UE-common TA and captures the delay/RTT of Sat-RP, which is common to all UEs. It is important to mention that $N_{TA,UE\_Specific}$ is acquired by every UE *autonomously*, given that NTN UEs are assumed to be equipped with GNSS receiver, and the broadcast of satellite ephemeris. On the other hand, the term N_{TA,UE_Common} has to broadcast to the UE. Currently, 3GPP WG RAN1 works on the signaling design of N_{TA,UE_Common}. In particular, some characteristics of the feeder link delay, e.g., the variations of the feeder link delay over time during the visibility window of a satellite, have been taken into account to reduce the signaling overhead for N_{TA,UE_Common}. We note that the choice of TA based on (14.6) for Msg1 (in 4-step RACH) or MsgA (in 2-step RACH) transmission of an NR NTN UE in idle/inactive mode leads to use the existing timing advance command 12-bit field in msg2 (for 4-step RACH) or msgB (for 2-step RACH) without any extension.

14.4.2.2 UL frequency synchronization

With respect to this issue, in 5G-NTN, two different approaches have been adopted for Doppler compensation of service link (between UE and satellite) and feeder link (between satellite and GW). In particular, 3GPP assumes that the Doppler compensation of the service link is performed by the UE while the Doppler compensation of the feeder link must be transparent to the network and the UE. Below, we briefly explain the UL frequency synchronization enhancement with respect to terrestrial 5G NR:

- **Service link:** It is agreed that an NR NTN UE in RRC_CONNECTED state shall be capable of at least using its acquired GNSS position and satellite ephemeris to perform frequency precompensation to counter shift the Doppler experienced on the service link.
- **Feeder link:** With respect to the compensation of Doppler on the feeder link, it is agreed that the Doppler shift over the feeder link and any transponder

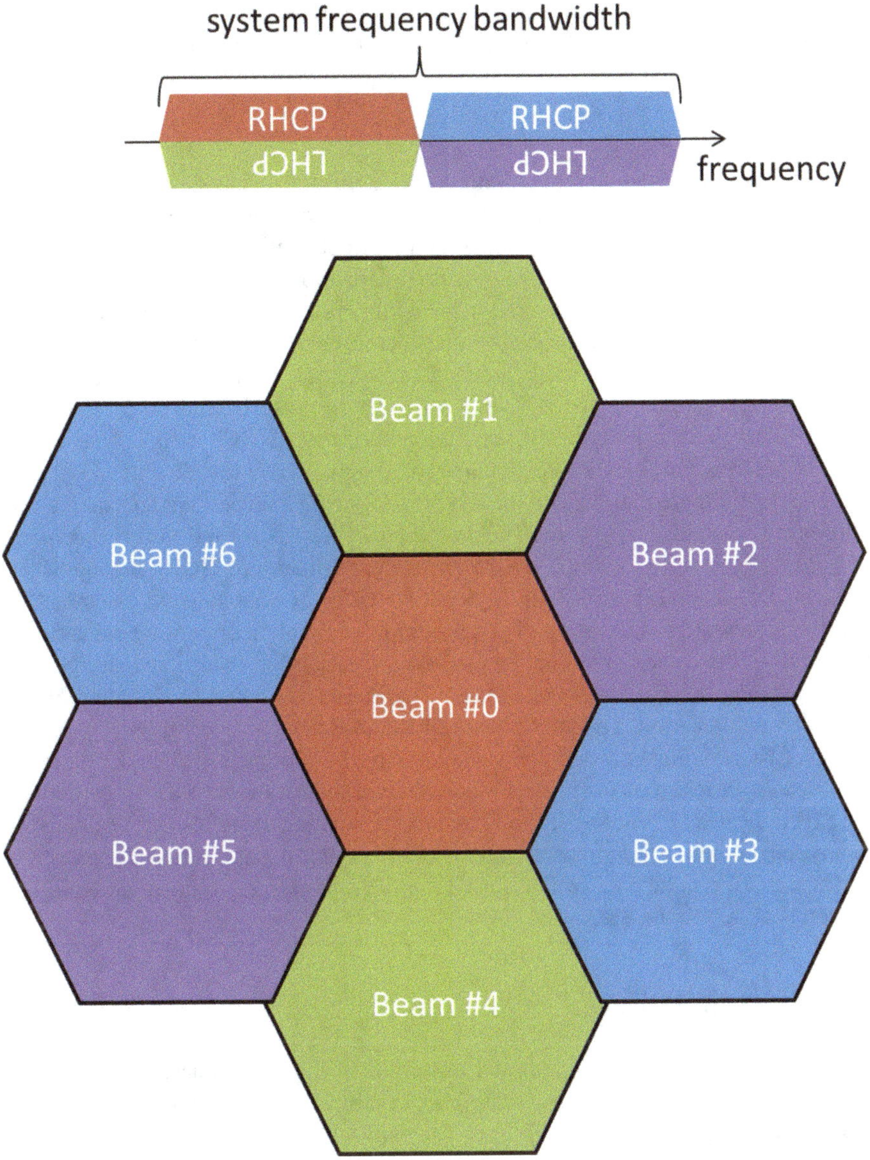

Figure 14.10 The so-called option 3 of beam-layout planning according to Reference 5

frequency error for both downlink and UL are compensated by the GW and satellite-payload without any specification impacts in Release 17.

14.4.3 Polarization signaling

In 3GPP Release 17, some enhancements with respect to the signaling of polarization have been introduced. Before we briefly review these enhancements, it is worth mentioning that in NTN polarization can optionally be used as part of the beam-frequency planning. In other words, polarization can be optionally enabled and jointly used with frequency reuse factor for beam-layout planning. However, enabling polarization reuse should be considered only, when circular polarization is supported by the terminal UE antenna. An example of beam-layout planning, when frequency reuse factor is 3, and polarization reuse is enabled is shown in Figure 14.10. Here, right-hand-side circular polarization (RHCP) and left-hand-side circular polarization (LHCP) are assumed to be supported at the UE antenna.

Due to the potential support of RHCP and LHCP, several enhancements are introduced in 3GPP Release 17. The first enhancement is related to the *indication* of polarization, which is done by the *network*. Furthermore, 3GPP distinguishes between DL and UL polarization. Later, two different options were discussed in 3GPP with respect to indication of polarization, i.e., implicit or explicit indications. Eventually, it is decided to adopt an explicit indication for a clean and straightforward signaling design. The second enhancement was related to the signaling of the polarization indication. The possibilities of Radio Resource Control (RRC) signaling, DCI-based indication, and SIB signaling have been discussed there. Eventually, a consensus is achieved to signal the polarization indication via SIB. In particular, the SIB indicates DL and/or UL polarization information using respective polarization type parameters to indicate RHCP or LHCP or linear polarization. Moreover, a UE assumes the same polarization for UL and DL, when the UL polarization information is absent in the SIB.

14.5 Conclusion

The 3GPP Release 17 is an important milestone in the 5G standardization groups in order to develop and to approve the TS to enable a direct access technology via satellite links for the first time. The goal is to deploy NTNs as part of 5G by approximately 2025 in order to meet the challenges of mobile network operators in terms of seamless coverage, availability, and resiliency. The satellite industry is gaining more and more interest in this emerging topic, and several companies even participate actively in the traditional terrestrial standards organization. Prestandard trials of Release 17 5G-NTN technology elements are already reported in Reference 19.

Since the cellular standard as specified by 3GPP continuously evolves, functional extensions were discussed for 5G-NTN in Release 18 as well. For 5G NR, multiple topics were discussed, which are summarized in Reference 20. In the same document, the potential extensions for IoT-NTN are described, on top of the specification as planned in Release 17 [9]. The final approved features for NR-NTN in Release 18 are NTN coverage enhancements, NR-NTN deployment in above 10 GHz frequency bands, NTN-TN and NTN-NTN mobility and service continuity enhancements as well as network-based UE location [21]. The approved objectives in Reference 22 for the LTE-based IoT-NTN in Release 18 are performance and mobility enhancements and potential enhancements for discontinuous coverage.

References

[1] 3GPP. TR 22.822. '3rd generation partnership project; study on using satellite access in 5G'. 2020.
[2] 3GPP. TR 23.737. '3rd generation partnership project; study on architecture aspects for using satellite access in 5G'. 2020.
[3] 3GPP SP-191335. *New WID: integration of satellite systems in the 5G architecture (proposed for Rel-17,* 5GSAT ARCH*)*. Sitges, Spain: SA2 WG2; 2019.
[4] 3GPP TR 38.811. '3rd generation partnership project; technical specification group radio access network; study on new radio (NR) to support non terrestrial networks (release 15)'. 2019.
[5] 3GPP TR 38.821. '3rd generation partnership project; technical reportgroup radio access network; study on solutions for NR to support nonterrestrial networks (NTN)'. 2020.
[6] 3GPP RP-213691. '*Solutions for NR to support non-terrestrial networks (NTN)*'. Thales. 2021.
[7] Sirotkin S. (ed.) *5G radio access network architecture*. Wiley; 2020.
[8] 3GPP. TR 36.763. '3rd generation partnership project; study on narrow-band internet of things (NB-IoT) / enhanced machine type communication (EMTC) support for non-terrestrial networks (NTN) Release (17)'. 2021.
[9] 3GPP RP-211601. 'WID on NB-IoT/EMTC support for NTN'. Mediatek, Eutelsat. 2021.
[10] 3GPP R1-2005548. 'NR-NTN: timing relationship enhancements'. Fraunhofer IIS, Fraunhofer HHI. 2020.
[11] 3GPP R1-2008722. 'Discussion on timing relationship enhancements for NTN'. Fraunhofer IIS, Fraunhofer HHI. 2020.
[12] 3GPP R1-2101694. 'Discussion on timing relationship enhancements for NTN'. Fraunhofer IIS, Fraunhofer HHI. 2021.
[13] 3GPP R1-2103655. 'Discussion on UL time synchronization for NTN'. Fraunhofer IIS, Fraunhofer HHI. 2021.
[14] 3GPP R1-2101694. 'Discussion on timing relationship enhancements for NTN'. Fraunhofer IIS, Fraunhofer HHI. 2021.
[15] 3GPP R1-2105272 'Discussion on UL time synchronization for NTN'. Fraunhofer IIS, Fraunhofer HHI. 2021.
[16] 3GPP. TR 38.214. '3rd generation partnership project; technical specification group radio access network; NR; physical layer procedures for data'. 2020.

[17] 3GPP. TR 38.133 '3rd generation partnership project; requirements for support of radio resource management'. 2020.
[18] 3GPP. TR 38.104. '3rd generation partnership project; base station (BS) radio transmission and reception'. 2020.
[19] *First software-defined 5G new radio demonstration over GEO satellite*. 2021. Available from https://www.iis.fraunhofer.de/en/pr/2021/20210312_5G_new_radio.html
[20] 3GPP RP-211658. 'Moderator's summary for discussion [ran93e-r18prep-08] NTN evolution, AT&T'. 2021.
[21] 3GPP RP-213690. 'NR NTN (non-terrestrial networks) enhancements'. Thales. 2021.
[22] 3GPP RP-213596. 'IoT NTN (non-terrestrial networks) enhancements'. Thales. 2021.

Chapter 15

Anti-jamming solutions for non-geostationary orbit satellite systems

Chen Han[1], *Xinhai Tong*[2], *and Liangyu Huo*[3]

The anti-jamming communication of the non-geostationary orbit (NGSO) satellite systems has drawn increasing attention due to the smart jamming and high dynamics caused by the satellite movement. This chapter investigates the anti-jamming scheme for NGSO satellite systems, with the aim of minimizing routing costs under jamming threats via Stackelberg game and reinforcement learning. Section 15.2 formulates the anti-jamming routing problem as a hierarchical anti-jamming Stackelberg game. It is proven that there is a Stackelberg equilibrium (SE) in the proposed game. Section 15.3 introduces the anti-jamming scheme for NGSO satellite systems, which consists of two stages: the available routing selection and the fast anti-jamming decision. To tackle the high dynamics caused by the intermittent interruptions and the unexpected congestion, a deep reinforcement learning-based routing algorithm (DRLR) is proposed to obtain an available routing subset. Furthermore, based on the available routing subset, a fast response anti-jamming algorithm (FRA) is proposed to make a fast anti-jamming decision. Satellites utilize DRLR and FRA algorithms to empirically analyze the jammer's strategies and adaptively make an anti-jamming decision according to the dynamic and unknown jamming environment. Section 15.4 shows that the proposed algorithm has lower routing cost and better anti-jamming performance than existing approaches, and the anti-jamming policies converge to the SE. The last section is the summary of this chapter.

15.1 Satellite routing

NGSO satellite systems can achieve wide-area, high-speed, and reliable transmission, and it is also the irreplaceable system to guarantee secure communication in defense applications. Thus, reliable transmission is an essential requirement of the

[1]Sixty-third Research Institute, National University of Defense Technology, Nanjing, China
[2]College of Communications Engineering, Army Engineering University, Nanjing, China
[3]School of Electronic and Information Engineering, Beihang University, Beijing, China

NGSO satellite systems. As artificial intelligence (AI) technology has been widely used in the wireless network domain, it provides jammers with more diverse and intelligent jamming attacks. Intelligent anti-jamming communication for NGSO satellite systems is an urgent and inevitable choice. Furthermore, due to the high dynamics caused by the intermittent interruptions and the unexpected congestion in the NGSO satellite systems [1], the difficulty of intelligent anti-jamming is further aggravated.

Satellites stand exposed to the malicious jamming attack [2] because of the periodic visibility and fixed orbits. As presented in Reference 3, a broadcasting jammer could launch chirp jamming attacks on the global navigation satellite systems. Some other common jamming, such as constant jamming, deceptive jamming, and flow jamming were discussed in Reference 4. In fact, there are many traditional anti-jamming methods, such as the frequency-based anti-jamming technologies including direct sequence spread spectrum (DSSS) and frequency hopping spread spectrum (FHSS), the space-based anti-jamming technologies, presenting as multi-beam antennas, adaptive nulling antennas, and self-adaption anti-jamming routing. However, these anti-jamming approaches cannot deal with the smart jamming which adjusts the jamming actions utilizing learning and reasoning, and seriously endangers the reliability of the NGSO satellite systems. For example, the smart jamming formulated in Reference 5 was able to automatically adjust the jamming channel. A novel smart jammer was proposed in Reference 6 to maximize jamming effects by adaptively adjusting the jamming power and jamming channel. To tackle the threat from smart jamming, the satellites also have to obtain the ability of learning and reasoning to achieve intelligent anti-jamming communications. The open literature for intelligent anti-jamming mostly focuses on the defense in time and frequency domain, while the space-based anti-jamming technologies, such as routing anti-jamming has drawn less attention.

Actually, many routing technologies have been proposed for the NGSO satellite systems, including the virtual node-based routing methods [7], the virtual topology-based routing methods [8], and the routing approaches which are originally used for the mobile Ad-Hoc network [9], such as the Ad-Hoc on-demand distance vector (AODV) algorithm [10], the independent zone routing algorithm (IZR) [11], and the optimized link state routing (OLSR) algorithm [12]. These approaches do not consider the anti-jamming defense, and cannot deal with the smart jamming. Moreover, these routing selection strategies mainly focus on the routing optimization for networks with known topology and do not address intermittent connectivity and unexpected congestion caused by satellite movement and the burst transmission mode [13–15]. The large-scale satellite network with high dynamics sharply increases the user's decision space. The uncertain burst traffic requests users to explore the unknown environment, and the smart jamming forces users to obtain the ability to learn, reason, and predict, which both aggravates the difficulty of intelligent anti-jamming routing.

This chapter proposes a spatial anti-jamming scheme (SAS) for NGSO satellite systems. The communication countermeasures between smart jammers and satellite users are modeled as a Stackelberg anti-jamming routing game. On the one hand, because of the sharply increasing decision space caused by the high dynamics of the NGSO satellite systems, deep learning (DL) technology can be utilized to extract effective environmental characteristics. On the other hand, due to the intelligence of the smart jammer, reinforcement learning (RL) techniques can be used to deal with the dynamic interactions between the satellites and the unknown jamming environment. Therefore, deep reinforcement learning (DRL) technology is adopted to solve the routing selection problem for the NGSO satellite systems. DRL technology is used to maintain an available routing subset to simplify the decision space for the Stackelberg anti-jamming routing game. Then, based on this routing subset, Q-Learning technology is used to respond quickly to the smart jamming and adjust anti-jamming strategies.

15.2 The problem of anti-jamming routing for NGSO satellite networks

In this chapter, the NGSO satellite system consists of multiple low-earth-orbit (LEO) satellite constellations. Due to the satellite movement, the inter-satellite distance and connectivity of the NGSO satellite system changes over time. The smart jammer considered in this chapter is power-limited. It divides the satellite communication network into several regions.

As shown in Figure 15.1, the smart jamming selects a region to launch jamming attacks. At this time, the channel rate of the *kth* link L_k from the source satellite node n_S to the destination satellite node n_D is given in Reference 16.

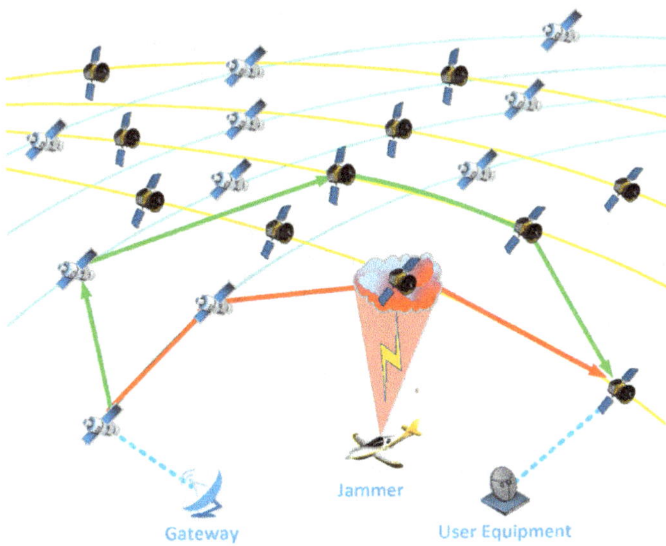

Figure 15.1 *The jamming model. The red line represents the jammed path, and the green line is the reconstructed path.*

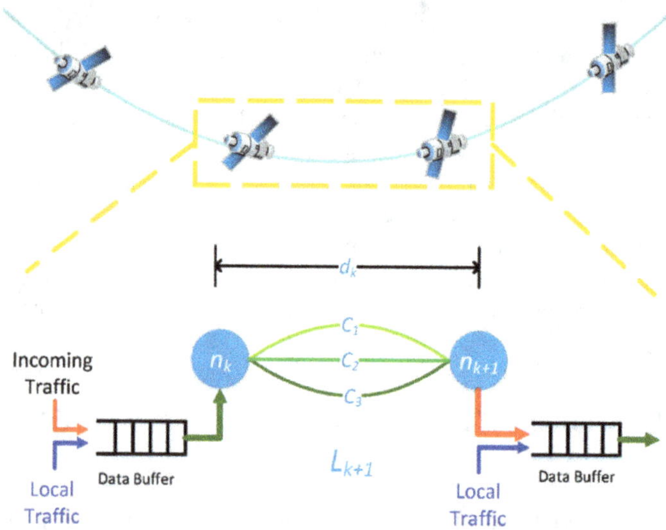

Figure 15.2 The transmission link model with existing traffic

$$C_k = B_k \log_2 \left(1 + \frac{p_k f_k^2 / d_k^2}{n_{0,k} B_k (4\pi / v_c)^2 + p_k^J f_J^2 / d_{J,k}^2} \right) \tag{15.1}$$

where B_k, p_k, $n_{0,k}$, and d_k, respectively, represent the channel bandwidth, transmission power, channel noise, and inter-satellite distance of L_k; $\sigma_{u,k}$, f_k, v_c are the fading exponent, communication frequency, and speed of light. p_k^J, $d_{J,k}$, and f_J are the jamming power, the jamming distance, and the jamming frequency in L_k. The incoming data composes of N_D bits, and the required process time of sending and receiving in the L_k is expressed as

$$t_k^1 = 2 \times \frac{N_D}{C_k} \tag{15.2}$$

As shown in Figure 15.2, the communication delay between n_k and n_{k+1} is not only related to the incoming traffic N_D, but also depends on the existing local traffic in n_k. The already existing traffic X_k in the current node follows the Poisson distribution with parameter λ_k [17]:

$$Pr(X_k = x_k) = \frac{(\lambda_k)^{x_k}}{(x_k)!} e^{-\lambda_k} \tag{15.3}$$

The queuing time before the kth transmission is given by

$$t_k^2 = \frac{x_k}{C_k} \tag{15.4}$$

Therefore, under the threat of smart jamming, the total communication delay of the jammed routing path is denoted by

$$\tau = \sum_{k=1}^{\zeta} \left(t_k^1 + t_k^2\right) \tag{15.5}$$

where ζ is the total hop count.

The existing routing works mostly adopt simple routing metrics, such as the least hop [14, 18, 19], the lowest delay [7, 20, 21], and the smallest congestion probability [22–24]. However, there are fewer works focusing on the multi-objective routing optimization. Considering the user's requirements, satellite communication services may be sensitive to channel rate and queuing delay. Thus, the cost function of multi-objective routing proposed in this chapter is presented as

$$c = \sum_{k=1}^{\zeta} ln\left(1 + w_1 t_k^1 + w_2 t_k^2\right) \tag{15.6}$$

where w_1 denotes the trend towards high data throughout. t_k^1 is only dependent on the channel rate, and the power consumption reduces with the improvement of the channel state; w_2 indicates the user's tolerance degree of queuing delay.

The anti-jamming routing problem in this chapter aims at minimizing the routing cost under jamming threat, which is divided into two subproblems: the routing selection problem and the fast response anti-jamming problem.

15.2.1 Routing selection problem for NGSO satellite networks

As for one transmission task, an agent usually used in the DRL paradigm is used to select nodes, which is hereafter referred to as the user. The node selection in the L_k only depends on the node selection in the L_{k-1}. Thus, the problem of node selection in L_k can be regarded as a Markov decision process (MDP). The user's state set S is $\{s_k \in S | S = n_1, n_2, \dots n_N\}$, where s_k is the source node of L_k. A_k is the set of possible action $\{a_k \in A_k | A_k = A^{s_k}\}$, where a_k is the destination node of L_k. State-action pair $\{a_k | s_k\}$ represents the selecting action a_k in the state s_k. Then, an immediate routing cost is denoted as

$$c_k = \text{In}\left(1 + w_1 t_k^{\prime} + w_2 t_k^{\prime\prime}\right) \tag{15.7}$$

The users aim to obtain an optimal policy π_U^* that probabilistically maps state s_k to action a_k. According to π_U^*, users can make the best decision $a_U^* \sim \pi_U^*$.

$$\pi_U^* = \left\{Pr\left(a_k^* | s_k\right) | k = 1, 2, \dots \zeta\right\} \tag{15.8}$$

$$a_U^* = \left\{a_1^*, a_2^* \dots a_\zeta^*\right\} \tag{15.9}$$

Then, the user can find the available routing set $\Re$ including all the routing links which meet the communication need R_0, i.e., $c \leq R_0$.

$$find\, \Re = \left\{a_U^0 \mid a_U^0 = \underset{a_U}{arg}\, \{c \leq R_0\}\right\} \tag{15.10}$$

15.2.2 Fast response anti-jamming problem

The network environment changes with the implementation of smart jamming. Thus, the users need to explore the new environment and corresponding anti-jamming routing. However, due to the limited power of the jammer, the changes caused by the smart jamming may just affect a few nodes, thus, the large-scale relearning is unnecessary and costly. Especially, against the threat of smart jamming, the fast response and quick decision is an important consideration of anti-jamming performance. Therefore, we formulate the fast response anti-jamming problem based on the available routing subset $\Re$.

Ω is a sub-network composed of all the nodes in $\Re$. As for smart jammer, only the nodes in Ω are jammed, the jamming aiming at user's routing is successful. Based on learning, the jammer reduces the valid jamming range to Ω. It launches jamming attacks on one node in Ω with a jamming power p_j to reduce its channel rate. A_J is the set of jamming nodes:

$$\{a_{J,m} \in \mathrm{A}_J | \mathrm{A}_J = a_{J,1}, a_{J,2} \ldots a_{J,G}\} \tag{15.11}$$

where G is the node count in Ω.

The jammer senses the status of the NGSO satellite systems, and estimates the transmission time τ. The utility function of the jammer is defined as

$$r_J = \vartheta \times (\tau - \tau_{min}) - \lambda_J \times p_{J,v} \tag{15.12}$$

where the minimum transmission time of user without suffering jamming is estimated as τ_{min}. But the jammer cannot obtain the accurate jamming feedback, and $\vartheta \sim N(\mu, \delta)$ is used to assess the incomplete estimation. $p_{J,v} \in [p_{J,1}, p_{J,2} \ldots p_{J,V}]$ is the available jamming power, and λ_J is the jamming cost per unit power.

The available routing path set of the user is $\left\{\Re = a^0_{U,1}, a^0_{U,2} \ldots a^0_{U,F}\right\}$, and F is the number of available routing paths in $\Re$. The user's utility function of routing selection is

$$r_U = c \tag{15.13}$$

Based on Q-learning, the user selects the best route to maximize the utility under the jamming threat. If $c_U > 1.2R_0$, the network state is considered to be changed. Then, the DRL process is restarted to update $\Re$.

15.2.3 Anti-jamming routing game

Inspired by Reference 6, the anti-jamming routing selection problem is modeled as a hierarchical Stackelberg routing game $\wp = \{J, U, \pi_J, \pi_U, r_J, r_U\}$, where J, U represent the jammer and user; π_J is expressed as jammer's mixed strategies of node selection and power selection, and π_U is user's mixed policies of routing selection; r_J, r_U are respectively denoted as the routing game utilities of the jammer and user.

The user and jammer employ mixed strategies, which define a probability distribution for all possible actions, including the routing path for the user, and the jamming node and jamming power for the jammer. According to their own strategies, the jammer chooses the best jamming node and power to maximize the jamming utility:

$$\left(a_J^*, p_J^*\right) = \arg \max_{a_{J,m}, p_{J,v}} \left\{r_J\left(a_{J,m}, p_{J,v}\right)\right\} \tag{15.14}$$

The user selects the best routing path to maximize user's routing utility:

$$a_U^{0*} = \arg \max_{a_{U,n}^0} \left\{r_U\left(a_{U,n}^0, a_{J,m}, p_{J,v}\right)\right\} \tag{15.15}$$

The expected utility is denoted as $r(\pi_U, \pi_J) = E\left[r|\pi_U, \pi_J\right]$, and the Stackelberg equilibrium (SE) is defined as follows.

Theorem 1: If the following conditions in (15.16) are met, the strategy profile $\left(\pi_U^*, \pi_J^*\right)$ constitutes the SE. Then, no player can increase its own utility by diverging unilaterally within the proposed game model.

$$\begin{aligned} \hat{r}_U(\pi_U^*, \pi_J^*) &\geq \hat{r}_U\left(\pi_U, \pi_J^*\right) \\ \hat{r}_J(\pi_U^*, \pi_J^*) &\geq \hat{r}_J\left(\pi_U^*, \pi_J\right) \end{aligned} \tag{15.16}$$

Theorem 2: In this game, there exist steady strategies of the user and jammer that constitute the SE.

Proof: According to Reference 25, the finite strategic game has a mixed strategy equilibrium. Thus, there exists a SE in the proposed game, in the meaning of the stationary policy. According to Theorem 1 and considering that the purpose of user is to maximize utility, the optimal strategy of the user is given by

$$\pi_U^* = \arg \max_{\pi_U} \{\hat{r}_U(\pi_U, \pi_J)\} \tag{15.17}$$

Then, the jammer's optimal strategy is obtained by

$$\pi_J^* = \arg \max_{\pi_J} \left\{\hat{r}_J\left(\pi_J, \pi_U\left(\pi_J\right)\right)\right\} \tag{15.18}$$

Therefore, $\left(\pi_J^*, \pi_U^*\left(\pi_J^*\right)\right)$ constitutes a steady SE.

15.3 Anti-jamming scheme for NGSO satellite networks

The proposed anti-jamming scheme for NGSO satellite networks consists of two algorithms. A DRLR is proposed to solve the routing selection problem in large NGSO satellite networks. A FRA is proposed to tackle the fast response problem.

15.3.1 Deep reinforcement learning-based routing algorithm

Specifically, as shown in Figure 15.3, for the routing selection in L_k, the SAS agent observes the position of the current node n_k as the state s_k, then selects the next node as the action a_k from the accessible satellite nodes A^{s_k}. Meanwhile, the satellite communication network, as known as the environment in DRL diagram, updates its internal status and response an environment reward r_k.

According to the ε-greedy exploration strategy, the agent randomly samples an action $a_t \sim \pi\left(s_k\right)$ and performs it in the environment. The satellite network updates

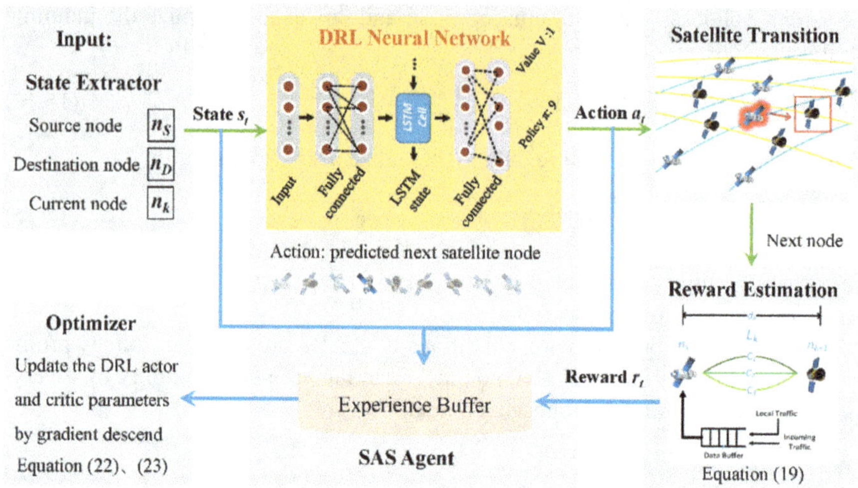

Figure 15.3 DRL modeling for routing selection

its internal status by the transition probability $Pr\left(s_{k+1}|s_k, a_k\right)$. Note that since the current position is directly used as the observed state in DRL, the state transition probability that is caused by the SAS agent is deterministic. On the other hand, for the unobservable transition rule in the internal status of the satellite network, the delay time is updated according to the current position and the corresponding channel information, while the queuing time is updated by re-generating a Poisson distribution at each time step. Recall that these parts of the transition of the internal satellite network are not observed by the SAS agent. Then the agent receives the instantaneous reward r_k to evaluate the efficiency of the yielded policy and then observes a new state s_{k+1} and makes the next choice.

The objective of the agent is to maximize a cumulated reward in each episode, which we refer to as episodic return. Spontaneously, we regard a completed routing path that is found by the SAS agent from the source node to the destination node as one episode in our DRL diagram. The instantaneous reward r_k can be defined as follows:

$$r_k\left(s_k, a_k, s_{k+1}\right) = -c_k \tag{15.19}$$

Accordingly, during one episode, the cumulated reward of policy π yielded by the SAS agent can be denoted as follows:

$$R\left(\pi\right) = \sum_{k=1}^{\xi} \eta^k r_k\left(s_k, a_k, s_{k+1}\right) \tag{15.20}$$

where $\eta \in \left(0, 1\right]$ is a discount factor indicating that the importance of future routing selection compared to that at the current moment. Generally, η is set as a decimal close to 1, which can force the agent to pay more attention to the decision at the current moment while remaining the ability of asking into account the total return of the entire path. Moreover, in the early stages of exploration, there is no guarantee that the destination node will be reached every time by the SAS agent. Thus, to accelerate the exploration to the destination node, an additional reward Γ is attached to the action which leads to the destination node as follows:

$$r_k = \begin{cases} r_k + \Gamma a_k = n_D \\ r_k \quad a_k \neq n_D \end{cases} \tag{15.21}$$

The episodic interaction finishes as soon as the current node is the destination node or the number of selected nodes exceeds the maximum path length. Furthermore, the agent records the best path $a_U^* = \{a_1, a_2 \ldots a_\xi\}$ and the corresponding routing cost c_U.

15.3.1.1 DRL training for routing selection

A deep neural network is utilized to extract features from previous experiences, regarding to the satellite communication network with varying spatial connectivity, and then generate the corresponding policy $\pi(s_k)$. The most popular DRL diagram actor-critic is used to train the agent. Specifically, actor-critic DRL is composed of two separated neural networks called actor and critic. The actor neural network is employed for generating the routing selection policy π, while the critic neural network is intended to evaluate the potential expected accumulated routing cost for each satellite node.

The structure of the deep neural network is also shown in Figure 15.3. In general, it contains two fully-connected layers (size: 128) and one layer of long- and short-term memory (LSTM) cell (size:128) to extract the feature of the satellites [26]. Sequentially, the output feature of LSTM (denoted by h_k) inputs to another fully-connected layer to produce the final policy $\pi(s_k)$ and value function$v(s_k)$. As depicted above, $\pi(s_k)$ is a probability distribution from a Softmax function over the accessible satellite node.

At the same time, the LSTM feature h_k is also output and saved. This is done to generate the policy for the next time the DRL neural network loads the temporally related information of the changing satellite network. Note that, the LSTM used here belongs to the recurrent neural network. In the NGSO satellite systems, the currently accessible state s_k only indicates the position of the current node, while it does not include utility information between different nodes of the entire satellite communication network. Hence, this process can be regarded as a partial observable MDP (POMDP). It has been shown that recurrent neural network performs well in POMDP [27]. Accordingly, to encourage the agent to consider varying spatial position and connectivity in successive time steps, LSTM is used to record the utility rule in different satellite nodes. Then the agent samples one node from the available neighboring nodes with respect to the yielded policy and performs it as the selected node in the environment followed by an instantaneous reward r_k. Finally, the interaction experience of DRL, which refers to $(s_k, a_k, s_{k+1}, r_k, h_k)$, is saved in the experience replay buffer. Finally, the neural network parameters are updated the experience via DRL training. The whole interaction is achieved through the following process.

1. At the kth routing selection, the current position of the satellite node is extracted as the state s_k. Then, the state combined with the temporal related feature at the last time step from the LSTM composes the input to DRL neural network.
2. The critic neural network of DRL model outputs a value function $v\left(s_k\right)$ to evaluate the expected reward at s_k. Meanwhile, the actor neural network outputs a corresponding policy $\pi\left(s_k\right)$.
3. For a given policy $\pi\left(s_k\right)$, the agent randomly samples an action a_k according to a standard disturbance variable ε. This is done to remain the potential of exploration to the other part of the satellite communication network.
4. The environment is updated with respect to the action a_k. Specifically, the current satellite node updates as $s_k \leftarrow a_k$. The observed state updates as $s_k \leftarrow s_{k+1}$. The cache of LSTM feature updates as $h_k \leftarrow h_{k+1}$. Moreover, the propagation time and queuing time update with respect to the adjacency of the current node and the re-generated Poisson distribution. Finally, an instantaneous reward r_k is responded.
5. The interaction experience $\left(s_k, a_k, s_{k+1}, r_k, h_k\right)$ is saved in the experience replay buffer.
6. Once time step k satisfies the termination condition, i.e., $s_k = n_D$or$k = \zeta_{max}$, the environment responds an additional reward and the interaction process will be interrupted. Substantially, the experience replay buffer yields a batch of randomly sampled experience into the optimizer to update the parameters in actor and critic neural networks.
7. Consequently, the agent records the completed path and the corresponding routing cost. After the whole training process terminates, the DRL model outputs a routing subset $\Re$ which meets the communication requirement, i.e., $c_U \leq R_0$.

15.3.1.2 DRL updating for routing selection

This section focuses on the updating diagram of the DRLR algorithm. Considering the characteristic of DRL, in general, directly using the interaction experience to update the parameters leads to lower data utilization and exploration efficiency. A3C [28] is a typical representative of actor-critic algorithm which combines the advantages of classical DQN [29] and policy gradient [30]. To further improve data utilization, trust region policy optimization (TRPO) [31] is proposed based on important sampling, which performs well on both discrete and continues state space.

In this chapter, the neural network parameters are trained and updated according to proximal policy optimization (PPO) [32]. PPO is a simplification of TRPO, which is able to reduce computational complexity while remaining the core advantage. To tackle this problem, a synchronous version of PPO algorithm is employed, which can alternate between the sampled experience through time-varying interacting policy.

Specifically, before the rollout in each episode, the sub-models copy parameters from the master model. Then the gradient will be computed in each sub-models and be uniformly collected in the master model. Under the principle of PPO, the current interaction policy with the satellite network should not change sharply from the old policy, which is denoted as $\pi_{old}\left(a_k|s_k\right)$. Thus, the KL divergence penalty is used to constrain the difference to compose a new surrogate objective. Moreover, this objective is further clipped for stable convergence. The clip function is omitted in this chapter, and details can be found in Reference 32. After processing by PPO, normalized gradients synchronously update the parameters. Note that the neural network and the environment in each workflow are independent of each other. θ_π and θ_v is used to parameterize the actor and critic neural networks, respectively. To maximize the DRL reward, the agent update θ_π and θ_v by stochastic gradient descent according to (15.19). Furthermore, due to unreachable conditions between some nodes, in our experiment, the DRL algorithm is easy to converge to local optimum often. Thus, the maximum entropy regularization is employed to encourage the exploration, while remaining the generalization ability under different utility situation of the satellite network. The parameters are updated according to (15.22) and (15.23).

$$\theta_v \leftarrow \theta_v + \phi_v \sum_{k=1}^{K} \nabla_{\theta_v} \left(r_k + \eta V\left(s_{k+1};\theta_v\right) - V\left(s_k;\theta_v\right)\right)^2, \tag{15.22}$$

$$\theta_\pi \leftarrow \theta_\pi + \phi_\pi \sum_{k=1}^{K} \left\{ \begin{array}{c} \nabla_{\theta_\pi} ln \frac{\pi(a_k|s_k)}{\pi_{old}(a_k|s_k)} \left(r_k + \eta V\left(s_{k+1};\theta_v\right) - V\left(s_k;\theta_v\right)\right) \\ -\beta \mathrm{KL}\left(\pi\left(s_k\right)|\pi_{old}\left(s_k\right)\right) + \gamma H\left(\pi\left(s_k;\theta_\pi\right)\right) \end{array} \right\} \tag{15.23}$$

where ϕ_π and ϕ_v are denoted as the gradient learning rates of actor and critic neural works. In addition, K_L and H are defined as the KL divergence and entropy function, while β and γ represent the corresponding coefficient.

15.3.2 *Fast response anti-jamming algorithm*

After obtaining the $\Re$, the user utilizes Q-learning to choose the reliable path from $\Re$ for anti-jamming defense. This design has many advantages: first, it can response quickly to smart jamming and makes an efficient anti-jamming decision; second, it can reduce network congestion caused by users' competition for the best route with the minimal routing cost. Because all the links in $\Re$ meet the communication requirements; thirdly, it is able to supervise the network dynamics, and if the routing cost of the current optimal selection still does not meet the requirements, the user restarts the DRL process and explore the new network state to update the $\Re$.

The Q-routing function of the user is defined as $Q_{\Re,t}\left(a^0_{U,n}\right)$, which is updated as follows:

$$Q_{\Re,t+1}\left(a^0_{U,n}\right) = (1-\alpha)\, Q_{\Re,t}\left(a^0_{U,n}\right) + \alpha \left(r_U + \eta \max_{n'} Q_{\Re,t}\left(a^0_{U,n'}\right)\right) \tag{15.24}$$

where $\alpha \in (0,1)$ is learning rate, $\alpha = \alpha_0 / \left(\omega\left(a\right) lg\left(\omega\left(a\right)\right)\right)$,

$$\sum_{l=0}^{\infty} \alpha_l = \infty, \sum_{l=0}^{\infty} \alpha_l^2 < \infty \tag{15.25}$$

where α_0 is the initial learning step size, and $\omega(a)$ is the times that the action a is selected.

$\pi_U(t) = \pi_{U,1}(t), \pi_{U,2}(t) \dots \pi_{U,F}(t)$ is denoted as the mixed strategy of routing selection, where $\sum_{n=1}^{F} \pi_{U,n}(t) = 1$, and $\pi_{U,n}(t)$ represents the probability to choose the routing path $a^0_{U,n} \in \left[a^0_{U,1}, a^0_{U,2} \dots a^0_{U,F}\right]$. Specially, the $\pi_{U,n}(t)$ is updated as follows:

$$\pi_{U,n}(t+1) = \frac{e^{Q_{\Re,t}\left(a^0_{U,n}\right)/\xi}}{\sum_{n'=1}^{F} e^{Q_{\Re,t}\left(a^0_{U,n'}\right)/\xi}} \tag{15.26}$$

where ξ is the parameter of the Boltzmann model.

$$\begin{array}{ll} \xi = \xi_0 e^{-\upsilon t} & \xi \geq \hat{\xi} \\ \xi = \hat{\xi} & \xi < \hat{\xi} \end{array} \tag{15.27}$$

ξ_0 is related to the exploration time, and ξ represents the ending condition in the exploration state. υ affects the transition from exploration to exploitation.

The Q function of the jammer $Q_{J,\varsigma}(a_{J,m}, p_{J,v})$ is updated as follows:

$$Q_{J,\varsigma+1}(a_{J,m}, p_{J,v}) = Q_{J,\varsigma}(a_{J,m}, p_{J,v}) + \alpha\left(r_J - Q_{J,\varsigma}(a_{J,m}, p_{J,v})\right) \tag{15.28}$$

At time ς, $\pi^J_p(\varsigma) = \left[\pi^J_{p,1}(\varsigma), \pi^J_{p,2}(\varsigma) \dots \pi^J_{p,V}(\varsigma)\right]$ and $\pi^J_n(\varsigma) = \left[\pi^J_{n,1}(\varsigma), \pi^J_{n,2}(\varsigma) \dots \pi^J_{n,G}(\varsigma)\right]$ are denoted as the jammer's mixed policies of jamming power and node. Similarly, $\pi^J_{p,v}(\varsigma)$ and $\pi^J_{n,m}(\varsigma)$ are the probabilities to choose the jamming power $p_{J,v} \in \left[p_{J,1}, p_{J,2} \dots p_{J,V}\right]$ and jamming node $a_{J,m} \in \left[a_{J,1}, a_{J,2} \dots a_{J,G}\right]$, which are updated as follows:

$$\pi^J_{p,v}(\varsigma+1) = \frac{e^{\frac{1}{G\xi}\sum_{m'=1}^{G} Q_{J,\varsigma}\left(a_{J,m'}, p_{J,v}\right)}}{\sum_{v'=1}^{V} e^{\frac{1}{G\xi}\sum_{m'=1}^{G} Q_{J,\varsigma}\left(a_{J,m'}, p_{J,v'}\right)}} \tag{15.29}$$

$$\pi^J_{n,m}(\varsigma+1) = \frac{e^{Q_{J,\varsigma}\left(a_{J,m}, p_{J,v}\right)/\xi}}{\sum_{m'=1}^{G} e^{Q_{J,\varsigma}\left(a_{J,m'}, p_{J,v}\right)/\xi}} \tag{15.30}$$

15.3.3 Analysis of the proposed scheme

In this chapter, the DRLR algorithm is used to get the available routing subset $\Re$ to simplify the decision space for anti-jamming game. If the smart jamming is launched before obtaining $\Re$, the impact of jamming is equivalent to changing the dynamics of the environment, and the user is still able to obtain $\Re$ by the DRLR algorithm. On the contrary, if the jamming is launched after obtaining $\Re$, the set $\Re$ will get smaller. However, it still contains all the available paths in the current status.

Furthermore, if the jamming does not attack the key nodes in the $\Re$, the jamming attacks are invalid. Instead, if the jammer attacks the node in the $\Re$, the jammed routing path will be deleted from the $\Re$, and the user will get the available routing set $\Re', \Re' \subset \Re$. After the dynamical anti-jamming game, $\Re'$would be further narrowed down until it converges to $\Re^*$, which is determined by the SE. However, as mentioned above, the environment changes caused by the smart jamming maybe just affect a few nodes. Thus the restart of the DRL process is unnecessary and costly, though it is indeed able to make it. In fact, the user only needs to maintain the validity of the available routing set $\Re^*$, then, the current communication requirements will be met well and the smart jamming can be tackled quickly. Therefore, based on $\Re$ obtained by DRLR algorithm, FRA is proposed to achieve fast and reliable anti-jamming routing. The analysis on the SE of the anti-jamming policies obtained by FRA is given as follows.

From the perspective of Stackelberg anti-jamming game, the jammer firstly determines the jamming policy, and automatically adjusts it according to the jamming effect. As for the user, it explores the jamming environment in the DRL process, and obtains $\Re$ which is the action set of the FRA algorithm for anti-jamming routing game. Due to the finite satellite nodes in $\Re$, the policies of the user and jammer are both finite. According to Theorem 2, there is a stable SE between the user and the smart jammer.

Based on Reference [33], the Q function can be described by the differential equation:

$$\frac{dQ(\varsigma+1)}{d\varsigma} = \alpha\left(r - Q\left(\varsigma\right)\right) \tag{15.31}$$

Substituting (15.31) into differential (15.30), and (15.32) is obtained. It has been proved in Reference 33 that the stable strategy of jamming power can be obtained by

$$\frac{d\pi^J_{n,m}(\varsigma)}{d\varsigma} = \pi^J_{n,m}\left(\varsigma\right)\frac{\alpha_j}{\xi_j}\left\{\left[r_{J,m}\left(\varsigma-1\right) - \sum_{m'=1}^{G}\pi^J_{n,m'}\left(\varsigma\right)r_{J,m'}\left(\varsigma-1\right)\right] - \xi_j\sum_{m'=1}^{G}\pi^J_{n,m'}\left(\varsigma\right)ln\left(\frac{\pi^J_{n,m}(\varsigma)}{\pi^J_{n,m'}(\varsigma)}\right)\right\} \tag{15.32}$$

$$\pi^{J*}_{n,s} = \frac{e^{r_{J,s}/\xi_j}}{\sum_{s'=1}^{G}e^{r_{J,s'}/\xi_j}} \tag{15.33}$$

As for the jamming power policy and the user's strategy, we can get similar results.

Denoting the strategy of user and jammer as $\pi\left(\varsigma\right) = \left(\pi_J\left(\varsigma\right), \pi_U\left(\varsigma\right)\right)$, the convergence of $\pi\left(\varsigma\right)$ can be analyzed by an ordinary differential equation (ODE). Set the right hand of (15.32) as $f(\pi)$. As $\alpha \to 0$, $\pi\left(\varsigma\right)$ is able to weakly converge to$(\pi^*_J, \pi^*_U\left(\pi^*_J\right))$, which is the solution of $d\pi/d\varsigma = f(\pi)$, with any initial value $\pi\left(0\right) = \pi_0$. According to Reference 6, the Q-Learning algorithm can obtain optimal strategy, if the learning rate satisfies (25). Thus, the proposed FRA algorithm can converge to an optimal strategy. Then, it can be proved by contradiction that the optimal strategy is a stable SE. Assuming the optimal strategy is not a SE point, according to Reference 34, Q-Learning process converges to a steady point, which is an ODE's solution. Thus, the no-SE strategy is steady, which is in contradiction to Theorem 1.

Therefore, the proposed scheme can obtain an optimal strategy which is the steady SE.

15.4 Experiments and discussions

Simulation experiments are conducted to demonstrate the performance of the proposed scheme. The NGSO satellite systems contain 120 satellite nodes, distributed in 2 circular polar-orbit constellation networks R_1, R_2, where R_1 includes 60 nodes numbered 1 to 60, R_2 consists of 60 nodes numbered 61 to 120. Other parameters are given in Table 15.1.

Table 15.1 Simulation parameters

Parameters	**Value**
Orbital inclination of R_1 and R_2	$\varphi = 90^\circ$
Orbital period of and Orbital period of R_1 and R_2	$T = 120$ min
Orbital altitude of f and R_1 and R_2	$H_1 = 800$ km, $H_2 = 1200$ km
Frequency used in R_1 and R_2	$f_1 = 1.5\text{GHz}, f_2 = 3\text{GHz}$
Poisson parameter in R_1	$\lambda_1 \sim U(800\text{M}, 900\text{M})$ bit
Poisson parameter in R_2	$\lambda_2 \sim U(300\text{M}, 400\text{M})$ bit
Channel noise of R_1	$n_{01} = -180$ dBmW
Channel noise of R_2	$n_{02} = -185$dBmW
Distance threshold	$d_{th} = 6500$km
Channel bandwidth	$B = 10$ MHz
Transmission power	$p_u = 1000$W
Transmission data	$N_D = 400$Mbit
Communication requirement	$R_0 = 2.2000$
Observation accuracy of jammer	$\vartheta \sim N(1, 0.5)$
Jamming cost per unit power	$\lambda_j = 0.001$
Discount factor	$\eta = 0.99$
Weight of the KL divergence penalty	$\beta = 0.1$
Weight of the entropy exploration	$\gamma = 0.1$
Learning rates in DRLR	$\phi_v = \phi_\pi = 0.0003$
Boltzmann coefficients $\xi_0, \hat{\xi}, \upsilon$	10^7, 0.1, 0.1

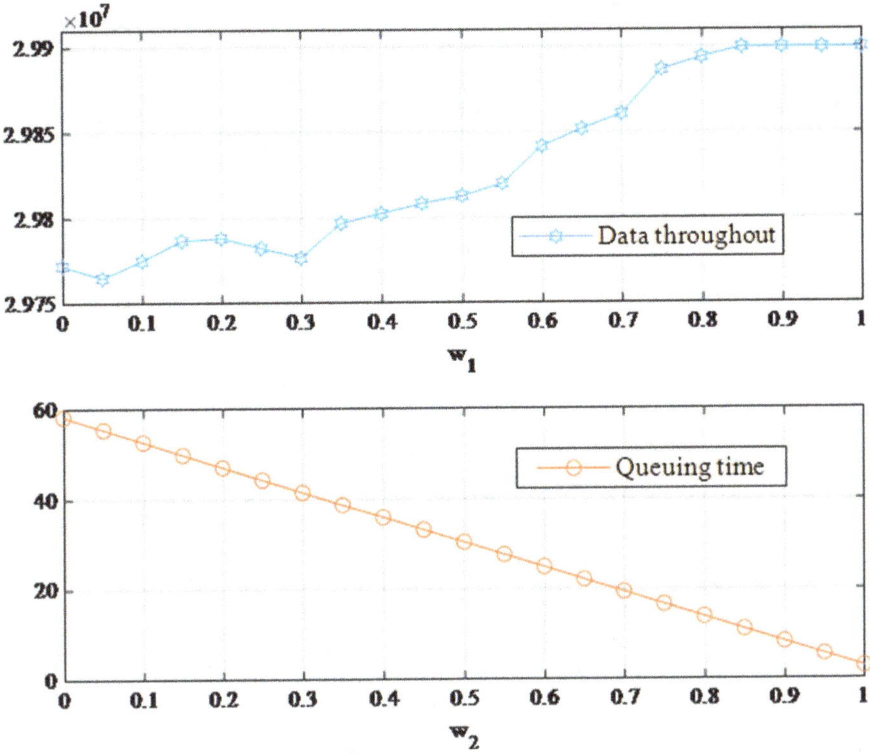

Figure 15.4 *The comparison experiments for weighted parameters. n_S and n_D are the 60th and 90th node, and the routing time is the 20th minute.*

15.4.1 The performance of multi-objective routing function

The weighted multi-objective routing cost function with parameters w_1, w_2 is given in (15.6). As shown in Figure 15.4, the routing performance changes with the different weighted parameters. It proves that data throughout increases as w_1 rises from 0 to 1. Because a larger w_1 indicates that the paths with larger channel rate should be given priority. Meanwhile, the queuing time reduces with the increasing ofw_2 $(w_2 = 1 - w_1)$, which reflects the user's tolerance degree of queuing time and guides to select the path with less local uncertain traffic.

15.4.2 The performance of the DRLR algorithm

As Ruiz-De-Azua *et al.* [35] remarked, the OLSR scheme could be well applied to NGSO satellite systems and had good routing performance. Thus, the proposed algorithm is compared to OLSR schemes with different routing metrics including the smallest hop counts and the shortest distance. The comparison of the routing performance from the 60th node to the 90th node between the proposed algorithm and the OLSR schemes is shown in Figure 15.5. The blue dotted line represents the minimal

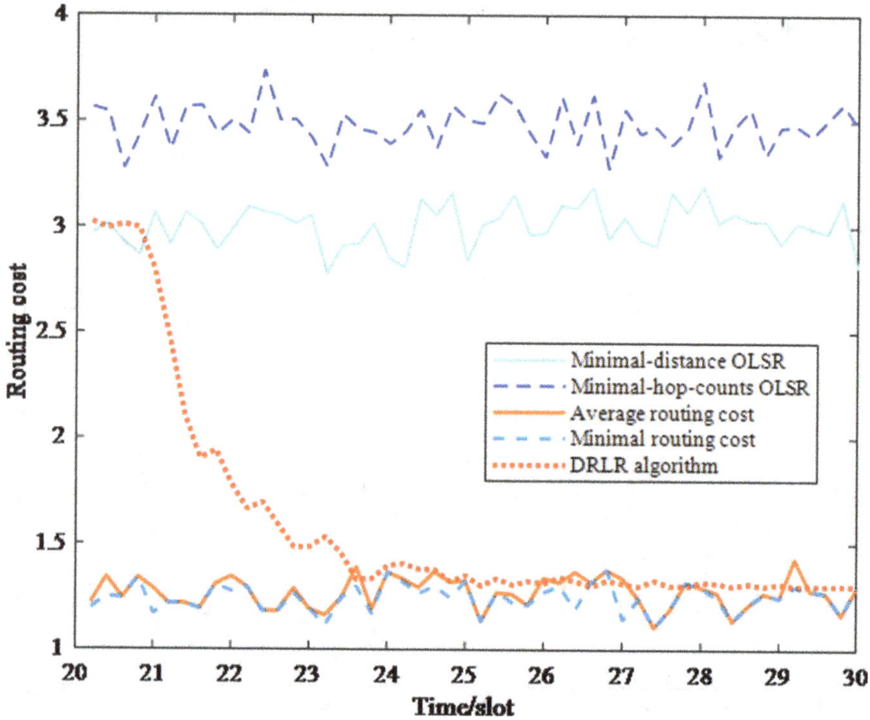

Figure 15.5 The comparison of routing performance from the 60th node to the 90th node, and the routing time is from the 20th minute to the 30th minute, $w_1 = w_2 = 0.5$

routing cost obtained by the optimal routing selection with complete information; the yellow triangle expresses the average routing cost considering the uncertain burst traffic; the magenta rectangle and the green dotted inverse triangle represent the routing cost of the OLSR algorithm with the metrics of minimal distance and minimal hop counts, respectively. As indicated in Figure 15.5, the minimal-distance OLSR scheme has the lower routing cost than that of the minimal-hop-count OLSR scheme, while both are above the theoretical value. But the proposed algorithm has significant performance improvement than the OLSR scheme and approximately converges to the mean value.

Figure 15.6 and Figure 15.7 show the routing cost comparison in a quarter period, as we can see that the performance of the proposed algorithm is always better than the OLSR scheme and gradually converges to the optimal value.

The available routing set $\Re$ obtained by the DRLR algorithm, from the 60th node to the 90th node with $w_1 = w_2 = 0.5$ in the 20th minute, is given in Table 15.2.

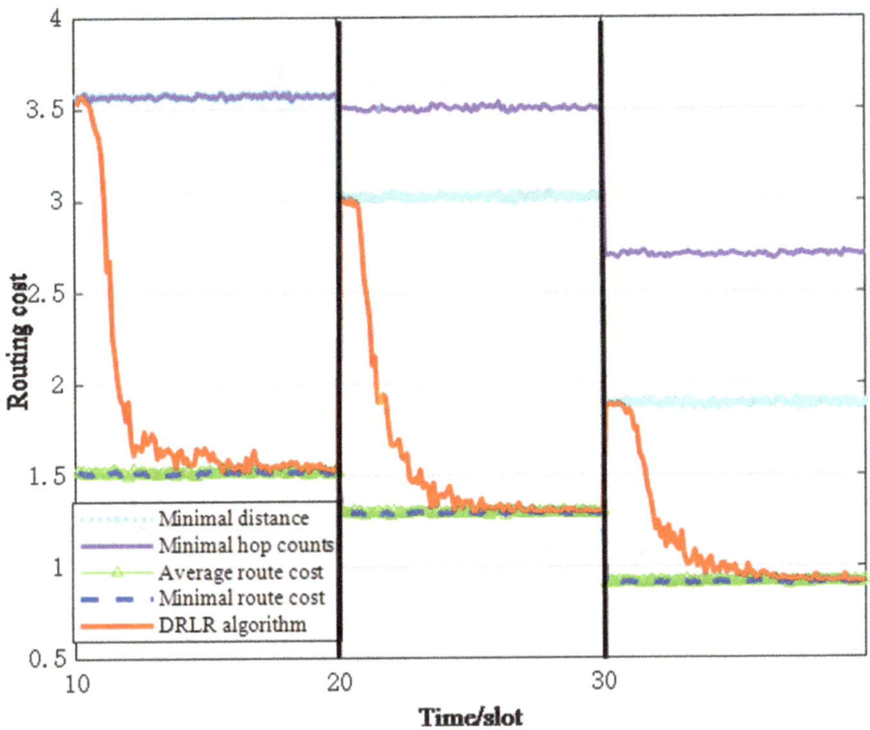

Figure 15.6 *The comparison of routing performance from the 60th node to the 90th node between proposed algorithm and the OLSR schemes,* $w_1 = w_2 = 0.5$

15.4.3 The performance of the FRA algorithm

As shown in the left picture of Figure 15.8, when the jamming aims at the 110th node, the agent automatically updates routing path according to the jammed network, and the routing cost rises from 1.30 to 1.97. Due to the limitation of the current network structure, almost all of the better paths contain the 110th node. If the 110th node is jammed, the suboptimal path is $60 \rightarrow 111 \rightarrow 101 \rightarrow 91 \rightarrow 81 \rightarrow 90$, and the routing cost increases approximately to 1.97. In the right picture of Figure 15.8, when the jammer launches jamming at the 40th node, the agent reselects routing path and the new utility is still about 1.3, because that if the 40th node is jammed, the current selected path $60 \rightarrow 110 \rightarrow 100 \rightarrow 90$ is nearly equal to the optimal path.

As for the user, the routing path selection is a stateless problem. The action set and reward are $\Re$ and r_U. The user wants to choose the better path with a lower routing cost. As for the jammer, the jamming power and jamming node are both the optimization objective, and the jammer aims to launch jamming attacks to the most crucial node with the lowest power. The state set of the user is $p_{j,v} \in [200\text{W}, 500\text{W}, 1000\text{W}, 1500\text{W}]$, and the action set is Ω which is given in Table 15.3, and the jammer's reward is r_J.

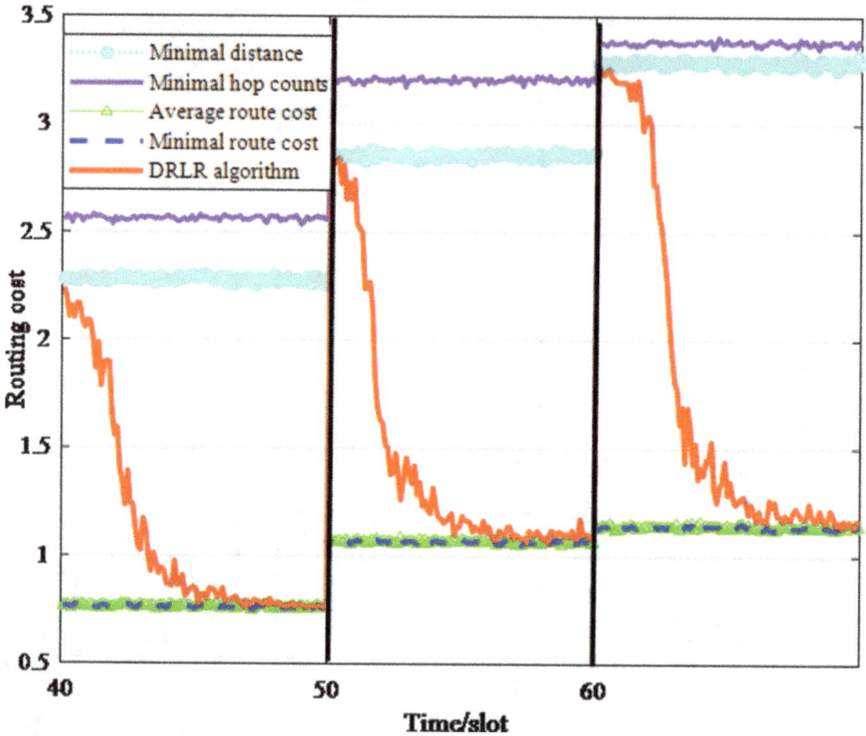

Figure 15.7 *The comparison of routing performance from the 1st node to the 30th node,* $w_1 = w_2 = 0.5$

Table 15.2 *The available routing set* $\mathfrak{R}$

Number	Path	Routing cost
1	60 → 110 → 40 → 90	1.2984
2	60 → 110 → 100 → 90	1.3056
3	60 → 110 → 50 → 100 → 90	1.4851
4	60 → 110 → 100 → 40 → 90	1.4882
5	60 → 120 → 110 → 100 → 90	1.4911
6	60 → 110 → 100 → 30 → 90	1.4926
7	60 → 110 → 21 → 90	1.5534
8	60 → 110 → 21 → 81 → 90	1.7335
9	60 → 110 → 91 → 81 → 90	1.8168
10	60 → 110 → 31 → 81 → 90	1.8183
11	60 → 111 → 101 → 91 → 81 → 90	1.9694
12	60 → 50 → 100 → 90	2.1457

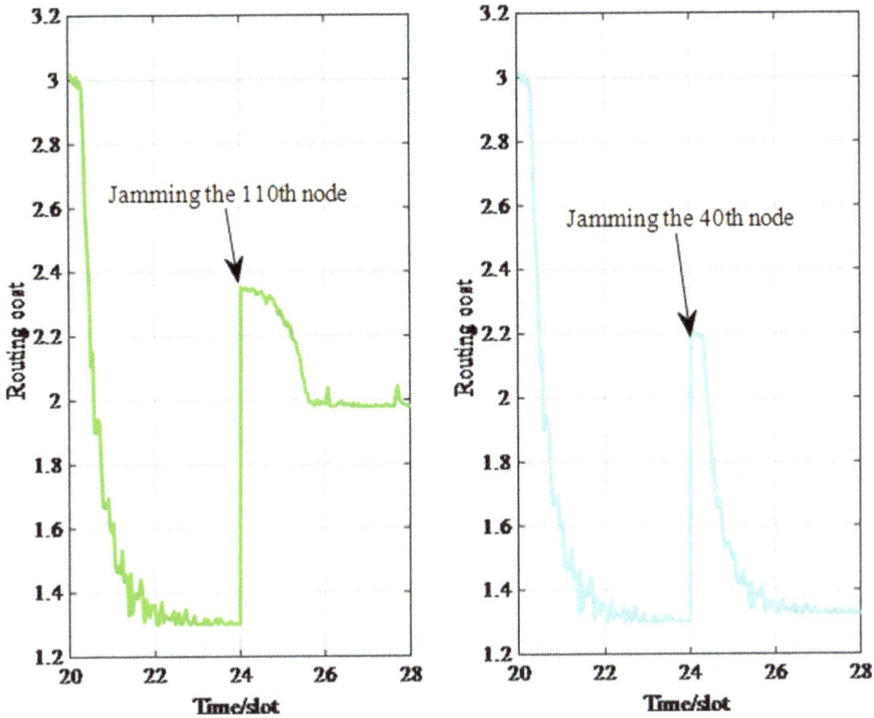

Figure 15.8 *The anti-jamming performance of the simple DRL-based algorithm. n_S and n_D are the 60th and the 90th node; the routing time is the 20th minute,* $w_1 = w_2 = 0.5$

The jamming power selection policies are shown in Figure 15.9. Considering both the jamming effects and power consumption, the power 2 (i.e., 500W) is the best power selection for the smart jammer. The jamming node selection policies are indicated in Figure 15.10. As previously mentioned, the node 10, i.e., the 110th node, is the key node, and the jammer chooses the 110th node to launch jamming so that the jammer could obtain the maximal jamming effect.

The user's routing path selection policies are shown in Figure 15.11. The jammer chooses the 110th node to launch jamming attacks, then, the top ten routing paths with lower routing cost are all interrupted, so the user reselects the path 11 for reliable communication. Because the routing cost of the path 12 is very close to

Table 15.3 *The action space of jamming nodes* Ω

	1	**2**	**3**	**4**	**5**	**6**	**7**	**8**	**9**	**10**	**11**	**12**
Node	21	30	31	40	50	81	91	100	101	110	111	120

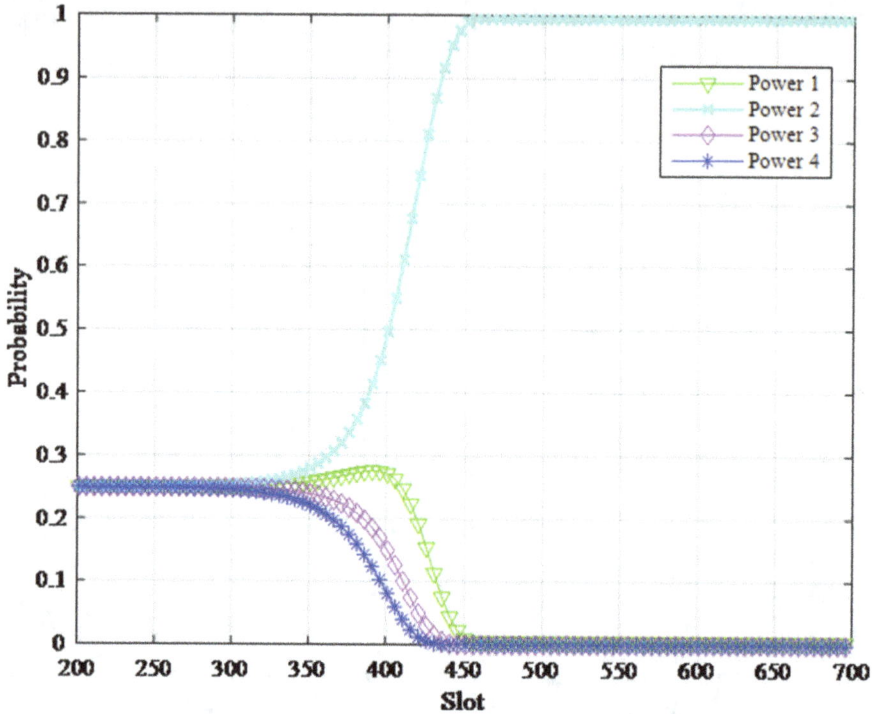

Figure 15.9 The jamming power selection policies of the jammer

that of the path 11, thus, the selection probability of the path 11 does not converge exactly to 1.

As indicated in Figure 15.12, during the anti-jamming game, the reward of both the jammer and user converges to the equilibrium point. As the jammer gradually converges to the strategies with the maximum jamming effect, the user's routing cost is forced to rise, but it still converges to the stable and best strategy in the current jamming environment. The jammer finds the optimal jamming policy and no longer diverges it. Then, the user makes the best and stable anti-jamming decision as well.

The routing cost comparison between the random selection anti-jamming (RSA) algorithm and the FRA algorithm is shown in Figure 15.13. The upper picture shows the routing cost comparison in the whole anti-jamming process, and the under picture elaborates on the comparison of routing cost after the jamming policy has converged to the equilibrium point. It has proved that compared with the RSA algorithm, the proposed FRA algorithm has the better anti-jamming performance with the lower routing cost and better convergence.

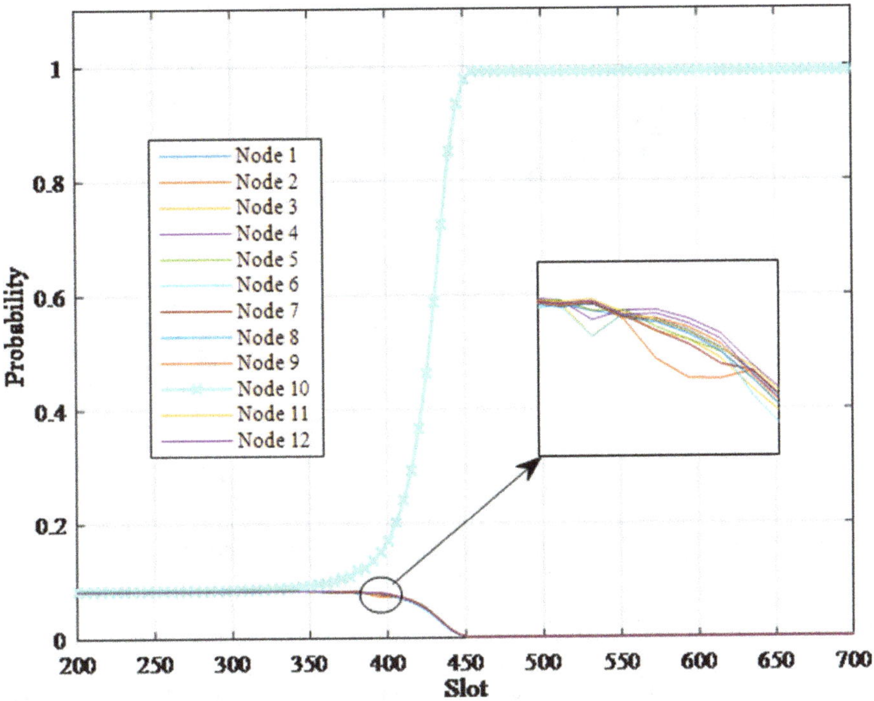

Figure 15.10 The jamming node selection policies of the jammer

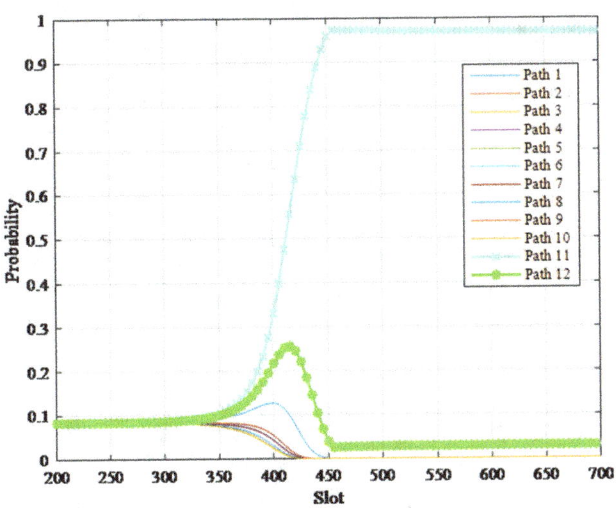

Figure 15.11 The routing path selection policies of the user

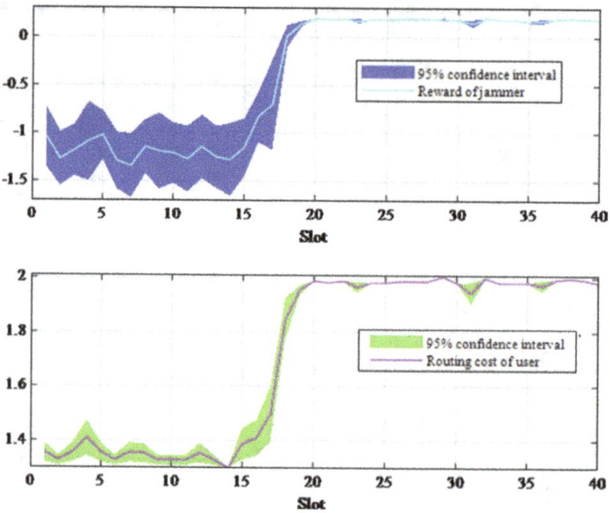

Figure 15.12 The convergence of the jammer and user

15.5 Chapter summary

This chapter investigates the anti-jamming routing selection problem for the NGSO satellites system. First, the anti-jamming routing selection problem between users and smart jammers is formulated as a hierarchical Stackelberg anti-jamming routing game. Second, DRLR is proposed to obtain an available routing subset. Based on

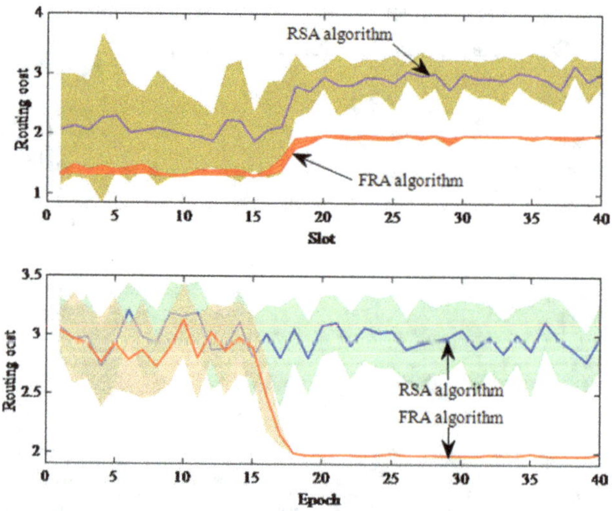

Figure 15.13 The routing cost comparison of the RSA algorithm and the FRA algorithm. The shaded areas represent 95% confidence intervals.

this subset, FRA is proposed to make a quick anti-jamming decision. The jammer can automatically adjust the targeted node and jamming power according to the jamming effect, and the user utilizes DRLR and FRA algorithm to actively explore the dynamic network, empirically analyze the jammer's strategies and adaptively make an anti-jamming decision. Finally, the simulations have proven that the proposed algorithm has better performance than existing approaches, and the anti-jamming policies converge to the Stackelberg equilibrium.

References

[1] Liu J., Shi Y., Fadlullah Z.M., Kato N. 'Space-air-ground integrated network: a survey'. *IEEE Communications Surveys & Tutorials*. 2019, vol. 20(4), pp. 2714–41.

[2] Ziebold R., Medina D., Romanovas M., Lass C., Gewies S. 'Performance characterization of GNSS/IMU/DVL integration under real maritime jamming conditions'. *Sensors (Basel, Switzerland)*. 2018, vol. 18(9), pp. 2954–74.

[3] Wang P., Wang Y., Cetin E., Dempster A.G., Wu S. 'GNSS jamming mitigation using adaptive-partitioned subspace projection technique'. *IEEE Transactions on Aerospace and Electronic Systems*. 2019, vol. 55(1), pp. 343–55.

[4] Zou Y., Zhu J., Wang X., Hanzo L. 'A survey on wireless security: technical challenges, recent advances, and future trends'. *Proceedings of the IEEE*. 2019, vol. 104(9), pp. 1727–65.

[5] Yao F., Jia L., Sun Y., Xu Y., Feng S., Zhu Y. 'A hierarchical learning approach to anti-jamming channel selection strategies'. *Wireless Networks*. 2019, vol. 25(1), pp. 201–13.

[6] Han C., Niu Y. 'Cross-layer anti-jamming scheme: A hierarchical learning approach'. *IEEE Access: Practical Innovations, Open Solutions*. 2011, vol. 6, pp. 34874–83.

[7] Fischer D., Basin D., Eckstein K., Engel T. 'Predictable mobile routing for spacecraft networks'. *IEEE Transactions on Mobile Computing*. 2011, vol. 12(6), pp. 1174–87.

[8] Lu Y., Sun F., Zhao Y. 'Virtual topology for LEO satellite networks based on earth-fixed footprint mode'. *IEEE Communications Letters*. 2006, vol. 17(2), pp. 357–60.

[9] Boukerche A., Turgut B., Aydin N., Ahmad M.Z., Bölöni L., Turgut D. 'Routing protocols in AD hoc networks: a survey'. *Computer Networks*. 2011, vol. 55(13), pp. 3032–80.

[10] Marina M.K., Das S.R. 'Ad hoc on-demand multipath distance vector routing'. *Wireless Communications and Mobile Computing*. 2006, vol. 6(7), pp. 969–88. Available from http://doi.wiley.com/10.1002/wcm.v6:7

[11] Samar P., Pearlman M.R., Haas Z.J. 'Independent zone routing: an adaptive hybrid routing framework for ad hoc wireless networks'. *IEEE/ACM Transactions on Networking*. 2004, vol. 12(4), pp. 595–608.

[12] Toutouh J., Garcia-Nieto J., Alba E. 'Intelligent OLSR routing protocol optimization for vanets'. *IEEE Transactions on Vehicular Technology*. 2004, vol. 61(4), pp. 1884–94.

[13] Radhakrishnan R., Edmonson W.W., Afghah F., Rodriguez-Osorio R.M., Pinto F., Burleigh S.C. 'Survey of inter-satellite communication for small satellite systems: physical layer to network layer view'. *IEEE Communications Surveys & Tutorials*. 2004, vol. 18(4), pp. 2442–73.

[14] Asadpour M., Hummel K.A., Giustiniano D., Draskovic S. 'Route or carry: motion-driven packet forwarding in micro aerial vehicle networks'. *IEEE Transactions on Mobile Computing*. 2004, vol. 16(3), pp. 843–56.

[15] Tang F., Mao B., Fadlullah Z.M. 'On removing routing protocol from future wireless networks: A real-time deep learning approach for intelligent traffic control'. *IEEE Wireless Communications*. 2004, vol. 25(1), pp. 154–60.

[16] Han C., Huo L., Tong X., Wang H., Liu X. 'Spatial anti-jamming scheme for internet of satellites based on the deep reinforcement learning and Stackelberg game'. *IEEE Transactions on Vehicular Technology*. 2018, vol. 69(5), pp. 5331–42.

[17] Moscholios I.D., Vassilakis V.G., Sarigiannidis P.G., Sagias N.C., Logothetis M.D. 'An analytical framework in LEO mobile satellite systems servicing batched poisson traffic'. *IET Communications*. 2018, vol. 12(1), pp. 18–25. Available from https://onlinelibrary.wiley.com/toc/17518636/12/1

[18] Korçak Ö., Alagöz F., Jamalipour A. 'Priority-based adaptive routing in NGEO satellite networks'. *International Journal of Communication Systems*. 2007, vol. 20(3), pp. 313–33. Available from http://doi.wiley.com/10.1002/dac.v20:3

[19] Kim H.-S., Paek J., Bahk S. 'QU-RPL: queue utilization based RPL for load balancing in large scale industrial applications'. *2th Annual IEEE International Conference on Sensing, Communication, and Networking (SECON)*; Seattle, WA, 2018. pp. 265–73.

[20] Kawamoto Y., Nishiyama H., Kato N., Kadowaki N. 'A traffic distribution technique to minimize packet delivery delay in multilayered satellite networks'. *IEEE Transactions on Vehicular Technology*. 2018, vol. 62(7), pp. 3315–24.

[21] Michael N., Tang A. 'HALO: hop-by-hop adaptive link-state optimal routing'. *IEEE/ACM Transactions on Networking*. 2015, vol. 23(6), pp. 1862–75.

[22] Taleb T., Mashimo D., Jamalipour A., Kato N., Nemoto Y. 'Explicit load balancing technique for NGEO satellite IP networks with on-board processing capabilities'. *IEEE/ACM Transactions on Networking*. 2009, vol. 17(1), pp. 281–93.

[23] Song G., Chao M., Yang B., Zheng Y. 'TLR: a traffic-light-based intelligent routing strategy for NGEO satellite IP networks'. *IEEE Transactions on Wireless Communications*. 2014, vol. 13(6), pp. 3380–93.

[24] Alagoz F., Korcak O., Jamalipour A. 'Exploring the routing strategies in next-generation satellite networks'. *IEEE Wireless Communications*. 2007, vol. 14(3), pp. 79–88.

[25] Han Z., Niyato D., Saad W., Başar T., Hjørungnes A. *Game theory in wireless and communication networks [online]*. Cambridge University Press; 2011 Oct 20. Available from https://www.cambridge.org/core/product/identifier/9780511895043/type/book

[26] Zhao Z., Chen W., Wu X., Chen P.C.Y., Liu J. 'LSTM network: a deep learning approach for short-term traffic forecast'. *IET Intelligent Transport Systems*. 1999, vol. 11(2), pp. 68–75. Available from https://onlinelibrary.wiley.com/toc/17519578/11/2

[27] Sutton R.S., Precup D., Singh S. 'Between MDPS and semi-MDPS: A framework for temporal abstraction in reinforcement learning'. *Artificial Intelligence*. 1999, vol. 112(1–2), pp. 181–211.

[28] Mnih V., Badia A.P., Mirza M., Graves A, *et al.* 'Asynchronous methods for deep reinforcement learning'. *Proc. ICML*; 2016. pp. 1928–37.

[29] Mnih V., Kavukcuoglu K., Silver D, *et al.* 'Human-level control through deep reinforcement learning'. *Nature*. 2015, vol. 518(7540), pp. 529–33.

[30] Schulman J., Moritz P., Levine S., Jordan M, *et al.* 'High-dimensional continuous control using generalized advantage estimation'. *ArXiv Preprint ArXiv:1506.02438*. 2015.

[31] Schulman J., Levine S., Moritz P., Jordan M.I, *et al.* 'Trust region policy optimization'. *Proceedings of ICML*; 2015. pp. 1889–97.

[32] Schulman J., Wolski F., Dhariwal P., Radford A, *et al.* 'Proximal policy optimization algorithms'. *ArXiv Preprint ArXiv:1707.06347*. 2017.

[33] Kianercy A., Galstyan A. 'Dynamics of Boltzmann Q learning in two-player two-action games'. *Physical Review E*. 2012, vol. 85(4), pp. 1574–604.

[34] Sastry P.S., Phansalkar V.V., Thathachar M.A.L. 'Decentralized learning of nash equilibria in multi-person stochastic games with incomplete information'. *IEEE Transactions on Systems, Man, and Cybernetics*. 1994, vol. 24(5), pp. 769–77.

[35] Ruiz-De-Azua J.A., Camps A., Calveras Auge A. 'Benefits of using mobile ad-hoc network protocols in federated satellite systems for polar satellite missions'. *IEEE Access: Practical Innovations, Open Solutions*. 1994, vol. 6, pp. 56356–67.

Chapter 16

Non-terrestrial network testbeds for 5G and beyond

Jorge Querol[1], Sumit Kumar[1], Oltjon Kodheli[1], Abdelrahman Astro[1], Juan Duncan[1], Mohammad Gholamian[1], Rakesh Palisetty[1], Symeon Chatzinotas[1], Thomas Heyn[2], Guido Casati[2], and Bo Zhao[2]

The fifth generation (5G) of wireless communications offers a plethora of new communications paradigms, scenarios, and services. Among them, the integration of terrestrial and non-terrestrial networks is one of the most relevant aiming at providing seamless ubiquity connectivity and economies of scale for satellite communications. This chapter presents the most relevant experimental testbeds for the demonstration of 5G and beyond 5G adaptations for their use in non-terrestrial networks. In the first sections, a state-of-the-art review is presented including a description of the principal hardware and software components, highlighting OpenAirInterface as the most prominent open-source 5G stack. The latest sections describe the features and capabilities of the most advanced 5G non-terrestrial network testbeds based on OpenAirInterface: the 5G-SpaceLab and the 5G-Lab.

16.1 State-of-the-art NGSO testbeds

16.1.1 Overview of NGSO testbeds

This section highlights some of the ongoing and completed non-geostationary orbit (NGSO) testbeds and their operational features. We discuss about hardware-based testbeds providing both over-the-air testing as well as satellite channel emulation capabilities. The testbeds under discussion are the following:

1. Sat5G Testbed [1]
2. University of Surrey 5G Testbed [2]
3. 5G Space Communication Lab [3]

[1]Interdisciplinary Centre for Security, Reliability and Trust, University of Luxembourg, Luxembourg
[2]Fraunhofer IIS, Erlangen, Germany

4. IoT-ANCSAT Testbed [4]
5. SATis5 Testbed [5]

16.1.1.1 Sat5G project

Sat5G project testbed brings Satcom into 5G by defining optimal satellite-based backhaul and traffic offloading solutions. It uses live *MEO* and emulated *GEO* for experimentation. It also facilitates end-to-end connection via a commercially available 5G core network. The testbed achieved 3GPP integration to satellite links so that the satellite gateway presents as gNB and satellite remote terminal presents as UE to a 5G core network. Among the many capabilities of the project include multicast over live GEO and MEO satellite links, with latency-reducing content fetching and broadcasting to network edge [6].

16.1.1.2 University of Surrey 5G testbed

University of Surrey 5G testbed provides backhauling through satellites and multi-linked connections via satellite and terrestrial links. The gateway and the user terminal in the testbed are 3GPP Rel 15/16 compliant; however, the air interface is not compliant and uses DVB-S2x. Nonetheless, research is in progress to implement the 5G NR as the air interface in the near future. Among the notable usages of the testbed include end-to-end 5G connection with Telesat Ka-band *LEO* satellite [7]. Additionally, the use-case for the 5G moving platform was demonstrated over SES's O3b *MEO* satellite system [8], using real terminals and a commercial 5G core network. The testbed has been used for demonstrating 5G backhauling, delivery of content to the edge, caching, and multi-linking using a combination of satellite and terrestrial networks.

16.1.1.3 5G Space Communication Lab

The 5G Space Communications Lab (5G-SpaceLab) at the University of Luxembourg allows to test, validate, and demonstrate space operations for two different scenarios: Earth-orbiting satellite communications and Earth-Moon communications. The testbed is under development and uses a software-defined radio (SDR)-based approach for the implementation of 5G and satellite nodes. The testbed uses open-source 5G protocol stack OpenAirInterface5G [9] for 5G RAN and is capable of emulating *MEO* and *GEO* satellite links using an in-house developed satellite channel emulator. The testbed is also capable of emulating Inter Satellite Links (ISLs).

16.1.1.4 IoT-ANCSAT testbed

IoT-ANCSAT testbed is an evolving testbed at the University of Luxembourg that will facilitate satellite-based IoT links. It covers all the three satellite orbit scenarios (*LEO*, *MEO*, and *GEO*). The testbed is developed with the help of OQtech [10] and will be a simulator-based testbed. One of the aims of the testbed is to come up with

5G-based Virtual Network Functions (VNF) that could comprise a satellite network slice to demonstrate the IoT satellite network slice in the testbed.

16.1.1.5 SATis5G testbed

The SATis5G project testbed mainly focuses on the convergence of satellite and terrestrial networks and provides the user to perform a comprehensive assessment of the capabilities of such convergence. The testbed uses *GEO* and *MEO* constellation of satellites for experimentation in two connectivity modes: backhaul connectivity for terrestrial 5G networks and direct 5G connection. Likewise SAT5G testbed, this testbed also uses a standard compliant 5G core network [11].

Apart from the efforts from academia, significant developments have been done by the industry for facilitating connectivity through LEO and MEO satellite constellations. However, such platforms are commercial in nature and generally not available for conducting tests or prototype validation. Table-16.1.1.5 lists some of such NGSO satellite constellations developed by industry. For details, the interested audience can go through the respective references.

Platform	Constellation	References
Oneweb	LEO	[12, 13]
Starlink	LEO	[14, 15]
O3b mPOWER	MEO	[8, 16]
Kuiper	LEO	[16, 17]
Inmarsat-Orchestra	LEO	[18]

16.1.2 Hardware components

NGSO testbeds allow designing, developing, testing, and validating non-terrestrial networks toward 5G and beyond. They may be using either dedicated hardware, usually when using a proprietary software stack, or SDRs, usually when using an open-source software stack. Software-defined components offer an agile approach to verify and validate current wireless technologies or design and develop completely new ones. This is because of their flexibility of having the PHY, MAC, and/or higher layers functions being software-defined. Accordingly, end-users and developers can have end-to-end control over all the software stack from the waveform up to the application layer.

A combination of off-the-shelf commercial SDR units and processing units (PUs) offers a turnkey solution to build a small form factor, low cost, and agile solution to support a wide spectrum of NGSO applications and use-cases.

16.1.2.1 Processing units

The PUs are responsible for running the different flavours of the 3GPP software stack on Intel-based architecture. Having PUs powered by Intel is a major requirement due to optimized DSP functions, which rely on Single Instruction, Multiple

Table 16.1 *Processing units components specifications*

	Processing units		
Specifications	**Precision 7920 Rack Workstation**	**Precision 3640 Tower Workstation**	**Z2 Mini G5 Workstation**
Vendor	Dell	Dell	HP
Processor	Intel Xeon Gold 6248 2.5GHz,(3.9GHz Turbo)	Intel Core i9-10900K 10th generation 3.7 GHz(5.3GHz Turbo)	Intel® Core™ i9-10850K 3.6 GHz (5.2 GHz Turbo)
Cores	16	10	10
Cache	27.5 MB	20 MB	20 MB
Memory	48GB DDR4	64GB DDR4	64GB DDR4
Hard Drive	512GB SSD	1.0TB SSD	1.0TB SSD
Interfaces	USB 3.2 Type A RJ45 Network Connection Serial RJ45 for iDRAC SFP+	USB 3.2, Type A USB 3.2, Type C RJ45 Network Connection Serial	USB 3.2, Type A USB 3.2, Type C RJ45 Network Connection

Data (SIMD) instructions (SSE, SSE2, SSS3, SSE4, and AVX2). However, these requirements can be also relaxed, according to the software portability, and modularity that can afford running the software stack on ARM-based architecture with some performance or function constraints. The overall performance of the hardware is a combination of different metrics, which includes the processor frequency, number of cores, number of threads, memory, cache size, and physical network interface. All metrics are sort of constraints that can be relaxed and adjusted according to the hardware function, requirements, and expected performance.

In Table 16.1, we provide a non-exhaustive list of the workstation's minimum requirements that can comply with the current 3GPP standards, specifically LTE and 5G-NR. For instance, choosing the minimalist criteria for 5G-NR gNodeB requires a minimum CPU frequency of 3 GHz, otherwise, it will struggle under heavy computational loads. Caching, number of cores, and threads are also impacting the multithreading and multiprocessing performance. In consequence, selecting the right platform guarantees the desired performance metrics such as downlink and uplink throughput that comply with 5G-NR requirements.

16.1.2.2 Software-defined radio units

The RF front-haul is based on SDR components. Different off-the-shelf commercial SDRs can be used from Ettus Research/National Instrument such as Universal Software Radio Peripheral (USRP) or space-ready SDR from GomSpace. Some of the RF units that can be deployed within the testbed may utilize the following models:

- Ettus Research-based SDRs (USRP N310, USRP B210, and USRP X310) [19–21]
- GomSpace-based Space ready SDR (Nanocom SDR) [22]

The requirements of the radio can be utilized according to use-case, hardware function, and application as previously mentioned. Accordingly, we can choose our radio frontend with respect to the maximum used bandwidth, the number of available RF channels, and the onboard FPGA. In Table 16.2, we summarize some of the tested SDRs specifications and their compatibility with the 3GPP standards 4G and 5G.

16.1.2.3 Field-programmable gate array units

Channel impairments such as Doppler, delay, link budget, and many others can be implemented on FPGA. Different FPGA-based development boards can be chosen to emulate the delay and Doppler effects from Xilinx such as Zynq UltraScale RFSoC family [23] that integrates the key subsystems for multiband, multi-mode cellular radios into an SoC platform that contains an Arm-based processing system. In addition to the channel emulation use-case, FPGA boards can be used as an accelerator by offloading processing hungry tasks within the physical layer from the Intel-based processors to FPGAs. Accordingly, the processing is distributed and could be load-balanced among the PUs and the FPGA boards, where the interconnection between them can be through any of the standard interfaces like PCIe, Ethernet, etc. On the other hand, FPGAs that are located inside the SDRs can be utilized to perform part of the Physical layer functions which could be low-phy or high-phy. As a consequence, no need for an additional FPGA board if its resources, number of logic cells, block RAM size, DSP blocks etc., are satisfying the requirements of the offloaded task.

16.1.2.4 Hardware integration

As illustrated in Figure 16.1, using the different combinations of the hardware, we can construct either a base station, user equipment, or a satellite according to the desired scenario, use-cases, and specifications.

For instance, a software-defined base station can be constructed by connecting an Intel Xeon-based dell server that runs an LTE or 5G-NR flavour of the 3GPP software stack, to the SDR USRPN310 via an SFP+ link. The SFP+ communication link between the server and USRP N310 has a bitrate up to 10 Gbps. Further, the USRP N310 is connected to the channel emulator to transmit and receive the analog waveform through the RF cables.

On the other hand, we can construct the software-defined user equipment as defined before for the base station, since the only difference is the software running on top of the server, which can control the functionality of the SDR which can be defined as a *Base-Station - BT* or *a User-Equipment - UE*. Similarly, a software-defined UE can be modelled using different hardware components with more relaxed requirements. This is achieved by connecting an Intel i9-based HP compact

Table 16.2 Software-defined radio units components specifications

Software-defined radio units				
Specifications	**USRP N310**	**USRP B210**	**USRP X310**	**NanoCom SR2000**
Vendor	Ettus Research	Ettus Research	Ettus Research	GOMSPACE
Driver	UHD - USRP Hardware Driver	UHD - USRP Hardware Driver	UHD - USRP Hardware Driver	IIO - Industrial I/O
RF Frequency Range	10 MHz – 6GHz	70 MHz – 6 GHz	DC – 6 GHz **depends on daughterboard*	70 Mhz – 6 GHz
Duplexing	FDD or TDD	FDD or TDD	FDD or TDD	FDD or TDD
Open-Source	FPGA/driver - UHD	FPGA/driver - UHD	FPGA/driver - UHD	FPGA/driver - IIO
MIMO	4×4 MIMO	2×1 MIMO 2×2 MIMO	2×2 MIMO	2×2 MIMO
Interfaces	USB Type A host portmicro-USB port (serial console, JTAG) RJ45 - 1 GbE SFP+	USB 3.0	USB Type A host port micro-USB port (serial console, JTAG) RJ45 - 1 GbE SFP+ PCIe Express	USB to UART CAN I2C RS422 LVDS (TR-600)
Communication	SFP+	USB 3.0	RJ45/SFP+/PCIe	CAN/UART
Compatibility	4G/5G (up to 100MHz)	4G/5G (40MHz with ¾ sampling)	4G/5G (80MHz with ¾ sampling)	N/A Transparent Payload

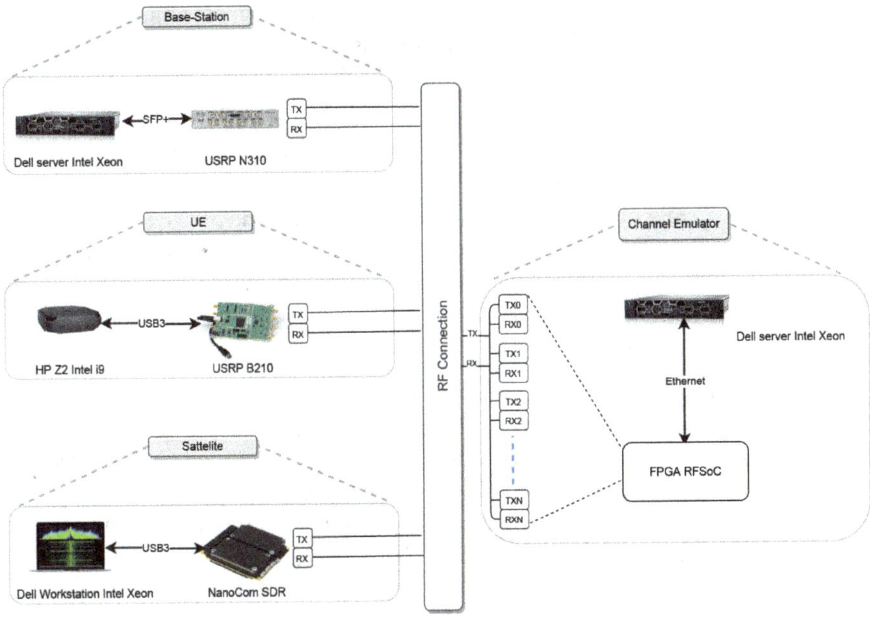

Figure 16.1 Interconnection and interfaces of the hardware components

workstation to the SDR USRPB210 via USB3. The communication link here is capable of streaming up to 56 MHz of real-time RF bandwidth. Furthermore, the USRPB210 is connected to the channel emulator via RF cables.

The communication link between the USRPs and the servers/workstations is achieved by the USRP Hardware Driver (UHD) [24]. The driver supports many standard interfaces such as USB3, Ethernet, PCIe, and SFP+. Accordingly, the software running on the servers can have control over all the receive and transmit signal chains using the UHD driver. This is achieved by deploying the UHD Application Programming Interface (API) functions within the software stack running on the servers.

Another use case is to design a software-defined satellite that can be achieved using the GomSpace SDR platform and is implemented as a ready-to-use S-band standalone radio. For instance, to create ISL for communication between satellites in orbit (ISL). The SDR deploys AD9361 transceiver from Analog Devices, a high-performance, highly integrated radio frequency (RF) agile transceiver. Unlike USRPs, the control of GomSpace SDR is done using the Industrial I/O subsystem (IIO) Linux subsystem driver [25].

16.1.2.5 Conclusion

In summary, we have defined various hardware components within the NGSO testbed. Most of the radios are software-defined, which have great flexibility and offers an agile solution to accelerate the development time by introducing multi-functional radios that can be used for 5G and beyond use-cases within the NGSO testbed.

16.1.3 Software stacks

An open-source software stack is essential to implement the enhancements which are in the standardization of cellular technologies in 3GPP. This section provides an overview of available open-source SDR platforms for 4G and 5G NR components. The main focus is on the UE and eNB and gNB components, but the availability of open source core network implementations is addressed as well.

16.1.3.1 GNU Radio

In the GNU Radio ecosystem, there are no open-source 5G NR implementations available, but there are two projects focusing on different parts of the LTE system: gr-lte (UE, published in github) and openLTE (eNodeB and EPC, hosted by SourceForge). GNU Radio provides a simple LTE UE implementation with gr-lte and LTE eNodeB and EPC implementations with openLTE. But there are no current activities, especially not with respect to 5G NR.

16.1.3.2 Software Radio Systems

Software Radio Systems Limited (https://www.softwareradiosystems.com/) is an Irish company providing open-source implementations of their srsUE, srsENB, and srsEPC. Additionally, they sell licenses for their AirScope tool building on these open-source developments. They also provide services by the means of consulting, training, and testbed development. The community for the open-source core can be found here: https://www.srslte.com/.

Software Radio Systems (SRS) provides an open-source implementation of several LTE features for UE, eNodeB and EPC, release 21.04 contains the first Open-Source 5G NSA UE, including 5G-NR PHY layer for x86 including SIMD-optimized LDPC encoder/decoder, compatible and tested with 3rd-party RAN/Core solutions and data traffic over Secondary Cell Group (SCG) bearer for NR. In October 2021, SRS will release version 21.10 that will contain the 5G NSA eNB/gNB application. Another release 22.04 is planned for April 2022 with the first elements for 5G SA.

16.1.3.3 O-RAN

The O-RAN Alliance (www.o-ran.org) was founded by operators to clearly define requirements and build a supply chain ecosystem to realize its objectives. Therefore, they specify an overall 5G Radio Access Network (RAN) Architecture as shown on the O-RAN website, not covering UE nor core network components. Besides the O-RAN Alliance, there is also the O-RAN Software Community (SC). This is a collaboration between the O-RAN Alliance and Linux Foundation with the mission to support the creation of software for the RAN.

The last three releases Amber (November 2019), Bronze (June 2020), and Cherry (December 2020) of the O-RAN Software Community (SC) focus on the RAN Intelligent Controller (RIC), but also for the other RAN components, software was released. In Amber release, the code for the O-RAN Central Unit (OCU) was mostly based on openLTE and contains the modules LTE S1AP, RRC, PDCP, RLC, and MAC. In Bronze release, these modules were replaced by an initial 5G user plane functions with 5G NR SDAP and PDCP layers. Additional, in Cherry release, 5G NR RRC for control plane functions has been added.

The O-RAN Distributed Unit (O-DU) module is split into O-DU High and O-DU Low modules. The O-DU High mainly contains the 5G NR MAC ad 5G NR RLC layers, whereas the O-DU Low contains the PHY-high layer. The interface between O-DU High and O-DU Low is the FAPI Interface.

In Amber release, the code for the O-RAN DU High component contains mainly first draft versions of the modules F1AP, 5G NR MAC, and RLC. In Bronze release, there were improvements on the modules O-DU High layers MAC, RLC and app, F1-U interface, and F1-C to support additional F1AP messages and basic FAPI messages. In Cherry release, additional goals have been achieved, among others support for 64 QAM in DL and 16QAM in UL, support for all short PRACH formats, integration of O-DU High with O-DU Low and establishing Netconf session for O1 interface for CM.

The Amber release of the O-RAN Distributed Unit (DU) Low component contains the Open Front Haul (O-FH) library implementation.

In Bronze release, new modules and extensions supporting O-RAN FrontHaul compliant Radio to Layer 1 interface and a FAPI compliant Layer 1 to Layer 2 interface have been implemented. So far no open-source Layer 1 implementation is available in O-RAN SC. Instead, the link to a high-performance Layer 1 stack is referencing to Intel's binary only FlexRAN solution.

In Cherry release, the O-DU Low and O-DU High integration, the integration of O-DU Low and O-RU/RRU emulator and E2E integration according to the RSAC and INT project alignment features and scope have been performed. Despite this, the O-DU Low has been integrated with 3rd party commercial SW to verify the UE attachment and traffic.

The O-RAN Distributed Unit (DU) seems to be most promising with the Open Front Haul Interface Library as well as the 5G NR MAC and RLC implementations and the F1 interface. Also, the O-RAN Centralized Unit (CU) is evolving with the first draft implementation of the SDAP and PDCP layers. And, as can be seen with the closed-source L1 implementation in the O-DU Low component, this project focuses more on the open implementation of the interfaces and less on the open implementation of the component's functionality.

16.1.3.4 FlexRAN

The name "FlexRAN" is used for at least two projects, the Mosaic5G FlexRAN and the Intel FlexRAN.

The Mosaic5G FlexRAN (http://mosaic-5g.io/flexran/) is a subproject of the Mosaic5G project and defines an interface similar to the O-RAN E2 interface for monitoring and controlling of RAN. Besides the interface specification itself, the project provides the FlexRAN real-time controller and the FlexRAN runtime. The FlexRAN runtime is integrated into each RAN module and the FlexRAN controller connects to these RAN modules. This FlexRAN is mainly used for RAN optimization. The LTE implementation of OpenAirInterface already integrates the FlexRAN runtime.

The Intel FlexRAN (https://software.intel.com/en-us/articles/flexran-lte-and-5g-nr-fec-software-development-kit-modules) is a set of libraries optimized for Intel processors for computation-intensive parts of the LTE and 5G NR FEC (Forward Error Correction) modules. The Intel FlexRAN library seems to be a promising solution for optimized PHY layer modules in the area of coding/decoding. It may be interesting to combine this with open source 5G NR stacks, for example, OpenAirInterface. Despite this, the Intel FlexRAN library is used for the closed-source L1 implementation used in O-RAN SC DU Low.

16.1.3.5 Nvidia

The Nvidia Aerial SDK (https://developer.nvidia.com/aerial-sdk) consists of two SDKs: cuVNF and cuBB. As both SDKs are targeting the 5G NR PHY layer (L1), located in the Distributed Unit, the "cu" prefix does not indicate Centralized Unit, but Cuda. The NVIDIA cuVNF SDK provides optimized input output (IO) with memory allocations and support for network data flow and processing use cases. cuVNF improves performance for signal processing with NVIDIA GPU multi-core compute capability. The NVIDIA cuBB SDK provides a fully offloaded 5G PHY layer processing pipeline (5G L1) that delivers unprecedented throughput and efficiency by keeping entire all processing within the GPU's high-performance memory. With 5G NR based uplink and downlink channels running on GPUs, cuBB SDK provides high performance for latency and bandwidth utilization. NVIDIA cuPHY SDK provides beamforming, LDPC encode/decode, and other functionalities for PHY pipeline. Currently, access to these SDKs is only granted in an early access program after signing an NDA.

16.1.3.6 free5GC

The free5GC (https://www.free5gc.org/) is an open-source project, but only for the 5G core network defined in 3GPP Release 15 and beyond. The 5G RAN is not in the scope of free 5GC.

16.1.3.7 open5GS

This open5GS project (https://open5gs.org/) implements 5GC and EPC networks for private LTE or 5G networks in C language. As the free5GC project, the RAN is not covered by the developments. A WebUI is implemented in Node.JS and React and is available for testing purposes.

16.1.3.8 UERANSIM

The open-source project UERANSIM (https://github.com/aligungr/UERANSIM) is licensed under GPL-3.0 and covers the 5G UE and the 5G standalone gNB-Central Unit (CU). According to the status in the repository, the physical, MAC, and SDAP layer implementations are pending.

16.1.3.9 OpenAirInterface

OpenAirInterface (OAI, https://www.openairinterface.org/) is an open-source software platform for simulation and emulation of 3GPP mobile networks. Contributions to this open-source project are provided by the members of the OpenAirInterface Software Alliance (OSA), which is a non-profit consortium fostering a community of industrial as well as academic contributors.

OAI provides a full experimental LTE implementation (Rel 8, partial Rel 10) in real time under Linux optimized for x86 and with interworking functions. It includes EUTRAN (eNodeB and UE) and EPC (MME, xGW, and HSS). Regarding 5G, OAI currently comprises software components for the 5Gcore (AMF, SMF, NRF, AUSF, UDM, UDR), for the gNB (5G Standalone software stack) as well as for the UE. Currently, the members of the OSA are working to evolve the software toward future 5G releases of 3GPP.

16.1.3.10 Conclusion

In conclusion, OAI is currently the most advanced open-source stack for 5G system implementations, which is also in accordance with Reference [26]. In the next chapter, we describe the modifications implemented in OpenAirInterface to support NGSO satellites in 5G.

16.2 OpenAirInterface modifications

The general principles of the 3GPP 5G NR updates are described already in the previous chapter on 3GPP integration. In this section, the modifications especially in OAI to support 5G-NTN are described.

16.2.1 Modifications on OAI PHY/MAC layers

16.2.1.1 Comparison between OAI implementation and 3GPP

3GPP has already provided some solutions for NR to support NTN [27] as well as several approved change requests packed in [RP-212969]. Currently, OAI does not cover all the changes on the PHY layer proposed by 3GPP for Release 17. However, some important modifications have been implemented in OAI. For instance, for the enhancement of transmission time adjustments the parameter k_{offset} was used in OAI via command line assuming that gNB and UE both have the information about the long delay. This k_{offset} can be configured in RRC configuration as proposed in [RP-212969]. Moreover, k_{offset} shall be further considered in the CSI reference resource definition, HARQ-ACK reporting, etc. in OAI to be in line with [RP-212969]. Current OAI has deactivated the HARQ process due to simplicity, that is, OAI uses one HARQ process for NTN transmission. If necessary, the number of HARQ processes can be extended from 16 to 32 as mentioned in [RP-212969].

The modifications to support communication via the LEO satellite implemented in OAI by Fraunhofer IIS are not limited to the proposals from 3GPP. For instance, in case UE could not acquire a global navigation satellite system (GNSS) position and satellite ephemeris, it must find its own way to initialize frequency synchronization. The modifications on the PHY/MAC layers in OAI with focus on time and frequency synchronization are described below.

16.2.1.2 Doppler shift pre-/post-compensation

As mentioned in Reference 27, for LEO satellite systems it is recommended to have pre-/post-compensation of a common frequency offset at the network side, conducted with respect to the spot beam center. The UE-specific frequency offset can be estimated and compensated by UE or indicated by the network. The common frequency offset is much larger compared with the remaining UE-specific frequency offset. Without pre-compensation for downlink, additional complexity is needed at UE receiver to achieve robust downlink initial synchronization performance based on Rel-15 SSB. To avoid the complexity at UE receiver, it is assumed that in OAI the common frequency offset is known to gNB and will be compensated at gNB.

In OAI gNB uses write/read functions to send/receive I/Q samples to/from either the RF simulator or the hardware (USRPs). The pre-/post-compensation is applied as follows:

1. The additive inverse of the common frequency offset is used to calculate the Doppler-like factor according to (16.1),

$$e^{j2\pi\frac{f_c}{f_s}n} = \cos(2\pi\frac{f_c}{f_s}n) + j\sin(2\pi\frac{f_c}{f_s}n) \tag{16.1}$$

   where f_c is the common frequency, f_s is the sampling frequency, and n is the sample index.
2. This factor is then multiplied with each I/Q sample immediately before the write function and after the read function.

Since the sine/cosine functions work in floating points and manipulation of each sample using sine/cosine functions is quite time-consuming, a look-up table for the sine/cosine functions is applied to accelerate the processing time when emulating using hardware, for example, USRPs and the channel emulator.

16.2.1.3 UE initial frequency synchronization

For initial frequency synchronization, in case that an exact pre-/post-compensation at the gNB side cannot be guaranteed or there is no pre-/post-compensation at the gNB side at all, UE needs to adjust its carrier frequency by measuring and searching for the synchronization signal from the gNB within a certain frequency band.

Starting from its carrier frequency UE continuously applies an increasing positive/negative frequency offset with a fixed step size (currently +/−40 kHz) to its received signal. If the synchronization signal can be successfully decoded, UE will apply the current frequency offset permanently and the residual frequency shift can be compensated by the following dynamic compensation algorithm.

16.2.1.4 Dynamic residual frequency shift compensation

The Doppler frequency shift is in the range of several hundred kilohertz in an LEO transmission system. Reference 27 assumes a common frequency offset at the network side, i.e., pre-/post-compensation shall be available at the satellite. This compensates for the major part of the Doppler frequency shift. However, there is still a residual frequency offset at the UE side if UE is away from the spot beam center. This frequency offset shall be estimated and compensated with the help of the DMRS symbols.

At the time of writing, the released OAI only implements a single initial frequency offset estimation based on SS Block. Frequency offset is not continuously tracked and compensated. With the following dynamic residual frequency shift estimation and compensation algorithm the maximum Doppler shift that can be compensated is about +/−1975 Hz.

1. Multiple DMRS symbols within a slot are mandatory to estimate the frequency shift caused by Doppler since the signal is always proceeded slot by slot. Assuming the channel is not varying within a slot, the channel estimates of different DMRS symbols should only have a phase rotation depending on the Doppler frequency shift, the sampling frequency, and the time interval. By measuring the phase difference between different DMRS symbols, the Doppler shift can be derived. If there are more than two DMRS symbols per slot, averaging should be used.
2. The average Doppler frequency shift is then given into a PI (proportional-integral) controller. The output of the PI controller is the estimated Doppler shift that will be used by the compensation at the UE side. The coefficients of the controller are based on the trial and error approach. The reason is that there is a delay between the estimation of Doppler and compensation. Without a proper controller the tracking of the frequency shift may fluctuate or keep a constant non-zero distance from the real value.
3. The estimated Doppler shift is then applied in time domain immediately after I/Q sample reception for DL and before I/Q sample transmission for UL.

16.2.1.5 Timing drift compensation and autonomous timing advance update

Due to Doppler effect UE will observe different frequencies from the transmit frequency of the satellite. It can be also seen that due to the high velocity of the LEO satellite the received signal at the UE side is squeezed/stretched. For terrestrial scenarios where the velocity is much lower, this squeeze/stretch effect is negligible and can be compensated not very often. For NTN if the sample offset compensation is not performed frequently enough, the timing drift/offset will accumulate and lead to loss of time synchronization.

Assuming the velocity of an LEO satellite is 7800 m/s, in an extreme case that the satellite moves directly toward the UE, the timing drift within a frame (10 ms) is 0.26 μs. This corresponds to about 16 samples in the exemplary configuration (FFT size 2048, normal cyclic prefix, 30720 samples per slot). Currently, OAI is able to shift only one sample per frame. In this case, the timing offset will accumulate and lead to synchronization error. Therefore, the estimation and compensation of the timing drift should be enhanced.

The enhancement of timing drift compensation comprises the following steps:

1. In the SS block the PBCH channel is estimated regularly (once per two frames) by the UE from the PBCH DMRS. The peak position of the channel estimates in time domain (channel impulse response) from the SS block provides the timing information, as shown in Figure 16.2.
2. Each time the peak of the channel estimates is calculated, it is compared with the target position and the difference (error) is calculated. This difference is then given as an input to a PI controller. The controller output is used by the UE to read/receive less or more samples. The reason is similar to the dynamic residual frequency shift compensation. Due to the delay between the estimation of the timing drift value and the compensation, fluctuation of the difference (error) or a constant non-zero difference might occur. To minimize the difference and stabilize the whole compensation procedure, the PI controller is chosen.

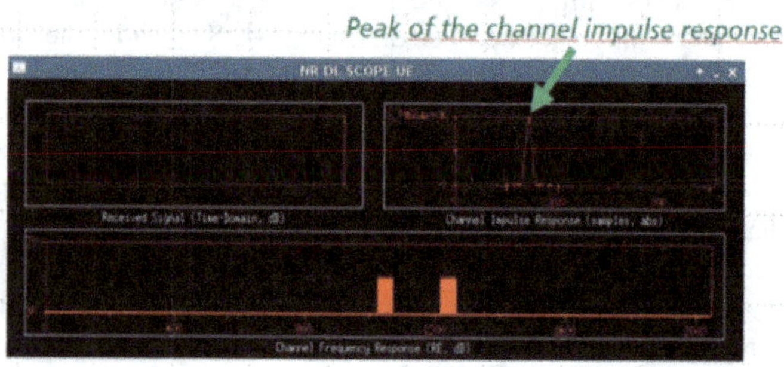

Figure 16.2 Peak of the channel impulse response of the PBCH channel in the OAI UE scope

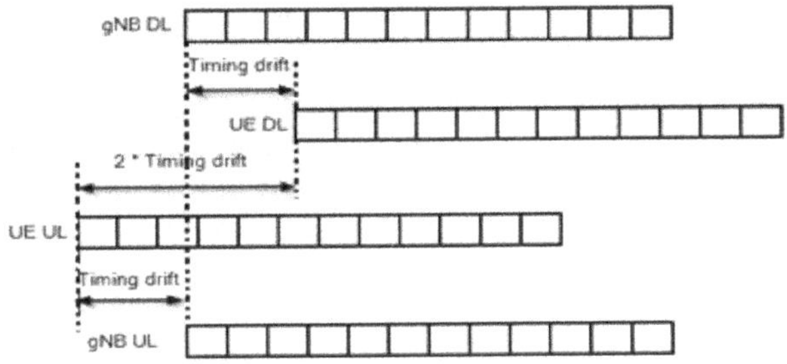

Figure 16.3 UL transmission compensation for the timing drift

3. The peak as well as the controller output is calculated once per two frames. In order to adapt the timing more frequently to avoid accumulation of the timing drift, we apply the compensation for each slot. The controller output value is applied evenly distributed at the end of each slot.
4. Since the timing drift occurs on both DL and UL, UE could also pre-compensate the timing drift for UL signal using the estimated offset from the PI controller. The UL transmission time is adjusted by two times of the compensated value for each slot, so that the signal arrived at the gNB is synchronized. This UL time adjustment can be seen as an autonomous timing advance update in addition to the normal TA update, as depicted in Figure 16.3.

16.2.1.6 Timing advance update

To guarantee that a new TA update is only calculated after the old TA update is applied, the TA update period is increased from 10 frames to 50 frames in OAI. The TA update period must be at least twice of the one-way delay between gNB and UE.

16.2.1.7 Random access procedure modifications

The random access (RA) procedure is a very important mechanism in 5G systems, which is mainly utilized for achieving uplink synchronization among users. It consists of a four-message exchange among the users and the base station briefly described as follows:

- Message 1 - When there is a Random Access Opportunity (RAO), the users send a preamble to the serving base station in order to initiate the RA procedure. This enables the estimation of the round trip time (RTT) at the base station side for every single user. At this step, all the users compete for the same radio resources, meaning that the RAO is the same for all the users, hence preamble collision may occur. To reduce the probability of collision, there exist a list of possible preamble sequences defined in the standard and the UEs randomly select one. Notably, in case two UEs randomly select the same preamble sequence to initiate the RA procedure, a collision will occur, leading to a failure of the RA procedure. In such a case, the UEs will try again to send the preamble after a back-off time and a power-ramping.

- Message 2 - The estimate of the RTT based on the time of arrival (ToA) of the preambles coming from various users, is utilized by the base station to compute the TA value. The TA is then reported to the users with a successful RA procedure in order to align their uplink data transmission in the subsequent message exchanges. In addition, information regarding the scheduling of Message 3 is provided.
- Message 3 - In this stage, a contention request is initiated by the users with the purpose of identifying themselves in the network and obtain a unique ID. This phase is also known as the contention resolution phase. Please note that in case of a contention-free RA Message 3 and Message 4 transmission are skipped. Users may also report their data volume status and power headroom to facilitate the scheduling and power allocation algorithms for subsequent transmissions.
- Message 4 - In this final step, the users are granted a permanent unique ID in the network, and the connection between the users and the base station is established.

When considering a non-terrestrial network, one of the main impairments to be taken into account is the increased RTT in the communication link, and the first procedure impacted is the RA procedure. This is due to the frame misalignment among the user and the base station that overcomes the subframe length. Figure 16.4 illustrates such problem, comparing the small frame misalignment that occurs in a Terrestrial Network (computed to be 0.67 ms for a 100 km cell) with that experienced over a nonterrestrial network with 600 km altitude of the NTN terminal. To counteract this challenge, modifications are required in the 3GPP protocol, mainly

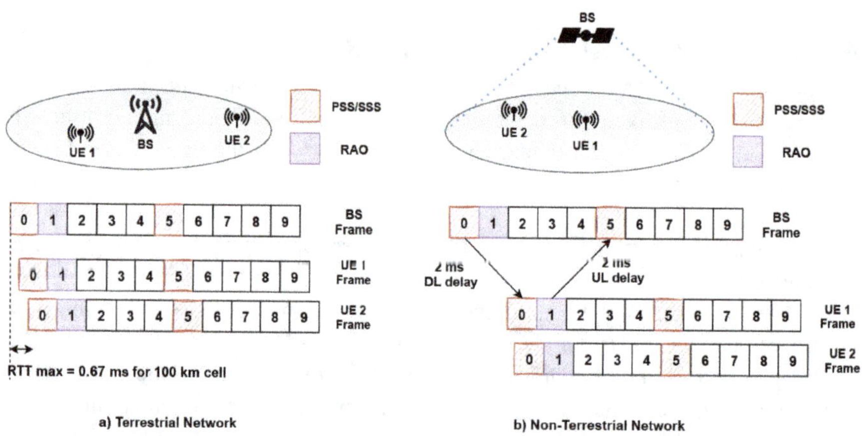

Figure 16.4 Frame misalignment among UE and gNB: a) Terrestrial network; b) Non-terrestrial network

divided in two branches: a) subframe-level TA at the user side and b) subframe-level timing delay (TD) at the base station side.

- Subframe-level TA: The TA concept, as it exists, can be implemented only at a sample level. Therefore, once the frame misalignment overcomes the supported limit by the terrestrial network, the typical TA fails to be implemented, leading to a failure of the RA procedure. To tackle this, a subframe-level TA can be applied by the users in order to align the frames even at high values of RTT in the communication link. To do so, two parameters are needed, the location estimate of the users and the satellite trajectory data. This enables a specific user on-ground to estimate the RTT and apply the subframe-level timing advance even before initiating the RA procedure.
- Subframe-level TD: An alternative way to solve the problem would be to apply a SF-level TD at the base station. Doing so, all the channels and procedures can be processed by the base station by taking into account the RTT present over the NTN channel. Clearly, the BS has to be aware of the NTN altitude and the RTT experienced at the center of its beam. Such information can be included in the deployment phase of the network since it will be fixed over time. The advantage of solving this issue at the base station side is that it relaxes the need for extra processing and algorithms by the users.

16.2.2 Modifications on OAI RLC/PDCP/RRC layers

Likewise the physical layer and the MAC layer, necessary modifications have to be done at both the gNB and UE sides, at both the user plane as well as the control plane, to cope up with the high RTT observed by the LEO and MEO satellites. The layers under consideration are RLC, PDCP, and SDAP on the user plane while RLC, PDCP, and RRC on the control plane. We will use the protocol stack as shown in Figure 16.5 as a reference for further discussions.

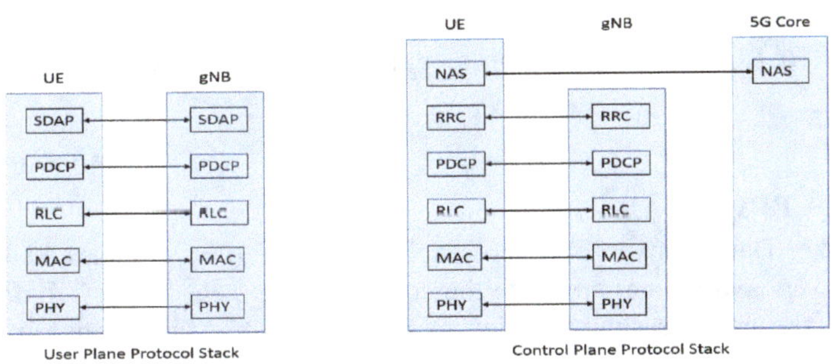

Figure 16.5 5G protocol stack for the user-plane and control-plane

16.2.2.1 RLC

The Radio Link Control (RLC) [28] layer takes SDU from PDCP and delivers them to the corresponding peer entity (UE or gNB). Three transmission modes are supported by the RLC: Transparent Mode (TM), Unacknowledged Mode (UM), and Acknowledged Mode (AM). Some of the major responsibilities of the RLC layer include loss detection and error correction through ARQ, RLC SDU segmentation, and re-assembly, RLC SDU discard. RLC configurations are agnostic of the 5G numerology under use.

- *Status Reporting*: A status report can be triggered by the polling procedure or by detection of reception failure of an Acknowledged Mode Data (AMD) PDU that is indicated by the expiration of the t-Reassembly timer. This timer is started when an AMD PDU segment is received from the lower layer, is placed in the reception buffer, at least one-byte segment of the corresponding SDU is not received and the corresponding timer is not running already. The procedure to detect loss of RLC PDUs at lower layers by expiration of timer t-Reassembly is used in RLC AM as well as in RLC UM. t-Reassembly timer can be configured to any value between 0 and 200 ms. For the terrestrial case, this timer covers the largest time interval in which the individual segments of the corresponding SDU have to arrive out of order at the receiver due to SDU segmentation and/or HARQ retransmissions before a status report and consequently an ARQ-retransmission is triggered. Moreover, if HARQ is enabled in NTN, the value of t-Reassembly timer will be required to be modified, because the timer should cover the maximum time allowed for HARQ transmission.
- *Mitigating HARQ/ARQ interaction*: In 5G-NR, HARQ of the physical layer and ARQ of the RLC layer in AM mode independently can perform re-transmission to mitigate transmission errors on the physical channel. Besides, HARQ and ARQ interact for improving re-transmissions. Feedback information/error received from the HARQ is reported to the ARQ function, this enables the ARQ to perform re-transmission and to cope with any HARQ feedback error. For NTN-based NR access, it is proposed to disable HARQ mechanisms [27], to face longer propagation delay on the NR user links. And hence, disabling HARQ also involves changing the HARQ/RLC interaction, to avoid RLC's useless re-transmitting due to feedback error received from HARQ. However, if the HARQ is not disabled, this modification is not required.

16.2.2.2 PDCP

The Packet Data Convergence Protocol (PDCP) [29] layer takes SDU from higher layers and delivers them to the corresponding peer entity (UE or gNB). Some of the major responsibilities of PDCP include Maintenance of PDCP Sequence Number (SN), Ciphering and Deciphering, Integrity protection and verification, Timer-based SDU discard, Reordering, in-order delivery, and

Duplicate detection. We enlist some of the modifications to be done in the PDCP layer to adapt it for the NTN operation.

- *Extension of SDU Discard Timer*: The transmitting PDCP entity shall discard the PDCP SDU when the discard Timer expires for a PDCP SDU or when a status report confirms the successful delivery [28]. The discard Timer can be configured between 10 ms and 1500 ms or can be switched off by choosing infinity [30]. The discard Timer mainly reflects the QoS requirements of the packets belonging to a service. However, by choosing the expiration time of the discard Timer or the QoS requirements, the RTD as well as the number of retransmissions on the RLC layer and/or HARQ shall be considered. By increasing the expiration time of discard Timer, one should keep in mind that extended timer values will increase the amount of required memory for the buffer.
- *Reordering and In-order Delivery*: To detect loss of PDCP Data PDUs, there is the timer t-Reordering which is started or reset when a PDCP SDU is delivered to upper layers [28]. The maximum configurable expiration time is 3000 ms; however, in the terrestrial network settings, this timer is configured according to the maximum RTT experienced. Hence, according to the LEO and MEO, this timer has to be adapted accordingly.

16.2.2.3 SDAP

The Service Data Application Protocol (SDAP) [31] is a new layer that has been introduced in 5G. Among many others, the most critical task of SDAP is mapping between a QoS flow and a data radio bearer. It has been found that there will be no impact on the SDAP layer due to the NTN operation, i.e., large propagation delay.

16.2.2.4 RRC

The Radio Resource Control (RRC) [30] is a control plane layer whose main roles include: Broadcast of system information related to Access Stratum (AM) and Non-Access Stratum (NAS), Establishment-Maintenance-Release of RCC connection (data and signaling bearers) between the UE and the gNB, security functions, handover, cell selection/re-selection, and Delivery of NAS messages from UE to the AMF.

The RRC layer is responsible for the reliable functioning of many procedures such as RRC-setup request, RRC-reestablishment, RRC-resume, RRC-suspend, etc. All these procedures are associated with timers, expiry of which lead toward restarting the procedures. These timers are in general started when any RRC message is sent from the UE to gNB (and vice versa) and the timer is stopped once the response is received from the peer. If the response is not received before the expiry of the timer, appropriate action is taken. A detailed list of such timers and the associated values can be found in Reference [30]. These timers have been set according to the RTT experienced in the terrestrial networks which are significantly lower compared to the RTT observed in satellite links. And hence, these timer values need to be extended for coping up with such a large RTT.

16.3 5G-SpaceLab testbed

16.3.1 Overview

The 5G Space Communications Lab (5G-SpaceLab) is an interdisciplinary experimental testbed combining the expertise, facilities, and infrastructure of multiple laboratories located at the Interdisciplinary Center for Security, Reliability and Trust (SnT) of the University of Luxembourg (UniLu). The 5G-SpaceLab is a unique integrated and interdisciplinary space communications and control emulation platform that allows testing, validating, and demonstrating the next generation of space applications. The main capabilities of the 5G-SpaceLab include 5G NTN communications, NGSO satellite and channel emulation, small satellite payload design and implementation, space-based edge computing, lunar rover control and teleoperation, AI-enhanced control and communications, and space-based Internet of Things (IoT) applications. A team of more than 20 researchers collaborate on a range of national and international projects with a high-TRL development component.

This section focuses on the 5G-SpaceLab components relevant to the emulation of 5G NTN and NGSO satellite communication channel and two case studies implemented and validated in the 5G-SpaceLab testbed.

16.3.2 SnT Satellite channel emulator

16.3.2.1 Overview

The NTN channel emulator is a hardware equipment developed by the SIGCOM group of the SnT—University of Luxembourg to replicate the effects of a satellite channel using a realistic transponder. One salient characteristic of the channel emulator is its capability to operate on a MIMO configuration. This feature gives the channel emulator the capability to mimic interference scenarios between the multiple carriers in multi-beam satellite systems. The channel emulator was designed and built using Software Defined Radio (SDR) tools to maximize the implementation flexibility. The current version of the channel emulator has eight channels in a full-MIMO configuration, and was implemented using two different hardware platforms: The Zynq® UltraScale+™ RFSoC ZCU111 evaluation kit, which features a Zynq UltraScale+ RFSoC supporting 8 12-bit 4.096GSPS ADCs, 8 14-bit 6.554GSPS DACs, and 4 GB DDR4 on PL; and the AMC574 FPGA board, which is a costumed industrial board based on RFSoC XCZU29DR Xilinx FPGA that support 16 channels of ADCs and DACs with 8GB DDR4 on PL.

The channel emulator implements the typical impairments of communication satellite payload. Figure 16.6 describes the different impairments implemented in the channel emulator.

The channel emulator operates with inputs and outputs at IF frequencies and generates digitally the impairments happening in the RF frequencies. With this approach, the emulator can replicate the end-to-end behavior from the IF interface at the transmitter side to the IF interface that comes out of the LNB on the receiver side

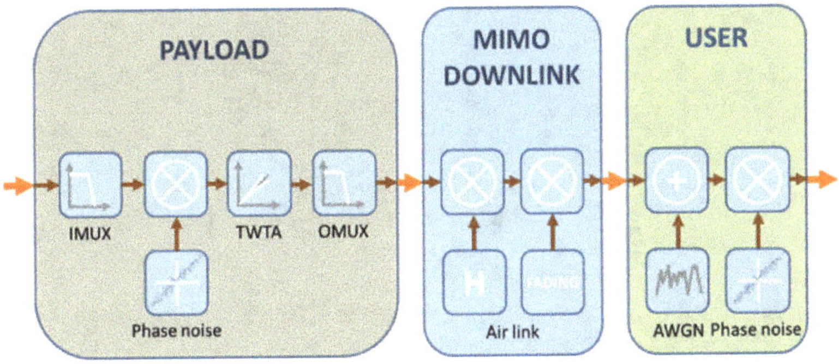

Figure 16.6 Functional block diagram of the satellite channel emulator

(RF impairments at the ground transmitter are considered negligible). The effects that occur in the actual RF bands are emulated in the payload emulator.

The channel emulator is therefore divided into three main components, as seen in Figure 16.6, the payload emulator, the MIMO downlink emulator, and the user equipment emulator. The payload block implements the linear and non-linear distortion to the input signal and applies the phase noise effects for a given frequency conversion in the transponder. The MIMO downlink block imposes the interference matrix between the different input streams and different fading patterns. Finally, the user emulator block implements the thermal and phase noise that typically affects the low-cost terminal RF equipment.

16.3.2.2 STK orbits

To design and emulate the satellite orbit we utilize the Systems Tool Kit (STK), which is a platform for analyzing and visualizing complex systems. Also, we are able to extract useful parameters in order to be used by other hardware or software components. For example, in case of an LEO satellite, one of the parameters that we change over time is the delay. This can be extracted directly from STK, and then acting as an input for the channel emulator that has to mimic the satellite channel impairments, as we will see in the following sections.

16.3.2.3 Delay emulation

The round trip time (RTT) of communication over a non-terrestrial channel is implemented by using a deep first in first out (FIFO) buffer, which utilizes an external Double Data Rate (DDR) Synchronous Dynamic Random-Access Memory (SDRAM) chipset. The implemented delay length is based on the frequency rate at which the data writes and reads to/from the Deep FIFO, and the depth of the FIFO buffer. By fixing the sampling frequency at the maximum performance of the hardware and changing the depth of Deep FIFO, we are able to emulate various delays. Different delay values can be emulated depending on the orbit of the targeted satellite ranging from few ms for LEO to hundreds of ms for GEO (~ 250ms RTT).

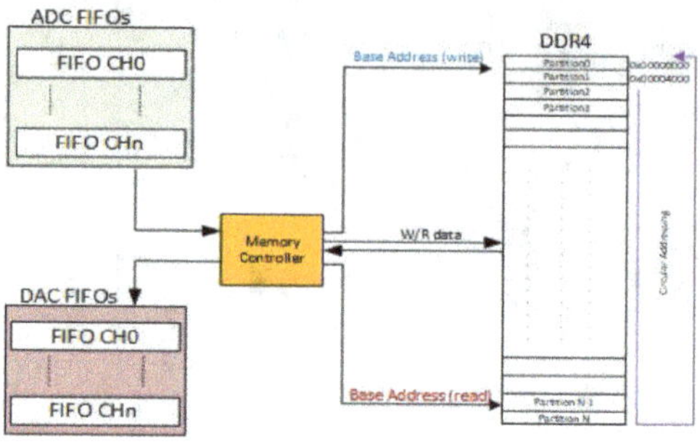

Figure 16.7 Block diagram of changing the DDR4 to the Deep FIFO

Furthermore, deep space communications cases are also accounted, for example, the RTT delay required to emulate a Lunar Relay Satellite orbiting the Moon L2 point is about ~ 400ms. The maximum aggregated delay for all channels with the current implementation is 1.4 s. Figure 16.7 shows the implementation of the Deep FIFO based on the external DDR4 memory.

16.3.2.4 Doppler emulation

Conversely to GEO satellites with constant propagation delay, NGSO involve movement, which translates into a variation of the propagation delay over time. In practice, these variations in NGSO propagation delay need to be tracked and compensated to avoid packet errors at the receiver side. Such delay variations are the cause of the undesired Doppler effect, which widens or narrows the spectrum of the original signal.

In this section, we explain the proposed "resampler" approach to emulate this varying delay propagation for NGSO (or Doppler effect), inspired by the compensation approach often used at the receiver side. The resampler implemented in the channel emulator applies a polynomial timing compensation that comprises five possible polynomial coefficients as shown in Table 16.3. This polynomial resampler is an adaption of the cubic polynomial re-sampler described in Reference [32]. The modification used in our actual implementation of the re-sampler allows handling positive and negative time compensations.

The coefficients are applied as the coefficients of an FIR filter to the incoming data samples as shown in Figure 16.8. To emulate these variable delays, the implementation approach employs a digital resampler as mentioned earlier. The first in first out (FIFO) memory can be employed for the integer value of the delays. The fractional delay is given as an input to the FIR coefficient module that generates the variable coefficients and applied to the incoming input data samples in the resampler module as shown in Figure 16.8. These filter coefficients change the location of the sample points and thereby applying the fractional delay. In this way, the variable fractional delay is offset with the help of FIR filter coefficients and resampler, and integer delay through the FIFO is performed in real time, thereby introducing the timing delay.

Table 16.3 Set of coefficients for polynomial resampler

	$\mu < 0$	$\mu \geq 0$
c_{-1}	$\frac{\mu^3 - \mu}{6}$	0
c_{-2}	$\frac{-\mu^3 + \mu^2 + 2\mu}{2}$	$\frac{-\mu^3 + 3\mu^2 - 3\mu}{6}$
c_{-3}	$\frac{\mu^3 - 2\mu^2 - \mu + 2}{2}$	$\frac{\mu^3 - 2\mu^2 - \mu + 2}{2}$
c_{-4}	$\frac{-\mu^3 + 3\mu^2 - 3\mu}{6}$	$\frac{-\mu^3 + \mu^2 + 2\mu}{2}$
c_{-5}	0	$\frac{\mu^3 - \mu}{6}$

The prototype uses NI USRP FPGA and Labview to implement a real-time doppler delay emulator. The communication between these blocks is governed through the AXI4-Stream protocol, and FIFOs are employed to process the input and receive the offset compensated samples in real time.

16.3.3 Case study: "Random Access Procedure over NTN"

By modifying the existing RA procedure of the 3GPP protocol so as to tolerate higher RTT values, as described in section 16.2.1.7, we have demonstrated at the 5G-SpaceLab a successful access phase and measured the time required by a single user to access the network under different values of NTN RTTs. The results are shown in Figure 16.9 for the two proposed approaches, the TA and TD. As it can be seen, the TD approach results in a step-function behavior because there are discrete values of the timers that are reported by the base station to the users. On the other hand, since the TA approach relies on the RTT estimate at the user side, the timers are modified in accordance with the exact value of the RTT.

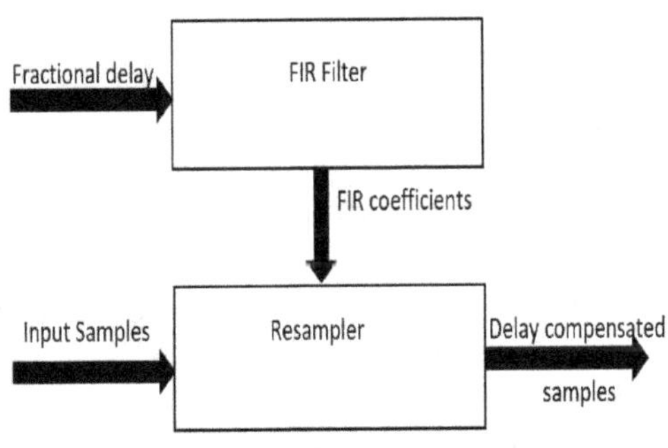

Figure 16.8 Block diagram of doppler delay emulator using resampler

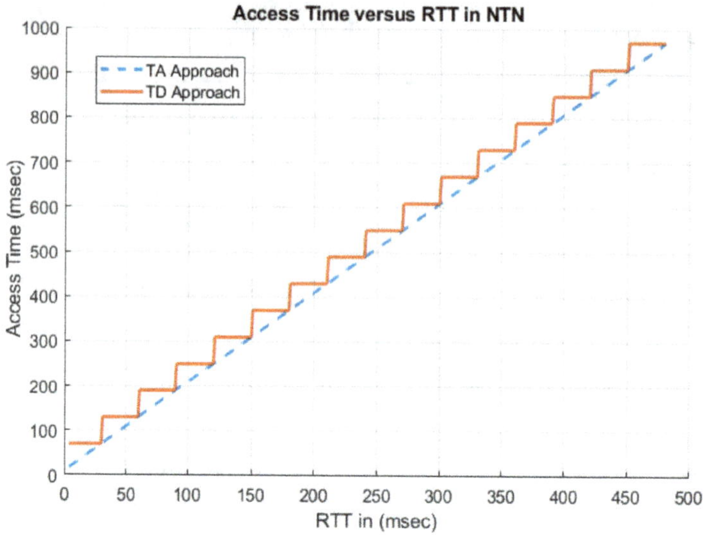

Figure 16.9 Single user access time

As can be observed from Figure 16.9, the access time for a single user can be as high as 970 ms in the worst case scenario of a GEO satellite with a transparent payload. Notably, these results represent only a lower bound since a single user is used for the experiment, thus no collision can occur. In a scenario where many users will try to access the network simultaneously, the presence of collisions in the RA procedure will naturally lead to a further increase in the user access times. Nevertheless, the obtained results are a big step forward toward NTN systems since the technology readiness level (TRL) is increased.

16.4 5G-Lab Fraunhofer

16.4.1 Laboratory environment

The 5G NR laboratory at Fraunhofer IIS consists of several hardware and software components which are enabling the emulation of a vast spectrum of 5G NR scenarios. The deployed testbed is suited to run two major SDR platforms for simulation and emulation of 3GPP mobile networks: Amarisoft [33], one of the most advanced commercial software implementation of 5G NR, and OAI, the most flexible open-source implementation of 5G NR already introduced in section 16.1.3.9. Both software options are enabling real-time emulation of 5G NR under Linux optimized for x86 architectures, with the physical layer running on COTS hardware. The emulation platform at Fraunhofer IIS and the relevant hardware components deployed for NGSO emulation are shown in Figure 16.10.

The operation of real-time 5G-NR modems requires high-performance hardware. SDR systems are able to perform most of the digital signal processing (DSP)

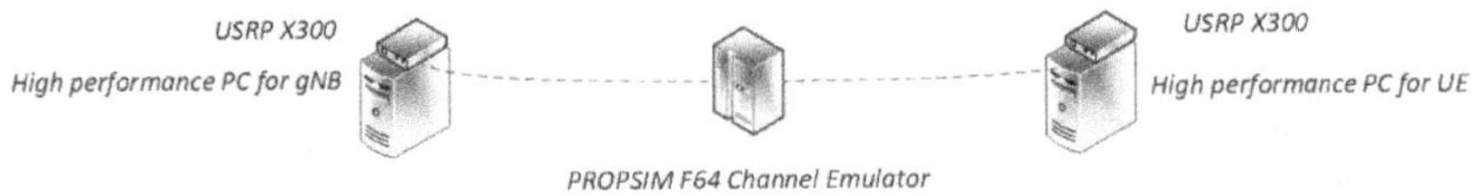

Figure 16.10 NGSO software-defined hardware components

using general-purpose computers, combined with dedicated hardware such as signal processors and/or field programmable gate arrays (FPGA) if required. The SDR radio is characterized by the fact that the received radio signal is digitized as far as possible after the antenna or mixer and further processed as a digital signal. The RF section with its analogue mixers and other RF circuits is implemented in digital design. Only the RF power amplifier on the transmitting side is still analogue. The key feature of SDRs is that the main parameters of the radio signal (e.g., waveform, modulation, coding, bandwidth) can be easily configured by implementing DSP procedures and switching from one to the other requires only minor changes to the code. These features of SDRs lead to an evident major advantage that is flexibility and low-cost upgrade of the code, which translates into fast adaptability to the changes in the market or in the radio communication standards. The aim here is to keep the hardware expenditure as low as possible and instead to carry out the entire baseband signal processing in software. Such systems are by no means limited to the transmitting side and can also be used as receivers in highly mobile applications. To complete the platform, high-performance PCs are required to cope with the real-time processing of high data rates.

The testbed deploys high-performance multi-core PCs hosting the SDR software and Ettus USRP X300, X310, and N310 [21] from National Instruments for the RF front-end. The high-performance PCs are equipped with Intel® Core™ i7-7820X (or i7-9800X) processors, containing eight cores with hyper-threading, running with clock frequencies up to 4.5 GHz. Supporting these processors, 32 GB high-speed RAMs are equipped. The X300 USRPs are equipped with two CBX-120 daughter boards. The CBX-120 daughterboard has the following key parameters:

- Frequency range: 1 200–6 000 MHz
- Bandwidth: up to 120 MHz
- Duplex mode: TDD or FDD
- Maximum output power: 22 dBm (below 3 GHz), 12–22 dBm (above 3 GHz)
- 1 TX/RX port, 1 RX-only port

As each USRP X300 contains two CBX-120 daughter boards, full 2x2 MIMO is possible with these devices. For maximum throughput and low latency between the high-performance PCs and the USRPs, four-lane PCIe connections are used. The X310 and N310 are compatible with the larger bandwidth of 5G NR. The X310 provides bandwidths of up to 160 MHz, but the maximum usable bandwidth for NR is only 80 MHz (217 PRB, which is due to the master clock rate of 184.32 Msps and requirement to use 3/4 sampling). Meanwhile, the N310 series provide maximum instantaneous bandwidth of up to 100 MHz per channel with configurable sample rates (e.g., 122.88, 125, and 153.6 Msps).

The emulation of the satellite channel is achieved by means of a channel emulator, namely the PROPSIM F64, engineered by Keysight Technologies [34]. This hardware unit enables bidirectional emulation of wireless radio channel propagation effects such as dynamic multipath propagation, pathloss, shadowing, fast fading, Doppler shift, noise, and interference in a controlled laboratory environment. A wide spectrum of 5G fast fading profiles and RF channel models are available to reproduce a vast array of different use cases and test scenarios (e.g., Constant, Rayleigh, Rice, Nakagami). Channel capacity and bandwidth are scalable and allow to emulate challenging 5G scenarios (e.g., up to 64 MIMO channels and 100 MHz). The NTN channel emulation is achieved through the Aerospace Option (ASO) feature which enables the emulation of SISO topologies with high Doppler shift (up to +/- 1.5 MHz), long propagation delay spread (up to 1.3 s) and high range rates. Although the input signal frequency range is limited to 6 GHz, the emulator allows to select an RF center frequency that can differ from the emulation center frequency. Each channel unit of the PROPSIM is equipped with RF input/output duplex ports and RF connectors that allow interconnection with third-party devices to be tested (e.g., SDRs), irrespective of the system technology or modulation. The emulator is also equipped with external local oscillators and interfaces to laboratory hardware (i.e., DVI display port, USB for external I/O) that is used to monitor and control purposes.

16.4.2 Case study: "5G lab tests with emulated LEO satellite"

Emulation of 5G communication under the LEO satellite channel has been successfully performed at Fraunhofer IIS by means of the ASO feature provided by the PROPSIM F64, which enables the emulation of LEO satellite channel models characterized by larger Doppler shift and longer propagation delay than in the terrestrial scenario. A block diagram of the emulation setup is shown in Figure 16.10.

In this case study, only one gNB and one UE are assumed. In order to establish a successful connection between the UE and gNB, the open-source platform OAI is running on both PCs. At the time of writing, the public release of OAI supports 5G compliant terrestrial communication. To facilitate satellite communications, modifications in the software stack were implemented by Fraunhofer IIS as described in section 16.2.1.

The purpose of this case study is to prove the feasibility of communication via the LEO satellite using OAI. The main challenge in such an emulation scenario is to check if the time and frequency synchronization are maintained under the LEO satellite channel condition. For this purpose, the emulated channel model reproduces the impairments caused by the LEO satellite, i.e., the large delay (up to several tens of milliseconds) and the time-variant Doppler shift (up to several hundred kilohertz). Table 16.4 shows the parameter settings of the exemplary LEO channel model. The center frequency is 3.61908 GHz, since the USRP X300 used in this setup only supports up to 6 GHz. This model emulates a transparent

Table 16.4 Configuration of the LEO channel model with PROPSIM

Model center frequency	**3.61908 GHz**			
Emulation RF frequency	**3.61908 GHz**			
Step	**Time(s)**	**Delay(s)**	**Doppler (kHz)**	**Gain (dB)**
1	0	0.013	900	0
2	30	0.013	900	0
3	180	0.0127	0	0
4	330	0.013	−900	0

satellite payload, meaning that the UE and gNB are located on the ground with a satellite altitude of about 1950 km. The channel emulator limits the maximum delay change to 0.003 s for each step. In order to create a model with large delay variation, a lot of steps have to be inserted. In this case study, we do not focus on the delay variations but more on the large delay. Therefore, the delay does not vary much with time. The Doppler shift keeps constant for the first 30 s and then decreases with a Doppler rate of −6 kHz/s. The maximal Doppler shift is +/-900 kHz. These values are compliant with the values given in Reference [27]. The emulation time is 330 s.

The test consisted in running the emulation of the channel model and the OAI gNB and UE sequentially. Once started, the gNB kept sending the synchronization signals and the UE could successfully decode it. Figure 16.11 shows the UE scope in OAI after decoding the synchronization signal. The transmitted QPSK symbols can be clearly seen on the UE side, meaning that the connection is stable. And during the whole emulation (330 s), the UE scope could keep showing the constellation and the peak of the received PBCH signal stably, indicating that the time and frequency synchronization can be maintained quite well.

This case study assumes a simple LEO satellite channel mode with high Doppler shift, Doppler rate, and large delay. To create and emulate a more accurate LEO channel model is for further study. Moreover, in the real satellite communication further adjustment of the Tx/Rx power gains and fine tuning are needed.

16.4.3 Conclusion

The advantages of the NGSO testbed deployed at the premises of Fraunhofer IIS are given by a combination of fast-prototyping and incremental development approach. The experimental implementation of 5G NR features to support NTN in OAI, occurring concurrently with the 3GPP standardization process, enables remarkable flexibility which allows the development team to test the features in a laboratory environment, ahead of the release of the standard. Once the laboratory tests are

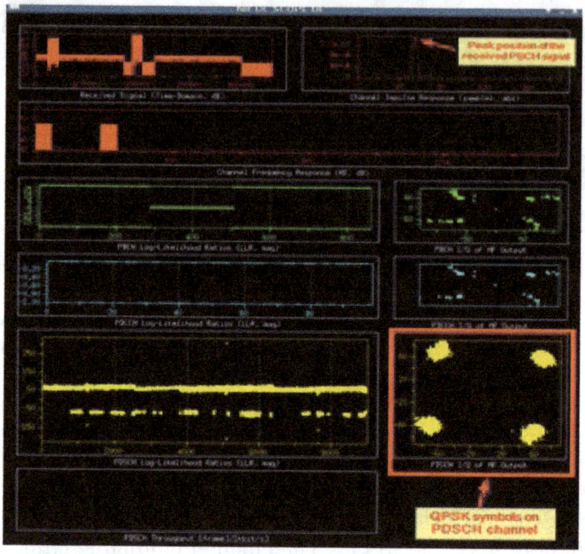

Figure 16.11 OAI UE scope under emulated LEO satellite channel conditions

successfully performed, the channel emulator can be easily unplugged and replaced by a real satellite channel by means of satellite hub equipment, including the up- and down-conversion to the operating band and satellite RF- and terminal equipment. This incremental approach reduces the risks to the minimum of going over the air with an unstable release of the code, thus allowing a cost-effective use of the expensive satellite resources.

References

[1] Available from https://www.sat5g-project.eu [Accessed 15 Oct 2021].

[2] Available from https://www.surrey.ac.uk/institute-communication-systems/facilities/satellite-networking-testbed [Accessed 15 Oct 2015].

[3] Available from https://wwwfr.uni.lu/snt/research/sigcom/5g_space_communications_lab [Accessed 15 Oct 2021].

[4] Available from https://wwwfr.uni.lu/snt/research/sigcom/projects/iot_anc-sat_agile_network_configuration_for_5g_internet_of_things_services_over_satellite

[5] Available from https://satis5.eurescom.eu [Accessed 15 Oct 2021].

[6] Liolis K., Geurtz A., Sperber R., *et al.* 'Use cases and scenarios of 5G integrated satellite-terrestrial networks for enhanced mobile broadband: the sat5g approach'. *International Journal of Satellite Communications and Networking*. 2019, vol. 37(2), pp. 91–112.

[7] Available from https://www.sstl.co.uk/media-hub/latest-news/2019/telesat-leo-phase-1-satellite-demonstrates-first-e [Accessed 15 Oct 2021].

[8] Available from https://www.ses.com/our-coverage/o3b-mpower [Accessed 15 Oct 2021].

[9] *OpenAirInterface project* [online]. 2021. Available from https://gitlab.eurecom.fr/oai/openairinterface5g/wikis/home

[10] Available from https://www.oqtec.space [Accessed 15 Oct 2021].

[11] Available from https://www.fokus.fraunhofer.de/en/fokus/news/open5gcore-campus-2020_06 [Accessed 15 Oct 2021].

[12] Henri Y. 'The oneweb satellite system", in handbook of small satellites: technology, design'. *Manufacture, Applications, Economics and Regulation*. 2020, pp. 1–10.

[13] Available from https://https://oneweb.net [Accessed 15 Oct 2021].

[14] McDowell J.C. 'The low earth orbit satellite population and impacts of the spacex starlink constellation'. *The Astrophysical Journal*. 2020, vol. 892(2), L36.

[15] Available from https://www.starlink.com [Accessed 15 Oct 2021].

[16] Huang J., Cao J. 'Recent development of commercial satellite communications systems'. *Artificial Intelligence in China*. 2020, pp. 531–36.

[17] Available from https://en.wikipedia.org/wiki/Kuiper_Systems [Accessed 15 Oct 2021].

[18] Available from https://www.inmarsat.com/en/about/technology/orchestra.html [Accessed 15 Oct 2021].

[19] *Ettus USRP N310, online*. 2021. Available from https://kb.ettus.com/N300/N310

[20] *Ettus USRP B210, online*. 2021. Available from https://kb.ettus.com/B200/B210/B200mini/B205mini

[21] *Ettus USRP X300, online*. 2021. Available from https://www.ettus.com/all-products/x300-kit

[22] *GomSpace nanocom*. 2021. Available from https://gomspace.com/shop/subsystems/payloads/software-defined-radio.aspx

[23] *Zynq ultrascale+ rfsoc*. 2021. Available from https://www.xilinx.com/products/silicon-devices/soc.html

[24] *uhd software API [online]*. 2021. Available from https://www.ettus.com/sdr-software/uhd-usrp-hardware-driver

[25] *Industrial I/O — the linux kernel documentation, online*. 2021. Available from https://www.kernel.org/doc/html/v5.0/driver-api/iio/index.html

[26] Sirotkin S. (ed.) *5G radio access network architecture*; 2020.

[27] '3GPP TR 38.821. '3rd generation partnership project; technical specification group radio access network; solutions for NR to support non-terrestrial networks (Ntn)'. *Release*. 2021, vol. 16.

[28] *Introduction of non-terrestrial network operation innr.*. 2021

[29] 'TS 38.323; NR; packet data convergence protocol (PDCP) specification'. *Release*, vol. 16.

[30] 'Ts 38.331; NR; radio resource control (RRC) protocol specification'. *Release*, vol. 16.

[31] *Radio Resource Control (RRC) Protocol Specification*, vol. 16.

[32] '3GPP. 'Ts 37.324; E-UTRA and NR; service data adaptation protocol (SDAP) specification'. *Release 16.*
[33] Mengali U. *Synchronization Techniques for Digital Receivers [Online]*. 2013. Available from https://www.amarisoft.com
[34] *Keysight technologies*. 2021. Available from https://www.keysight.com

Chapter 17

Conclusion and future perspectives

Eva Lagunas Symeon Chatzinotas[1], *Kang An*[2], *and Bassel F. Beidas*[3]

Higher broadband speed, lower latency, and expanded coverage are the key characteristics behind the popularity of non-geostationary orbit (NGSO) satellite constellations. While NGSO satellite constellations for broadband are in the initial stages of development and deployment, it is expected that in the coming decades, we will witness a substantial increase in the number of NGSO satellites launched to space. In this book, we have provided an overview of the main uncertainties pose by such imminent deployment and for a successful and efficient operation of such megaconstellations. Some of them include their coexistence and/or integration with legacy satellites and terrestrial wireless communication systems, flexible radio resource allocation and interference management, constellation design and reliability, etc. These open challenges crucially interconnected puzzle pieces that shall fit together to unleash the full potential of NGSO communication systems.

Nevertheless, the way how NGSO operators will achieve profitability is still under discussion. While the evolution of the spectrum regulations may shape the competitive landscape, the hardware for the NGSO end-user remains expensive nowadays, and the service rates offered by operators are far from the ones offered by terrestrial competitors. The latter is justified by the infrastructure that is needed to coordinate and operate such complex mesh networks, which are NGSO constellations. Only in rural or remote accessible areas, NGSO broadband could potentially compete price-wise with the terrestrial alternatives.

On the other hand, most of the envisaged large constellations of NGSO satellites consider a fully meshed intersatellite network, generally combined with a complex ground network with a significant number of gateways. Clearly, global coverage generally requires that each satellite has "at least" one gateway within its coverage area. However, this is not always possible. Therefore, in parallel with the advances on the space segment, also the ground segment is at the center of the discussion.

[1]Interdisciplinary Centre for Security, University of Luxembourg, Luxembourg
[2]National University of Defense Technology, Sixty-third Research Institute, Changsha, Hunan, China
[3]Hughes, Advanced Development Group, Germantown, MD, USA

Link availability, number of handovers, proximity to cloud-based services, and sovereignty of the geographical location are different factors from a complex equation that needs to be resolved to determine the ground segment dimensioning of NGSO constellations. The potential advent of feeder links in Q/V band and the subsequent link blockage due to weather impairments would require advance weather statistical and prediction models, combined with automated gateway switching for traffic off-loading.

All in all, NGSO constellations will pave the way for revolutionary use cases, which come along with the corresponding challenges. Below, we shortlisted some of these novel use cases that would definitely need a careful study in the near future:

- **LEO missions as NGSO internet users**: Usually, NGSO communication systems are conceived as Internet providers. However, assuming a scenario where Low-Earth Orbit (LEO) satellites (space missions in general) can access the Internet via a space-based Internet provider, then the satellites can be connected to the network permanently. This is certainly a game changer that will cause a paradigm shift from relying to on-ground stations to downlink data from the LEO satellite (or send data to the satellite) once per orbit, to accessing the data whenever needed (24 h × 7 days). The ability to communicate with the satellite (either downlink or uplink) on-demand through the Internet can improve several important aspects, such as: (i) throughput; (ii) real-time tasking; (iii) timeliness of data; (iv) selective downlink, and (v) operation cost.
- **Gateway network as a service**: Following the trends in terrestrial service model named Network as a Service, the ground segment of NGSO constellations can also benefit of such model in which the infrastructure is owned by third parties, and the feeder link service is rented to satellite operators from infrastructure owners. This approach is known as Ground Stations as a Service and its main benefits include: (i) data latency improvement by avoiding delays related to the use of few ground stations; (ii) reduction of cost in infrastructure deployment and maintenance; and (iii) increased availability needed to better support impairments and satellite dysfunction management.
- **Open-RAN for space**: The Open radio access network (RAN) architecture has become a popular approach used in cellular networks to virtualize parts of the network that are traditionally handled by specialized hardware and software. Open RAN shall ease the reliance on specific vendors in delivering communications infrastructure and overcome the lack of supplier diversity. The key point of this technology for space is its flexibility to draw on the innovations of multiple suppliers to upgrade their infrastructure with the latest technology. If implemented in the next generation of cellular networks, Open RAN may enable increased opportunities and streamline the ecosystem for integration of nonterrestrial systems into the network.
- **Satellite Internet of Things (IoT)**: IoT is undoubtedly one of the key use cases of excellence of NGSO communication systems. In many cases, IoT devices are distributed in large and remote areas, where it is difficult to have direct access from terrestrial networks. IoT space networks collect a considerable amount of

information, which can be multisource data (from multiple satellites/sensors) and sometime redundant or with low entropy. Data fusion and advanced analytics will definitely be needed to efficiently operate such systems.

- **Distributed satellite systems**: NGSO constellations are by definition distributed systems, i.e. several satellites are jointly operated to achieve a common goal. The conception of the first constellations has disregard the opportunities of an optimized distributed system, where self-organization, formation flying adjustment, coherent communications and beamforming, and joint attitude and communications optimization can be envisaged, among others.
- **Space edge processing**: Connected to the previous points, space edge processing or the capabilities of processing data on-board, is gaining momentum. Preparing data, compressing/fusing data, and performing certain calculations on-board before downloading to Earth is becoming a reality for the new satellite processors operated in low orbits.
- **Quantum for space**: Quantum technology provides a secure means of information sharing between two transceivers located far away from each other. While fiber optics offer a good solution for low-range communications, quantum satellite communications appear to be an opportunity for long-range wide-coverage communications. While preliminary tests have been conducted successfully by the Chinese Quantum Experiments at Space Scale or Micius, there are high expectations for quantum cryptography and quantum computers (with superior speeds and parallel processing) which will surely become popular in the coming years for the space community.

Index

www.ingramcontent.com/pod-product-compliance
Lightning Source LLC
LaVergne TN
LVHW011556200926
848484LV00001B/320